281/9221146 *
13.7.93

20 010 036 76

AF606455

University of Northumbria at Newcastle
LIBRARY

TRANSITION METAL HYDRIDES

TRANSITION METAL HYDRIDES

Edited by

A. Dedieu
Laboratoire de Chimie Quantique
Université Louis Pasteur
4 Rue Blaise Pascal
67008 Strasbourg Cedex France

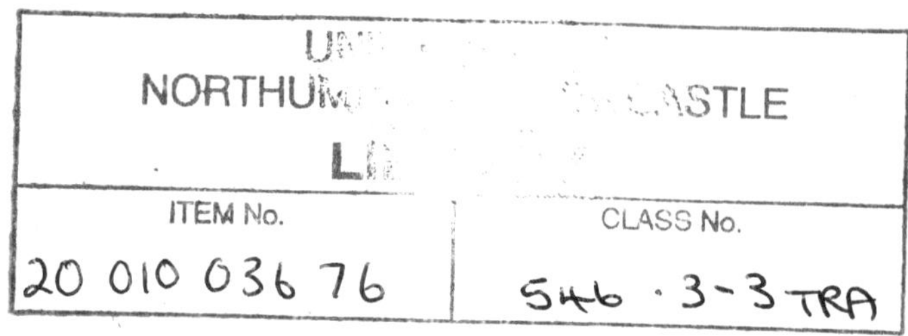
NORTHUM... ...ASTLE
L...
ITEM No.
20 010 036 76
CLASS No.
546 · 3-3 TRA

A. Dedieu
Laboratoire de Chimie Quantique
Université Louis Pasteur
4 Rue Blaise Pascal
67008 Strasbourg Cedex France

Library of Congress Cataloging-in-Publication Data

Transition metal hydrides/edited by A. Dedieu.

p. cm.
Includes bibliographical references and index
ISBN 0-89573-781-7
1. Transition metal hydrides. I. Dedieu. A. (Alain)
QD172.T6T733 1991
547'.056—dc20 91-29176
CIP

British Library Cataloging in Publication Data

Dedieu. A.
Transition metal hydrides.
I. Title
546.6

ISBN 0-89573-781-7
ISBN 3-527-27985-7

© 1992 VCH Publishers, Inc.
This work is subject to copyright.
All rights reserved, whether the whole or part of the material is concerned, specifically those of translation, reprinting, re-use of illustrations, broadcasting, reproduction by photocopying machine or similar means, and storage in data banks.
Registered names, trademarks, etc., used in this book, even when not specifically marked as such, are not to be considered unprotected by law.

Printed in the United States of America
ISBN 0-89573-781-7
ISBN 3-527-27985-7

Printing History:
10 9 8 7 6 5 4 3 2 1

Published jointly by

VCH Publishers, Inc.
220 East 23rd Street
Suite 909
New York, New York 10010

VCH Verlagsgesellschaft mbH
P.O. Box 10 11 61
D-6490 Weinheim
Federal Republic of Germany

VCH Publishers (UK) Ltd.
8 Wellington Court
Cambridge CB1 1HW
United Kingdom

PREFACE

Transition metal hydrides are critically involved in many stoichiometric or catalytic processes of organometallic chemistry. They display a broad reactivity pattern, some of them behaving as hydride donors whereas others may act as protonating reagents or may transfer their hydrogen atom through a radical mechanism. The recent recognition of the so-called nonclassical hydride complexes, containing the dihydrogen ligand, has opened a new field of investigations and calls perhaps for a reconsideration of some reaction mechanisms.

There has been, therefore, an upsurge in the number of experimental studies devoted to transition metal hydrides, either in the gas phase or in solution. Alongside this, the development of modern computer programs and very fast computers has allowed quite accurate calculations to be performed on such systems. In fact, for the smaller "naked" transition metal hydrides, calculated data can be safely compared to experimental data obtained from gas-phase studies, which in turn may be linked with solution studies. New refinements in the methodologies and techniques used to delineate reaction mechanisms in solution have appeared recently. A review of the recent advances made in the understanding of the structure and reactivity of transition metal hydrides seems therefore quite appropriate. It is our hope that the contributions of this book will provide a sound basis for discussing these topics and therefore will lead to new and fruitful cooperative work.

Strasbourg, December 1990

CONTENTS

TRANSITION METAL HYDRIDES

CHAPTER 1

Gas-Phase Organometallic Chemistry of Transition Metal Hydrides

P. B. ARMENTROUT and L. S. SUNDERLIN

Department of Chemistry, University of Utah, Salt Lake City, Utah 84112

1.1. INTRODUCTION AND SCOPE

The simplicity of gas-phase transition metal hydrides makes them an appropriate starting point for a review of such species. In solution and on surfaces, the possible variations in environment (solvent, ligation, nearest neighbors, and support) are responsible for both the staggering array of chemical transformations that are known and the staggering complexity that their study entails. Gas-phase studies are free of solvent and support interactions, allowing the metal–hydrogen interaction to be studied in isolation. In addition, by working in the gas phase, transition metal hydrides that are free of the influence of other ligands (including other metals) can be examined. Thus, species that are not stable in condensed phases and would react in any other environment (with the possible exception of a rare-gas matrix) can be studied in detail. Because this isolation can be achieved, gas-phase studies are an ideal starting point for elucidation of the bonding and reactivity of metal hydrides. In essence, studies in the gas phase provide a key link between theoretical work on transition metal hydrides and the technologically important "real world" environments of transition metal hydrides. The purpose of this chapter is to examine the isolated metal–hydrogen interaction and its chemistry in detail, to help provide a basis for understanding the more-complicated systems discussed later in this book.

In this chapter, we attempt to cover comprehensively all relevant work on *gas-phase* transition metal hydrides. We restrict ourselves to discussion of species in which metal–hydrogen bonds are known (or likely) to occur. Explicitly excluded are those studies for which it is known that the hydrogen does not interact directly with the metal. Condensed-phase results and studies in rare-gas matrices are mentioned only where necessary to understand the gas-phase results. These restrictions, although sometimes arbitrary, are necessary in order to limit the scope of this chapter and to avoid extensive overlap with this material in the remaining chapters.

This chapter is broken down into several major sections. Section 1.2 covers the enormous amount of data available for the simplest transition metal hydrides, the diatoms. Results for the diatomic metal hydrides relate nicely to theory, both for high-level calculations and for more-qualitative considerations that are most apparent for these simple systems. Section 1.3 discusses the slightly more complicated homoleptic metal hydrides, species that contain two or more metal–hydrogen bonds but no other ligands. Section 1.4 extends the review to transition metal complexes that can have any ligands surrounding the metal center. This section includes information on both coordinatively saturated and unsaturated complexes. Next, we discuss recent work on the reactions of clusters of transition metals with hydrogen in Section 1.5. Bare metal clusters are unique species that have been studied intensely for only a decade. This work promises significant insight into the reactivity of metal surfaces. In the final section, the available experimental work is examined for its shortcomings, and research areas ripe for further exploitation are identified.

1.2. DIATOMIC TRANSITION METAL HYDRIDES

A number of studies of the diatomic transition metal hydrides involve work in which they are simply observed. Such observations include the use of a field evaporation ion source and a magnetic sector mass spectrometer to observe the singly charged monohydride cations of Ni, Pd, Cu, and Au and the doubly charged monohydride cations of Ti, Fe, Zr, Nb, and Mo [1]. The singly charged cationic metal monohydrides of Mn [2], Fe, Co, and Ni [3] also have been observed in hydrocarbon flames. Cesium sputter sources have been used to produce the diatomic transition metal monohydride anions of Sc, Ti, V, Mn, Fe, Co, Ni, Y, Zr, Nb, Mo, Re, Rh, Pd, La, Ta, and Pt [4]. This work attests to the stabilities of the diatomic transition metal hydrides in a wide variety of environments, but does little to characterize these species quantitatively.

More useful are the many experiments that are designed to measure the properties (molecular constants, reactivities, and thermodynamic stabilities) of these molecules. In the following subsections, the spectroscopy of the diatomic transition metal hydrides is reviewed, including both optical studies of the neutrals and photoelectron spectroscopy of the anions. Next, the thermodynamic stabilities of the neutral, anionic, and cationic metal hydrides are critically evaluated and analyzed to understand the periodic variations in these quantities. Finally, studies that probe the reactivity of these species are discussed, including both reactions that form diatomic transition metal hydrides and reactions of these species.

1.2.1. Spectroscopy

Two types of spectroscopic experiments have been performed with the diatomic transition metal hydrides: optical spectroscopy on the neutral species and photoelectron spectroscopy on the anions. These experiments are reviewed later. At present, there is no spectroscopic information available regarding cationic transition metal hydride diatoms.

1.2.1.1. ABSORPTION AND EMISSION SPECTROSCOPY

Considering the simplicity of the transition metal hydride diatoms, it seems reasonable to expect that detailed spectroscopic information would long have been available. Two factors have frustrated such efforts. First, formation of the metal hydrides requires extreme conditions. Shock tubes, King furnaces, hollow cathode discharges, arcs, and flames are needed to generate these species. This leads to a broad manifold of populated rotational and vibrational states, although free-jet expansions have been used recently to cool NiH rotationally [5, 6]. Second, ready interpretation of most of the spectra that have been obtained is hindered by their complexity. The manifold of electronic states and spin–orbit levels means that transitions can interleave and that perturbations are the rule rather than the exception.

Previous reviews of the optical spectroscopy of transition metal hydrides include work of Cheetham and Barrow [7], Smith [8], Scott and Richards [9], and the compilation of Huber and Herzberg [10]. In many cases, the information provided in these reviews is still remarkably up-to-date and the reader is referred to these for more-detailed information. Optical spectroscopy is available for all of the first-row transition metals but only the spectroscopy of NiH and CuH are really well-developed. For the second- and third-row transition metal hydrides, experimental results exist for only the group III, X, and XI metals. In no cases are there optical spectra for ionic transition metal hydrides, either anions or cations [11].

ScH. Only two observations of the scandium hydride molecule have been made. Using a shock tube, Smith [8] saw a complex absorption band extending from 17,700 to 18,350 cm^{-1}. No analysis was performed although deuterium labelling studies indicate that the band probably involves (0, 0) vibrational transitions. A band in this same wavelength region and one in the near infrared (11,800–12,300 cm^{-1}) were observed in emission by Bernard, Effantin, and Bacis [12]. Again, no analysis could be performed.

TiH. By using shock-heated hydrogen and titanium dust, Smith and Gaydon [13] were able to observe three complex absorption bands of TiH and TiD in the region 18,200–21,300 cm^{-1}. Deuterium labelling studies indicate that they probably involve (0, 0) vibrational transitions, but it was not until later that Gaydon [14] performed a partial rotational analysis on the lowest energy band. He obtained a rotational constant of ~ 5 cm^{-1} for the lower state and concluded that a $^4\Delta$–$^4\Phi$ transition was consistent with the observed spectrum, assuming that the ground state is the theoretically predicted $^4\Phi$ [15, 16].

VH. There has been only a single experimental observation of vanadium hydride. Smith [8] observed a band extending from 20,800 to 22,700 cm^{-1} that deuterium labelling suggests is a (0, 0) vibrational transition. No analysis of this spectrum has been published.

CrH. Chromium hydride is the first transition metal hydride for which spectroscopic information is available. Two bands in the ultraviolet (27,000–28,000 cm^{-1} and 30,070–30,500 cm^{-1}) have been observed but not analyzed [8, 17]. The former has been attributed to a $B\,^6\Pi$–$X\,^6\Sigma^+$ transition [8]. Another band in the infrared has been assigned to the $X\,^6\Sigma^+$–$A\,^6\Sigma^+$ transition at ν_{00} = 11,552 cm^{-1} [18, 19]. The upper state is highly perturbed. The vibrational and rotational structure of this band has been analyzed in detail and the constants derived are given in Table 1.1.

MnH. Several electronic bands have been reported in the literature (near 11,200–14,500, 17,600, 20,900, 22,000, and 23,600 cm^{-1}) [20]. For many years, only one of these (ν_{00} = 17,666 cm^{-1}) had been analyzed in detail [20, 21]. Assigned to $A\,^7\Pi$–$X\,^7\Sigma$, this transition has been reanalyzed [22] to obtain the $v = 0$ constants for both the lower and upper states as listed in Table 1.1.

Very recently, the identification of a low-lying quintet state of MnH via photoelectron spectroscopy [23, 24] (described later) motivated Balfour [25] to analyze an emission spectrum straddling 11,800 cm^{-1}. This was successfully interpreted as a $^5\Sigma$–$^5\Sigma$ transition. It was also suggested that the 20,900 cm^{-1} band could be attributed to a $^5\Pi$–$^5\Sigma$ transition. Rotational constants for all three quintet states were derived and are listed in Table 1.1 along with approximate excitation energies. Theoretical work [26] suggests that the 20,900 cm^{-1} band might also correspond to another $^5\Sigma$–$^5\Sigma$ transition and that a band at 10,650 cm^{-1} is a $b\,^5\Pi$–$a\,^5\Sigma$ transition.

FeH. Three bands of iron hydride (10,100, 18,600–19,200, and 20,000–21,000 cm^{-1}) have been observed in emission and absorption, suggesting that the lower state is the ground state of the molecule [27–29]. Although the latter two bands have not been analyzed, Balfour, Lindgren, and O'Connor [28] have assigned 1000 lines in the lowest band of the FeD spectrum. They identify this as a $^4\Delta$–$^4\Delta$ transition and its $\Omega = \frac{7}{2}$–$\Omega = \frac{7}{2}$, $\Omega = \frac{5}{2}$–$\Omega = \frac{5}{2}$, $\Omega = \frac{3}{2}$–$\Omega = \frac{3}{2}$, and $\Omega = \frac{1}{2}$–$\Omega = \frac{1}{2}$ subsystems. Analysis yields rotational constants for both the upper and lower states and vibrational constants for the upper state only. An even more complete analysis of this $^4\Delta$–$^4\Delta$ transition is provided by Phillips et al. [29], who obtain molecular

Table 1.1

Molecular Constants for Diatomic Transition Metal Hydrides[a]

Molecule	State	T_0	ω	$\omega_e x_e$	B	r
TiH[b]	$^4\Phi$				5.0	1.8
CrH^-[c]	$^7\Sigma^+$	0			5.57	1.75
	$^5\Sigma^+$	810	1125		5.57	1.75
CrH[d]	$^6\Sigma^+$	0	1581	32	6.22	1.66
	$^6\Sigma^+$	11,552	1479	22	5.34	1.79
MnH^-[e]	$^6\Sigma$ ($^6\Delta$)[c]	0	1050		5.1	1.82
MnH	$^7\Sigma$	0	1548[f]	29[f]	5.61[g]	1.74
	$^5\Sigma$	1,725[e]	1720[e]	70[e]	6.34[h]	1.64
	$^7\Pi$	17,667[g]	1623[d]	33[d]	6.34[g]	1.64
	$^5\Sigma$	13,500[h]			6.22[h]	1.66
	$^5\Pi$	22,600[h]			6.08[h]	1.67
FeH^-[e]	$^5\Delta$	0	1300		5.3	1.79
FeH	$^4\Delta_{7/2}$[i]	0	1827	32	6.59	1.61
	$^6\Delta$[e]	1,945			5.44	1.77
	$^6\Sigma^+$[e]	4,925			5.3	1.8
	$^4\Delta_{7/2}$[i]	9,929	1499	37	5.94	1.69
CoH^-[c]	$^4\Phi$	0	1300		6.1	1.67
CoH	$^3\Phi_4$[j,k]	0	1925	35	7.39	1.52
	$^5\Phi$[c]	6,625			6.1	1.67
	$^3\Phi_4$[j,k]	22,367	1656	64	6.67	1.60
NiH^-[c]	$^3\Delta$	0	1430		6.57	1.61
NiH	$^2\Delta_{5/2}$	0	2007[l]	40[k,l]	7.89[m]	1.47
	$^4\Delta$[c]	11,610	1650		7.4	1.52
	$^4\Pi_{5/2}$[n]	15,520[d]	1508[o]		6.42[p]	1.63
	$^2\Delta_{5/2}$[q]	15,977	1571	35		
	$^2\Delta_{5/2}$[r]	23,761	1780	340	6.16	1.66
CuH[s]	$^1\Sigma^+$	0	1941	37	7.94	1.46
YH[t]	$^1\Sigma^+$	0	1530	20	4.58	1.92
	$^1\Sigma$	14,295	1330	16	4.23	2.00
PdH[d,k]	$^2\Sigma^+$	0	2035	39	7.23	1.53
AgH[d]	$^1\Sigma^+$	0	1760	34	6.45	1.62
LaH[d]	$^1\Pi$				3.78	2.11
	$^3\Delta$				3.84	2.09
PtH[d]	$^2\Delta_{5/2}$	0	2295	46	7.20	1.53
AuH[d]	$^1\Sigma^+$	0	2305	43	7.24	1.52

[a] All values in cm^{-1} except for r, which is in angstroms. Vibrational constants are ω_e if $\omega_e x_e$ is provided and ν_{01} if not. Rotational constants are generally equilibrium values.

[b] Reference 14. [c] Reference 24. [d] Reference 10.

[e] Reference 23. [f] Reference 21. [g] Reference 22.

[h] Reference 25. [i] Reference 29. [j] Reference 32.

[k] Converted from experimental values for the metal deuteride.

[l] Calculated from data in reference 41.

[m] Equilibrium value derived from reference 41.

[n] State assignment from reference 6.

[o] Reference 40.

[p] Equilibrium values calculated from data in reference 40.

[q] Reference 8. [r] Reference 36. [s] Reference 44.

[t] Reference 45.

constants for both the upper and lower states (Table 1.1). They also determine the splittings between the different Ω levels for each state and give the (0, 0) origin for the lowest transition ($\Omega = \frac{7}{2}$–$\Omega = \frac{7}{2}$) as 9929 cm^{-1}. Finally, Beaton et al. [30] observed the $J = \frac{11}{2} \leftarrow \frac{9}{2}$ transition in the $\Omega = \frac{7}{2}$ level of the $^4\Delta$ ground state of FeH in the far infrared via laser magnetic resonance.

CoH. The earliest observation of cobalt hydride was made by Heimer [31], with later additions by Smith [8] and Klynning and Kronekvist [32]. Several bands of CoH and CoD have been characterized in both emission [31] and absorption [32] spectra, which suggests that the lower state is probably the ground state of the molecule. Several of these features have been identified as a $^3\Phi$–$^3\Phi$ transition and its $\Omega = 4$–$\Omega = 4$ and $\Omega = 3$–$\Omega = 3$ subsystems [32]. (Herzberg [33] suggested that this state could be $^1\Gamma$ based on the electronic configuration given by Heimer; however, the ground state is undoubtedly the $^3\Phi$ state identified theoretically [16].) These bands have been analyzed for the molecular constants given in Table 1.1 (in most cases, these are converted from CoD values). The spitting between the $^3\Phi_4$ and $^3\Phi_3$ states has been estimated as about 800 cm^{-1} by Klynning and Kroneqvist. A transition in the far infrared identified as $J = 6 \leftarrow 5$ transition in the $\Omega = 4$ level of the $^3\Phi$ ground state of CoH has also been observed by Beaton et al. [30] via laser magnetic resonance.

NiH. The spectroscopy of NiH has seen the most activity of any transition metal hydride during the last decade. Early contributions came from Gaydon and Pearse [34], Heimer [35], Åslund et al. [36], and Smith [8]. In 1979, when Huber and Herzberg's compilation was published, transitions from the $^2\Delta$ ground state to three excited states (called the A, B, and C states, and all identified as $^2\Delta$) had been observed and partially analyzed for molecular constants. In several cases, however, the spectra are sufficiently perturbed that the molecular constants may not be reliable. Åslund et al. identified both the $\Omega = \frac{5}{2}$–$\Omega = \frac{5}{2}$ and $\Omega = \frac{3}{2}$–$\Omega = \frac{3}{2}$ subsystems in the $C\,^2\Delta$–$X\,^2\Delta$ transition and found that the spin–orbit splitting for the ground state was -490 cm^{-1}.

In 1982, Scullman, Löfgren, and Kadavathu [37] reexamined this system because theory [38, 39] indicated that there could not be two $^2\Delta$ states in the region of 15,000 cm^{-1}, where the A and B states were observed. They found a $^2\Delta_{3/2}$–$^2\Delta_{5/2}$ transition in this energy region and also a $^2\Phi_{7/2}$–$X\,^2\Delta_{5/2}$ transition. Further, perturbations in the $v = 1$ level of the $X\,^2\Delta_{5/2}$ state show that there is another state at about 2800 cm^{-1}. In 1985, Field and co-workers examined this question by using Stark and Zeeman spectroscopy. They found that the Stark spectrum of jet-cooled NiH at about 17,463 cm^{-1} could be described as a $B\,^2\Delta_{5/2}$–$X\,^2\Delta_{5/2}$ transition [5]. Later, they used Zeeman spectroscopy to verify that the B state is indeed $^2\Delta$ but that the A state cannot be $\Lambda = 2$, $\Sigma = \frac{1}{2}$ [6]. Rather, the A state is a $^4\Pi_{5/2}$ and the spin-forbidden A–X transition borrows intensity from the B–X transition via spin–orbit interactions.

Most recently, Kadavathu, Löfgren, and Scullman [40] reanalyzed several of these transitions to obtain additional molecular parameters (although they still refer to the A state as $^2\Delta$). In addition, the vibration–rotation spectroscopy of NiH and NiD has been examined via laser magnetic resonance [30, 41]. The various molecular constants that have been derived are listed in Table 1.1.

CuH. The spectroscopy of copper hydride is the best-developed of the transition metal hydrides. Even a decade ago, results including molecular parameters were available for 11 electronic states of CuH. These were compiled by Huber and Herzberg [10], who relied greatly on the work of Ringström [42]. More recently, Brown and Ginter [43] observed four new transitions between 47,600 and 53,000 cm^{-1}. Bernath and coworkers [44] used Fourier transform methods to improve the spectroscopic constants for the $X\,^1\Sigma^+$ and $A\,^1\Sigma^+$ states. The reader is referred to these works for information about all but the ground state, which is contained in Table 1.1.

YH and LaH. Spectroscopy by Bernard and Bacis [45] provides the information available for yttrium hydride. The six bands observed in emission are identified as a $^3\Phi$–$^3\Delta$ transition and several transitions between various singlet states: $^1\Pi$, $^1\Sigma \rightarrow {}^1\Sigma$ and $^1\Sigma$, $^1\Delta \rightarrow {}^1\Pi$. Rotational constants for all of these states and vibrational constants for two of the $^1\Sigma$ states are obtained. Based on early ab-initio calculations of ScH [46, 47], Bernard and Bacis assign the ground state as $^3\Delta$. More recent theoretical work [48] establishes a $^1\Sigma^+$ state as the ground state and obtains molecular constants in very good agreement with those obtained experimentally. Table 1.1 lists the molecular constants obtained experimentally for the $^1\Sigma^+$ ground state and the one excited state for which complete information also is available. Based on the calculated excitation energies for the $A\,^1\Pi$ and $a\,^3\Delta$ states, an approximate reordering and reassignment of the states of YH is $X\,^1\Sigma^+$ (0 cm^{-1}), $a\,^3\Delta$ (6940 cm^{-1}), $A\,^1\Pi$ (10,900 cm^{-1}), $^1\Sigma$ (14,300 cm^{-1}), $^1\Pi$ (19,600 cm^{-1}), $^1\Delta$ (22,300 cm^{-1}), $^3\Phi$ (25,000 cm^{-1}), and $^1\Sigma$ (30,800 cm^{-1}).

Bernard and Bacis [49] also performed the only spectroscopy available for lanthanum hydride. Three transitions ($^3\Phi$–$^3\Delta$ and $^1\Delta$, $^1\Sigma$–$^1\Pi$) are observed in both emission and absorption, indicating that the $^1\Pi$ and $^3\Delta$ states lie within several thousand wave numbers of the ground state. These authors later assigned the $^3\Delta$ state as the ground state [45], in analogy with their treatment of YH. In contrast, Huber and Herzberg [10] assigned the $^1\Pi$ and $^3\Delta$ states as the A and a states, respectively, presumably in anticipation of an unobserved $^1\Sigma$ ground state. The rotational constants derived experimentally for these two states are listed in Table 1.1.

PdH and PtH. The only work concerning palladium hydride includes a short paper by Lagerqvist, Neuhaus, and Scullman [50] and more-detailed work by Malmberg, Scullman, and Nylen [51]. In the latter paper, 12 absorption bands of PdH and PdD are observed in the region of 22,000–25,000

cm^{-1}. These are attributed to transitions between the $^2\Sigma^+$ ground state and five excited electronic states, all of $^2\Sigma$ character. Because of strong perturbations, only the PdD spectrum could be analyzed and that for only the ground electronic state. The resulting molecular constants have been converted to values for PdH and these are listed in Table 1.1.

Spectroscopic data for platinum hydride come largely from the work of Scullman and coworkers [52] and are summarized by Huber and Herzberg [10]. Like NiH, the ground state of PtH is $^2\Delta$. Transitions between two subsystems of this state and two excited states (both identified as $^2\Delta$) have been observed in both emission and absorption, and an additional state ($^2\Phi$) has been observed in emission only. Molecular constants are available for all of these states [10], and those for the ground state are listed in Table 1.1. Based on the preceding discussion for NiH, it seems likely that one of the excited $^2\Delta$ states is inappropriately labelled. Transitions between several higher-lying states (identified as $^2\Sigma$) also have been observed.

AgH and AuH. The spectroscopies of silver and gold hydride are well-developed and are described in detail by Huber and Herzberg [10], who take much of their data from Ringström and Åslund [53] and Ringström [54]. Both molecules have a $^1\Sigma^+$ ground state with the molecular parameters given in Table 1.1. Nine excited electronic states are known for AgH and six for AuH. For both molecules, the lowest is a $^1\Sigma^+$ state (although a 0^+ designation is more appropriate for AuH), approximately 28,000 cm^{-1} above the ground state. The remaining excited states are over 38,000 cm^{-1} above the ground state. The reader is referred elsewhere [10] for detailed information on these states.

1.2.1.2. PHOTOELECTRON SPECTROSCOPY

Stevens, Feigerle, and Lineberger [23, 24] systematically studied five of the first-row diatomic transition metal hydride anions. Not only do these studies provide the electron affinities of the metal hydrides (as listed in Table 1.2), but they provide spectroscopic data for both the anionic and neutral species and thereby complement the optical spectroscopy studies discussed previously. In the case of FeH, the photoelectron spectroscopy (PES) data resolved a long-standing question regarding the ground state of this molecule. In other cases, excited states of the neutral molecule that differ in spin from the ground state (and therefore are not observed optically) can be identified.

The diatomic transition metal hydride anions were produced in a discharge source containing ammonia and the metal carbonyls $Cr(CO)_6$, $Fe(CO)_5$, $Co(CO)_3NO$, and $Ni(CO)_4$. For MnH^-, the discharge contained only $HMn(CO)_5$.

MnH^-. The photoelectron spectrum of MnH^- exhibits six peaks [23]. Deuterium labelling studies show that three of these correspond to (0, 0) vibrational transitions whereas the others involve vibrational excitations. The

Table 1.2
Electron Affinities (in Electron Volts) of Transition Metal Hydrides

M	EA(M)	EA(MH)	$EA(MH_2)$[a]
Cr	0.666 (0.012)[b]	0.563 (0.010)[c]	$\geq$ 2.5
Mn	$\leq$ 0	0.869 (0.010)[e]	0.444 (0.016)
Fe	0.151 (0.003)[f]	0.934 (0.011)[e]	1.049 (0.014)
Co	0.662 (0.003)[f]	0.671 (0.010)[c]	1.450 (0.014)
Ni	1.157 (0.010)[g]	0.481 (0.007)[c]	1.934 (0.008)

[a] Reference 107. [b] Reference 103. [c] Reference 24.
[d] Reference 104. [e] Reference 23. [f] Reference 105.
[g] Reference 106.

lowest-energy peak shows little vibrational excitation, establishing that it involves removal of a nonbonding electron. It therefore is assigned to a transition between the ground state of MnH, known by optical spectroscopy to be a $^7\Sigma^+(3d^5\sigma^2\sigma^{*1})$, and the ground state of MnH^-. (The σ and σ^* molecular orbitals are largely the bonding and antibonding combinations of the metal 4s orbital and the H atom 1s orbital.) Originally, the MnH^- ground state was suggested to be $^6\Delta(3d^6\sigma^2\sigma^{*1})$ [23]; however, later studies [24] suggested it could also be $^6\Sigma$ (if the excited quintet state of MnH was found to be $^5\Sigma$, as was recently confirmed spectroscopically, see the next paragraph).

The second peak occurs 0.213 $\pm$ 0.006 eV higher in energy and shows a long vibrational progression, indicating removal of a bonding or antibonding electron. This is assigned to formation of MnH in an excited quintet state. Originally suggested to be $^5\Delta(3d^6\sigma^2)$ [23], later studies found a likely interpretation to be a $^5\Sigma$ state due to a mixture of $(3d^6\sigma^2)$ and $(3d^5\sigma^2\sigma^{*1})$ electron configurations. The recent spectroscopic analysis of Balfour [25] confirms the latter assignment. Franck–Condon analyses of the vibrational progression in this band yield a vibrational frequency for MnH($^5\Sigma$) of 1720 $\pm$ 55 cm^{-1}, higher than that of the $^7\Sigma^+$ ground state, and a bond length of 1.60 $\pm$ 0.02 Å, shorter than that of the ground state. The weak third electronic band occurs at 2.27 $\pm$ 0.03 eV and falls near the energy cutoff of the instrument.

FeH$^-$. The photoelectron study of FeH^- is particularly interesting because it established the ground state of FeH for the first time [23]. Both the $^6\Delta(3d^6\sigma^2\sigma^{*1})$ state and the $^4\Delta(3d^7\sigma^2)$ state were possibilities and theory found them close in energy. The PES spectrum of FeH^- is fairly complex, although deuterium labelling studies show that there are four electronic origins present [23]. The lowest shows both hot band structure and a vibrational progression in the neutral. By comparison with the MnH^- spectrum, this transition was assigned to removal of the antibonding σ^* electron from $FeH^-(^5\Delta, 3d^7\sigma^2\sigma^{*1})$ to form $FeH(^4\Delta, 3d^7\sigma^2)$. Careful consideration of other possible assignments and a systematic variation of the FeH^- source conditions were used to verify this assignment.

The second band in the PES spectrum showed no vibrational progression and was assigned to formation of $FeH(^6\Delta)$, the other plausible ground state. This lies only 0.24 ± 0.01 eV above the $^4\Delta$ ground state. Another band, 0.61 eV above the ground state, is assigned to the $^6\Sigma^+$ state of FeH, which has the same electron configuration as the $^6\Delta$ state. Franck–Condon analysis shows that both of these states have a bond length comparable to that of $FeH^-(^5\Delta)$ (Table 1.1). A third band at 1.24 eV (9965 ± 155 cm^{-1}) was also observed but unassigned; however, this energy corresponds exactly with the spectroscopically identified $^4\Delta$–$^4\Delta$ transition at 9929 cm^{-1} [29], discussed in Section 1.2.1.

CoH⁻. Two bands are observed in the PES spectrum of CoH^- [24]. The lowest shows a vibrational progression with a frequency that matches the value determined by optical spectroscopy (Table 1.1). This is consistent with formation of ground state $CoH(^3\Phi, 3d^8\sigma^2)$ by detaching the antibonding electron from $CoH^-(^4\Phi, 3d^8\sigma^2\sigma^{*1})$. The second band shows no vibrational excitation and therefore is assigned to loss of a nonbonding $3d$ electron to form $CoH(3d^7\sigma^2\sigma^{*1})$, probably a $^5\Phi$ state. This state is found to lie 0.82 ± 0.01 eV above the ground state and to have a bond length comparable to CoH^- (Table 1.1). These results are directly analogous to those for FeH except that the splitting between the low-spin and high-spin states of the neutral is much larger in the case of cobalt.

NiH⁻. Five electronic bands are observed in the PES of NiH^- [24]. The lowest shows a vibrational progression and is therefore assigned to a $NiH^-(^3\Delta, 3d^9\sigma^2\sigma^{*1})$ to $NiH(^2\Delta, 3d^9\sigma^2)$ transition. The other strong peak observed is the highest in energy and is assigned to formation of the high-spin $NiH(^4\Delta, 3d^8\sigma^2\sigma^{*1})$ state, 1.44 eV above the $^2\Delta$ ground state. Assignments for the three other bands (0.28, 0.45, and 0.92 eV above the ground state) are not obvious; however, the first two bands occur at an energy where strong perturbations have been observed in the optical spectrum [37] and where theory suggests the $^2\Pi$ and $^2\Sigma^+$ states of NiH lie [39, 55]. Because detachment to form these doublet states is a two-electron process from $NiH^-(^3\Delta)$, these states must be highly mixed.

CrH⁻. The PES spectrum of CrH^- is discussed last because it is the most complex. Five sharp bands are observed [24], although the two at highest energy are barely resolved. The bands fall into two groups that are spaced by an energy close to the excitation energy for the $A\,^6\Sigma^+$ state of CrH, 1.432 eV, as determined by optical spectroscopy [19]. Indeed, the first and third bands are separated by 1.428 eV and the second and fifth bands are separated by 1.426 eV. This suggests that the first and second bands correspond to formation of CrH($X\,^6\Sigma^+, 3d^5\sigma^2$), whereas the third and fifth bands correspond to formation of CrH($A\,^6\Sigma^+, 3d^4\sigma^2\sigma^{*1}$). The 0.10 ± 0.01-eV splitting between the first and second peaks and the third and fifth peaks is then assigned to the energy difference between two states of the anion. Because the ground state is almost certainly the $^7\Sigma^+(3d^5\sigma^2\sigma^{*1})$ state, a plausible excited state is the $^5\Sigma^+(3d^5\sigma^2\sigma^{*1})$. The isoelectronic MnH is observed to have a similar low-lying quintet excited state at 0.21 eV, as discussed previously. The fourth band in the CrH^- PES is unassigned.

1.2.2. Thermochemistry

Examination of the thermochemical stability of the diatomic transition metal hydrides has involved several different techniques. Probably the best and most complete thermodynamic information exists for the cationic hydrides. This comes from studies of the interaction of M^+ with H_2. Bond energies for the neutral diatoms have been derived by using several different methods, but for many transition metals this thermochemistry is not yet well-established. Given the neutral thermochemistry, the stabilities of the anions are straightforwardly calculated from the electron affinity measurements described previously. These various experiments are described later, and the periodic trends in the thermochemistry are discussed.

We have chosen to tabulate the thermochemistry in terms of the bond dissociation energies of the diatomic transition metal hydrides, $D(\text{M—H})$, rather than as heats of formation. This allows the periodic trends in this thermochemistry to be examined easily. Further, the bond energies are given at a temperature of 298 K. Because many values in the literature are given at 0 K, these are converted to 298-K values by assuming that the system is an ideal gas with only translational and rotational degrees of freedom. This means that $D(\text{M—H}) \equiv D_{298}(\text{M—H}) = D_0(\text{M—H}) + \frac{3}{2}kT$, where $\frac{3}{2}kT = 3.7\ \text{kJ mol}^{-1}$ at 298 K. This approximation is almost certainly accurate to within 1 kJ mol^{-1}.

1.2.2.1. CATIONS

Ion Beam Experiments. With few exceptions, bond energies for diatomic transition metal hydride cations have been measured by using ion beam mass spectrometry to examine the kinetic energy dependence of reaction (1.1),

$$M^+ + H_2 \rightarrow MH^+ + H \qquad (1.1)$$

and its isotopic analogues with D_2 and HD. The kinetic energy threshold E_T for this reaction is related to the metal hydride cation bond energy by

$$D_0(\text{M}^+\text{—H}) = D_0(\text{H}_2) - E_T - E_{\text{rot}} - E_{\text{el}} \qquad (1.2)$$

where E_{rot} is the rotational energy of the H_2 reactant, E_{el} is the electronic energy of the metal ion reactant, and D_0 indicates that the bond energies are at 0 K. Results for reaction with HD and D_2 also yield $D_0(M^+—H)$ after correcting for zero-point energy effects. Equation (1.2) implicitly assumes that there are no barriers to reaction in excess of the endothermicity. This is commonly true for ion–molecule reactions and there is no evidence that such barriers exist in these systems [56].

The technically difficult aspects of these studies are the interpretation of the threshold and control of the electronic energy of the reactant ion. The former difficulty arises because both reactants have a distribution of kinetic energies that must be accounted for in determining the true threshold. Methods for handling this are now reasonably well developed and are described in the literature [57]. The efficacy of these methods for reaction (1.1) has been shown in test cases such as M = Si. In this study [58], the SiH^+ bond energy derived from the ion beam experiments, 312 ± 3 kJ mol^{-1}, is in excellent agreement with values obtained from a detailed spectroscopic study, 311 ± 3 kJ mol^{-1} [59]. Control of the metal-ion electronic energy is a more subtle problem, but one that is experimentally tractable [60, 61].

Armentrout and Beauchamp were the first to study reaction (1.1) with transition metals. They reported detailed data for Ni^+ [62] and Co^+ [63] and later the thermochemical results for Cr^+, Mn^+, and Fe^+ [64]. Later work by Beauchamp and co-workers includes detailed data for Sc^+ [65], Fe^+ [66], Ru^+, Rh^+, and Pd^+ [67]. In all cases, the metal ions are produced by surface ionization (in which the desired ions boil off a rhenium filament heated resistively to ~ 2500 K and exposed to a compound containing the metal). There are good reasons to believe that this process produces ions with a Maxwell–Boltzmann distribution of the accessible electronic states at the filament temperature such that ground-state species are preferentially produced (kT = 21 kJ mol^{-1} at 2500 K) [68]. Hence, these studies made no correction for E_{el} in eq. (1.2).

This work has now been superseded and extended by studies of Elkind and Armentrout using an improved *guided* ion beam apparatus [57]. This instrument allows a more precise and accurate determination of the cross sections for reaction (1.1) and thereby yields improved precision and accuracy in the resulting thermochemistry. In addition, Elkind and Armentrout explicitly considered the effects of electronic excitation of the metal (generally by producing the metal ions both by surface ionization *and* other techniques such that the electronic energy content of the ions is systematically varied). These results now include detailed reports of the cross sections and thermochemistry for all first-row transition metals (Sc [69], Ti [70], V [71], Cr [72], Mn [73], Fe [74], Co, Ni, and Cu [75]), Y, La, and Lu [69]. These authors have also reported thermochemical results for all second-row metals except Tc, Ru, and Rh [76], although the effects of electronic excitation have *not* been treated in detail for these metals. Unpublished work [77] on Rh and Pd has been completed also. The thermochemical results of these various studies are

Table 1.3
Experimental Bond Dissociation Energies of Diatomic Transition Metal Hydrides[a]

M	$D(M^+—H)$[b]	$D(M—H)$[c]	$D(M^-—H)$[d]
Sc	239(9)[e]	205(17)	
Ti	227(11)[f]	205(9)	
V	202(6)[g]	165(13)	
Cr	136(9)[h]	223(16)	213(16)
Mn	203(14)[i]	126(18)	> 210(18) [258][j]
Fe	208(6)[k]	157(8)	235(17)
Co	195(6)[l]	194(13)	195(13)
Ni	166(8)[l]	252(8)	187(8)
Cu	92(13)[l]	258(11)	
Y	260(6)[e]		
Zr	230(13)[m]		
Nb	225(13)[m]		
Mo	175(13)[m]	207(19)	
Ru	171(13)[n]	234(21)	
Rh	151(13)[o]	247(21)	
Pd	197(13)[o]	234(25)	
Ag	67(13)[m]	220(8)	
La	243(9)[e]		
Pt		< 336	
Au		303(13)	

[a]Values (in kilojoules per mole) are for 298 K. Uncertainties are in parentheses.

[b]In some cases, these values have been adjusted from 0-K values as described in the text (by adding $\frac{3}{2}kT = 3.7$ kJ mol^{-1} at 298 K).

[c]Recommended values from Table 1.4.

[d]Calculated by using eq. (1.14), the neutral bond energies listed here, and the data in Table 1.2.

[e]Reference 69. [f]Reference 70. [g]Reference 71.

[h]Reference 72. [i]Reference 73.

[j]Estimate. See text for a discussion of this number.

[k]Reference 74. [l]Reference 75. [m]Reference 76.

[n]Reference 67. [o]Reference 77.

listed in Table 1.3. This work has been reviewed elsewhere [61, 78] and the thermochemical values have been verified by similar experiments with hydrocarbons [79–84] and ammonia [85].

The earlier results of Beauchamp and co-workers agree with those of Elkind and Armentrout within experimental error, except for Fe. In the case of iron, the discrepancy arises because the threshold observed in the cross section for reaction (1.1) is due primarily to reaction of the ^4F first excited state of Fe^+, 0.3 eV above the ^6D ground state [74]. This was not accounted for in the bond energy derived by Halle, Klein, and Beauchamp [66]. Correcting for this electronic excitation brings the values into agreement.

Deprotonation Studies. The other experimental means that has been used to determine the bond energies of the diatomic transition metal hydride cations is the examination of reaction (1.3),

$$MH^+ + B \rightarrow M + BH^+ \tag{1.3}$$

where B is an appropriate base. This type of experiment has been utilized only twice, for MnH^+ and FeH^+.

Stevens and Beauchamp [86] formed MnH^+ by electron impact ionization and fragmentation of $HMn(CO)_5$ in an ion cyclotron resonance (ICR) mass spectrometer. In the following discussion, we have updated the proton affinity values (PA) used by these authors. In particular, they assigned $PA(NH_3)$ as 866 kJ mol^{-1}, whereas the best value is now considered to be 853.5 $\pm$ 12 kJ mol^{-1} [87]. Stevens and Beauchamp found that MnH^+ undergoes reaction (1.3) efficiently with 2-propyl cyanide (PA = 813 kJ mol^{-1}) and stronger bases, reacts inefficiently with dimethyl ether (PA = 804 kJ mol^{-1}), and does not react at all by proton transfer with methyl cyanide (PA = 788 kJ mol^{-1}) and weaker bases. Based on these observations, they assigned $PA(Mn) = 805 \pm 21$ kJ mol^{-1} and converted this to the homolytic bond energy by using

$$D\left(M^+{-}H\right) = PA(M) + IE(M) - IE(H) \tag{1.4}$$

where IE are ionization energies taken from Lias et al. [88]. This yields a $Mn^+{-}H$ bond energy of 210 $\pm$ 21 kJ mol^{-1}, in very good agreement with the value derived from beam studies of reaction (1.1), 203 $\pm$ 14 kJ mol^{-1} (Table 1.3). Note that this latter value suggests $PA(Mn) = 798 \pm 14$ kJ mol^{-1}, consistent with a near-thermoneutral proton transfer reaction with dimethyl ether.

Halle et al. [66] performed a similar experiment with FeH^+ produced by electron impact ionization and fragmentation of $Fe(C_5H_4CH_3)_2$ in an ion beam machine. They then measured the cross sections for the proton transfer reactions (1.3) with a series of bases at a low kinetic energy (0.5 eV). It was observed that bases having large proton affinities generally had larger reaction cross sections (although there were large variations and NH_3, which had the largest PA of the bases studied, had a very small cross section). A plot of these cross sections versus $PA(B)$ was extrapolated to zero cross section to determine the PA of Fe, 787 $\pm$ 8 kJ mol^{-1} relative to $PA(NH_3) = 853.5$ kJ mol^{-1}. This was in good agreement with the value determined from their beam experiments of reaction (1.1), 795 $\pm$ 21 kJ mol^{-1}, but disagrees with the more recent determination, 758 $\pm$ 6 kJ mol^{-1}, calculated from the value listed in Table 1.3 [74]. A reexamination of the data of Halle et al. finds that reaction (1.3) is exothermic with CH_3CHO ($PA = 781 \pm 8$ kJ mol^{-1}) [87] suggesting that PA(Fe) is less than this value. With CH_3OH ($PA = 761 \pm 8$ kJ mol^{-1} [87], reaction (1.3) is observed to be "slightly endothermic," suggesting that PA(Fe) is somewhat greater than this value. Consequently, the observations of Halle et al. actually are consistent with the most recent determination of $D(Fe^+—H)$. The confusion demonstrates that slow reactions cannot necessarily be equated with endothermic processes.

1.2.2.2. NEUTRALS

Four techniques have been instrumental in determining the bond energies of the neutral transition metal hydride diatoms. Spectroscopy and high-temperature mass spectrometry were the first to be utilized. More recently, these bond energies have been bracketed by studying the reactions of M^- with appropriate proton donors or the reactions of M^+ with appropriate hydride donors. Similar experiments in which the kinetic energy threshold for the latter reaction is identified provide a more precise determination of the bond energies. These methods are described in the following subsections.

Spectroscopy. Spectroscopically determined bond energies can be the most precise and accurate of thermochemical values, but for the transition metal hydrides, most "spectroscopic" values are linear Birge–Sponer extrapolations (LBSE). Such values depend on a Morse potential providing an accurate representation of the metal hydride out to the dissociation limit. If this is the case, then $D_e = (\hbar\omega_e)^2/4\hbar\omega_e x_e$ and $D_0 = D_e - \hbar\omega_e/2 + \hbar\omega_e x_e/4$. The utilization of the LBSE method to obtain dissociation energies has been discussed in detail by Gaydon [89], who concludes that the method generally (but not always) provides a reasonable upper limit to the true bond energy. For metal hydrides where the true bond energy is known (mainly those of groups I, XI, and XIII, the LBSE values are about 8% too high. This value is sometimes used to "correct" LBSE values for other metal hydrides.

Included among the LBSE values found in the literature are bond energies for CrH, MnH, FeH, CoH, and PdH. For chromium hydride, Gaydon recommends a value of $D_0 = 275 \pm 48$ kJ mol^{-1}, whereas the original literature citation [18] suggests $D_e = 26{,}000$ cm^{-1}, which can be corrected to $D_0 = 302$ kJ mol^{-1} by using the data in Table 1.1. For MnH, an LBSE yields $D_0 = 2.5$ eV, as cited by both Gaydon [89] and Huber and Herzberg [10]. Gaydon also notes the possible presence of an accidental predissociation that suggests D_0(MnH) < 2.4 eV. He recommends a value of 2.4 ± 0.3 eV $= 232 \pm 29$ kJ mol^{-1}. For cobalt hydride, Huber and Herzberg [10] cite $D_0 = 3.2$ eV, a value that is apparently from an LBSE with no correction. For palladium hydride, Malmberg et al. [51] perform an LBSE to obtain D_e(PdH) $= 3.3$ eV, which can be converted to D_0(PdH) $= 3.2_2$ eV $=$ 311 kJ mol^{-1} by using the data in Table 1.1.

Among the most notorious and least accurate bond energies is the oft-cited LBSE "spectroscopic" value for FeH. Dendramis, Van Zee, and

Weltner [90] measured the infrared spectrum of FeH and FeD in an argon matrix and determined $\omega_e = 1747\ \text{cm}^{-1}$ and $\omega_e x_e = 43\ \text{cm}^{-1}$. They then corrected these matrix values to the "gas phase" values of 1764 and 46 cm^{-1}. An LBSE was then used to obtain $D_0(\text{FeH}) = 193\ \text{kJ mol}^{-1}$, which they arbitrarily decreased by 20% to yield a suggested bond energy of 164 kJ mol^{-1}. The true accuracy of this whole procedure is demonstrated by using the vibrational frequency and anharmonicity listed in Table 1.1 for ground-state FeH. These values, which are very different than the matrix values, yield $D_0(\text{FeH}) = 301\ \text{kJ mol}^{-1}$ from an LBSE.

Slightly more advanced spectroscopic methods have been used to determine bond energies for the group XI metal hydrides. In the case of copper hydride, Ringström [42] does both a Birge–Sponer and Rydberg extrapolation of seven vibrational levels of the $A\,^1\Sigma^+$ state to obtain a dissociation limit of $33{,}200 \pm 500\ \text{cm}^{-1}$. By assuming that the A state dissociates to $\text{Cu}(^2D_{5/2}) + \text{H}(^2S)$, an asymptote that lies 11,203 cm^{-1} above the ground-state atoms, he finds that $D_0(\text{CuH}) = 22{,}000 \pm 500\ \text{cm}^{-1} = 263 \pm 6$ kJ mol^{-1}. This value is in agreement with an upper limit of 280 kJ mol^{-1} determined by Herzberg and Mundie [91] from an observed predissociation in the A state. (An LBSE yields a bond energy of 289 kJ mol^{-1}, 10% higher than the actual value.)

For silver hydride, both Gaydon [89] and Huber and Herzberg [10] cite a graphical Birge–Sponer extrapolation of 12 vibrational levels in the ground state, which yields $D_0(\text{AgH}) = 2.28 \pm 0.1\ \text{eV} = 220 \pm 10\ \text{kJ mol}^{-1}$. For gold hydride, Ringström [54] finds that the a, b, c, and B states all dissociate to a common level, $\text{Au}(^2D_{3/2}) + \text{H}(^2S)$. LBSEs for all four states lead to a dissociation limit that is $47{,}500 \pm 1000\ \text{cm}^{-1}$ above the ground state of AuH. Because the $\text{Au}(^2D_{3/2})$ lies 21,435.3 cm^{-1} above ground-state $\text{Au}(^2S)$, he estimates that $D_0(\text{AuH}) = 26{,}000 \pm 1000\ \text{cm}^{-1} = 311 \pm 12\ \text{kJ mol}^{-1}$.

More-detailed information is also available for the group X transition metal hydrides. For NiH, predissociation in the C state has lead Åslund et al. [36] to suggest an upper limit to $D_0(\text{NiH})$ of $26{,}000\ \text{cm}^{-1} = 3.22$ eV, whereas Huber and Herzberg [10] give a limit of 3.07 eV. Gaydon cites an LBSE value of 3.2 eV, although the more recent spectroscopic constants listed in Table 1.1 suggest 3.0 eV. The most accurate value is probably a graphical Birge–Sponer extrapolation of four vibrational levels in the $^2\Delta$ excited state [89]. By assuming that dissociation is to the $\text{Ni}(^3D) + \text{H}(^2S)$ asymptote, a bond energy of $2.6 \pm 0.3\ \text{eV} = 251 \pm 29\ \text{kJ mol}^{-1}$ is derived. For PtH, Kaving and Scullman [52] observe a predissociation in the $v = 0$ level of the B state. Assuming that the ground state of PtH is the $^2\Delta_{5/2}$ leads to an upper limit for the dissociation energy of $27{,}750\ \text{cm}^{-1} = 332\ \text{kJ mol}^{-1}$. This agrees with the value of $347 \pm 38\ \text{kJ mol}^{-1}$ cited by Gaydon [89] and taken from an LBSE of the $A\,^2\Delta$ state.

High-Temperature Mass Spectrometry. Kant and Moon [92, 93] have examined the gas-phase equilibrium in eq. (1.5) for several metals by using high-temperature mass spectrometry,

$$M + \tfrac{1}{2}H_2 = MH \tag{1.5}$$

This method determines the abundance of each chemical constituent (corrected for an estimate of the relative detection efficiency, i.e., the ionization cross section) as a function of temperature (typically over a range of 200 K in the region of 1000–2000 K). The equilibrium constants so derived are related to the heat of reaction at each temperature via a "third-law" relationship. This depends on knowing the vibrational frequency, bond length, and electronic degeneracy of the MH molecule (see Section 1.2.1.1). A "second-law" method (which uses a plot of $\ln K$ vs. $1/T$) is also used, although these values are generally considered to be less reliable because they involve an extrapolation to low temperatures.

Bond energies were determined by Kant and Moon for the transition metal hydrides of Sc, Ni, Cu, Ag, and Au. Second- and third-law values differ from one another by an average of 8 ± 4 kJ mol^{-1}, with the largest discrepancy being less than 16 kJ mol^{-1}. This agreement demonstrates that equilibrium in eq. (1.5) is achieved. The values suggested by Kant and Moon have been adjusted to 298 K values and are listed in Table 1.4, for all but ScH. In this case, accurate molecular parameters were unavailable to Kant and Moon. In retrospect, the bond length and vibrational frequency chosen by Kant and Moon are reasonable compared with calculated values [16], but they chose an electronic degeneracy of 2.45 for ScH. The ground state of ScH has been calculated to be $^1\Sigma^+$ with a low-lying $^3\Delta$ state [16]. If the third-law bond energy is reevaluated assuming a $^1\Sigma^+$ ground state, the 0-K bond energy can be revised to 212 kJ mol^{-1} (as noted by Kant and Moon); however, including the low-lying $^3\Delta$ state in the calculations lowers the bond energy back to $\sim$ 201 kJ mol^{-1}. After adjusting to 298 K, this is the bond energy listed in Table 1.4.

Kant and Moon [93] also tried to make measurements for the hydrides of Cr, Mn, and Fe, but either observed nothing (Fe) or very small amounts of the hydrides (Cr and Mn). As a consequence, they calculated upper limits to the bond energies from upper limits to the hydride ion currents at the experimental temperatures and hydrogen pressures. This, of course, assumes that the failure to observe these species is attributable only to thermodynamic considerations (rather than kinetic factors or spurious experimental conditions). Given this proviso, these upper limits (adjusted to 298 K) are also listed in Table 1.4. Also, Kant and Moon assumed that the ground state of FeH was $^6\Sigma$. As discussed in Section 1.2.1, FeH is now known to have a $^4\Delta$ ground state with a low-lying $^6\Delta$ state. Reevaluation of the third-law value therefore suggests that the upper limit could be even lower than the value given by Kant and Moon (although insufficient information is available to obtain a more exact value).

Table 1.4
Experimental Bond Dissociation Energies of Diatomic Transition Metal Hydride Neutrals[a]

M	Spectrosc.[b,c]	KM[c,d]	SLSF[e]	TB[f]	Beam
Sc		[205(17)]			
Ti					205(9)[i]
V			159(13) [> 146]		170(17)[g]
Cr	{279(48)}[h]	< 192	172(13) [> 159]		223(16)[i]
Mn	{235(29)}[h]	< 138			126(18)[j]
Fe	{168(29)}[k]	< 184	124(13) [> 109][l]	180(25) [> 155]	157(8)[m]
Co	{312}[n]	192(13)	177(13) [> 163]	226(42) [> 184]	195(13)[o]
Ni	255(29)[h]	252(8)		272(25) [> 247]	249(15)[o]
Cu	267(6)[p]	248(8)			258(18)[o]
Mo			192(13) [> 180]	222(21) [> 201]	
Ru				234(21) [> 205]	
Rh				247(21) [> 226]	
Pd	{314}[q]			234(25) [> 209]	
Ag	224(10)[h,n]	215(8)			
Pt	< 336[r]				
Au	315(12)[s]	292(8)			

[a]Values (in kilojoules per mole) are for 298 K. Uncertainties are in parentheses. Values in square brackets are obtained from a more-conservative interpretation of the data (see text).

[b]Values in curly brackets indicate that they were derived from a linear Birge–Sponer extrapolation and are unreliable.

[c]These values have been adjusted from 0-K values as described in the text (by adding $\frac{3}{2}kT = 3.7$ kJ mol^{-1} at 298 K).

[d]Kant and Moon [92, 93].

[e]Sallans, Lane, Squires, and Freiser [95].

[f]Tolbert and Beauchamp [96].

[g]Reference 98. [h]Reference 89. [i]Reference 100.

[j]Reference 99. [k]Reference 90.

[l]This value has been adjusted for a small difference in *EA*(Fe) compared to the one used by Sallans et al. [95].

[m]Reference 97. [n]Reference 10. [o]References 82 and 84.

[p]Reference 42. [q]Reference 51. [r]Reference 52.

[s]Reference 54.

Reactions of Anions. Sallans et al. [94, 95] have used ion cyclotron resonance (ICR) mass spectrometry to study reaction (1.6),

$$M^- + AH \rightarrow MH + A^- \tag{1.6}$$

where AH is a suitable proton donor having a gas-phase acidity of ΔH_a(AH) = $D(A^-—H^+)$. By examining a series of acids, the acidity of the metal hydride, $D(M^-—H^+)$, was estimated for M = V, Cr, Fe, Co, and Mo. This acidity is then converted to the homolytic bond energy of MH via eq. (1.7),

$$D(\mathrm{M{-}H}) = D(\mathrm{M^-{-}H^+}) + EA(\mathrm{M}) - IE(\mathrm{H}) \tag{1.7}$$

The metal anions are generated via several steps. First, the appropriate metal carbonyl reacts with low-energy electrons via dissociative electron attachment. The metal carbonyl anions so generated are then accelerated in the presence of a rare gas. Collisions induce dissociation and this process is continued until the bare anion remains. Experiments were attempted with Mn and W but insufficient quantities of the atomic metal anion were produced for these metals. Sallans et al. found that all five metal anions would undergo reaction (1.6) with 5,5-dimethylcyclohexane-1,3-dione, $\Delta H_a = 1418 \pm 8$ kJ mol^{-1}. Only Co^- and Mo^- would react with $C_6H_5COCH_2COCH_3$, $\Delta H_a = 1422 \pm 8$ kJ mol^{-1}. Mo^- reacts with p-$H_2NC_6H_4COOH$, $\Delta H_a = 1427 \pm 8$ kJ mol^{-1}, but not with $CH_3COCH_2COCH_3$, $\Delta H_a = 1438 \pm 8$ kJ mol^{-1}.

The failure to observe a reaction can be caused by kinetic rather than thermodynamic restrictions. For example, in complex systems like these, it is possible that competing reaction channels might suppress an otherwise exothermic proton transfer reaction (although Sallans et al. report no other reactions) or barriers along the reaction path may exist. Therefore, strictly speaking, these results provide only lower limits to the acidities of the metal hydrides (although even this can be in error because slightly endothermic reactions can be observed under ICR conditions). Sallans et al. assumed that the reactions are under thermodynamic control (an assumption that is often true for ion–molecule reactions) and thereby assigned ΔH_a(VH, CrH, FeH) $= 1420 \pm 13$ kJ mol^{-1}, ΔH_a(CoH) $= 1425 \pm 13$ kJ mol^{-1}, and ΔH_a(MoH) $= 1433 \pm 13$ kJ mol^{-1}. By using eq. (1.7), these values were converted to the neutral bond energies listed in Table 1.4.

Reactions of Cations. In a similar type of experiment, Tolbert and Beauchamp [96] used an ion beam instrument to examine reaction (1.8),

$$\mathrm{M^+ + BH \rightarrow MH + B^+} \tag{1.8}$$

where BH is a suitable hydride donor. These authors measured the kinetic-energy-dependent cross section for reaction (1.8) to determine which reactions were exothermic for a series of bases and for M = Fe, Co, Ni, Ru, Rh, Pd, and Mo. As did Sallans et al., Tolbert and Beauchamp reasonably assumed that these reactions are under thermodynamic control and therefore bracketed $D(\mathrm{M^+{-}H^-})$ between the heterolytic bond energies $D(\mathrm{B^+{-}H^-})$ of the bases that did undergo reaction (1.8) and those that did not or that were endothermic. The heterolytic metal hydride bond energies derived in this way were converted to the homolytic bond energy $D(\text{M—H})$ by using

$$D(\text{M—H}) = D(\text{M}^{+}\text{—H}^{-}) - IE(\text{M}) + EA(\text{H}) \tag{1.9}$$

The resultant values are listed in Table 1.4.

One potential problem in these studies is the presence of excited states of the atomic metal cations. In all cases, M^+ is generated by surface ionization on a hot filament at 2500 K. This produces less than 2% excited states in the cases of Ni^+, Ru^+, Rh^+, Pd^+, and Mo^+, but 24 and 19% excited states for Fe^+ and Co^+, respectively. This effect was considered carefully by comparing the magnitudes of the experimental cross sections with the theoretical collision cross sections. This comparison was used to establish whether an observed exothermic process was due exclusively to excited-state species.

In a variation on this experiment, Armentrout and Beauchamp [63] examined the reactions of Co^+ with hydrocarbons. Included among these were the endothermic processes (1.10) and (1.11),

$$M^{+} + RH \rightarrow MH^{+} + R \tag{1.10}$$

$$M^{+} + RH \rightarrow MH + R^{+} \tag{1.11}$$

They found that reaction (1.10) had a larger cross section than reaction (1.11) when RH was propane, and therefore they concluded that $IE(\text{CoH}) < IE(\text{R})$, where R = 2-$C_3H_7$. Likewise, reaction (1.10) had a smaller cross section than reaction (1.11) when RH was cyclopentane, suggesting that $IE(\text{CoH}) > IE(c\text{-}C_5H_9)$. Combined with the bond energy for CoH^+ determined via reaction (1.1), this led to $D(\text{Co—H}) = 163 \pm 25$ kJ mol^{-1}. More recently, Georgiadis, Fisher, and Armentrout [82] have reexamined reactions (1.10) and (1.11) for Co^+ and propane. Because of better collection of the alkyl product ions, they find that formation of $C_3H_7^+$ clearly begins at a lower kinetic energy than formation of CoH^+, establishing that $IE(\text{CoH}) > IE(2\text{-}C_3H_7)$, in direct contrast to the conclusions of Armentrout and Beauchamp. This result sets a lower limit of 146 kJ mol^{-1} for $D(\text{Co—H})$.

This leads directly to the third type of cation experiment used to measure the bond energies for neutral diatomic transition metal hydrides. Georgiadis et al. [82] and Fisher and Armentrout [84] have examined the kinetic-energy dependence of reaction (1.11) for a series of hydrocarbons and M = Co, Ni, and Cu. The presence of excited states is explicitly discussed and treated. As with reaction (1.1), the kinetic-energy threshold is measured for these reactions and assumed to be equal to the endothermicity of the overall reaction. This assumes that there are no barriers in excess of the endothermicity, as is commonly true for ion–molecule reactions. This endothermicity, E_T, is then related to the desired bond energy by

$$D(\text{M—H}) = D(\text{R—H}) + IE(\text{R}) - IE(\text{M}) - E_T - E_{\text{el}} \tag{1.12}$$

where the quantities D(R—H), IE(R), and IE(M) are well-known. Compared with the bracketing techniques described previously for metal anions and cations, this method has the advantage of providing a more-precise determination of the metal hydride bond energy. Values for the systems where RH = ethane, propane, isobutane [82], and cyclopropane [84] were consistent for all three metals, and the averages of these results are given in Table 1.4.

Results of a similar study are also available for Fe, although here the interpretation is less straightforward. Schultz, Elkind, and Armentrout [81] examined the state-specific reactivity of Fe^+ with propane and found that reaction (1.11) was observed for the ^4F excited state of Fe^+ but was very inefficient for the ^6D ground state. The threshold measured for this reaction was corrected for the excitation energy of $Fe^+(^4F)$, 0.28 eV, and the iron hydride bond energy calculated by using eq. (1.12). The accuracy of the value obtained, 151 ± 10 kJ mol^{-1}, was questioned because the threshold for the directly competing reaction (1.10) was observed to be higher than the thermodynamic value by 0.42 ± 0.12 eV (41 ± 12 kJ mol^{-1}), a result that was believed to be due to competition with very efficient exothermic reactions. Another approach to obtaining this thermochemistry was to measure the *relative* thresholds of the directly competing reactions (1.10) and (1.11). If the barriers in both reaction channels are the same, this provides the ionization energy of FeH relative to the propyl radical. The result was $IE(\text{FeH}) = IE(2\text{-C}_3\text{H}_7) - 0.33 \pm 0.10$ eV $= 7.69 \pm 0.10$ eV, which is converted to the neutral bond energy by using

$$D(\text{M—H}) = D(\text{M}^+\text{—H}) + IE(\text{MH}) - IE(\text{M}) \qquad (1.13)$$

At the time, the resultant bond energy, $D(\text{FeH}) = 191 \pm 13$ kJ mol^{-1}, was considered to be the more-accurate value, although it is technically only an upper limit to the true bond energy. In recent work, Schultz and Armentrout [97] have examined reaction (1.11) in four additional systems (CH_3CHO, *n*-butane, cyclopentane, cyclopropane) and found bond energies that are in agreement with the 151 kJ mol^{-1} result from the propane system. The good agreement for all five systems suggests that barriers are *not* a problem in this reaction. The value for the average FeH bond energy obtained from these five systems is 157 ± 8 kJ mol^{-1}.

Similar studies also have been performed in our laboratories for Ti, V, Cr, and Mn. Aristov [98] observed the endothermic reaction (1.11) between V^+ and isobutane. After correcting for electronic excitation, the threshold of 2.39 ± 0.17 eV corresponds to a VH bond energy of 170 ± 17 kJ mol^{-1}. In the case of Mn, Sunderlin and Armentrout [99] found that ground-state $Mn^+(^7S)$ does not react with isobutane via reaction (1.11), although the excited ^5S and ^5D states do. Measurement of the threshold for this reaction was straightforward although the relative contributions of the ^5S and ^5D

states could not be determined unequivocally. Two possible interpretations of the data lead to $D(\mathrm{Mn{-}H}) = 177 \pm 7\ \mathrm{kJ\ mol^{-1}}$ or $126 \pm 18\ \mathrm{kJ\ mol^{-1}}$. Based on a comparison with the upper limit to the bond energy given by Kant and Moon [93] (See Table 1.4), the later value is chosen as the correct interpretation. Work in progress in our laboratory [100] on the reactions of Ti^+ and Cr^+ with methyl amines leads to preliminary values for $D(\mathrm{Ti{-}H}) = 205 \pm 9\ \mathrm{kJ\ mol^{-1}}$ and $D(\mathrm{Cr{-}H}) = 223 \pm 16\ \mathrm{kJ\ mol^{-1}}$.

Comparison of Values. In evaluating the reliability of the thermochemistry of the transition metal hydrides, we start by considering the cases where the spectroscopic bond energies are reliable (NiH, CuH, AgH, and AuH). In all four cases, there is reasonable agreement with the values of Kant and Moon ([92], hereafter referred to as KM). For NiH and CuH, the beam results of Armentrout and coworkers ([82, 84], hereafter referred to as GFA) also agree nicely. In these cases, the average values for these measurements are the recommended values listed in Table 1.3. In these four cases, the only other experimental value is that derived by Tolbert and Beauchamp (TB) for NiH [96]. Although the specific value obtained by TB is much higher than the others, the bracketing method used by these authors (and also by Sallans et al. [95], hereafter referred to as SLSF) provides only a definitive lower limit and a likely upper limit to the bond energy. In this context, the value of TB is in satisfactory agreement with the other values. For CoH, the high-temperature value from KM and the beam results of GFA agree nicely and the average is given in Table 1.3. Again the value of TB is consistent with this value.

The case of FeH is relatively controversial because there is clearly no consensus among the many different measurements. Based largely on the ability of the beam results to reproduce the spectroscopic values and those of KM in the cases of CoH, NiH, and CuH, we recommend $D(\mathrm{FeH}) = 157 \pm 8\ \mathrm{kJ\ mol^{-1}}$ [97]. This is consistent with the upper limit of KM and with the work of SLSF and TB when these bracketing experiments are conservatively viewed as deriving lower limits to the true bond energy. Justification for this value comes from ab-initio theoretical work. Bauschlicher et al. [101] suggest that $D_0(\mathrm{FeH}) = 158\ \mathrm{kJ\ mol^{-1}}$, which is equivalent to $D(\mathrm{FeH}) = 162\ \mathrm{kJ\ mol^{-1}}$, whereas Sodupe et al. [102] obtain $D_e(\mathrm{FeH}) = 148\ \mathrm{kJ\ mol^{-1}}$, which we adjust to $D(\mathrm{FeH}) = 141\ \mathrm{kJ\ mol^{-1}}$.

Less information is available for the other transition metal hydrides. In none of these cases are the spectroscopic values reliable. In the cases of VH and MoH, there is reasonable agreement between the two available determinations, so the average value is taken as the recommended bond energy. For CrH, the value obtained by SLSF is considerably below the beam determination, which we choose to list in Table 1.3. In the remaining cases, only one specific value is available and this is listed in Table 1.3. These values must be considered to be tentative until they are reproduced by other techniques.

1.2.2.3. ANIONS

The bond dissociation energies of the diatomic transition metal hydride anions are easily calculated from the electron affinities *EA* of the atomic metal and its hydride and the bond dissociation energy of the neutral hydride. This relationship is given by

$$D(\mathrm{M^-{-}H}) = D(\mathrm{M{-}H}) + EA(\mathrm{MH}) - EA(\mathrm{M}) \qquad (1.14)$$

The electron affinities come from the photoelectron spectroscopy of Lineberger and co-workers [23, 24, 103–106] and are listed in Table 1.2. Combined with the neutral bond energies given in Table 1.3, these values provide the anionic bond energies, also listed in Table 1.3.

In the case of manganese, the metal electron affinity is less than or equal to zero. This establishes that $D(\mathrm{Mn^-{-}H}) \geq 210 \pm 18$ kJ mol^{-1}. A speculative value for this bond energy can be estimated by noting that Miller, Feigerle and Lineberger [107] find a good linear correlation between $EA(\mathrm{MH_2})$ and $EA(\mathrm{M})$ for M = Fe, Co, and Ni. By using the measured value for $EA(\mathrm{MnH_2})$ in this correlation, one can estimate that $EA(\mathrm{Mn}) \approx -0.5$ eV for the purposes of these simple metal hydride species. Using this value in eq. (1.14) predicts that $D(\mathrm{Mn^-{-}H}) \approx 258$ kJ mol^{-1}.

1.2.2.4. PERIODIC TRENDS

The variations in the transition-metal-diatom bond energies with metal identity have been discussed extensively for the cationic [64, 76, 78, 108–110] and neutral [109–111] species. One particularly useful way of understanding these variations is in terms of promotion energy arguments. This work suggests that there exists an "intrinsic" metal hydride bond energy that represents the most favorable bond that can be formed between a metal and a hydrogen atom. For the first-row diatomic metal hydrides, such a bond can be formed between a 4*s* orbital on the metal and a 1*s* orbital on the hydrogen atom. Because the ground states of the metals do not have this $4s^13d^{n-1}$ configuration, energy is required to promote the atom to this configuration. This "promotion" energy should include the energy needed to spin-decouple the 4*s* electron from the 3*d* electrons [108]. The extent to which this idea accurately reproduces the experimental bond energies is shown in Figure 1.1, which shows the bond energies of Table 1.3 plotted against promotion energies taken from the compilation of Carter and Goddard [112]. Clearly, the bond energy and the promotion energy are strongly correlated. The zero-promotion-energy intercept of this plot defines the intrinsic bond energy for the first-row transition metal hydride cations as 234 kJ mol^{-1}.

Figure 1.1 also shows data for neutral first-row metal hydrides. It can be seen that the periodic trends in the neutral bond energies are in reasonable accord with those for the cationic bond energies. Indeed, the neutral data follow a very similar quantitative correlation (not shown in the figure, for

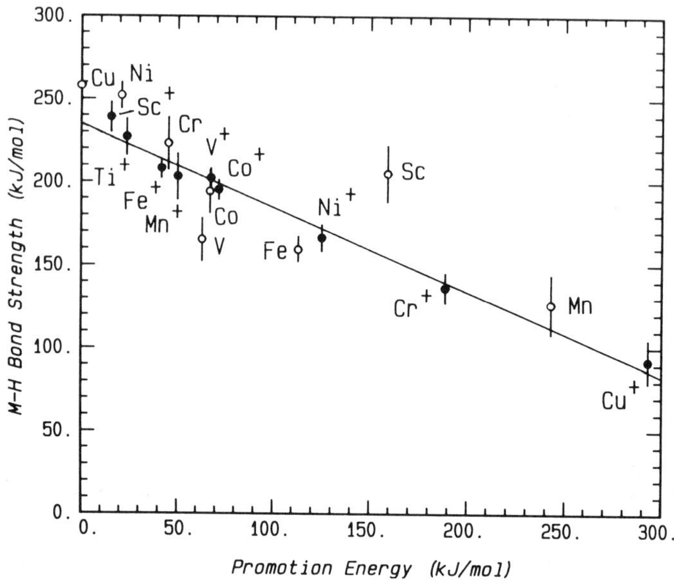

FIGURE 1.1
First-row transition metal hydride bond energies versus atomic metal ion promotion energy to a $4s3d^{n-1}$ spin-decoupled state (see text). Closed and open symbols show data for cationic and neutral species, respectively. The line is a linear regression fit to the cationic data.

clarity) with a comparable intrinsic bond energy of 248 kJ mol^{-1}. This indicates that the electronic configuration of the metal rather than its charge is the key factor that determines the strength of the metal–hydrogen bond.

Figure 1.2 shows a similar plot for the second- and third-row transition metal hydrides. In general, bond energies for the second-row neutral and cationic species also correlate with promotion energy, although Pd^+ clearly deviates from the correlation. Pd^+ is believed to be an exception to the M(s)—H(1s) bonding scheme because the promotion energy to a $5s4d^8$ configuration is very high (361 kJ mol^{-1}) and the palladium ground state has a $4d^9$ configuration with one singly occupied $4d$ orbital. Consequently, hydrogen bonds to Pd^+ by using this $4d$ orbital (with zero promotion energy) rather than bonding to the $5s$ orbital and paying the large promotion energy [76]. This has been confirmed by ab-initio calculations, which also show that a similar situation may hold for Ru^+ and Rh^+ [113, 114]. A fit to the data for the second-row cations (excluding Pd^+) gives an intrinsic bond energy of 261 kJ mol^{-1}, somewhat higher than for the first-row metals.

An alternate means of understanding the periodic trends in the *neutral* transition metal hydride bond energies has been suggested by Squires [115].

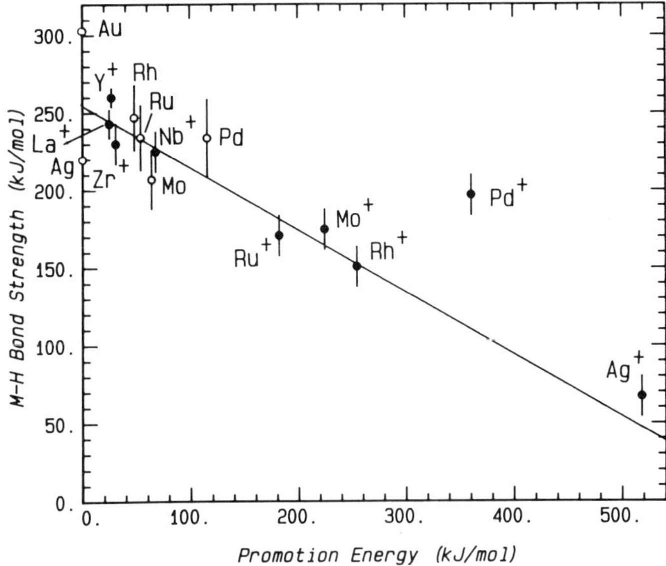

FIGURE 1.2
Second- and third-row transition metal hydride bond energies versus atomic metal ion promotion energy to an sd^{n-1} spin-decoupled state (see text). Closed and open symbols show data for cationic and neutral species, respectively. The line is a linear regression fit to the second row cationic data (excluding Pd^+).

He found that the acidities of MH, $D(M^-—H^+)$, were approximately constant, which means that D(M—H) and the electron affinity EA of M are directly related. For the data given in this review (Table 1.3), Figure 1.3 shows that there is a reasonable correlation between these quantities, especially for the best-known D(MH) values: Co, Ni, and Cu. The "constant" acidity, which was found to be 1427 ± 21 kJ mol^{-1} by Squires [115], is now 1448 ± 30 kJ mol^{-1} with the updated data. (This is an average of all neutral metal hydride data in Table 1.3 except for Mn, an element for which the anion is unstable, and Pt, for which only a limit to the bond strength is known. Electron affinities are taken from Mead, Stevens, and Lineberger [116].) A line corresponding to this acidity is shown in Figure 1.3. For the neutral metal hydrides, the promotion energy and electron affinity correlations are related because they are both measures of the energy required to make MH with a $d^{n-1}\sigma^2$ electron configuration. We note, however, that the acidities of the cationic transition metal hydrides MH^+ are not constant (ranging from 648 to 1014 kJ mol^{-1}), such that this correlation appears to be less general than the one involving promotion energy.

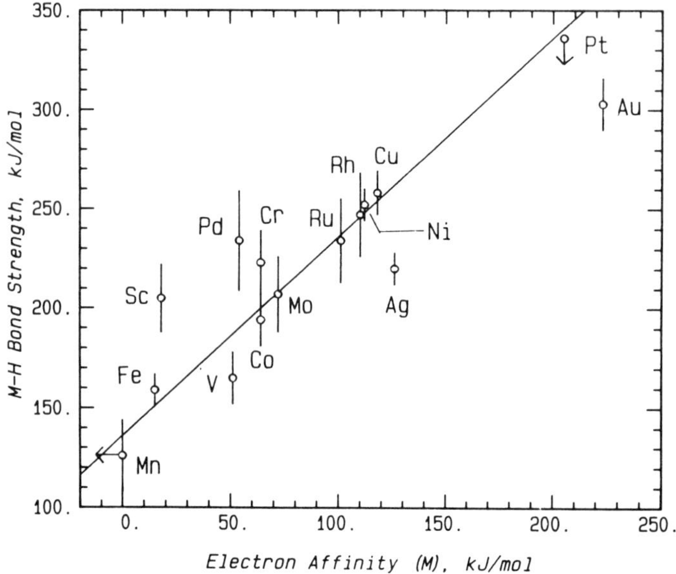

FIGURE 1.3
Transition metal hydride neutral bond energies versus atomic metal electron affinity. The line corresponds to the average acidity of the data shown, $D(M^{-}—H^{+}) = 1448$ kJ mol^{-1}. The arrow on the platinum point indicates an upper limit to the bond energy. The arrow on the manganese point indicates an unstable atomic anion.

1.2.3. Reactivity

1.2.3.1. REACTIONS TO FORM DIATOMIC TRANSITION METAL HYDRIDES

One of the interesting aspects of the study of reaction (1.1) is the variation in reactivity observed for different transition metal ions. The studies of these reactions include detailed examination of the effect of electronic excitation on the reactivity and mechanistic studies by looking at the changes in reactivity for H_2, HD, and D_2. These aspects of these studies have been reviewed previously [61, 78, 117] and are summarized briefly here.

Three categories of reactivity appear to exist based on the electron configuration and spin state of the atomic metal ion. Experimental results that illustrate these categories are shown in Figure 1.4.

(1) If the $4s$ and $3d\sigma$ orbitals are unoccupied, the systems react efficiently. The metal ion reacts with HD to form nearly equal amounts of MH^+ and MD^+. Overall, the process occurs via a statistically behaved intermediate. $Ti^+(^4F, 3d^3)$ [70] and $V^+(^5D, 3d^4)$ [71] (shown in Figure 1.4) are examples of this type.

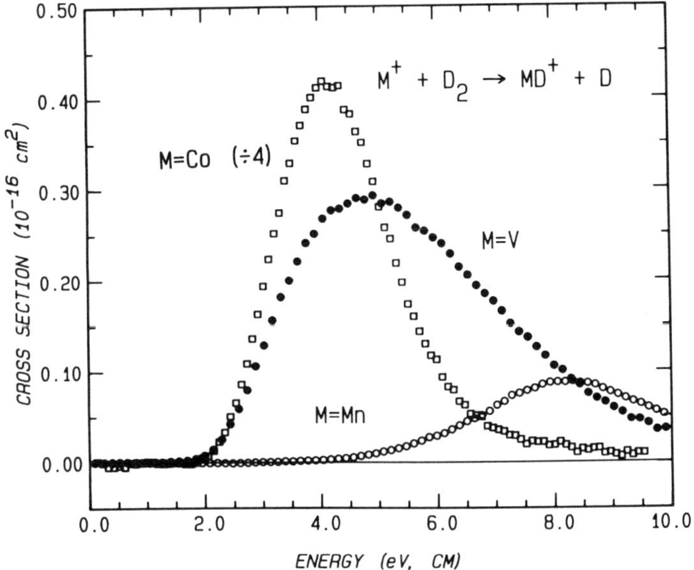

FIGURE 1.4
Cross sections for reaction of V^+, Mn^+, and Co^+ with D_2 as a function of relative translational energy. In all cases, the ions are produced by surface ionization such that the data correspond closely to reactivity of only the ground electronic states, $V^+(^5D, 3d^4)$, $Mn^+(^7S, 4s3d^5)$, and $Co^+(^3F, 3d^8)$.

(2) If either the $4s$ or $3d\sigma$ orbital is occupied, the systems can react efficiently if they have low spin. The branching ratio in reaction with HD is $\approx 3:1$ in favor of the MH^+ product. This indicates that the reaction is more direct (such that angular momentum conservation dictates the branching ratio). For metals on the left side of the Periodic Table, such configurations are all excited states with $4s3d^{n-1}$ configurations, for example, $Ti^+(^2F, 4s3d^2)$ [70], $V^+(^3F, 4s3d^3)$ [71], $Cr^+(^4D, 4s3d^4)$ [72], and $Mn^+(^5S, 4s3d^5)$ [73]. For metals on the right side of the Periodic Table, this is the $3d^n$ configuration. Examples are the $Co^+(^3F, 3d^8)$ (shown in Figure 1.4), $Ni^+(^2D, 3d^9)$, and $Cu^+(^1S, 3d^{10})$ ground states [75] and the $Fe^+(^4F, 3d^7)$ [74] and $Mn^+(^5D, 3d^6)$ [73] excited states.

(3) If either the $4s$ or $3d\sigma$ orbital is occupied and the ion has a high spin, the systems react inefficiently at the thermodynamic threshold and via an impulsive (hard-sphere-like) mechanism at elevated energies. The branching ratio in reaction with HD favors production of the MD^+ product at low energies. $Mn^+(^7S, 4s3d^5)$ [73] (shown in Figure 1.4) and $Fe^+(^6D, 4s3d^6)$ [74] are typical.

Although these guidelines are attractive in their simplicity, they do not predict the behavior of all valence states of all ions. For example, the $Sc^+(^3D, 4s3d)$ and $Ti^+(^4F, 4s3d^2)$ ground states (which both have high-spin $4s3d^{n-1}$ configurations of category 3) react as suggested by category 1. The explanation is that the potential energy surfaces evolving from these states mix with those from more reactive excited states [the $Sc^+(^3F, 3d^2)$ and $Ti^+(^4F, 3d^3)$, which have the same spin and are in category 1].

Molecular Orbital and Spin Considerations. Activation of H_2 can be understood via the simple molecular orbital interactions shown in Figure 1.5. In essence, the bonding electrons of H_2, $\sigma_g(H_2)$, are donated to an acceptor orbital on the metal center (here, the s orbital of the atomic metal ion, with contributions from the $d\sigma$). In turn, a donor orbital on the metal (here a $d\pi$ orbital) donates electrons into the antibonding orbital of H_2, $\sigma_u^*(H_2)$. Both interactions serve to weaken the H_2 bond while building electron density between the metal and the H atoms. Interactions of H_2 with metal atoms [118, 119], metal complexes [120, 121], metal clusters (see Section 1.5), and metal surfaces [120, 122] all have been discussed theoretically in such terms.

Clearly, the most favorable electron configuration for a reactant metal is to have the acceptor orbital empty (as for the category 1 ions discussed previously) and the donor orbital fully occupied. If the acceptor orbital is occupied (as for categories 2 and 3), formation of an MH_2^+ intermediate is impeded and the reaction is forced to occur via a more direct pathway. Categories 2 and 3 are distinguished by the spin of the reactants, which directly influences the repulsiveness of the M^+–H_2 interactions. High spin prevents efficient bonding interactions involving the nonbonding electrons on the metal. This is illustrated by noting that the ground states of all MH_2^+ intermediates have low spin.

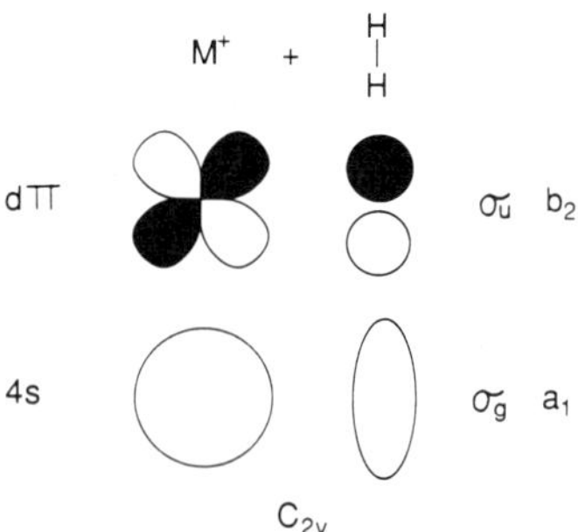

FIGURE 1.5
Molecular orbital interactions for a perpendicular approach of atomic transition metal ions to H_2.

1.2.3.2. REACTIONS OF DIATOMIC TRANSITION METAL HYDRIDES

There are few studies of the reactions of transition metal hydride diatoms in the gas phase. No work on neutral and anionic species has been performed and only the cationic hydrides of Mn, Fe, Co, and Ni have been studied. The reason for this paucity of information is that in almost all of the experiments discussed previously, the metal hydrides are the *products* of reactions. As a consequence, either the product yield is insufficient or the experimental apparatus is ill-equipped to examine the subsequent reactivity of these species.

In 1981, Stevens and Beauchamp [86] examined the reactivity of MnH^+ via ICR mass spectrometry, largely to determine the $Mn^+—H$ bond energy by examining the deprotonation reaction (1.3), as described in Section 1.2.2.1. Two additional reactions were observed in addition to the simple deprotonation processes. MnH^+ reacts with the parent complex, $HMn(CO)_5$, to eliminate H_2 and CO to form $Mn_2(CO)_4{}^+$. With acetaldehyde, MnH^+ reacts to form $MnCH_3{}^+ + H_2 + CO$ (H_2CO formation is also possible and has nearly the same energetics). This interesting process is suggested to proceed via the sequence,

$$MnH^+ + CH_3CHO \rightarrow H_2MnCOCH_3{}^+ \rightarrow MnCOCH_3{}^+ + H_2 \rightarrow$$

$$(CO)MnCH_3{}^+ + H_2 \rightarrow MnCH_3{}^+ + CO + H_2$$

although there is no direct evidence for any of the proposed intermediates.

Carlin et al. [123] have used ICR to examine the reactivity of FeD^+, CoD^+, and NiD^+. These species were generated by reaction of the atomic metal cation with d_3-methylnitrite followed by collisional activation (with an unspecified partner, designated here as Rg).

$$M^+ + CD_3ONO \rightarrow MOCD_3{}^+ + NO$$

$$MOCD_3{}^+ + Rg \rightarrow MD^+ + CD_2O + Rg$$

In a related study, Halle et al. [66] generated FeH^+ by electron impact ionization of $(CH_3C_5H_4)_2Fe$ and examined its reactivity by using an ion beam apparatus. Studies involving proton and hydride transfer reactions of this species are described in detail in Sections 1.2.2.1 and 1.3.1, respectively.

Carlin et al. found that the thermoneutral H–D exchange reaction

$$MD^+ + H_2 \rightarrow MH^+ + HD \qquad (1.15)$$

was observed for CoD^+ and NiD^+, but not for FeD^+. This is consistent with the observations of Halle et al., who found that FeH^+ exchanges with D_2 but with an activation energy of 84 ± 29 kJ mol^{-1}. Only NiD^+ was observed to react with methane, and it exchanged D for H (to form $NiH^+ + CH_3D$) and D for CH_3 (to form $NiCH_3^+ + HD$). With ethane, both CoD^+ and NiD^+ react to form $MC_2H_5^+ + HD$. At elevated kinetic energies, FeH^+ reacts with ethane to form $FeCH_3^+ + CH_4$. With larger alkanes (RH), all three metal hydrides react to form MR^+, and this product undergoes varying amounts of subsequent decomposition by losing H_2 or CH_4. The extent of decomposition increases from Fe to Co to Ni, consistent with the reactivity observed with the smaller alkanes. All three MD^+ ions react with ethene in an H–D exchange reaction, process (1.16), and NiD^+ and CoD^+ also react to lose HD to form the metal vinyl ion, reaction (1.17):

$$MD^+ + C_2H_4 \rightarrow MH^+ + C_2H_3D \quad (1.16)$$

$$MD^+ + C_2H_4 \rightarrow MC_2H_3^+ + HD \quad (1.17)$$

The exchange process is facile and probably proceeds via a reversible equilibrium between a metal hydride ethene complex, $DM(C_2H_4)^+$, and the metal ethyl ion, $MCH_2CH_2D^+$. Additional chemistry with propene, benzene [123], and various oxygen bases [66] also has been examined.

An interesting observation in these studies is that the reactivity of the metal hydride cations increases from Fe to Co to Ni, whereas the reactivity of the bare atomic ions is generally considered to have the opposite order. The reactivity of MH^+ inversely correlates with its bond energy (Table 1.3), suggesting that the reactivity of these species is directly related to their stability. Another way of thinking about this is to examine the relative stabilities of the HMR_2^+ species (where R = H or an alkyl), the intermediate formed by oxidative addition of a covalent bond to the metal hydride ion. This can be estimated from the energies required to promote MH^+ to a state that can strongly bond two additional covalent ligands. Based on the calculations of Carter and Goddard [112], these promotion energies are 219, 139, and 49 kJ mol^{-1}, respectively. Clearly, these values correlate nicely with the relative reactivities of the metal hydride cations.

1.3. HOMOLEPTIC TRANSITION METAL HYDRIDES

Very little experimental work has been performed on the homoleptic hydrides of transition metals, MH_x ($x > 1$), in the gas phase. Field evaporation mass spectrometry has been used to observe MH_2^+ (M = Ni, Pd, Au), CuH_x^+ ($x = 2$–5, where the even numbers are much more abundant than the odd), and such exotic species as MH_2^{2+} (M = Nb, Mo, Rh), TaH_2^{3+}, TiH_x^{2+} ($x = 2$–5), and ZrH_x^{2+} ($x = 2$–4) [1]. Cesium sputter sources have been used to produce the following: dihydride anions of Sc, Ti, V, Cr, Fe, Y, Zr, Nb, Mo, Rh, La, and Ta; the trihydride anions of Sc, Ti, V, Y, Mo, and La; and tetrahydride anions of Sc, Y, and La [4, 124]. No gas-phase optical spectroscopy has been performed on the homoleptic transition metal hydrides, although work in rare-gas matrices exists [125]. Metal dihydride cations are formed transiently in the experiments, described in Section 1.2.3.1, on the reactions of atomic metal cations with H_2, but no quantitative data regarding these intermediates are obtained in these studies.

Quantitative experiments on the homoleptic transition metal hydrides include production of MH_2^+ (Sc, Y, La) and FeH_2 in ion beam experiments and the photoelectron spectroscopy of the dihydride anions of Mn, Fe, Co, and Ni [107]. These studies are described in more detail in the following subsections.

1.3.1. Ion Beam Experiments

Metal Dihydride Cations (Sc, Y, La). Tolbert and Beauchamp [65] were the first to observe that Sc^+ reacts with ethane and formaldehyde to form the metal dihydride cation product, reactions (1.18) and (1.19).

$$M^+ + C_2H_6 \rightarrow MH_2^+ + C_2H_4 \tag{1.18}$$

$$Sc^+ + CH_2O \rightarrow ScH_2^+ + CO \tag{1.19}$$

The first process was found to be endothermic whereas the latter process is exothermic. These results show that 436 kJ mol^{-1} $< D(Sc^+\text{—}H) + D(HSc^+\text{—}H) <$ 572 kJ mol^{-1}. A more specific value was later obtained by Sunderlin et al. [80], who made a more-detailed study of reaction (1.18). In addition, these authors found that Sc^+ reacted with propane, cyclopentane, and cyclohexane to form ScH_2^+ in endothermic processes. Determination of the endothermicities for all four reactions yielded an average value for $D(Sc^+\text{—}H) + D(HSc^+\text{—}H)$ of 483 ± 13 kJ mol^{-1}, in agreement with the limits set by Tolbert and Beauchamp.

YH_2^+ and LaH_2^+ have been observed by Huang et al. [126] as products in the high energy collision induced dissociation of the complexes of these metal ions with alkanes and alkenes. A quantitative characterization of these species has been provided by Sunderlin and Armentrout [83], who determined that Y^+ and La^+ form the dihydride cations via reaction (1.18). Analyses of these endothermic processes yield the bond energies in Table 1.5. Observations of metal dihydride cations also have been made for two Lanthanide elements (Gd^+ via reaction (1.19) [127] and Lu^+ via reaction (1.18) [83]). To date, no other transition metal ions have been observed to form such metal dihydride cations in reactions with hydrocarbons, formaldehyde, or other organic molecules. Estimates of the thermochemistry of such MH_2^+ species have been made [128] and propagated [129], but are not reported here because they are not based on direct experimental information.

Table 1.5
Experimental Bond Dissociation Energies of Transition Metal Dihydrides[a]

M	D(M—H)[b]	D(HM—H)	Sum
Sc^+	239(9)	244(15)	483(13)[c]
Y^+	260(6)	273(9)	533(7)[d]
La^+	243(9)	266(12)	509(8)[d]
Fe	157(8)	> 313 [> 271]	> 478 [> 436][e]
Fe^-	234(17)	> 314 [> 272]	> 565 [> 523][f]

[a]Values (in kilojoules per mole) are for 298 K. Uncertainties are in parentheses.

[b]Values from Table 1.3. [c]Reference 80. [d]Reference 83.

[e]Derived from results of reference 66. Values in brackets are obtained from a more-conservative interpretation of the data (see text).

[f]Derived from results for FeH_2 and data in Table 1.2 (see text).

It can be seen in Table 1.5 that the second metal hydride bond energy, $D(HM^+—H)$ for M = Sc, Y, and La, is slightly larger than the first for all three metals. This is reasonable because these three metal ions have only two valence electrons, such that formation of the second bond requires no promotion energy. Because the first metal hydride bond energy, $D(M^+—H)$, is near the maximum for all three metals (see Figures 1.1 and 1.2), the sum of the two bond energies is approximately twice the intrinsic metal hydride bond energy determined in Section 1.2.2.4. On this basis, these values are probably reasonable estimates for the maximum stability of metal dihydride species. Because the H_2 bond energy is 436 kJ mol^{-1}, it can be seen that these MH_2^+ species are stable to reductive elimination of H_2 by 47–97 kJ mol^{-1}.

Iron Dihydride. One intriguing study, which has unfortunately not been repeated for other metals, was performed by Halle et al. [66]. They examined the reactions of FeH^+ (generated as described in Section 1.2.2.1) with a series of hydride donors to form FeH_2, by process (1.20):

$$MH^+ + BH \rightarrow MH_2 + B^+ \tag{1.20}$$

This reaction occurred at low kinetic energies with CH_3CHO but not with $(CH_3)_2O$ and was apparently close to thermoneutral with C_2H_5OH. Based on the organic thermochemistry chosen by these authors, they concluded that $D(HFe^+—H^-) \geq D(C_2H_4OH^+—H^-) = 971$ kJ mol^{-1} although thermochemistry listed by Lias et al. [88] suggests the limit is ≥ 962 kJ mol^{-1}. The heat of formation of FeH_2 was then derived by using

$$\Delta_f H(MH_2) \leq \Delta_f H(MH^+) + \Delta_f H(BH) - \Delta_f H(B^+) \tag{1.21}$$

They found $\Delta_f H(FeH_2) < 324$ kJ mol^{-1} although this calculation utilized a bond energy for FeH^+ of 247 kJ mol^{-1}. Using the value listed in Table 1.3 and the thermochemistry from Lias et al. [88] provides $\Delta_f H(FeH_2) <$ 372 kJ mol^{-1}. This means that FeH_2 is stable relative to Fe + H_2 by > 42 kJ mol^{-1} or that the sum of the two Fe—H bond energies is > 478 kJ mol^{-1}. Given the value for D(Fe—H) listed in Table 1.3, the second hydride bond energy can be calculated as D(HFe—H) > 321 ± 8 kJ mol^{-1} [and if D(Fe—H) were as low as 126 kJ mol^{-1}, D(HFe—H) would be > 352 kJ mol^{-1}]. This is an exceptionally strong bond compared with any other gas-phase metal hydride bond energy (cation, neutral, or anion) given in Table 1.3.

A more-conservative interpretation of the data of Halle et al. considers the possibility that the FeH^+ reactant has a small amount of internal energy. This could mean that reaction (1.20) with CH_3CHO and C_2H_5OH is actually endothermic. The reactant with the next-highest hydride affinity examined by Halle et al. was C_2H_5CHO. Reaction (1.20) with this species was clearly exothermic, suggesting that $D(HFe^+—H^-) \geq D(C_2H_5CO^+—H^-) = 920$ kJ mol^{-1} [88], which in turn means that $\Delta_f H(FeH_2) < 414$ kJ mol^{-1} or that D(Fe—H) + D(HFe—H) > 436 kJ mol^{-1}. This yields an estimated D(HFe—H) bond, > 279 ± 8 kJ mol^{-1}, which is more in keeping with the other metal hydride bond energies in Table 1.3. Also, this result is consistent with theoretical calculations, which find that formation of FeH_2 from Fe and H_2 is a nearly thermoneutral process [119], implying that D(Fe—H) + D(HFe—H) ≈ 436 kJ mol^{-1}.

These new values also suggest an alternative interpretation of the experiment of Halle et al., namely, that the process observed is *not* reaction (1.20) but process (1.22), in which the iron dihydride is not formed at all.

$$FeH^+ + BH \rightarrow Fe + H_2 + B^+ \tag{1.22}$$

Because the ion beam experiments detect only the ionic products, reactions (1.20) and (1.22) can be differentiated only by their thermochemistry. Given the bond energy for FeH^+ listed in Table 1.3, reaction (1.22) should be exothermic for species where $D(B^+—H^-) < 916 \pm 8$ kJ mol^{-1}. Although there is no direct evidence that this alternate interpretation of the data is more valid than the original interpretation, more-definitive experiments that rule out the possibility of internally excited FeH^+ must be performed before the thermochemistry of FeH_2 as derived by Halle et al. can be considered unequivocal.

1.3.2. Photoelectron Spectroscopy

The PES spectra of MnH_2^-, FeH_2^-, CoH_2^-, and NiH_2^- are all rather simple, and as a consequence no detailed spectroscopic data are obtained [107]. No vibrational progressions are observed except in the case of NiH_2^-, where there is a possible vibrational band of the neutral at $\sim$ 2200 cm^{-1}. All other peaks correspond to electronic states of the neutral, as demonstrated by deuterium labelling experiments. Excited states of the neutrals are observed at 1.182 $\pm$ 0.022 eV for FeH_2, 0.066 $\pm$ 0.011 eV and $\sim$ 0.32 eV for CoH_2, and $\sim$ 0.2 eV for NiH_2. The energy of the first peak is used to determine the electron affinities of the dihydrides listed in Table 1.2. CrH_2^- was also studied but could not be photodetached with 2.54- or 2.71-eV photons.

Although spectroscopic information is not obtained directly from this work, Miller et al. [107] argue that the failure to observe extensive vibrational progressions demonstrates that the dihydride anions and neutrals are both linear. (If either the anion or the neutral or both had a bent structure, it seems likely that a progression in the bending mode would be excited upon photodetachment.) This is consistent with theoretical calculations for the structures of the neutral transition metal dihydrides [119, 130, 131]. These observations indicate that photodetachment removes a nonbonding electron from the anion. Thus, the ground-state electron configurations of the dihydrides are likely to be $MH_2^-(3d^{n+1}\sigma^2\sigma^2)$ and $MH_2(3d^n\sigma^2\sigma^2)$, where σ represents the two MH bonding orbitals involving M(4*s*), M(4*p*), and H(1*s*) orbitals.

Finally, because there is an estimate for the thermochemistry of the neutral FeH_2 molecule, the electron affinity of FeH_2 can be used to derive the thermochemistry of the FeH_2^- species. This is listed in Table 1.5. It can be seen from this data that the second bond energy, $D(\mathrm{HFe^-{-}H})$, is toward the high side of the simple metal hydride bond energies listed in Tables 1.3 and 1.5. This suggests that the lower value listed in brackets may be a more-reasonable estimate of the true bond energy. Even this value is quite high, equalling $D(\mathrm{HY^+{-}H})$, which is expected to be one of the strongest bonds. This observation is reasonable in light of the electron configurations for FeH_2^- and FeH^- proposed by Miller et al. [23, 107]. Binding another hydrogen atom to $FeH^-(3d^6\sigma^2\sigma^{*1})$ leads directly to the ground-state configuration of $FeH_2^-(3d^6\sigma^2\sigma^2)$. Thus, the $\mathrm{HFe^-{-}H}$ bond should be strong because the FeH^- configuration is ideally suited to binding the second hydrogen atom, that is, no promotion is required.

1.4. TRANSITION METAL HYDRIDE COMPLEXES

Gas-phase studies of transition metal hydride complexes (species where both a hydrogen ligand and other types of ligands are present) again encompass both neutral and ionic species. Studies of the neutral complexes are largely restricted to stable closed-shell (18-electron) complexes for obvious reasons. These species have been examined by many techniques, including mass spectrometry, photoelectron spectroscopy, ultraviolet spectroscopy, electron diffraction, and photolysis–transient-IR spectroscopy. The methods used to study ionic transition metal complexes are such that coordinatively unsaturated species are easily included. Such work includes flow-tube studies of anionic transition metal hydride complexes, determinations of the proton affinities or hydride affinities of metal complexes, various reactions of metal-complex ions, and collision-induced dissociation experiments. In these ionic experiments, the metal hydride complexes are identified mass spectrometrically, that is, exclusively by mass-to-charge ratio. Thus, definitive proof for a metal–hydrogen bond is lacking although the chemical behavior of such species can be used to make such a linkage highly plausible.

1.4.1. Neutral Transition Metal Hydride Complexes

1.4.1.1. IONIZATION STUDIES

Mass spectrometry has been used to identify transition metal hydrides since the 1960s. The method was important because of the difficulty in detecting hydrides via optical spectroscopies. Most of the work has been reviewed previously [132, 133] and involves simple observation of such species by electron impact ionization and fragmentation of the complexes at electron energies of 70 eV. The general trend observed for hydride carbonyl complexes is that the hydride ligand is lost competitively with carbonyl ligands for monomeric metal hydride complexes and polymetallic complexes with terminal hydride ligands. In contrast, bridging hydrides are not lost unless several carbonyl ligands are also lost [132].

1.4.1.2. SPECTROSCOPY

Spectroscopic studies of a number of volatile transition metal hydride complexes have been performed. Many of these involve ultraviolet photoelectron spectroscopy (PES) but X-ray PES and electron diffraction also have been applied to gas-phase metal hydride compounds.

The most extensive work involves the four complexes, $HMn(CO)_5$, $HRe(CO)_5$, $H_2Fe(CO)_4$, and $HCo(CO)_4$. The UV PES of $HMn(CO)_5$, [134–136], its second-row analog $HRe(CO)_5$ [135, 136], and $HCo(CO)_4$ [135] were interpreted by Cradock, Ebsworth, and Robertson [135], who assigned bands at 10.6, 10.5, and 11.5 eV, respectively, to ionizations from the M—H bonding orbital. Guest et al. [137] measured the UV PES of $H_2Fe(CO)_4$ and assigned two bands at 10.95 and 11.3 eV to ionizations from the M—H bonding orbitals. Chen et al. [138] used X-ray PES to study the manganese, iron, and cobalt complexes and concluded that the hydride ligands carry charges of -0.8, -0.3, and -0.75, respectively.

Ultraviolet optical spectra of $HMn(CO)_5$ and $HRe(CO)_5$ yield results concerning the electronic levels that are consistent with the PES studies [139]. Stobart [140] measured the gas-phase IR spectrum of $H_2Fe(CO)_4$ and $GeH_3(H)Fe(CO)_4$ and tentatively assigned bands in the spectrum of $H_2Fe(CO)_4$ at 1895 (stretch), 800, 693, and 585 cm^{-1} to metal–hydrogen interactions. The gas-phase IR spectrum of $HMn(CO)_5$ has been measured and discussed several times, with bands at 1794 (stretch), 731, 612, and 362 cm^{-1} attributed to vibrations involving the hydride ligand [141]. Electron diffraction results [142] indicate that $HMn(CO)_5$, $H_2Fe(CO)_4$, and $HCo(CO)_4$ have metal hydride bond lengths of 1.58, 1.56, and 1.56 ± 0.02 Å, respectively, and that the H—Fe—H angle is $100° \pm 10°$. The manganese result is in good agreement with a neutron diffraction measurement on a crystal of $HMn(CO)_5$, $r(\text{Mn—H}) = 1.60 \pm 0.02$ Å [143], but differs from an earlier gas-phase electron diffraction measurement of 1.43 ± 0.05 Å [144].

Ultraviolet photoelectron studies of the metallocene derivatives [$(C_5H_5)_2ReH$, $(C_5H_5)_2MoH_2$, and $(C_5H_5)_2WH_2$ [145]] and the niobium and tantalum compounds [$(C_5H_5)_2MH_3$ [145, 146], $(C_5H_5)_2M(CO)H$ [146], and $(C_5H_5)_2M(C_2H_4)H$ [146]] all show ionization from the M—H bonding orbitals in the region from 8 to 11 eV. Lichtenberger et al. [146] note that, for the monohydrides, the ionization energies from M—H bonding orbitals are higher than those from the M–alkyl bonding orbitals in the analogous monoalkyl complexes. This is interpreted to indicate that the M—H bond is stronger than the M—CH_3 bond.

An unusual case involves studies of cyclohexenyl manganese tricarbonyl, η^3-$C_6H_9Mn(CO)_3$, a molecule that exhibits agostic [147] metal–CH bonding in solution [148]. Gas-phase UV PES studies [149] find no evidence for donation from the C—H bonding orbital to the metal, although there is some evidence for donation from the metal into a C—H antibonding orbital.

1.4.1.3. PHOTOLYSIS REACTIONS

Following the discovery of condensed-phase transition metal complexes involving an η^2-dihydrogen ligand [150], such complexes have also been studied in the gas phase in two instances. Fletcher and Rosenfeld [151] photolyzed $Cr(CO)_6$ to form the unsaturated $Cr(CO)_4$ complex. The disappearance of this species was monitored by transient infrared spectroscopy [152]. In the presence of large amounts of He and small quantities of H_2 or D_2, this species was found to decay in first order with respect to the dihydrogen pressure at a rate about one-tenth the collision rate. No data regarding the product of this reaction was obtained, but it seems likely to be the simple association complex, $Cr(CO)_4H_2$.

Ishikawa et al. [153] performed a similar experiment by photolyzing $W(CO)_6$ to form $W(CO)_4$ and $W(CO)_5$. In the presence of dihydrogen, these unsaturated complexes formed $W(CO)_4H_2$ transiently and the stable (lifetimes exceeding 1 ms) saturated complexes, $W(CO)_5H_2$ and $W(CO)_4(H_2)_2$. Transient infrared spectroscopy was used to monitor the disappearance of reactants, to monitor the appearance of products, and to aid in product identification. The data were interpreted to indicate weak W—H_2 electron back-donation, and the lifetime suggests W—H_2 bond energies of > 67 kJ mol^{-1}.

1.4.2. Anionic Transition Metal Hydride Complexes

1.4.2.1. PRODUCTION AND CHARACTERIZATION OF ANIONIC TRANSITION METAL HYDRIDES

Several methods have been used to generate transition metal hydride anion complexes. The major routes include the reaction of anionic complexes with hydrogen atoms and hydrogen molecules, and the reactions of hydride anions with transition metal compounds.

Reactions with Hydrogen Atoms. Conceptually, one of the simplest ways to form metal hydride complexes is the reaction of hydrogen atoms with an appropriate precursor complex. Two such studies have been undertaken, both by McDonald and co-workers, who utilized a flowing afterglow apparatus. The H atoms were produced by passing H_2 through a microwave discharge. In the first study, McDonald, Chowdhury, and Schell [154] observed that $Fe(CO)_4^-$ reacts with hydrogen atoms to form both $Fe(CO)_3^-$ and $HFe(CO)_3^-$. Later, McDonald and Chowdhury [155] examined the reactions of H and D atoms with (1,3-*c*-C_6H_8)$Fe(CO)_3^-$. The major products were again $Fe(CO)_3^-$ and $HFe(CO)_3^-$, but (*c*-C_6H_8)$Fe(CO)_2(H)^-$ also was formed in small amounts. This latter species did not react further with H_2, suggesting it is a stable 18-electron complex, the η^4-diene hydride. Results for the reaction with D atoms were used to support the contention that this reaction occurs primarily via initial formation of an Fe—H bond.

Reactions with Hydride Ion. Lane, Sallans, and Squires [156] observed that the hydride ion H^- would react with $Fe(CO)_5$ under high-pressure-flow conditions to form $HFe(CO)_3^-$ and $HFe(CO)_4^-$. The latter ion was found to be relatively unreactive, which supports its identification as the 18-electron tetracarbonyl iron hydride anion (rather than a 16-electron tricarbonyl iron formyl anion). The tricarbonyl hydride anion was observed to react further with $Fe(CO)_5$ to form $HFe_2(CO)_6^-$ and $HFe_2(CO)_7^-$, and also with CO_2 to form an adduct.

With benzene chromium tricarbonyl, the H^- is observed to deprotonate the benzene ring, yielding a $C_6H_5Cr(CO)_3^-$ product, but equal amounts of $HCr(CO)_3^-$ are observed [157]. Several labelling studies were used to show that initial attack of the hydride ion is at the benzene ring and that hydrogen migration to the metal is the slow step in the latter reaction.

Reactions with Dihydrogen. In the first of the flow-tube studies involving anionic transition metal hydride complexes, McDonald et al. [154] examined the reactions of H_2 with $Fe(CO)_4^-$, $Fe(CO)_3^-$, and $HFe(CO)_3^-$. Whereas $Fe(CO)_4^-$ (a 17-electron complex) did not react, the other two complexes (15- and 16-electron complexes, respectively) formed the adducts $H_2Fe(CO)_3^-$ and $H_3Fe(CO)_3^-$, respectively, via termolecular processes in

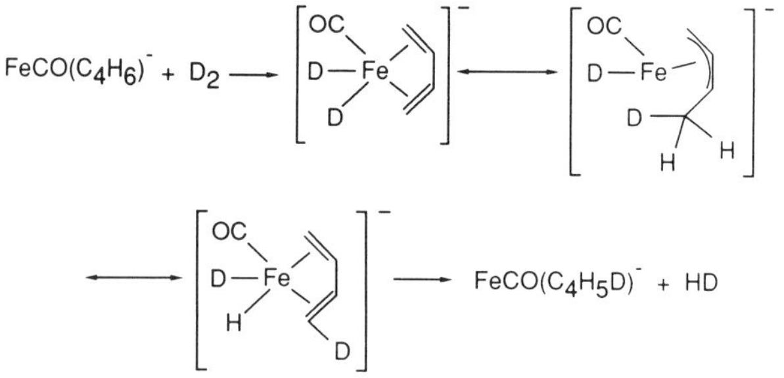

$FeCO(C_4H_6)^- + D_2 \longrightarrow$ [OC, D—Fe, D]$^-$ ⟷ [OC, D—Fe, D, H, H]$^-$ ⟷ [OC, D—Fe, H, D]$^-$ $\longrightarrow FeCO(C_4H_5D)^- + HD$

Scheme 1.1

the high-pressure flow of He (0.5 torr). Likewise, the unsaturated $Fe(CO)_2^-$, $Mn(CO)_3^-$, and $(C_4H_6)Fe(CO)^-$ complexes undergo similar condensation reactions with H_2 [158, 159]. These termolecular reactions are inefficient (0.4–3.0% under these flow conditions) [158–160]. In most cases, these studies present no evidence that addresses the specific structure of the complexes, that is, whether they are dihydride and trihydride species or whether they contain an η^2-H_2 ligand. However, in the case of $(C_4H_6)Fe(CO)^-$, reaction with D_2 results in exchange of up to four hydrogens in the diene ligand, indicating that only the terminal CH bonds are involved. This can be explained by the mechanism shown in Scheme 1.1 which involves an equilibrium between the diene ligand and a π-allyl metal hydride complex.

Reactions with Other Hydride Donors. McDonald, Chowdhury, and Jones [158] also observed that $Fe(CO)_2^-$ and $Mn(CO)_3^-$, but not $Fe(CO)_3^-$ and $Mn(CO)_4^-$, react with H_2O, H_2S, NH_3, and PH_3 to displace a CO ligand as the primary reaction channel. In the case of H_2S and PH_3, dehydrogenation is also observed, indicating that the SH and PH bonds have been activated. Further, products like $Fe(CO)(H_2O)^-$ will react with another water molecule to form $Fe(CO)(OH)_2^-$, also indicating that activation of the OH bond has taken place. The kinetic isotope effects of deuterium substitution are consistent with direct bond activation. With the exception of the $Mn(CO)_3(NH_3)^-$ species, the various reaction products observed by McDonald and co-workers are assigned to the oxidative addition products in which a metal–hydrogen bond is formed.

One rather unique route to the formation of anionic transition metal hydride complexes has been explored in the literature. Originally noted by Lane et al. [161], a more-complete paper by Lane and Squires [162] shows that $Fe(CO)_5$ reacts with $OH(NH_3)^-$ to form $HFe(CO)_4^-$ as a primary product (presumably accompanied by NH_3 and CO_2). Although no specific studies supporting the hydride structure were performed, the tetracarbonyl iron hydride is an 18-electron complex that is presumably more stable than the alternative structure, the 16-electron iron–formyl complex, $(CO)_3FeCHO^-$. The fascinating thing about this process is that when $Fe(CO)_5$ reacts with OH^-, the only product is $HOFe(CO)_3^- + 2CO$ [161] (in contrast to the solution-phase chemistry, where $HFe(CO)_4^- + CO_2$ is formed via a mechanism that is in dispute [163]). Thus, the NH_3 solvent molecule is an active participant in generation of the CO_2 molecule in this reaction. In a related experimental observation, Lane, Sallans, and Squires [164] found that HCO_2^- reacts with $Fe(CO)_5$ to form $HFe(CO)_4^-$ as well as the adduct and the formyl anion, $(CO)_4FeCHO^-$.

Transient Generation of Anionic Transition Metal Hydride Complexes. McDonald and Jones [165] suggest that metal hydrides are formed in the reactions of the unsaturated $Mn(CO)_3^-$ with several alkanes. They observe that this complex will dehydrogenate ethane, propane, isobutane, and cyclopentane. In the latter three cases, they contend that the final product structures are the 18-electron hydrido π-allyl species, for example, $(CO)_3Mn(H)(C_3H_5)^-$ in the case of the propane system. This conclusion is based on the failure of these anions to react further with D_2, H_2S, and $(CH_3)_3SiH$, processes that are generally observed for unsaturated metal complexes. In the case of the ethane system, the evidence for metal hydride bonds is more compelling. McDonald and Jones found that $(CO)_3Mn(C_2H_4)^-$ would react with D_2 to exchange all four hydrogens on the ethene ligand. This must occur via an intermediate like $(CO)_3Mn(H)(C_2H_5)^-$, a species that is presumably formed directly by reaction of C_2H_6 with $Mn(CO)_3^-$. In reaction with methanol the $Mn(CO)_3^-$ complex will eliminate CO or H_2 [166]. Kinetic isotope effects are used to support the contention that these products involve initial OH and CH bond activation, respectively.

In similar chemistry, McDonald, Reed, and Chowdhury [167] found that $Fe(CO)_2^-$ reacts with neopentane to form the adduct anion and that this adduct reacts with D_2 to exchange a single hydrogen atom. This suggests that the structure of the adduct is $(CO)_2Fe(H)(CH_2C(CH_3)_3)^-$. The $Fe(CO)_2^-$ complex was also observed to form an adduct with dimethyl ether and to dehydrogenate this molecule. Isotope labelling experiments showed that the latter reaction was a 1,1-dehydrogenation and that D_2 would react with the product to exchange one hydrogen atom. Thus, the dehydrogenation reaction probably proceeds via a $(CO)_2Fe(H)(CH_2OCH_3)^-$ intermediate to form $(CO)Fe{=}CHOCH_3^-$.

1.4.2.2. REACTIONS OF ANIONIC TRANSITION METAL HYDRIDES

Isotopic exchange reactions have been demonstrated for two metal hydride anion complexes. McDonald et al. [154] found that addition of D_2 to the flow downstream of the formation of $HFe(CO)_3^-$ resulted in the formation of $HD_2Fe(CO)_3^-$ and $D_3Fe(CO)_3^-$, as well as the probable formation of other deuterated species. Although the former product can be generated by simple D_2 addition, production of the latter species requires that $DFe(CO)_3^-$ be formed first via an isotopic exchange reaction between $HFe(CO)_3^-$ and D_2.

Similarly, Lane and Squires [157] found that $HCr(CO)_3^-$ undergoes H–D exchange in the reaction with D_2. Other reactions of $HCr(CO)_3^-$ include those with alkenes containing allylic CH bonds. The result is to eliminate H_2 and form what are probably π-allyl complexes of the tricarbonyl chromium anion. Ethene reacts with the chromium hydride complex to form an adduct, but acetylene reacts to form an adduct and to eliminate H_2, forming $(CO)_3CrC_2H^-$. Additional chemistry of the $HCr(CO)_3^-$ complex is described in a review by Squires [168] where he notes that C_2D_4 reacts with the complex to exchange the hydride ligand and that dienes and aromatic compounds react exclusively by addition to the complex. Water, ammonia, and alcohols (ROH) react via a "σ-metathesis" pathway to eliminate H_2 and form $(CO)_3CrX^-$, where X can be OH, NH_2, and OR, respectively.

McDonald and Schell [169] also observed reactions of the 17-electron species $HMn(CO)_4^-$, [generated by dissociative electron attachment of $HMn(CO)_5$] with various neutral molecules and noted reactions such as adduct formation, carbonyl ligand displacement, and electron abstraction. Reaction with O_2 produces a more-varied list of products, all manganese oxides.

1.4.3. Cationic Transition Metal Hydride Complexes

1.4.3.1. REACTIONS OF ATOMIC METAL CATIONS WITH ALKANES

A vast amount of work on the reactions of atomic transition metal cations with hydrocarbons and related species has been carried out since the 1970s. These reactions have been the subject of several recent reviews, and the reader is directed to these for further information [117, 168, 170–175]. It is generally believed that the reactions of the atomic transition metal ions with alkanes and substituted alkanes occurs via metal hydride intermediates. For example, the commonly postulated mechanism for 1,2-dehydrogenation is exemplified in reaction (1.23),

$$M^+ + C_2H_5R \rightarrow HM^+\text{—}CH(CH_3)R \rightarrow$$

$$H_2M^+\text{—}CH_2{=}CHR \rightarrow M(CH_2{=}CHR)^+ + H_2 \qquad (1.23)$$

(where R = H or an alkyl). In most cases, little or no direct information is obtained in these studies concerning the presumed metal hydride intermediates. Because neither the reactants nor the products in these systems are believed to contain metal hydrogen bonds, the large bulk of this type of study will not be discussed further. The following two sections discuss studies that provide specific evidence for cationic complexes that contain transition metal hydride bonds.

1.4.3.2. COLLISION-INDUCED-DISSOCIATION EXPERIMENTS

Collision-induced dissociation (CID) has been used as a structural diagnostic for transition metal complexes [174, 176]. In the first of this type of study on metal–alkane cation complexes, Freas and Ridge [177] generated $MC_4H_{10}^+$ complexes (M = Fe and Cr) by reaction of $M(CO)^+$ with *n*-butane and 2-methylpropane. These complexes were then collisionally activated by He at a laboratory kinetic energy of 6 keV. In the case of Cr^+, the dominant CID product was Cr^+, suggesting that the $CrC_4H_{10}^+$ complexes are simply a loose association of the butane with the metal ion. For Fe^+, more-complicated decompositions (including production of FeH^+) were observed, and the two different isomers of C_4H_{10} led to different results. However, there is a good correspondence between the products observed in CID and those seen in the chemistry occurring between Fe^+ and the butanes in ion beam experiments [175, 178].

Freas and Ridge interpreted the fragmentation patterns to mean that species like $H_2Fe(C_4H_8)^+$ and $CH_3(H)Fe(C_3H_6)^+$ were the dominant structures with contributions from a variety of other species (e.g., $H—Fe^+—C_4H_9$ and $CH_3—Fe^+—C_3H_7$). This conclusion is far from secure, however, as the experimental observations are also consistent with activation of the association species, $Fe^+ \cdot C_4H_{10}$, which then decomposes via the various pathways observed in the interaction of Fe^+ with butanes. In this scenario, the preferred pathways for decomposition are not controlled by the *structure* of the complex (other than the structure of the intact alkane) but rather by the excitation energy given to the complex in the collision. Although high-energy CID generally is presumed to provide direct structural information, this should only be true if the decomposition occurs rapidly compared to energetically more-favorable isomerization reactions. In systems as complex as these, it is unlikely that this can be taken for granted.

Such complications are very definitely present in low-energy CID processes. In one such study, Huang et al. [126] formed $MC_2H_6^+$ for M = Sc, Y, and La from the reaction of the atomic metal ions with *n*-butane in an ICR cell. Low-energy CID yielded M^+, MCH_3^+ and MCH_2^+ as the dominant products for all three metals. MH^+, MH_2^+, and $MC_2H_4^+$ also were observed for M = Y and La (but not for Sc). The authors suggest that the structure of $MC_2H_6^+$ is the dimethyl metal ion in most cases, but for M = Y and La, either this species can isomerize to $H_2MC_2H_4^+$ and $HMC_2H_5^+$ or these intermediates are also present initially. The decomposition products observed here again correlate nicely with the reactivity observed for the atomic metal ions with ethane (M = Sc [65, 80], Y, and La [83]).

Further studies of the CID of iron-containing cationic complexes have been carried out by Larson and Ridge [179]. One of the most interesting studies concerning metal hydrides involves $FeCH_2O^+$. Whereas CID of

$FeCH_2O^+$ formed in reaction (1.24) gives predominantly Fe^+, CID of $FeCH_2O^+$ formed in reaction (1.25) yields mostly FeH_2^+ and some Fe^+.

$$FeCO^+ + CH_2O \rightarrow FeCH_2O^+ + CO \tag{1.24}$$

$$Fe^+ + (CH_3)_2CO \rightarrow FeCH_2O^+ + C_2H_4 \tag{1.25}$$

This suggests that the product of reaction (1.25) is the iron dihydride carbonyl ion, H_2FeCO^+, whereas the product of reaction (1.24) is the adduct, $Fe^+—H_2CO$, formed by ligand displacement [179].

Rather than using CID to provide structural information, Forbes, Lech, and Freiser [180] also used this technique to provide thermochemical data. For example, in CID of $HSc^+—C_6H_6$ (formed by condensation of benzene with ScH^+), "loss of benzene is observed, suggesting that the bond strength of the $Sc^+—H$ is greater than for Sc^+–benzene." However, this conclusion is only valid if the kinetics associated with H-atom loss and benzene loss are similar, that is, the frequency or preexponential factors for these processes are comparable. This is unlikely for a case such as this, in which losses of a very light ligand and a much heavier ligand are competing [181]. Despite the appealing simplicity of such experiments, competitive CID is unlikely to yield reliable thermochemistry *unless* a more-detailed kinetic analysis is applied.

1.4.3.3. ISOTOPE EXCHANGE REACTIONS

As with negative-ion complexes, one method used to determine the nature of the bonding in transition metal cationic complexes is to react such species with deuterated compounds to see how many hydrogens will exchange. Several studies of H–D exchange reactions that suggest metal hydride structures are discussed in reference 172. Here we mention several illustrative examples. Related experiments are included in Section 1.2.3.2.

Beauchamp, Stevens, and Corderman [182] observed that $(C_5H_5)Rh^+—C_3H_5$ reacts with D_2 to exchange four of the five allyl hydrogens. The authors inferred the reaction mechanism shown in Scheme 1.2. Further reaction with D_2 eventually leads to the four hydrogens on the terminal

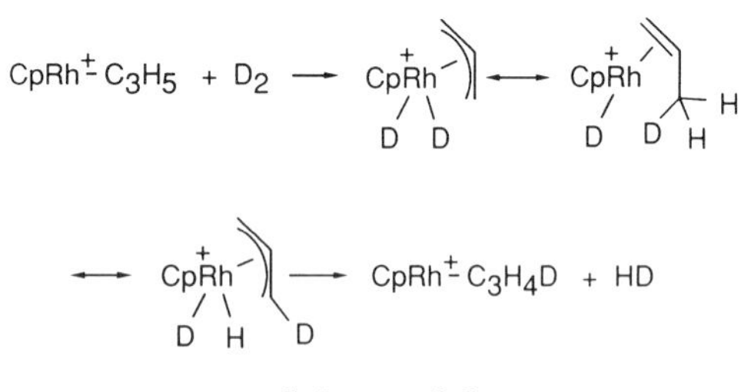

$CpRh^+—C_3H_5 + D_2 \longrightarrow$ CpRh(+)(D)(D)(allyl) $\longleftrightarrow$ CpRh(+)(D)(D)(CH$_2$=CH–CH$_2$) $\longleftrightarrow$ CpRh(+)(D)(H)(CH$_2$=CH–CH$_2$D) $\longrightarrow CpRh^+—C_3H_4D + HD$

Scheme 1.2

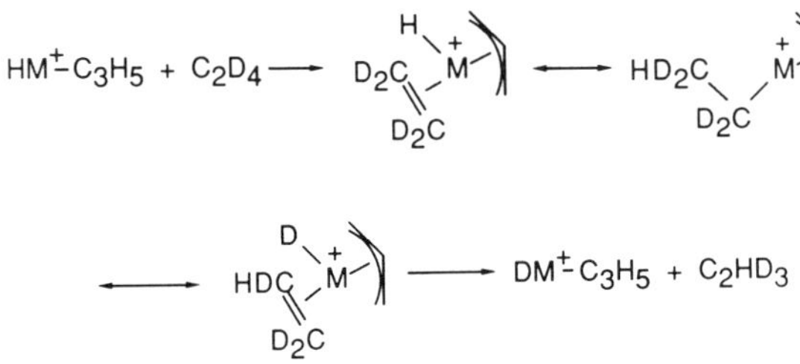

$HM^{+}\text{–}C_3H_5 + C_2D_4 \longrightarrow \cdots \longrightarrow DM^{+}\text{–}C_3H_5 + C_2HD_3$

Scheme 1.3

carbons being replaced by deuterons, whereas the one hydrogen on the central carbon remains unaffected.

In a related study, Byrd and Freiser [183] found that $RhC_3H_6^{+}$ (formed by the dehydrogenation of propane by Rh^{+}) reacts with D_2 to exchange five of six hydrogen atoms. They inferred that a metal–propene structure, $M^{+}—C_3H_6$, and a hydrido–allyl structure, $HM^{+}—C_3H_5$, are both accessible (as in Scheme 1.2), such that only the single hydrogen on the secondary carbon is inaccessible to exchange. Surprisingly, when M = Fe, Co, or Ni, $MC_3H_6^{+}$ will *not* exchange hydrogens with D_2 [184], but will with C_2D_4 [184]. This is believed to occur through the mechanism shown in Scheme 1.3, where the hydrido–allylic species interconverts with a metal–propene complex (similar to Scheme 1.2). This mechanism is analogous to that proposed to explain scrambling in the reactions of Co^{+} with 1,1,1,4,4,4-d_6-butane [63]. The difference in reactivity between D_2 and C_2D_4 has been attributed to the extra energy made available to the system by the complexation of an alkene to the metal. A related consideration is the ability of the metal complex to oxidatively add D_2, a step that is not required for association with ethene.

1.4.4. Thermochemistry

The thermochemistry for metal hydride bonds in transition metal complexes is still relatively sparse. The most extensive thermochemical data available come from measurements of the proton affinities PA of stable metal complexes, PA(MC) = $D(MC—H^{+})$. Gas-phase acidities $D(MC^{-}—H^{+})$ of several neutral transition metal hydrides are also available, but only two neutral metal-complex–hydrogen bond energies $D(MC—H)$ have been determined directly in the gas phase, although others have been estimated. These experiments and values are discussed later. Other relevant work includes measurements of the hydride affinities $D(MC—H^{-})$ of several metal carbonyls by Squires [168]. However, this work is not discussed in detail here because there is no metal hydrogen bond formed; rather a metal formyl anion complex is generated.

1.4.4.1. PROTON AFFINITIES

Many of the available proton affinity measurements come from work by Stevens and Beauchamp [185] who used ICR techniques to bracket the PAs of 13 organometallic compounds and listed those for an additional 7 compounds taken from previous work [186]. Since that time, Meckstroth and Ridge [187] have measured the PAs of $Mn_2(CO)_{10}$, $MnRe(CO)_{10}$, and $Re(CO)_{10}$, again by using ICR methods. Most recently, $PA(FeCp_2)$ has been remeasured with high-pressure mass spectrometry by two groups [188, 189]. All of this work has been compiled and discussed recently by Simões and Beauchamp [129].

Homolytic metal–hydrogen bond energies can be calculated from the PAs by using eq. (1.4) (where M now represents the entire metal complex). These values are listed in Table 1.6. The accuracy of these bond energies are limited by several considerations. The accuracy of the required ionization energy of the complex, IE(M), is clearly a factor. The recent high-pressure studies of ferrocene have pointed out that values taken from the ICR measurements do not include consideration of entropy effects upon protonation. Thus, the high-pressure PA values are about 15 kJ mol^{-1} lower than the ICR value. For our purposes here, an important consideration in these studies is whether the metal or one of the ligands is the site of protonation, that is, is a metal–hydrogen bond formed? This is difficult to ascertain and so Table 1.6 includes all complexes except those for which ligand protonation

Table 1.6
Experimental Bond Dissociation Energies of Transition Metal Hydride Cation Complexes

Complex (MC)	$D(MC^+—H)^a$	Complex (MC)	$D(MC^+—H)^a$
$V(CO)_6$	220(14)		
$Cr(CO)_6$	230(10)	$Mo(CO)_6$	260(9)
$Cr(CO)_3C_6H_6$	221(15)	$W(CO)_6$	257(9)
$Mn(CO)_5$	172(10)		
$Mn_2(CO)_{10}$	204(13)	$Re_2(CO)_{10}$	246(7)
$MnRe(CO)_{10}$	< 267(7)		
$Mn(CO)_5H$	349(11)[b]		
$Mn(CO)_5CH_3$	267(11)	$Re(CO)_5CH_3$	294(13)
$Mn(CO)_3(C_5H_4CH_3)$	283(13)		
$Fe(CO)_5$	299(15)		
$Fe(C_5H_5)_2$	200(11)	$Ru(C_5H_5)_2$	271(15)
$Fe(CO)_2(C_5H_5)CH_3$	210(14)		
$Co(CO)_2(C_5H_5)$	245(12)	$Rh(CO)_2(C_5H_5)$	287(12)
$Ni(CO)_4$	248(9)		

[a]Values (in kilojoules per mole) are for 298 K and are taken from reference 129. Uncertainties are parentheses.
[b]Probably protonates at the H ligand (see text).

has been demonstrated. It seems likely that some of the high values in this table may be due to protonation of the ligands. Indeed the highest value in Table 1.6, $D[Mn(CO)_5H^+—H] = 349 \pm 11$ kJ mol^{-1}, is probably protonated at the hydrogen ligand such that the structure is $(CO)_5Mn^+—H_2$ with an η^2-dihydrogen ligand [129].

Discounting the very high value of $D[Mn(CO)_5H^+—H]$, the highest values in Table 1.6 average 291 ± 11 kJ mol^{-1}. If we adjust this downward by 15 kJ mol^{-1}, to account approximately for the entropy effects mentioned previously, the "maximum" strength metal–hydrogen bond in a saturated metal complex cation is similar to that found from the periodic trends analysis of the simple metal hydride diatoms (Section 1.2.2.4). Further, it can be seen that $D(MC^+—H)$ values for second- and third-row metals are generally higher than those for the first row, again consistent with trends in the metal hydride diatom thermochemistry.

1.4.4.2. NEUTRAL BOND STRENGTHS

$HMn(CO)_5$ and $HCo(CO)_4$ are the only metal hydride complexes for which the M—H bond energies have been measured in the gas phase. The values have been measured differently in each case and are not well-established. The heats of formation of the closed-shell Mn and Co complexes are reasonably well-known, but the metal–hydrogen bond strengths are still controversial because of uncertainties in the heats of formation of the metal carbonyl radicals formed after metal–hydrogen bond cleavage.

Connor et al. [190] used microcalorimetry to measure the exothermicity of displacement by iodine of the hydride ligand in $HMn(CO)_5$ in benzene solution. Combining these values with other thermochemistry detailed previously [190] (including $\Delta H_{vap}[Mn(CO)_5H] = 37.8 \pm 0.8$ kJ mol^{-1} [191]) allows the gas-phase heat of formation of $Mn(CO)_5H$ to be derived as -740 ± 10 kJ mol^{-1}. These experiments were carried out near room temperature so there is no need to extrapolate the data to standard temperature. The $H—Mn(CO)_5$ bond strength can be derived by using this heat of formation in combination with the gas-phase heat of formation of the $Mn(CO)_5$ radical. This latter value in turn depends upon $D[(CO)_5Mn—Mn(CO)_5]$, which has been the subject of extensive study and debate. As recently summarized by Simões and Beauchamp [129], values from a low of 79 ± 14 kJ mol^{-1} [192] to a high of ≥ 176 kJ mol^{-1} [193] have been measured for this bond strength. Simões and Beauchamp choose a best value of 159 ± 21 kJ mol^{-1}, based on photoacoustic measurements of Goodman, Peters, and Vaida [194], which leads to $\Delta_f H[Mn(CO)_5, g] = -713 \pm 11$ kJ mol^{-1}. Combined with $\Delta_f H(H) = 218$ kJ mol^{-1}, this gives a Mn—H bond strength of 245 ± 15 kJ mol^{-1}.

Bronshtein et al. [195] measured the equilibrium of Co(s), H_2, and CO with $HCo(CO)_4(g)$ at elevated temperatures (≥ 450 K) and pressures (≥ 100 atm). The amount of $HCo(CO)_4$ observed was proportional to $[H_2]^{0.53}[CO]^{3.82}$, in reasonable agreement with the expected $[H_2]^{1/2}[CO]^4$ proportions. The thermodynamic data were then corrected to 298 K by using literature information to derive $\Delta_f H_{298}[HCo(CO)_4] = -569.2$ kJ mol^{-1}. No error limits were provided for these values, although Pilcher and Skinner [196] cite ± 2.2 kJ mol^{-1}, which seems rather small given the required temperature extrapolation. To derive the metal hydride bond energy in this complex, the heat of formation of the $Co(CO)_4$ radical is required, and this in turn depends on the Co—Co bond energy of $Co_2(CO)_8$. The only direct experimental measurements for this bond energy come from Bidinosti and McIntyre, who obtained 48 ± 19 kJ mol^{-1} [197] and 54 ± 13 kJ mol^{-1} [198]. Estimates of 88 kJ mol^{-1} [199] and 86 kJ mol^{-1} [200] for the Co—Co bond energy have been obtained indirectly by assuming that all terminal metal–ligand, all bridging metal–ligand, and all metal–metal bonds in $Co_2(CO)_8$ and $Co_4(CO)_{12}$ are the same. Simões and Beauchamp [129] choose an intermediate value of 64 kJ mol^{-1}, which we use here after assigning an uncertainty of ± 35 kJ mol^{-1}. Combined with $\Delta_f H_{298}[Co_2(CO)_8] = -1185 \pm 6$ kJ mol^{-1} [200], the most direct measure for the gas-phase heat of formation of this complex, this bond energy yields $\Delta_f H[Co(CO)_4] = -560 \pm 18$ kJ mol^{-1}. Combining this value with $\Delta_f H(H) = 218$ kJ mol^{-1} gives $D[(CO)_4Co—H] = 227 \pm 18$ kJ mol^{-1}. This is comparable to $D[(CO)_4Co—H] = 229$ kJ mol^{-1} measured in heptane solution [201], suggesting that the gas-phase derivation is not greatly in error.

No measurements of the gas-phase metal–hydrogen bond strengths in $H_2Fe(CO)_4$ have been made, although an estimate for the mean bond energy has been cited [129]. This is derived from the measured enthalpy of reaction (1.26),

$$H_2Fe(CO)_4 \rightarrow Fe(CO)_4 + H_2 \tag{1.26}$$

in solution (109 ± 8 kJ mol^{-1} [129]) and the gas-phase H_2 bond strength (436 kJ mol^{-1}). This gives an average Fe—H bond strength of $(436 + 109)/2 = 272 \pm 4$ kJ mol^{-1} [129].

The H—$Mn(CO)_5$ and H—$Co(CO)_4$ bond strengths are listed in Table 1.7. [The mean value for H_2—$Fe(CO)_4$ estimated from solution data is not included because the relationship of this value to the gas-phase bond energy is unknown.] The two bond energies (245 and 227 kJ mol^{-1}, respectively) are relatively close to the "intrinsic" first-row metal–hydrogen bond strengths discussed in Section 1.2.2.4, 234 and 248 kJ mol^{-1}. Both complexes should have little steric hindrance, and no promotion energy is involved in forming the metal–hydrogen bond. Thus, they are appropriate tests of whether the intrinsic bond-strength–promotion-energy model, which was derived from analysis of metal hydride diatoms, is applicable to coordinatively saturated metal complexes. The closeness of the results to the predicted bond strength indicates that the model is relatively independent of other ligands and can apply to closed-shell species.

Table 1.7
Experimental Bond Dissociation Energies of Neutral Transition Metal Hydride Complexes[a]

Complex (MCH)	D(MC—H)	$D(\mathrm{MC^-—H^+})$
$HRe(CO)_5$		1383(5)[b]
$H_2Fe(CO)_4$		1334(21)[c]
$HMn(CO)_5$	245(15)[c]	1333(6)[b]
$HCo(CO)_4$	227(18)[c, d]	≤ 1314[c]
$HMn(CO)_3(PF_3)_2$		≤ 1289[b]
$HMn(CO)_2(PF_3)_3$		≤ 1289[b]
$HIr(PF_3)_4$		≤ 1289[b]
$HCo(PF_3)_4$		≤ 1289[b]

[a]Values (in kilojoules per mole) are for 298 K. Uncertainties are in parentheses.
[b]Reference 203.
[c]Reference 129.
[d]See text.

1.4.4.3. *ACIDITIES*

Gas-phase acidities of several transition metal hydride complexes, $D(\mathrm{MC^-—H^+})$, have been determined by examining the deprotonation of neutral metal hydride complexes in an ICR mass spectrometer. Stevens [129, 202] has bracketed the acidities of $HMn(CO)_5$, $H_2Fe(CO)_4$, and $HCo(CO)_4$ and checked these results by studying the dissociation behavior with zero-energy electrons. Further work using a selected ion flow tube (SIFT) apparatus has provided limits on the acidities of $HRe(CO)_5$, $HMn(CO)_3(PF_3)_2$, $HMn(CO)_2(PF_3)_3$, $HCo(PF_3)_4$, and $HIr(PF_3)_4$ [203]. In this work, Stevens and co-workers note that substituting PF_3 for CO increases the acidity of a compound, such that $HCo(PF_3)_4$ is the strongest gas-phase acid presently known. A summary of these acidity results is given in Table 1.7. such values could be converted to the neutral bond energies via eq. (1.7) (where M^- would represent the deprotonated metal complex) but in general the electron affinity of the metal complex is unknown. Indeed the only exception is $Fe(CO)_4$. In the cases of $HCo(CO)_4$ and $HMn(CO)_5$, because the metal hydride bond energy is known, eq. (1.7) can be used to derive the electron affinities of $Mn(CO)_5$ and $Co(CO)_4$, 2.32 ± 0.17 and $\geq 2.3 \pm 0.2$ eV, respectively.

1.4.4.4. *UNSATURATED COMPLEX BOND STRENGTHS*

Very few metal hydride bond energies have been measured for unsaturated metal ligand complexes. One species for which data do exist is $HMCH_3^+$ (M = Sc and V), which has been seen as an endothermic product in the reactions of Sc^+ with propane and 2-butene [204] and in the reaction of V^+

with isobutane [98]. In both cases, kinetic and thermodynamic arguments are used to suggest that the product is the hydridomethyl–metal compound rather than a metal–methane complex, $M^+—CH_4$. Measurements of the reaction thresholds lead to $D(CH_3Sc^+—H) = 256 \pm 15$ kJ mol^{-1} and $D(CH_3V^+—H) = 180 \pm 26$ kJ mol^{-1}. The scandium value is comparable to $D(HSc^+—H) = 245 \pm 15$ kJ mol^{-1} (Table 1.5).

1.5. HYDRIDES OF TRANSITION METAL CLUSTERS

The most recent addition to experimental studies of transition metal hydrides concerns clusters of transition metals. Reactions of transition metal clusters with molecular hydrogen are the most intensely studied cluster reactions, and results now exist for several metals over large ranges of cluster sizes. In addition, studies that examine the effect of charge (both positive and negative) on such reactions have been performed. Interested readers are referred to other recent reviews of cluster chemistry for additional information [205, 206].

1.5.1. Reactions of Hydrogen with Neutral Transition Metal Clusters

In these studies, neutral transition metal clusters are formed by laser vaporization of a solid sample of the metal followed by clustering in a flow of He. The clusters are reacted with hydrogen in a fast-flow reactor by mixing the reactant gas with the He downstream of the condensation region. After expanding into vacuum, neutral reactant and product intensities are determined via photoionization–time-of-flight mass spectrometry.

Iron. The earliest study of the reactions of dihydrogen with neutral transition metal clusters was a short report by Riley et al. [207] on iron clusters. Several more detailed studies followed in rapid succession. These include work by Richtsmeier et al. [208], Whetten et al. [209], and Morse et al. [210]. Coupled with subsequent work by these groups, the reactivity of iron clusters with hydrogen is one of the best-explored metal cluster reactions.

In these studies, iron clusters react to adsorb several hydrogen molecules. None of these studies sees any evidence for odd numbers of hydrogen atoms adsorbed, consistent with dissociative chemisorption and no H-atom loss (a strongly endothermic process). Dissociative chemisorption is also indicated by the observation that fully hydrogenated iron clusters react with D_2 to undergo H–D exchange for clusters with fewer than ≈ 10 metal atoms or at elevated temperatures for other clusters [207, 211]. All three groups find that the reactivity of iron clusters with H_2 and D_2 shows a strong dependence on cluster size [208–210], as shown in Figure 1.6. Minima in the reactivity occur

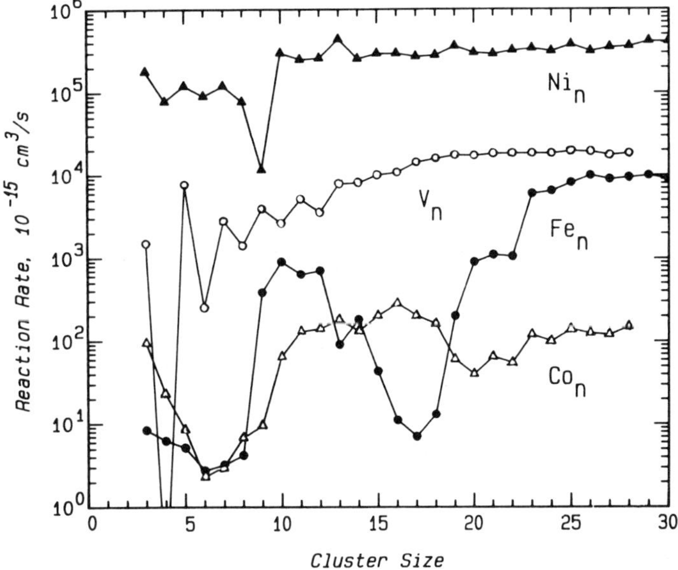

FIGURE 1.6
Relative D_2 chemisorption reactivity as a function of cluster size for neutral clusters of vanadium, iron, cobalt, and nickel, taken from references 206, 208, 219, and 224, respectively. For small nickel ($n = 3-6$) and iron ($n = 3-7$) clusters, the data are taken from reference 210 and are scaled to match the data of references 208 and 224 at $n = 7$ and 8, respectively. The data for Fe and Ni are shown on the same absolute reactivity scale. The relative reactivities of V and Co clusters are shown on the same scale but the absolute magnitudes of these rates are unknown.

for clusters with $n \approx 7$ and 17 atoms, whereas the most-reactive clusters are those where $n = 10–12$ and ≥ 23. Absolute rates determined by Richtsmeier et al. [208] show that the largest rate constants (those for $n \geq 23$) are only about 3% of the hard-sphere cross section, comparable to the sticking coefficient of H_2 on iron surfaces. Although the *pattern* of reactivity is nearly identical in all three studies, the *magnitude* of the reactivity scale differs somewhat. This is explained by the observation of Richtsmeier et al. [208] that increased temperature tends to wash out differences in reactivity.

A particularly important and influential observation was made by Whetten et al. [209]. They noted that iron clusters in the range $n = 8–25$ show a strong correlation between reactivity and ionization energy (IE), such that reactive clusters have low ionization energies. The close correlation of these properties suggested that hydrogen chemisorption on the clusters can be understood as electron donation from the cluster into the antibonding orbital of H_2, a concept discussed in Section 1.2.3.1. Further, Zakin et al.

[212] find that the ionization energy of hydrogenated iron clusters increases, consistent with electron donation from the metal to hydrogen. Morse et al. [210] argue that this cannot be the only effect because this correlation does *not* hold for small iron clusters ($n < 8$) and because copper clusters with low IEs do not react with D_2. We anticipate that a more-complete understanding of hydrogen chemisorption on transition metal clusters requires all of the concepts discussed in Section 1.2.3.1. Namely, the most-reactive clusters will have a donor orbital (of low IE) to interact with the antibonding orbital of H_2 *and* an empty acceptor orbital to interact with the bonding orbital of H_2. Similar ideas have been discussed by Trevor and Kaldor [205].

Continued work on the reactions of iron clusters with hydrogen has been carried out by Riley and co-workers. They find that up to $n = 30$, $\sim 1:1$ Fe–H ratios are observed when the hydrogenation reaction goes to completion [213]. For clusters up to $n \approx 80$, the composition of the cluster becomes deficient in hydrogen but tracks with the number of iron atoms on the *surface* of a globular cluster. For $n = 80$–260, less than full surface coverage is seen for iron clusters [214]. These results largely rule out open (planar or linear) structures of iron clusters. Multiphoton ionization of the hydrogenated iron clusters results in loss of hydrogen molecules at higher power, leading to an estimate of 1.3 ± 0.1 eV for the desorption energy of H_2 [215]. This is close to the 1.2-eV value found for bulk iron and implies that Fe_n—H bond energies are about 280 ± 10 kJ mol^{-1} and invariant with cluster size over the range of $n = 10$–32. This value is well above $D(\text{Fe—H}) = 157 \pm 8$ kJ mol^{-1} (Table 1.3), but is comparable to the intrinsic bond energy discussed in Section 1.2.2.4.

Recent work by the Riley group has returned to examine further the reasons for the strong dependence on cluster size in the reaction of iron clusters with hydrogen [216]. Here, it is proposed that changes in cluster structure may account for the large changes in reactivity. Evidence for structural changes are given for n between 14–15, 18–19, and 22–23, regions that correspond to large changes in the H_2 reactivity. Recent quantitative measurements of the stabilities of iron clusters by Loh et al. [217] and Lian and Armentrout [218] are consistent with such structural changes. Continued work is needed to elucidate further this interesting behavior, but it should be clear that the effects of physical and electronic structure are intimately tied together. For example, one can imagine that clusters with exposed atoms (and hence dangling bonds) will have low ionization energies and will be more reactive, while stable, symmetric structures might have relatively high IEs and be less reactive because most of the electron density will be engaged in metal–metal bonding.

Niobium. Niobium is another metal whose clusters have been studied actively. In one of the earliest cluster reactivity studies, Geusic, Morse, and Smalley [219] found that the reactions of niobium clusters with hydrogen also show a dramatic dependence on cluster size. Monomers, dimers, and the

$n = 8$, 10, and 16 clusters were found to be much less reactive than the surrounding clusters [210, 219]. Whetten et al. [220] then measured the ionization energies of niobium clusters and found that the unreactive $n = 8$, 10, and 16 clusters showed maxima in the IE values. Further, these authors found that clusters of $n = 23$–26 were found to have relatively high IEs and low reactivity with D_2, although this effect is not as pronounced as for the smaller clusters [221].

Further studies of this system by Zakin et al. [221] and Hamrick and co-workers [222, 223] found that niobium clusters of $n = 9$, 11, and 12 react with D_2 and N_2 to exhibit biexponential decay. Although several explanations of this behavior were considered [222], the evidence points to structural isomers, where the unreactive isomers have reactivities similar to $n = 8$ and 10. Zakin et al. also find that the reactions of niobium clusters are temperature-sensitive, indicating that D_2 chemisorption is an activated process.

Nickel. In contrast with the observations made for iron and niobium, the reactivity of nickel clusters with D_2 shows little variation with cluster size from $n = 3$ to $n = 20$ [210]. This is only partially confirmed by the studies of Hoffman et al. [224], who found that, for $n \geq 14$, the rate constant for H_2 and D_2 chemisorption scales smoothly as $n^{2/3}$, presumably in accordance with the hard-sphere cross section of the cluster. For these larger clusters, the reaction probability per collision was measured to be approximately 0.6 for H_2 and 0.3 for D_2. For smaller clusters, $n = 7$–14, a strong dependence of reactivity on size was observed with the $n = 9$ cluster being particularly unreactive (Figure 1.6). The discrepancy with the earlier work is probably explained by differences in cluster temperature. Parks et al. [214] have found that nickel clusters react with D_2 to maintain a 1 : 1 ratio of hydrogen atoms to surface metal atoms on the surface of the cluster for $n = 40$–250. This can be contrasted with the situation observed for iron clusters (discussed previously), which exhibit a coverage of about 80% [214].

Vanadium. Studies of the reactions of D_2 with vanadium clusters [205, 223, 225] show a pronounced oscillation in the reactivity of odd and even clusters from $n = 4$–13 such that the even clusters are much less reactive and V_5 is the most reactive cluster in this range (Figure 1.6). Such oscillations in the IEs of vanadium clusters are observed but are much less pronounced [205, 225]. For $n = 13$–18, the reactivity of the vanadium clusters increases smoothly and reaches a plateau for $n = 18$–28 [206]. vanadium clusters saturate with about 1.4 deuterium atoms per surface vanadium atom, assuming globular clusters [205], which is substantially higher than the coverages observed for Fe and Ni clusters.

Other Transition Metals. Less-extensive work also exists for several other metals. Smalley and co-workers examined the reactivity of copper and cobalt clusters with hydrogen [210, 219]. They find that Cu_n for $n = 1$–19 are unreactive with D_2, whereas the reactivities of Co_n for $n = 3$–28 show dramatic dependences on cluster size (similar to that for iron) such that the monomers, dimers, and clusters in the vicinity of $n = 6$ and 19 are much less reactive (Figure 1.6). Klots et al. [226] have elucidated the structures of hydrogenated cobalt clusters ($n = 40$–160). Rhodium ($n = 3$–17) clusters react readily with D_2 but no large variations are observed with cluster size [227]. Like vanadium, tantalum clusters with even numbers of atoms are much less reactive than the odd clusters and Ta_5 is the most reactive in the range $n = 3$–12 [206].

Palladium clusters also vary greatly in their reactivity with D_2 [228]. Reactive minima are observed for $n = 1, 3, 4, 9, 17$–19 and reactive maxima occur for $n = 7, 8, 10$–15, 22–25. Hydrogen coverage generally decreases with cluster size from a D–Pd ratio of 3 : 1 for $n = 4$ to about 1.5 for $n = 21$, but there is an abrupt shift upward at $n = 15$, suggesting the possibility of a structural feature. Platinum ($n < 12$) clusters react only slowly with H_2 to achieve fairly low coverages [229].

Effect of Adsorbates. Several studies have also examined how various adsorbates affect the reactions of transition metal clusters with hydrogen. Hoffman, Parks, and Riley [230] find that ammonia and hydrogen apparently bind to different sites on iron clusters. Bare iron clusters that are unreactive toward hydrogen have enhanced reactivity when ammoniated. The reactivities of other clusters are only slightly affected, such that they increase somewhat when 1–3 molecules of NH_3 are added and generally decrease when saturated with ammonia.

Hamrick and Morse [223] have compared the reactivity of D_2 with bare vanadium clusters to that for clusters that have a single carbon or oxygen atom adsorbed. Strong effects are observed and depend on the identity of the adsorbate. For example, V_5C is much less reactive than V_5 whereas V_5O has comparable reactivity to the bare cluster. In contrast, V_8O is one of the most-reactive oxygenated clusters whereas V_8 is one of the least-reactive bare clusters and V_8C is of moderate reactivity. Above about $n = 13$, the adsorbate effects are nearly gone. Previous work of Zakin et al. [212] showed that, for every atomic oxygen ligand, vanadium clusters ($n = 4$–8) would bind one less hydrogen molecule, a clear example of site blocking.

Nonose et al. [231] recently observed that the addition of aluminum atoms to small clusters of cobalt, vanadium, and niobium generally increases their reactivity, although the reactivities of niobium clusters of $n = 4$–8 are decreased. Similar work on mixed Co–V clusters shows that the addition of up to three vanadium atoms to Co clusters generally increases reactivity, but that more vanadium does not [232]. A special case is $Co_{12}V$, which is found to be distinctly less reactive than Co_{13}, a result taken to indicate that the V atom is in the center of an icosahedral structure.

1.5.2. Reactions of Hydrogen with Transition Metal Cluster Ions

Several methods have been used to examine the reactivity of ionic transition metal clusters with hydrogen. Experiments directly analogous to those described previously for neutral clusters can be carried out in fast-flow reactors. More-sophisticated techniques that permit a single-sized cluster ion to be isolated before reaction with a single reagent molecule include ICR methods and ion beam techniques. These latter methods avoid possible ambiguities in the observations resulting from fragmentation of clusters either during reaction or during the photoionization step necessary to detect neutral cluster species.

1.5.2.1. FAST-FLOW REACTOR STUDIES

Two studies have compared directly the reactivity of neutral clusters and ionic clusters with hydrogen in a fast-flow reactor. Zakin et al. [221] found that neither a positive nor a negative charge changed the maximum uptake of D_2 by niobium clusters over the range examined ($n \leq 28$). Neither did charge affect the reactivity by more than a factor of 2.5 compared to the neutral clusters, as illustrated by the data reproduced in Figure 1.7. Although the overall pattern of reactivity is clearly the same regardless of charge state, some differences are observed. The cations of $n = 7$–9 and the anions of $n = 12$ and 16 are about an order of magnitude more reactive than their neutral counterparts, whereas the cations of $n = 11$ and 14 and the anions of $n = 9$ and 11 are much less reactive. Evidence for an isomer of $Nb_{12}{}^+$ was found, but not for $Nb_9{}^+$ or $Nb_{11}{}^+$, the other clusters where neutral isomers were observed [221–223].

Similar experiments for positively charged iron clusters were performed by the same authors [233]. Although there are similarities in the size-dependent reactivity patterns for cationic and neutral iron clusters, the positive charge is found to exert much more of an influence than observed in the niobium case. Here, the reactivity of cations of $n = 4$–6 and 18 is greatly enhanced, that for $n \geq 19$ is moderately enhanced, and that for $n = 3$ and 9–14 is diminished. These changes are then interpreted by incorporating electrostatic interactions into a frontier orbital model similar to that discussed in Section 1.2.3.1.

Because positively charged clusters should be much more reticent to donate an electron to the hydrogen antibonding orbital and negatively charged clusters more willing to make this donation, the observations that the reactivities of charged and neutral clusters show striking similarities is reasonable evidence that a simple electron donation model for hydrogen chemisorption is inadequate. As discussed in Section 1.2.3.1, the interactions of atomic metal cations with H_2 can only be understood by understanding both the donor and acceptor abilities of the metal center. Presumably, similar

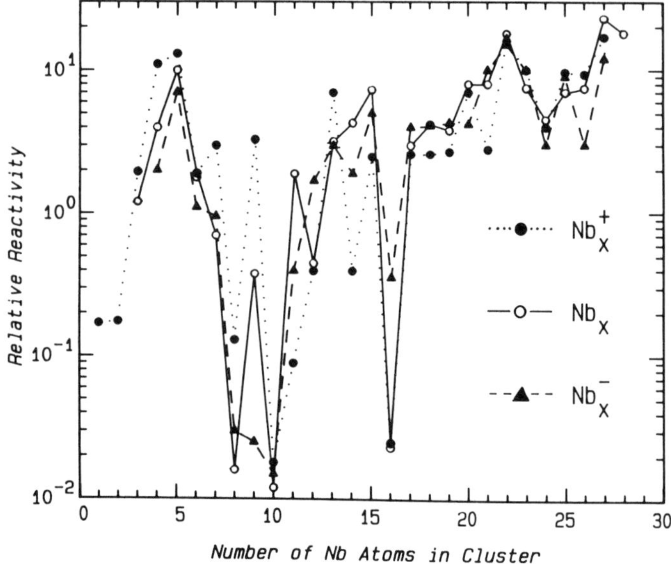

FIGURE 1.7
Relative D_2 chemisorption reactivity as a function of cluster size for cationic, neutral, and anionic niobium clusters. The data are taken from reference 221. In cases where the rates showed biexponential behavior (Nb_9, Nb_{11}, and $Nb_{12}{}^+$), only the more reactive data point is indicated.

considerations should hold for clusters. Although charge should change the relative contributions of the donor–acceptor interactions, apparently it does not affect the gross electronic structure of the clusters appreciably.

1.5.2.2. ION CYCLOTRON RESONANCE EXPERIMENTS

Smalley and co-workers also have examined the reactions of cationic niobium clusters with hydrogen by using ICR techniques [234]. In these studies, cold metal cluster ions are generated by using a laser-vaporization–supersonic-expansion source as per the neutral experiments. These positively charged metal clusters are then injected into an ICR cell, where their reactions with hydrogen can be monitored as a function of time. The sensitivity of this apparatus allows absolute reaction rates to be measured over a range of five orders of magnitude.

Results were in qualitative agreement with the results of Zakin et al. [221] and both groups agree that the trends in reactivity as a function of cluster size for the positively charged niobium clusters mimic the trends for the neutral clusters. Elkind et al. [234] find that saturation coverage varies with cluster size but is usually at an H–Nb ratio of ~ 1.3. Ion clusters of $n = 3$, 10, and 12 were found to be inert and those for $n = 4$, 5, 8, and 16 were relatively unreactive. Two forms of Nb_{19}^+ were found, one inert and the other over four orders of magnitude more reactive. the authors suggest that this implies the existence of two stable geometric isomers.

A qualitatively different ICR experiment has been pursued by Freiser and co-workers. Small metal cluster ions (dimers and trimers) are generated by reaction of atomic metal ions with volatile metal carbonyls [e.g., $Fe^+ + Fe(CO)_5 \rightarrow Fe_2(CO)_4^+ + CO$]. The resultant product ions are then activated by collisions to knock off the carbonyl ligands, leaving behind the bare metal cluster ion. This method is likely to form cluster ions with excess internal and kinetic energy. Freiser and co-workers have examined an extensive set of chemistries via such techniques, but very few involve transition metal cluster hydrides [172, 235]. One exception is the study of reaction (1.27) [180].

$$Fe_2^+ + C_2H_6 \rightarrow Fe_2H^+ + C_2H_5 \tag{1.27}$$

This endothermic process was observed by kinetically exciting the reactant ions by irradiation at their cyclotron frequency. The threshold for this reaction gave an estimated Fe_2^+—H bond strength of 218 ± 67 kJ mol^{-1}, which can be compared with $D(Fe^+—H) = 208 \pm 6$ kJ mol^{-1} (Table 1.3).

In related work, Jacobson has generated Fe_2H^+ and Fe_2D^+ and examined their reactions with ethene and propene at thermal energies [236]. H–D exchange is observed in reactions with ethene whereas propene reacts to form the Fe_2^+-allyl by H_2 elimination. $(C_5H_5)_2Fe_2H^+$ exchanges one hydrogen with C_2D_4 [237], suggesting a monohydride structure. $(C_5H_5)_2Fe_2D^+$ also exchanges once with C_2H_2.

1.5.2.3. *ION BEAM EXPERIMENTS*

The third method that can be used to study the reactions of transition metal ions is the ion beam technique introduced in Section 1.2.2.1. In the earliest such study [238], Mn_2^+ was generated by low-energy electron impact on $Mn_2(CO)_{10}$, a process that may leave the ion with small amounts of internal excitation (several tenths of an electron volt). Reaction of this ion with ethane was examined over a broad kinetic-energy range and only endothermic processes were observed. The two dominant products were Mn_2H^+ and MnH^+. The threshold of the former product indicated that $D(Mn_2^+—H) = 203 \pm 29$ kJ mol^{-1}, nearly identical with $D(Mn^+—H) = 203 \pm 14$ kJ mol^{-1}, (Table 1.3). A similar study of the reaction of D_2 with Co_2^+ [generated by low-energy electron impact on $Co_2(CO)_8$] finds that the reaction cross section and magnitude for production of $Co_2D^+ + D$ is nearly the same as that for the similar reaction of atomic Co^+ [56]. This implies that $D(Co_2^+—H)$ is nearly the same as $D(Co^+—H)$ (Table 1.3).

Finally, Loh, Lian, and Armentrout [239] examined the reactions of D_2 with Fe_2^+ and Fe_3^+ generated in a laser-vaporization–supersonic-expansion source. They find that the magnitude and shape of the cross sections for the endothermic production of Fe_nD^+ + D are very similar to that observed for $Fe^+(^4F)$ but distinct from that of ground state $Fe^+(^6D)$ (see Section 1.2.3.1). This implies that the electronic character of the dimer and trimer ions are more similar to the $^4F(3d^7)$ state than to the $^6D(4s3d^6)$ state. As discussed previously, the $^4F(3d^7)$ state is more reactive than the $^6D(4s3d^6)$ state because the 4*s* acceptor orbital is unoccupied in the former. This suggests that Fe_2^+ and Fe_3^+ also have an unoccupied acceptor orbital capable of efficient interaction with H_2.

Preliminary analysis of the thresholds for these processes indicates that $D(Fe_2^+—H) = 133 \pm 19$ kJ mol^{-1} and $D(Fe_3^+—H) = 163 \pm 19$ kJ mol^{-1}, which can be compared with $D(Fe^+—H) = 208 \pm 6$ kJ mol^{-1} (Table 1.3). The discrepancy with the work of Forbes et al. [180], $D(Fe_2^+—H) = 218 \pm 67$ kJ mol^{-1}, can be explained by internally excited ions in the ICR study.

1.6. SUMMARY AND OUTLOOK

It is clear from a look at the tables in this review that, although much is known about gas-phase transition metal hydrides, there are significant gaps in our knowledge. Progress in both optical and photoelectron spectroscopy of diatomic hydrides could be made, not only because of more sensitive techniques, but also because theoretical calculations recently have reached the level where they can assist in the assignment of spectroscopic features. Given current experimental technology, it should be possible to obtain a complete set of the electronic states and their vibrational and rotational frequencies (giving bond lengths) for the transition metal monohydrides. Such a data set would provide a firm basis for understanding the isolated metal–hydrogen bond and, in particular, the complicated electronic considerations that transition metal species entail. Extensions to more second- and third-row transition metal hydrides would provide useful comparisons to their first-row counterparts. Spectroscopic data on *any* cationic and additional anionic metal hydrides would augment our understanding of the influence of charge on these simple species.

The strengths of diatomic metal hydrogen bonds also deserve further work. For the first-row transition metal hydrides, there is a clear deficiency in the data for the neutral and anionic metal hydrides (Table 1.3). For the second- and third-row transition metal hydrides, the methods already in place need to be extended to all charge states of these species. Even in cases where the bond energy has been determined, further experiments using alternate techniques would be valuable to test the previous work and to enhance the precision of the bond strengths.

The homoleptic transition metal hydrides are very poorly characterized compared to the monohydrides. Here the methods for formation of the hydrides are less general and the experiments more complex. Although there are obvious experiments that presently can be performed [for example, reactions (1.20) for other metals and photoelectron spectroscopy of additional species], new approaches and techniques are needed in order to study these species systematically. Here again, theoretical calculations will be useful in interpreting the complex results that can be expected.

Although significantly more is known about transition metal hydrogen complexes compared with the homoleptic hydrides, the information available is haphazard. A particularly obvious deficiency is that only two M—H homolytic bond strengths are known for saturated complexes in the gas phase. The main problem is the difficulty in determining the heats of formation of the radical complexes formed upon scission of the metal hydrogen bond. Classical thermodynamic methods may not be the answer, but the application of newly developed techniques should be capable of providing such data. For example, we hope to ascertain the $H—Mn(CO)_5$ bond energy by examining the kinetic energy dependence of the endothermic reaction (1.28) for a series of atomic metal ions for which $D(MH^+)$ is well-characterized.

$$M^+ + HMn(CO)_5 \rightarrow MH^+ + Mn(CO)_5 \quad (1.28)$$

Extensions of this experiment to other volatile transition metal hydride complexes should be straightforward.

For the clusters, a comparatively large amount of quantitative information on reactivity has been gathered in a short amount of time. Needed are quantitative metal cluster hydrogen bond strengths and cluster geometries. Present technology probably is capable of providing this information, and recent advances have been made in both areas.

ACKNOWLEDGEMENTS

Our work in the area of transition metal hydrides has been supported by the National Science Foundation. Funds from the Alfred P. Sloan Foundation, the Camille and Henry Dreyfus Foundation, the National Science Foundation Presidential Young Investigator Program, and the National Science Foundation Predoctoral Fellowship Program (LSS) are also gratefully acknowledged.

REFERENCES

1. Kapur, S.; Müller, E. W. *Surf. Sci.* **1977**, *66*, 45 and references therein.
2. Tran, Q.; Karellas, N. S.; Goodings, J. M. *Can. J. Chem.* **1988**, *66*, 2210.
3. Goodings, J. M.; Tran, Q.; Karellas, N. S. *Can. J. Chem.* **1988**, *66*, 2219.
4. Middleton, R. *Nucl. Instrum. Methods* **1977**, *144*, 373, and references therein.
5. Gray, J. A.; Rice, S. F.; Field, R. W. *J. Chem. Phys.* **1985**, *82*, 4717.
6. Gray, J. A.; Field, R. W. *J. Chem. Phys.* **1986**, *84*, 1041.

7. Cheetham, C. J.; Barrow, R. F. *Adv. High Temp. Chem.* **1967**, *1*, 7.
8. Smith, R. E. *Proc. Roy. Soc. London A* **1973**, *332*, 113.
9. Scott, P. R.; Richards, W. G. *Chem. Soc. Spec. Period. Reports* **1976**, *4*, 70.
10. Huber, K. P.; Herzberg, G. *Molecular Spectra and Molecular Structure. Constants of Diatomic Molecules*; Van Nostrand Reinhold: New York, 1979.
11. Mahanti (Mahanti, P. C. *Nature* **1931**, *127*, 557) claims to have observed bands of CuH^+, but this observation has never been repeated, suggesting that the observed bands are erroneously identified.
12. Bernard, A.; Effantin, C.; Bacis, R. *Can. J. Phys.* **1977**, *55*, 1654.
13. Smith, R. E.; Gaydon, A. G. *J. Phys. B* **1971**, *4*, 797.
14. Gaydon, A. G. *J. Phys. B* **1974**, *7*, 2429.
15. Scott, P. R.; Richards, W. G. *J. Phys. B* **1974**, *7*, 500.
16. Chong, D. P.; Langhoff, S. R.; Bauschlicher, C. W.; Walch, S. P.; Partridge, H. *J. Chem. Phys.* **1986**, *85*, 2850.
17. Gaydon, A. G.; Pearse, R. W. B. *Nature* **1937**, *140*, 110. Kleman, B; Liljeqvist, B. *Ark. Fys.* **1955**, *9*, 345. O'Connor, S. *J. Phys. B* **1969**, *2*, 541.
18. Kleman, B.; Uhler, U. *Can. J. Phys.* **1959**, *37*, 537.
19. O'Connor, S. *Proc. Roy. Irish Acad. A* **1967**, *65*, 95.
20. Nevin, T. E. *Proc. Roy. Irish Acad.* **1942**, *48*, 1. Nevin, T. E.; Stephens, D. V. *Proc. Roy. Irish Acad.* **1953**, *55*, 109.
21. Hayes, W.; McCarvill, P. D.; Nevin, T. E. *Proc. Phys. Soc. London A* **1957**, *70*, 904.
22. Kovacs, I.; Pacher, P. *J. Phys. B* **1974**, *8*, 796.
23. Stevens, A. E.; Feigerle, C. S.; Lineberger, W. C. *J. Chem. Phys.* **1983**, *78*, 5420.
24. Stevens-Miller, A. E.; Feigerle, C. S.; Lineberger, W. C. *J. Chem. Phys.* **1987**, *87*, 1549.
25. Balfour, W. J. *J. Chem. Phys.* **1988**, *88*, 5242.
26. Langhoff, S. R.; Bauschlicher, C. W.; Rendell, A. P. *J. Mol. Spectrosc.* **1989**, *138*, 108.
27. Carroll, P. K.; McCormack, P. *Astrophys. J. Lett.* **1972**, *177*, L33. Carroll, P. K.; McCormack, P.; O'Connor, S. *Astrophys. J.* **1976**, *208*, 903.
28. Balfour, W. J.; Lindgren, B.; O'Connor, S. *Chem. Phys. Lett.* **1983**, *96*, 251; *Phys. Scr.* **1983**, *28*, 551.
29. Phillips, J. G.; Davis, S. P.; Lindgren, B.; Balfour, W. J. *Astrophys. J. Suppl. Ser.* **1987**, *65*, 721.
30. Beaton, S. P.; Evenson, K. M.; Nelis, T.; Brown, J. M. *J. Chem. Phys.* **1988**, *89*, 4446.
31. Heimer, A. *Z. Phys.* **1937**, *104*, 448.
32. Klynning, L.; Neuhaus, H. *Z. Naturforsch. A* **1963**, *18*, 1142. Klynning, L.; Kronekvist, M. *Phys. Scripta* **1972**, *6*, 61; **1973**, *7*, 72; **1981**, *24*, 21.
33. Herzberg, G. *Molecular Spectra and Molecular Structure I. Spectra of Diatomic Molecules*; Van Nostrand Reinhold: New York, 1950.
34. Gaydon, A. G.; Pearse, R. W. B. *Proc. Roy. Soc. London A* **1934**, *148*, 312.
35. Heimer, A. *Z. Phys.* **1937**, *105*, 56.
36. Åslund, N.; Neuhaus, H.; Lagerqvist, A.; Andersen, E. *Ark. Fys.* **1965**, *28*, 271.
37. Scullman, R.; Löfgren, S.; Kadavathu, S. A. *Phys. Scripta* **1982**, *25*, 295.
38. Bagus, P. S.; Björkman, C. *Phys. Rev. A* **1981**, *23*, 461.
39. Blomberg, M.; Siegbahn, P.; Roos, B. *Mol. Phys.* **1982**, *47*, 127.
40. Kadavathu, S. A.; Löfgren, S.; Scullman, R. *Phys. Scripta* **1987**, *35*, 277.
41. Nelis, Th.; Bachem, E.; Bohle, W.; Urban, W. *Mol. Phys.* **1988**, *64*, 759. Lipus, K.; Simon, U.; Bachem, E.; Nelis, Th.; Urban, W. *Mol. Phys.* **1989**, *67*, 1431.
42. Ringström, U. *Ark. Fys.* **1966**, *32*, 211; *Can. J. Phys.* **1968**, *46*, 2291.
43. Brown, C. M.; Ginter, M. L. *J. Mol. Spectrosc.* **1980**, *80*, 145.

44. Ram, R. S.; Bernath, P. F.; Brault, J. W. *J. Mol. Spectrosc.* **1985**, *113*, 269. Fernando, W. T. M. L.; O'Brien, L. C.; Bernath, P. F. *J. Mol. Spectrosc.* **1990**, *139*, 461.
45. Bernard, A.; Bacis, R. *Can. J. Phys.* **1977**, *55*, 1322.
46. Scott, P. R.; Richards, W. G. *J. Phys. B* **1974**, *7*, 1679.
47. Kunz, A. B.; Guse, M. P.; Blint, R. J. *J. Phys. B* **1975**, *8*, 1358.
48. Langhoff, S. R.; Pettersson, L. G. M.; Bauschlicher, C. W.; Partridge, H. *J. Chem. Phys.* **1987**, *86*, 268.
49. Bernard, A.; Bacis, R. *Can. J. Phys.* **1976**, *54*, 1509.
50. Lagerqvist, A.; Neuhaus, H.; Scullman, R. *Proc. Phys. Soc.* **1964**, *83*, 498.
51. Malmberg, C.; Scullman, R.; Nylen, P. *Ark. Fys.* **1969**, *39*, 495.
52. Neuhaus, H.; Scullman, R. *Z. Naturforsch. A* **1964**, *19*, 659. Scullman, R. *Ark. Fys.* **1964**, *28*, 255. Kaving, B.; Scullman, R. *Can. J. Phys.* **1971**, *49*, 2264; *Phys. Scripta* **1974**, *9*, 33.
53. Ringström, U.; Åslund, N. *Ark. Fys.* **1966**, *32*, 19.
54. Ringström, U. *Ark. Fys.* **1964**, *27*, 227.
55. Dunlap, B. I.; Yu, H. L. *Chem. Phys. Lett.* **1980**, *73*, 525.
56. Armentrout, P. B. In *Structure/Reactivity and Thermochemistry of Ions*; Ausloos, P., Lias, S. G., Eds.; D. Reidel: Dordrecht, 1987; p. 97.
57. Ervin, K. M.; Armentrout, P. B. *J. Chem. Phys.* **1985**, *83*, 166.
58. Elkind, J. L.; Armentrout, P. B. *J. Phys. Chem.* **1984**, *88*, 5454.
59. Carlson, T. A.; Copley, J.; Duric, N.; Elander, N.; Erman, P.; Larsson, M.; Lyyra, M. *Astron. Astrophys.* **1980**, *83*, 238.
60. Armentrout, P. B. *Ann. Rev. Phys. Chem.* **1990**, *41*, 313.
61. Armentrout, P. B. In *Gas Phase Inorganic Chemistry*; Russell, D. H., Ed.; Plenum: New York, 1989; p. 1.
62. Armentrout, P. B.; Beauchamp, J. L. *Chem. Phys.* **1980**, *50*, 37.
63. Armentrout, P. B.; Beauchamp, J. L. *J. Am. Chem. Soc.* **1981**, *103*, 784.
64. Armentrout, P. B.; Halle, L. F.; Beauchamp, J. L. *J. Am. Chem. Soc.* **1981**, *103*, 6501.
65. Tolbert, M. A.; Beauchamp, J. L. *J. Am. Chem. Soc.* **1984,** *106*, 8117.
66. Halle, L. F.; Klein, F. S.; Beauchamp, J. L. *J. Am. Chem. Soc.* **1984**, *106*, 2543.
67. Mandich, M. L.; Halle, L. F.; Beauchamp, J. L. *J. Am. Chem. Soc.* **1984**, *106*, 4403.
68. For a discussion of this process, see Sunderlin, L. S.; Armentrout, P. B. *J. Phys. Chem.* **1988**, *92*, 1209.
69. Elkind, J. L.; Sunderlin, L. S.; Armentrout, P. B. *J. Phys. Chem.* **1989**, *93*, 3151.
70. Elkind, J. L.; Armentrout, P. B. *Int. J. Mass Spectrom. Ion Processes* **1988**, *83*, 259.
71. Elkind, J. L.; Armentrout, P. B. *J. Phys. Chem.* **1985**, *89*, 5626.
72. Elkind, J. L.; Armentrout, P. B. *J. Chem. Phys.* **1987**, *86*, 1868.
73. Elkind, J. L.; Armentrout, P. B. *J. Chem. Phys.* **1986**, *84*, 4862.
74. Elkind, J. L.; Armentrout, P. B. *J. Am. Chem. Soc.* **1986**, *108*, 2765; *J. Phys. Chem.* **1986**, *90*, 5736.
75. Elkind, J. L.; Armentrout, P. B. *J. Phys. Chem.* **1986**, *90*, 6576.
76. Elkind, J. L.; Armentrout, P. B. *Inorg. Chem.* **1986**, *25*, 1078.
77. Elkind, J. L.; Armentrout, P. B. Unpublished work.
78. Elkind, J. L.; Armentrout, P. B. *J. Phys. Chem.* **1987**, 91, 2037.
79. Aristov, N.; Armentrout, P. B. *J. Am. Chem. Soc.* **1986**, *108*, 1806. Aristov, N.; Armentrout, P. B. *J. Phys. Chem.* **1987**, *91*, 6178. Sunderlin, L. S.; Armentrout, P. B. *J. Phys. Chem.* **1988**, *92*, 1209. Georgiadis, R.; Armentrout, P. B. *J. Phys. Chem.* **1988**, *92*, 7067. Sunderlin, L. S.; Armentrout, P. B. *Int. J. Mass Spectrum. Ion Processes* **1989**, *94*, 149.
80. Sunderlin, L.; Aristov, N.; Armentrout, P. B. *J. Am. Chem. Soc.* **1987**, *109*, 78.

81. Schultz, R. H.; Elkind, J. L.; Armentrout, P. B. *J. Am. Chem. Soc.* **1988**, *110*, 411.
82. Georgiadis, R.; Fisher, E. R.; Armentrout, P. B. *J. Am. Chem. Soc.* **1989**, *111*, 4251.
83. Sunderlin, L. S.; Armentrout, P. B. *J. Am. Chem. Soc.* **1989**, *111*, 3845.
84. Fisher, E. R.; Armentrout, P. B. *J. Phys. Chem.* **1990**, *94*, 1674.
85. Clemmer, D. E; Sunderlin, L. S.; Armentrout, P. B. *J. Phys. Chem.* **1990**, *94*, 208. Clemmer, D. E.; Sunderlin, L. S.; Armentrout, P. B. *J. Phys. Chem.* **1990**, *94*, 3008. Clemmer, D. E.; Armentrout, P. B. *J. Phys. Chem.* in press.
86. Stevens, A. E.; Beauchamp, J. L. *Chem. Phys. Lett.* **1981**, *78*, 291.
87. Lias, S. G.; Liebman, J. F.; Levin, R. D. *J. Phys. Chem. Ref. Data* **1984**, *13*, 695.
88. Lias, S. G.; Bartmess, J. E.; Liebman, J. F.; Holmes, J. L.; Levin, R. D.; Mallard, W. G. *J. Phys. Chem. Ref. Data* **1988**, *17*, 1 (Suppl. 1).
89. Gaydon, A. G. *Dissociation Energies and Spectra of Diatomic Molecules*, 3d ed.; Chapman Hall: London, 1968.
90. Dendramis, A.; Van Zee, R. J.; Weltner, W. *Astrophys. J.* **1979**, *231*, 632.
91. Herzberg, G.; Mundie, L. G. *J. Chem. Phys.* **1940**, *8*, 263.
92. Kant, A.; Moon, K. A. *High Temp. Sci.* **1979**, *11*, 52.
93. Kant, A.; Moon, K. A. *High Temp. Sci.* **1981**, *14*, 23.
94. Sallans, L.; Lane, K.; Squires, R. R.; Freiser, B. S. *J. Am. Chem. Soc.* **1983**, *105*, 6352.
95. Sallans, L.; Lane, K.; Squires, R. R.; Freiser, B. S. *J. Am. Chem. Soc.* **1985**, *107*, 4379.
96. Tolbert, M. A.; Beauchamp, J. L. *J. Phys. Chem.* **1986**, *90*, 5015.
97. Schultz, R. H.; Armentrout, P. B. *J. Chem. Phys.* **1991**, *94*, 1150.
98. Aristov, N. Ph.D. Thesis, University of California, Berkeley, 1986.
99. Sunderlin, L. S.; Armentrout, P. B. *J. Phys. Chem.* **1990**, *94*, 3589.
100. Chen. Y.; Clemmer, D. E.; Armentrout, P. B. Work in progress; *J. Chem. Phys.* in press.
101. Bauschlicher, C. W.; Langhoff, S. R.; Partridge, H.; Barnes, L. A. *J. Chem. Phys.* **1989**, *91*, 2399. This value is increased by 8.4 kJ mol^{-1} from the value directly calculated in [16] to account for basis-set incompleteness.
102. Sodupe, M.; Lluch, J. M.; Oliva, A.; Illas, F.; Rubio, J. *J. Chem. Phys.* **1990**, *92*, 2478.
103. Feigerle, C. S.; Corderman, R. R.; Bobashev, S. V.; Lineberger, W. C. *J. Chem. Phys.* **1981**. *74*, 1580.
104. Hotop, H.; Lineberger, W. C. *J. Chem. Phys. Ref. Data* **1985**, *14*, 731.
105. Leopold, D. G.; Lineberger, W. C. *J. Chem. Phys.* **1986**, *85*, 51.
106. Corderman, R. R.; Engelking, P. C.; Lineberger, W. C. *J. Chem. Phys.* **1979**, *70*, 4474.
107. Miller, A. E. S.; Feigerle, C. S.; Lineberger, W. C. *J. Chem. Phys.* **1986**, *84*, 4127.
108. Mandich, M. L.; Halle, L. F.; Beauchamp, J. L. *J. Am. Chem. Soc.* **1984**, *106*, 4403.
109. Armentrout, P. B.; Georgiadis, R. *Polyhedron* **1988**, *7*, 1573.
110. Armentrout, P. B. *ACS Symposium Series*, **1990**, *428*, 18.
111. Squires, R. R. *J. Am. Chem. Soc.* **1985**, *107*, 4385.
112. Carter, E. A.; Goddard, W. A. *J. Phys. Chem.* **1988**, *92*, 5679.
113. Schilling, J. B.; Goddard, W. A.; Beauchamp, J. L. *J. Am. Chem. Soc.* **1987**, *109*, 5566.
114. Pettersson, L. G. M.; Bauschlicher, C. W.; Langhoff, S. R.; Partridge, H. *J. Chem. Phys.* **1987**, *87*, 481.
115. Squires, R. R. *J. Am. Chem. Soc.* **1985**, *107*, 4385.
116. Mead, R. D.; Stevens, A. E.; Lineberger, W. C. In *Gas Phase Ion Chemistry*, Vol. 3; Bowers, M. T., Ed.; Academic: Orlando, FL, 1984; p. 213.

117. Armentrout, P. B. In *Selective Hydrocarbon Activation: Principles and Progress*, Davies, J. A., Watson, P. L., Liebman, J. F., Greenberg, A., Eds.; VCH: New York, 1990; p. 467.
118. Ruiz, M. E.; Garcia-Prieto, J.; Novaro, O. *J. Chem. Phys.* **1984**, *80*, 1529. Sevin, A.; Chaquin, P. *Nouv. J. Chimie* **1983**, *7*, 353.
119. Siegbahn, P. E. M.; Blomberg, M. R. A.; Bauschlicher, C. W. *J. Chem. Phys.* **1984**, *81*, 1373.
120. Saillard, J.; Hoffman, R. *J. Am. Chem. Soc.* **1984**, *106*, 2006.
121. Brothers, P. J.; *Prog. Inorg. Chem.* **1981**, *28*, 1. Noell, J. O.; Hay. P. J. *J. Am. Chem. Soc.* **1982**, *104*, 4578.
122. Shustorovich, E.; Baetzold, R. C.; Muetterties, E. L. *J. Phys. Chem.* **1983**, *87*, 1100. Siegbahn, P. E. M.; Blomberg, M. R. A.; Bauschlicher, C. W. *J. Chem. Phys.* **1984**, *81*, 2103.
123. Carlin, T. J.; Sallans, L.; Cassady, C. J.; Jacobson, D. B.; Freiser, B. S. *J. Am. Chem. Soc.* **1983**, *105*, 6320.
124. Alton, G. D. *Nucl. Instrum. Methods* **1986**, *A244*, 133.
125. Van Zee, R. J.; DeVore, T. C.; Wilkerson, J. L.; Weltner, W. *J. Chem. Phys.* **1978**, *69*, 1869. Van Zee, R. J.; DeVore, T. C.; Weltner, W. *J. Chem. Phys.* **1979**, *71*, 2051. Ozin, G. A.; McCaffrey, J. G. *J. Am. Chem. Soc.* **1984**, *106*, 807.
126. Huang, Y.; Wise, M. B.; Jacobson, D. B.; Freiser, B. S. *Organometallics* **1987**, *6*, 346.
127. Schilling, J. B.; Beauchamp, J. L. *J. Am. Chem. Soc.* **1988,** *110*, 15.
128. Halle, L. F.; Crowe, W. E.; Armentrout, P. B.; Beauchamp, J. L. *Organometallics* **1984**, *3*, 1694.
129. Martinho, Simões, J. A.; Beauchamp, J. L. *Chem. Rev.* **1990**, *90*, 629.
130. Demuynck, J.; Schaefer, H. F. *J. Chem. Phys.* **1980**, *72*, 311.
131. Blomberg, M. R. A.; Siegbahn, P. E. M. *J. Chem. Phys.* **1983**, *78*, 5682.
132. Johnson, B. F. G. In *Mass Spectrometry of Metal Compounds*; J, Charalambous, Ed.; Butterworths: London, 1975; Chapter 7, and references therein.
133. Johnson, B. F. G.; Lewis, J.; Robinson, P. W. *J. Chem. Soc. A* **1970**, 1684.
134. Evans, S.; Green, J. C.; Green, M. L. H.; Orchard, A. F.; Turner, D. W. *Discuss. Faraday Soc.* **1969**, *47*, 112.
135. Cradock, S.; Ebsworth, E. A. V.; Robertson, A. *J. Chem. Soc. Dalton Trans.* **1973**, 22.
136. Hall, M. B. *J. Am. Chem. Soc.* **1975**, *97*, 2057.
137. Guest, M. F.; Higginson, B. R.; Lloyd, D. R.; Hillier, I. H. *J. Chem. Soc. Faraday Trans. 2* **1975**, *71*, 902.
138. Chen, H.-W.; Jolly, W. L.; Kopf, J.; Lee, T. H. *J. Am. Chem. Soc.* **1979**, *101*, 2607.
139. Blakney, G. B.; Allen, W. F. *Inorg, Chem.* **1971,** *10*, 2763.
140. Stobart, S. R. *J. Chem. Soc., Dalton Trans.* **1972**, 2442.
141. Edgell, W. F.; Fisher, J. W.; Asato, G.; Risen, W. M. *Inorg. Chem.* **1969**, *8*, 1103, and references therein.
142. McNeill, E. A.; Scholer, F. R. *J. Am. Chem. Soc.* **1977**, *99*, 6243.
143. La Place, S. J.; Hamilton, W. C.; Ibers, J. A.; Davison, A. *Inorg. Chem.* **1969**, *8*, 1928.
144. Robiette, A. G.; Sheldrick, G. M.; Simpson, R. N. F. *J. Chem. Soc., Chem. Commun.* **1968**, 506.
145. Green, J. C.; Jackson, S. E.; Higginson, B. *J. Chem. Soc., Dalton Trans.* **1975**, 403.
146. Lichtenberger, D. L.; Darsey, G. P.; Kellogg, G. E.; Sanner, R. D.; Young, V. G.; Clark, J. R. *J. Am. Chem. Soc.* **1989**, *111*, 5019.
147. Brookhart, M.; Green, M. L. H. *J. Organometallic Chem.* **1983**, *250*, 395.

148. Lamanna, W.; Brookhart, M. *J. Am. Chem. Soc.* **1981**, *103*, 989. Brookhart, M.; Lamanna, W.; Humphrey, M. B. *J. Am. Chem. Soc.* **1982**, *104*, 2117.
149. Lichtenberger, D. L.; Kellogg, G. E. *J. Am. Chem. Soc.* **1986**, *108*, 2560.
150. Kubas, G. J. *Acc. Chem. Res.* **1988**, *21*, 120.
151. Fletcher, T. R.; Rosenfeld, R. N. *J. Am. Chem. Soc.* **1986**, *108*, 1686.
152. Weitz, E. *J. Phys. Chem.* **1987**, *91*, 3945.
153. Ishikawa, Y.; Weersink, R. A.; Hackett, P. A.; Rayner, D. M. *Chem. Phys. Lett.* **1987**, *142*, 271. Ishikawa, Y.; Hackett, P. A.; Rayner, D. M. *J. Phys. Chem.* **1989**, *93*, 652.
154. McDonald, R. N.; Chowdhury, A. K.; Schell, P. L. *Organometallics* **1984**, *3*, 644.
155. McDonald, R. N.; Chowdhury, A. K. *Organometallics* **1986**, *5*, 1187.
156. Lane, K. R.; Sallans, L.; Squires, R. R. *Organometallics* **1985**, *4*, 408.
157. Lane, K. R.; Squires, R. R. *J. Am. Chem. Soc.* **1985**, *107*, 6403.
158. McDonald, R. N.; Chowdhury, A. K.; Jones, M. T. *J. Am. Chem. Soc.* **1986**, *108*, 3105.
159. McDonald, R. N.; Schell, P. L.; Chowdhury, A. K. *J. Am. Chem. Soc.* **1985**, *107*, 5578.
160. McDonald, R. N.; Chowdhury, A. K.; Schell, P. L. *J. Am. Chem. Soc.* **1984**, *106*, 6095.
161. Lane, K. R.; Lee, R. E.; Sallans, L.; Squires, R. R. *J. Am. Chem. Soc.* **1984**, *106*, 5767.
162. Lane, K. R.; Squires, R. R. *J. Am. Chem. Soc.* **1986**, *108*, 7187.
163. See [161] and references therein for a discussion.
164. Lane, K. R.; Sallans, L.; Squires, R. R. *J. Am. Chem. Soc.* **1986**, *108*, 4368.
165. McDonald, R. N.; Jones, M. T. *J. Am. Chem. Soc.* **1986**, *108*, 8097.
166. McDonald, R. N.; Jones, M. T. *Organometallics* **1987**, *6*, 1991.
167. McDonald, R. N.; Reed, D. J.; Chowdhury, A. K. *Organometallics* **1989**, *8*, 1122.
168. Squires, R. R. *Chem. Rev.* **1987**, *87*, 623.
169. McDonald, R. N.; Schell, P. L. *Organometallics* **1988**, *7*, 1820.
170. Allison, J. *Prog. Inorg. Chem.* **1986**, *34*, 627.
171. Beauchamp, J. L. *ACS Symposium Series* **1987**, *333*, 11. Van Koppen, P. A. M.; Bowers, M. T.; Beauchamp, J. L.; Dearden, D. V. *ACS Symposium Series* **1990**, *428*, 34.
172. Freiser, B. S. *Chemtracts—Analytical and Physical Chemistry* **1989**, *1*, 65.
173. This field is nicely represented in the recent book, *Gas Phase Inorganic Chemistry*; Russell, D. H., Ed.; Plenum: New York, 1989.
174. Schwarz, H. *Acc. Chem. Res.* **1989**, *22*, 282.
175. Armentrout, P. B.; Beauchamp, J. L. *Acc. Chem. Res.* **1989**, *22*, 315.
176. MacMillan, D. K.; Gross, M. L. In *Gas Phase Inorganic Chemistry*; Russell, D. H., Ed.; Plenum: New York, 1989; p. 369.
177. Freas, R. B.; Ridge, D. P. *J. Am. Chem. Soc.* **1980**, *102*, 7129.
178. Halle, L. F.; Armentrout, P. B.; Beauchamp, J. L. *Organometallics* **1982**, *1*, 963.
179. Larson, B. S.; Ridge, D. P. *J. Am. Chem. Soc.* **1984**, *106*, 1912.
180. Forbes, R. A.; Lech, L. M.; Freiser, B. S. *Int. J. Mass Spectrom. Ion Processes* **1987**, *77*, 107.
181. This can be thought of in terms of conservation of angular momentum. This will tend to favor formation of [$ScH^+ + C_6H_6$], which has a reduced mass of 28.9 amu, instead of [$ScC_6H_6{}^+ + H$], with a reduced mass of only 1.0 amu.
182. Beauchamp, J. L.; Stevens, A. E.; Corderman, R. R. *Pure. Appl. Chem.* **1979**, *51*, 967.
183. Byrd, G. D.; Freiser, B. S. *J. Am. Chem. Soc.* **1982**, *104*, 5944.
184. Jacobson, D. B.; Freiser, B. S. *J. Am. Chem. Soc.* **1985**, *107*, 72.
185. Stevens, A. E.; Beauchamp, J. L. *J. Am. Chem. Soc.* **1981**, *103*, 190.
186. Stevens, A. E.; Beauchamp, J. L. *J. Am. Chem. Soc.* **1979**, *101*, 245. Fernando,

J.; Faigle, G.; Ferreira, A. M. da C.; Galembeck, S. E.; Riveros, J. M. *J. Chem. Soc., Chem. Commun.* **1978**, 126. Foster, M. S.; Beauchamp, J. L. *J. Am. Chem. Soc.* **1975**, *97*, 4808, 4814.
187. Meckstroth, W. K.; Ridge, D. P. *J. Am. Chem. Soc.* **1985**, *107*, 2281.
188. Ikonomou, M. G.; Sunner, J.; Kebarle, P. *J. Phys. Chem.* **1988**, *92*, 6308.
189. Meot-Ner (Mautner), M. *J. Am. Chem. Soc.* **1989**, *111*, 2830.
190. Connor, J. A.; Zafarani-Moattar, M. T.; Bickerton, J.; El Saied, N. I.; Suradi, S.; Carson, R.; Takhin, G. A.; Skinner, H. A. *Organometallics* **1982**, *1*, 1166.
191. Hieber, W.; Wagner, G. Z. *Z. Naturforsch. B* **1958**, *13*, 339.
192. Bidinosti, D. R.; McIntyre, N. S. *J. Chem. Soc., Chem. Commun.* **1966**, 555.
193. Smith, G. P. *Polyhedron* **1988**, *7*, 1605.
194. Goodman, J. L.; Peters, K. S.; Vaida, V. *Organometallics* **1986**, *5*, 815.
195. Bronshtein, Y. E.; Gankin, V. Y.; Krinkin, D. P.; Rudkovskii, D. M. *Russian J. Phys. Chem.* (Engl. Ed.) **1966**, *40*, 802.
196. Pilcher, G.; Skinner, H. A. In *The Chemistry of the Metal Bond*; Hartley, F. R.; Patai, S., Eds.; Wiley: New York, 1982; Chapter 2.
197. Bidinosti, D. R.; McIntyre, N. S. *J. Chem. Soc., Chem. Commun.* **1967**, 1.
198. Bidinosti, D. R.; McIntyre, N. S. *Can. J. Chem.* **1970**, *48*, 593.
199. Pilcher, G.; Skinner, H. A. In *Chemistry of the Metal–Carbon Bond*, Hartley, F. R., Patai, S., Eds.; Wiley: New York, 1982; p. 43.
200. Gardner, P. J; Carter, A.; Cunninghame, R. G.; Robinson, B. H. *J. Chem. Soc., Dalton Trans.* **1975**, 2582.
201. Alemdaroglu, N. H.; Penninger, J. M. L.; Oltay, E. *Monatsh. Chem.* **1976**, *107*, 1043.
202. Stevens, A. E. Ph.D. Thesis, California Institute of Technology, 1981.
203. Miller, A. E. S.; Kawamura, A. R.; Miller, T. M. *J. Am. Chem. Soc.* **1990,** *112*, 457.
204. Sunderlin, L. S.; Armentrout, P. B. *Organometallics* **1990**, *9*, 1248.
205. Trevor, D. J.; Kaldor, A. *ACS Symposium Series* **1987**, *333*, 43.
206. Kaldor, A.; Cox, D. M.; Zakin, M. R. *Adv. Chem. Phys.* **1988**, *70*, 211.
207. Riley, S. J.; Parks, E. K.; Pobo, L. G.; Wexler, S. *Ber. Bunsenges. Phys. Chem.* **1984**, *88*, 287.
208. Richtsmeier, S. C.; Parks, E. K.; Liu, K.; Pobo, L. G.; Riley, S. J. *J. Chem. Phys.* **1985**, *82*, 3659.
209. Whetten, R. L.; Cox, D. M.; Trevor, D. J.; Kaldor, A. *Phys. Rev. Lett.* **1985**, *54*, 1494.
210. Morse, M. D.; Geusic, M. E.; Heath, J. R.; Smalley, R. E. *J. Chem. Phys.* **1985**, *83*, 2293.
211. Riley, S. J.; Parks, E. K. In *Physics and Chemistry of Small Clusters*; Jena, P., Rao, B. K., Khanna, S. N., Eds.; Plenum: New York, 1987; p. 727.
212. Zakin, M. R.; Cox, D. M.; Whetten, R. L.; Trevor, D. J.; Kaldor, A. *Chem. Phys. Lett.* **1987**, *135*, 223.
213. Parks, E. K.; Liu, K.; Richtsmeier, S. C.; Pobo, L. G.; Riley, S. J. *J. Chem. Phys.* **1985**, *82*, 5470.
214. Parks, E. K.; Nieman, G. C.; Pobo, L. G.; Riley, S. J. *J. Phys. Chem.* **1987**, *91*, 2671.
215. Liu, K.; Parks, S. C.; Richtsmeier, S. C.; Pobo, L. G.; Riley, S. J. *J. Chem. Phys.* **1985**, *83*, 2882.
216. Parks, E. K.; Weiller, B. H.; Bechthold, P. S.; Hoffman, W. F.; Nieman, G. C.; Pobo, L. G.; Riley, S. J. *J. Chem. Phys.* **1988**, *88*, 1622.
217. Loh, S. K.; Hales, D. A.; Lian, L.; Armentrout, P. B. *J. Chem. Phys.* **1989**, *90*, 5466.
218. Lian, L.; Armentrout, P. B. Manuscript in preparation.
219. Geusic, M. E.; Morse, M. D.; Smalley, R. E. *J. Chem. Phys.* **1985**, *82*, 590.

220. Whetten, R. L.; Zakin, M. R.; Cox, D. M.; Trevor, D. J.; Kaldor, A. *J. Chem. Phys.* **1986**, *85*, 1697.
221. Zakin, M. R.; Brickman, R. O.; Cox, D. M.; Kaldor, A. *J. Chem. Phys.* **1988**, *88*, 3555.
222. Hamrick, Y.; Taylor, S.; Lemire, G. W.; Fu, Z.-W.; Shui, J.-C.; Morse, M. D. *J. Chem. Phys.* **1988**, *88*, 4095.
223. Hamrick, Y. M.; Morse, M. D. *J. Phys. Chem.* **1989**, *93*, 6494.
224. Hoffman, W. F.; Parks, E. K.; Nieman, G. C.; Pobo, L. G.; Riley, S. J. *Z. Phys. D* **1987**, *7*, 83.
225. Cox, D. M.; Whetten, R. L.; Zakin, M. R.; Trevor, D. J.; Reichmann, K. C.; Kaldor, A. *AIP Conf. Proc.* **1986**, *146*. Cox, D. M.; Zakin, M. R.; Kaldor, A. In *Physics and Chemistry of Small Clusters*; Jena, P., Rao, B. K., Khanna, S. N., Eds.; Plenum: New York, 1987; p. 741.
226. Klots, T. D.; Winter, B. J.; Parks, E. K.; Riley, S. J. *J. Chem. Phys.* **1990**, *92*, 2110.
227. Zakin, M. R.; Cox, D. M.; Kaldor, A. *J. Chem. Phys.* **1988**, *89*, 1201.
228. Fayet, P.; Kaldor, A.; Cox, D. M. *J. Chem. Phys.* **1990**, *92*, 254.
229. Kaldor, A.; Cox, D. M.; Trevor, D. J.; Zakin, M. R. *Z. Phys. D* **1986**, *3*, 195.
230. Hoffman, W. F.; Parks, E. K.; Riley, S. J. *J. Chem. Phys.* **1989**, *90*, 1526.
231. Nonose, S.; Sone, Y.; Onodera, K.; Sudo, S.; Kaya, K. *Chem. Phys. Lett.* **1989**, *164*, 427.
232. Nonose, S.; Sone, Y.; Onodera, K.; Sudo, K.; Kaya, K. *J. Phys. Chem.* **1990**, *94*, 2744.
233. Zakin, M. R.; Brickman, R. O.; Cox, D. M.; Kaldor, A. *J. Chem. Phys.* **1988**, *88*, 6605.
234. Alford, J. M.; Weiss, F. D.; Laaksonen, R. T.; Smalley, R. E. *J. Phys. Chem.* **1986**, *90*, 4480. Elkind, J. L.; Weiss, F. D.; Alford, J. M.; Laaksonen, R. T., Smalley, R. E. *J. Chem. Phys.* **1988**, *88*, 5215.
235. Buckner, S. W.; Freiser, B. S. In *Gas Phase Inorganic Chemistry*; Russell, D. H., Ed.; Plenum: New York, 1989; p. 279.
236. Jacobson, D. B. *Organometallics* **1988**, *7*, 568.
237. Jacobson, D. B. *J. Am. Chem. Soc.* **1989**, *111*, 1626.
238. Armentrout, P. B. In *Laser Applications in Chemistry and Biophysics* El-Sayed, M., Ed.; *Proc. SPIE* **1986**, *620*, 38.
239. Loh, S. K.; Lian, L.; Armentrout, P. B. Manuscript in preparation.

CHAPTER 2

Matrix Studies of Transition Metal Hydrides

RAY L. SWEANY

Department of Chemistry, University of New Orleans, New Orleans, Louisiana 70148

Matrix isolation refers to a spectroscopic technique for observing an analyte embedded in a solid material. First practiced by George Pimentel, it was hoped that the structure and spectroscopic behavior of atoms and molecules in inert-gas solids would duplicate behavior of the truly isolated species in the gas phase [1]. Although this ideal is never realized, valuable information has still been obtained by the use of this technique. Matrix materials that are least capable of perturbing an analyte are also incapable of strongly interacting with each other. Thus, matrix isolation is almost always practiced at the low temperature at which the matrix material is itself a solid. The purpose of this chapter is to review the use of the technique in studies of transition metal hydrides, pointing out how the hydrides are perhaps the most successfully studied metal complexes of any type. The technique has been useful in elucidating the spectrum and structure of transition metal hydrides and their photochemistry. Also, a number of hydrides, both classical and nonclassical, have been formed in matrices, many of them too unstable for conventional studies. The use of matrix isolation in general is reviewed periodically, most recently in a book edited by Andrews and Moskovits [2].

In developing this topic, I will begin by describing experiments in which hydrides were constituents in the gas mixture that was condensed to form the matrix. These matrices yield important information concerning the spectrum and photochemistry of molecules that could be characterized by other experimental methods. Thence, I will describe hydrides that are formed in situ by bimolecular reactions involving a coordinatively unsaturated metal center with a nearby hydrogen-containing molecule. Finally, I will describe reactions in which hydrides are formed by an intramolecular process.

2.1. MATRIX ISOLATION OF STABLE HYDRIDES

2.1.1. Techniques

There are a limited number of transition metal hydrides that can be introduced into matrices by conventional techniques because of the necessity of preparing matrices from gas mixtures. For the production of gas matrices, the analyte may be mixed with the matrix gas prior to deposition in order to achieve an accurate estimate of the host–guest ratio. Alternatively, the mixture may be prepared just prior to deposition by subliming the analyte into a flow of the matrix gas, but there is virtually no way to know the host–guest ratio by this method. Lacking this knowledge, infrared band shapes have been used to determine whether the guest is sufficiently dilute that it can be considered truly isolated. Poor isolation usually results in tailing of an absorption on the low-frequency side. This tailing results from an environment of inhomogeneous polarizability. As the polarizability of the environment increases, there is an increasingly large red shift in the positions of infrared absorptions [3]. Assuming that the polarizability of an analyte is greater than that of the host, infrared absorptions of species in contact with other analyte molecules will be red-shifted relative to those absorptions of well-isolated molecules. Thus, the envelope of all absorptions of molecules, both well and poorly isolated, will exhibit tailing on the low-frequency side of the absorption, the degree of tailing correlating with the number of molecules that are poorly isolated. Some care must be exercised in applying this criterion because of the possibility of nearby absorptions of rarer isotopomers. Figure 2.1 shows several spectra made of $CH_3WCp(CO)_3$ exhibiting varying degrees of tailing. The use of premixed gases is precluded when the analyte is insufficiently stable for long-term storage at ambient conditions or when it is insufficiently volatile to be handled reliably in a vacuum line. Because of these difficulties, the only published work involving matrices formed from premixed components were those reporting the photochemistry of $HMn(CO)_5$ and $HRe(CO)_5$ [4].

Less conventionally, paraffin hydrocarbons or polymer films have been used in place of frozen gases as matrices [5, 6]. Using these methods, the analyte is dissolved in the polymer from which a film is made. When cooled to liquid nitrogen temperatures these samples exhibit many of the same phenomena as can be observed in frozen-gas matrices. A principle advantage of these techniques is the avoidance of the necessity of vaporizing the isolate in order to form a matrix. However, the vibrational spectrum of the polymer film does interfere with observations in some regions of the infrared spectrum. Although coordinatively unsaturated molecules have been formed reversibly in these environments, it is reasonable to expect significantly stronger host–guest interactions than are found in rare-gas matrices.

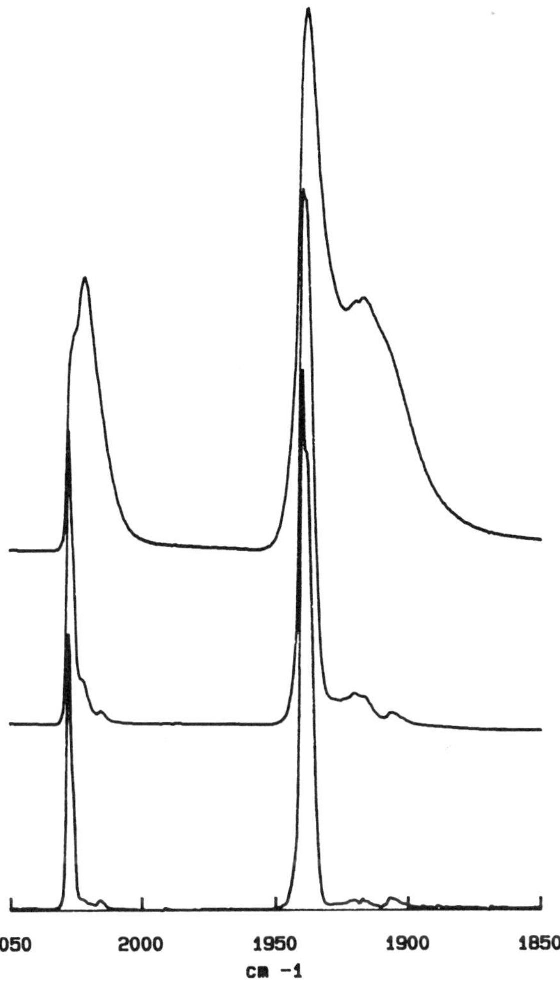

FIGURE 2.1
Spectra made of $CH_3WCp(CO)_3$ using sublimation. The upper two spectra exhibit various degrees of tailing, which suggests that some of the tungsten complexes occupy additional positions in the matrix [128].

2.1.2. Vibrational Spectroscopy of Hydrides

Before undertaking a detailed description of the use of matrix isolation in the study of the hydrides, it is important to discuss the means by which hydrides are characterized in matrices. Lacking the ability to make facile magnetic resonance measurements in matrices, the infrared spectrum is nearly the sole technique that gives direct evidence of the hydride ligand. Classical hydrides are characterized by the M—H stretching and bending vibrations. The characteristics of these modes have been long-appreciated. Typically, the M—H stretch appears in the region of 1700–2300 cm^{-1} and the bending vibration occurs in the region between 700 and 950 cm^{-1} [7]. The corresponding M—D absorptions are shifted dramatically to lower frequency. The ratio of $\bar{\nu}_{M-H}/\bar{\nu}_{M-D}$ is found to be around 1.35 [7, 8]. Exceptions occur when a hydrogen-containing vibrational mode is close in frequency to other modes of the same symmetry. In those instances, the modes are coupled and absorptions that are observed cannot be interpreted without a normal coordinate analysis in which the contributions from the hydrogen motions can be unravelled from the other motions [9–11].

Absorptions due to M—H stretching vibrations can be quite weak in the infrared spectrum, whereas absorptions due to bending vibrations are often strong. For example, the Co—H bending vibration of $HCo(CO)_4$ is approximately 10% of the intensity of the most intense carbonyl mode whereas the M—H stretching vibration is rarely seen [11]. Perutz has undertaken a detailed study of the hydrogen vibrational coordinates of a series of complexes, H_nMCp_2. In addition to being quite weak, the width of the absorptions due to the M—H stretch has been found to be strongly dependent on the identity of the host. In methane, the bands due to the W—H stretching vibrations of H_2WCp_2 are approximately 25 cm^{-1} full width at half-maximum (fwhm), whereas the same features are only 6 cm^{-1} fwhm in Ar. The modes are also split into components because of the existence of several types of trapping sites [12]. While making these generalizations it is important to acknowledge the exceptions. For example, the M—H deformation modes have not been observed in the spectra of $HMCp(CO)_3$ for M = Mo and W [13]. Although the M—H stretching vibrations are often weak in the spectrum of stable 18-electron metal hydrides, they are rarely missed in the spectra of simple complexes of the general formulation HMX, about which more will be written later. Raman spectroscopy rarely has been attempted in matrix studies of hydrides although it should readily detect the M—H stretching vibration for concentrations that are greater than 0.5 mol %.

Fewer infrared observations have been made of nonclassical hydrides [14]. There are six vibrational modes that arise from the incorporation of H_2 on anything except a naked metal atom. They are shown in Figure 2.2. In a simple triatomic system, the three lower illustrations describe simple rotations of the system. In this context, it is important to point out that rotation of η^2-H_2 on a single metal center is a rotation of the molecule. In condensed

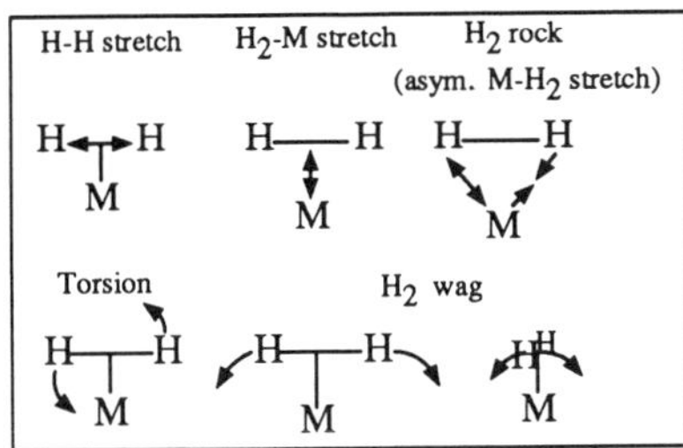

FIGURE 2.2
Valence coordinates of coordinated dihydrogen.

media, the barrier to such a rotation is intermolecular rather than intramolecular. The modes portrayed in Figure 2.2 are somewhat different than they were presented by Kubas et al. [15]. In the Kubas system, the M—H_2 stretch and the H_2 rock were described as symmetric and antisymmetric M—H stretches, analogous to simple dihydrides. The presentation in Figure 2.2 emphasizes the integrity of the H_2 moiety: It is depicted as either rocking in place or vibrating as a unit along the M—H_2 bond vector. It should be pointed out, however, that there is no large conceptual difference between the display of Figure 2.2 and one that describes the valence coordinates of a simple $M(H)_2$ fragment in a molecule.

The excitation of the H—H stretching vibration is never a formally forbidden infrared transition in dihydrogen complexes, but it is polarized along the direction of the M—H_2 bond in highly symmetric compounds. Thus, intensity only arises from the coupling of the H—H stretching vibration with other modes of the same symmetry. Of those, the M—H_2 stretching vibration would seem the most likely to couple strongly, but that mode itself does not produce a great deal of intensity in the infrared spectrum. Thus, one should not expect the H—H stretching vibration of an η^2-H_2 ligand ever to be very intense. The H—H stretching vibration has been observed in the region of 2600–3300 cm^{-1} [14]. The range may extend to as low as 2200 cm^{-1} [16, 17]. The M—H_2 stretch has been found between 975 and 850 cm^{-1}. These positions are based on very few reports and may be revised in the future as the spectra of more compounds become known.

Rotation of the hydrogen molecule on an axis normal to the H—H bond vector is intrinsically related to the torsional vibration. For small barriers of rotation, the spectrum of the torsional mode is complicated by the possibility of tunnelling [18]. As a consequence, a splitting of the torsional mode is expected, the size of which is inversely correlated with the barrier height. There are two wave functions that describe molecules that contain the MH_2 moiety [19]; they are identical except for the interchange of the hydrogen nuclei. For large barriers to rotation, the energies of the two states are degenerate. For smaller barriers, there is a splitting of the two states, the magnitude of which depends on the barrier to rotation and the torsional vibrational state. In the limit of free rotation, the states correlate with rovibrational states. As a result of hindered rotation, the torsional mode of hydrogen in $(H_2)W(CO)_3(PR_3)_2$ was found to be 347 cm^{-1} with a splitting of 45 cm^{-1} [20]. This splitting comes from the fact that quantum-mechanical splitting is strongly a function of the librational quantum number. The smaller splitting of the ground librational state is directly accessed by the inelastic neutron scattering experiment. From observations of the splittings, the barrier to rotation was determined to be 762 cm^{-1} [20].

Splittings have been observed in the $M—H_2$ stretching vibration of both $(H_2)Cr(CO)_5$ [21] and $Pd(H_2)$ [22] and in the H—H stretching vibration of $(H_2)Cr(CO)_5$ [21] in matrices and in $(H_2)Co(CO)_2NO$ [23] in liquid xenon. *The reason for the splittings is unclear*. Only the torsional mode of η^2-H_2 should be split by the mechanism just described. Although it is true that the H—H stretching vibration is a segment of a pathway that exchanges the hydrogen nuclei, the potential barrier for the exchange will be far greater than is the barrier for rotation of the hydrogen. Thus, one expects splitting of energy levels to be independent of the H—H stretching quantum number. In this context, it is important to point out that this analysis of quantum-mechanical doubling in infrared spectra is identical in many respects to the analysis of exchange coupling in the NMR spectrum of polyhydrides [24].

2.1.3. Electron Spin Resonance Characterization of the Hydrides

Very few hydrides have been studied by electron spin resonance (ESR) spectroscopy; therefore, I will not treat the subject in depth, although the ESR experiment is potentially as useful for the study of hydrides in matrices as the proton magnetic resonance experiment is for the study of hydrides in fluid media. The coupling of the free electron to a hydrogen nucleus can be large if it resides on the hydrogen atom 100% of the time as in the hydrogen atom, for which the isotropic hyperfine coupling constant is 508 G [8]. In transition metal hydrides, the coupling is never so large because atomic orbitals of hydrogen do not participate extensively in the singly occupied molecular orbital (SUMO). The M—H bonding orbital is largely hydrogen in character, but this orbital is completely filled and, thus, contributes nothing to the magnitude of the hyperfine coupling. To the extent that coupling is observed, the SUMO must be antibonding with respect to the M—H bond. Also, the covalency in the M—H bond must be fairly high so that there will be extensive participation of the hydrogen 1s atomic orbital in the antibonding orbital. In the limit of a saline hydride, the M—H antibonding orbital is entirely metal-centered and will yield no hyperfine coupling to the hydrogen. For example, the spin density on the hydrogen atom of $HNi(CN)_4^{2-}$ is less than 33% [25], whereas it is less than 20% on the hydrogen atom of the corresponding platinum complex [26]. In the diatomic PdH, the spin density on hydrogen is less than 10% [27]. Because the bonding orbital is dominated by the hydrogen 1s orbital, it follows that hydrogen bears a partial negative charge, a subject that will be treated in a later section.

2.1.4. Photochemistry of Hydrides

The most frequent use of matrix isolation techniques for studies of 18-electron transition metal hydrides arises from the characterization of photochemistry. In general, the hydrides that have been studied in matrices have fit the predictions of Geoffroy and co-workers [28] that are summarized in Scheme 2.1. Because of the necessity of having a volatile, stable parent complex, there has not been a wide range of ligand types explored. The only complex hydrides that have been matrix-isolated have been those containing CO and/or cyclopentadienyl ligands. When present in the coordination sphere, the loss of the carbonyl ligand is the most probable route of photodecomposition in nearly every instance. The lone exceptions to these generalizations are $H_2Fe(CO)_4$ and $H_2IrCp(CO)$, which eliminate H_2 upon photolysis [29, 30]. Typically, metal–hydrogen homolysis occurs as a minor reaction. Lacking a carbonyl ligand, hydrogen is lost; the monohydrido complex $HReCp_2$ yields hydrogen atoms and $ReCp_2$ upon ultraviolet irradiation in either CO or N_2 matrices [31]. In addition to the homolysis products, $(\eta^3\text{-}C_5H_5)(\eta^5\text{-}C_5H_5)ReLH$ forms, where L = N_2 or CO. With continued photolysis in CO, $(\eta^1\text{-}C_5H_5)(\eta^5\text{-}C_5H_5)Re(CO)_2H$ forms [32]. In argon matrices, no reaction of any kind is observed. It is unlikely that loss of a cyclopentadienyl group ever will occur without there being an abundance of free ligands in the matrix that can coordinate after the cyclopentadienyl ligand has been backed off, but it is intriguing that homolysis is not observed in argon (*vide infra*). Complexes containing *cis*-hydrides are predicted to lose H_2 [28]; experiments in matrices

Monohydrides

$$HML_n \xrightarrow[\text{inert gas}]{h\nu} HML_{n-1} + L \quad \text{major path A}$$

$$HML_n \xrightarrow[\text{inert gas}]{h\nu} ML_n + H \quad \text{minor path B}$$

Polyhydrides

$$H_mML_n \xrightarrow[\text{inert gas}]{h\nu} H_{m-2}ML_n + H_2 \quad (m \geq 2) \quad \text{path C}$$

Scheme 2.1
Photodecomposition paths involving hydrogen loss.

Table 2.1
Principal Photodecomposition Paths of Transition Metal Hydrides in Matrices

Precursor	Principal Path[a]	Ref.
$HMn(CO)_5$	A	[4, 33]
$HRe(CO)_5$	A	[4]
$H_2Fe(CO)_4$	C	[29]
$HCo(CO)_4$	A	[10, 34]
$HMo(\eta^5\text{-}C_5H_5)(CO)_3$	A	[35]
$HW(\eta^5\text{-}C_5H_5)(CO)_3$	A	[35]
$H_2Ir(\eta^5\text{-}C_5H_5)CO$	C	[30]
$H_3Nb(\eta^5\text{-}C_5H_5)_2$	C	[36]
$H_3Ta(\eta^5\text{-}C_5H_5)_2$	C	[36]
$H_2Mo(\eta^5\text{-}C_5H_5)_2$	C	[37]
$H_2W(\eta^5\text{-}C_5H_5)_2$	C	[37]
$HRe(\eta^5\text{-}(C_5H_5)_2$	B	[31]

[a] Letters refer to photochemical paths in Scheme 2.1.

conform to the prediction. Table 2.1 lists all complexes that have been matrix-isolated from stable precursors to date, showing which of the equations in Scheme 2.1 best describes the observed photochemistry in rare-gas matrices.

The homolysis of the metal–hydrogen bond can be observed exclusively when the loss of CO is frustrated by the presence of an abundance of CO in the matrix.

$$HM(CO)_n \xrightarrow[\text{CO matrix}]{h\nu} HCO + M(CO)_n$$

Thus, in CO matrices, the metal radical has been observed in every instance [33, 35, 38]. There are three important questions, which can be only partly elucidated by the existing experiments. First, how much more probable is CO loss than M—H bond homolysis? Second, what is the role of the CO in abstracting the hydrogen atom? Finally, can the M—H bond of the coordinatively unsaturated complex undergo homolysis? The answers to the questions interrelate.

Photolyses of carbonyl-containing hydrides in inert matrices usually give infrared evidence of only $HM(CO)_{n-1}$ [4, 11, 35]. Because most of these hydrides will undergo metal–hydrogen homolysis in CO matrices, one wonders whether the presence of CO is required for a metal–hydrogen bond to homolyze. $HCo(CO)_4$ and $HMn(CO)_5$, when irradiated with 193-nm light, do produce radical products [33, 34, 38, 39]. In these two instances, the CO loss processes are reversed, even in argon matrices, so that relatively large quantities of the parent hydrides are exposed to repeated hits by ultraviolet photons. In the case of $HCo(CO)_4$, so little $HCo(CO)_3$ accumulates in the matrix that the characterization of $HCo(CO)_3$ is nearly impossible.

$$HCo(CO)_4 \underset{\text{Ar matrix}}{\overset{h\nu}{\rightleftarrows}} HCo(CO)_3 + CO$$

It seems clear in these two instances that the homolysis of the metal–hydrogen bond is a minor path to photodecomposition, products from which can be observed if the saturated hydride has a long enough lifetime in the matrix. There is no suggestion that the coordinatively unsaturated complexes that are formed by CO loss are capable of losing hydrogen atoms upon subsequent irradiation [40]. In fact, to the extent that 16-electron products form, homolysis is shut off. For example, when the formation of $HMn(CO)_4$ is most efficient by irradiating with 229-nm light, there is little evidence of homolysis, although the process is detectable using the more-sensitive ESR spectroscopy [47].

While arguing that the coordinatively unsaturated complexes $HCo(CO)_3$ and $HMn(CO)_4$ cannot undergo homolysis, it is important to recall the behavior of $HReCp_2$ [31]. In argon matrices, no homolysis is noted. The fact that homolysis only occurs in CO and N_2 matrices suggest one of several possibilities.

1. The homolysis really is the result of the irradiation of $(\eta^3\text{-}C_5H_5)(\eta^5\text{-}C_5H_5)ReLH$, where $L = N_2$ or CO. [For this to be true, $(\eta^3\text{-}C_5H_5)(\eta^5\text{-}C_5H_5)ReL$ would have to rearrange rapidly to $(\eta^5\text{-}C_5H_5)_2Re$ because it is not one of the observed products.]

2. The reversal of homolysis is efficient in argon [31]. The observation of homolysis of the Re—H bond in N_2 and CO suggested that vibrationally more-complicated hosts were required to thermalize the radical products so that they would not recombine before the excess energy of the photon was dissipated. It should be noted that the homolysis of Mn—H and Co—H bonds is observed in argon. In fact, the excess energy of the photon was thought to be essential in powering the expulsion of the hydrogen atom beyond the immediate vicinity of the metal-centered radical [38].

3. As noted by Chetwynd-Talbot et al. [31], the inability to form species like HCO in Ar cannot explain the absence of homolysis in Ar. Homolysis *is* noted in N_2, but HN_2 is not especially stable. Thus, the formation of a host–H adduct cannot explain the absence of observed homolysis in argon.

There has only been one attempt to compare the quantum yields of CO loss with that of H-atom loss [34]. The estimate was based on observations of the rate of loss of $HCo(^{12}CO)_4$ in ^{12}CO and ^{13}CO matrices. In ^{12}CO matrices, the only reaction that yields new products is

$$HCo(CO)_4 \xrightarrow{h\nu} Co(CO)_4 + HCO$$

In ^{13}CO matrices, homolysis will continue to occur with nearly the same rate, while there is the additional possibility for CO exchange,

$$HCo(^{12}CO)_4 + n\,^{13}CO \xrightarrow{h\nu} HCo(^{12}CO)_{4-n}(^{13}CO)_n + n\,^{12}CO$$

From an analysis of the initial rates, it was estimated that eight CO-loss events occurred for every H—Co homolysis, assuming that homolysis was never reversed. This assumption is certainly open to question. If the hydrogen atom is intercepted by a near-neighbor CO, there is a strong thermodynamic driving force for the H atom to return to cobalt [48]. Moreover, HCO is itself photosensitive so that the hydrogen can be thought of as a hot potato, never spending very much time on any one CO. Thus, the estimate of eight losses of CO for every H-atom loss represents an upper limit.

It is intriguing to speculate on how much less is the relative quantum yield for M—H homolysis in the other monohydridocarbonyl systems that were studied. Those systems differ from $HCo(CO)_4$ in being able to form efficiently the metastable, coordinatively unsaturated complex. Having formed $HM(CO)_{n-1}$ from $HM(CO)_n$, homolysis is not a likely event [40]. If one assumes that neither the photolyzed CO nor the H atom returns to the metal, then the product yields due to CO loss and M—H homolysis are in the same ratio as the relative quantum yields. Thus, for there to be so few radical products observed in the typical example, the quantum yield for homolysis of the other M—H bonds must be much less than that of $HCo(CO)_4$.

It is worth noting that the homolysis of the metal–hydrogen bond is probably observable only because of the extreme mobility of the hydrogen atom. With the excess energy of the photon that induces the cleavage of the hydrogen–metal bond, the hydrogen atom can be ejected out of the vicinity of its partner-radical, thus preventing recombination. More-bulky radicals have not been shown to separate sufficiently. There is good evidence to suggest that irradiation of a variety of metal–methyl complexes will give radical products in fluid media [46], but the analogous processes in rigid matrices never show the formation of methyl radicals [41, 42]. If methyl radicals do form, they readily recombine with the partner-radical. The photolysis of $Co_2(CO)_8$ in CO matrices did result in the formation of $Co(CO)_4$ [50]. In this instance, the only particle that must diffuse is the cobalt atom. Once the metal–metal bond is broken, the separation can be increased by exchanging the original carbonyl groups with carbon monoxide from more remote regions of the matrix. It should be noted in this context that $Cp_2Rh_2(CO)_3$ is cleaved in CO matrices, but the product is not a 17-electron radical, but rather the stable, 18-electron $CpRh(CO)_2$ [51].

Irradiation of polyhydride complexes yields H_2, although the mechanism of H_2 loss is not always clear. One can envision either a sequential loss of hydrogen atoms analogous to the minor path of photodecomposition of the monohydrides or a concerted loss yielding H_2. There is no evidence of H· that can be detected by ESR in the photolysis of $H_2Fe(CO)_4$ [29], although

HCO can be observed when H_2WCp_2 is photolyzed in CO-containing matrices [37].

Interestingly, the loss of H_2 from $H_2Fe(CO)_4$ can be reversed when matrices are exposed to the visible light of the Nernst glower [29]. Even with irradiation, the reaction of $Fe(CO)_4$ and H_2 does not go to completion unless extra dihydrogen is doped into the matrix. Presumably, some of the excess energy of the photon causes H_2 to become separated from $Fe(CO)_4$ by too great a distance for them to recombine readily. The extra hydrogen makes it more probable that $Fe(CO)_4$ has a near-neighbor H_2. Similarly, IrCpCO, which is produced from the parent dihydride, also oxidatively adds H_2. In an analogous fashion, it also oxidatively adds CH_4 [30]. By contrast, $Fe(CO)_4$ does not add CH_4. There is a strong likelihood that $Fe(CO)_4$ forms a complex with methane, although it was not recognized as such at the time [52]. The oxidative addition of C—H bonds is unlikely for a first-row transition element, especially with carbonyl ligation (*vide infra*).

By contrast to $H_2Fe(CO)_4$ and $H_2IrCpCO$, the reversal of the loss of H_2 from H_2MCp_2 has not been observed, although irradiation of these hydrides does not go to completion [37]. Such behavior might suggest the existence of a photostationary state. There is good evidence to suggest that the rings of the photoproduct MCp_2 become parallel after H_2 is lost. The reverse reaction would probably require a substantial movement of the cyclopentadienyl ligands in order for hydrogen to add oxidatively. This requirement need not preclude reaction at such low temperature if an electronic excited state can be populated, which would force the complex to distort. On the other hand, the arrangement of the carbonyl groups of $Fe(CO)_4$ is not much different from that of $H_2Fe(CO)_4$ [52, 53]. Still, the reaction of H_2 and $Fe(CO)_4$ requires photoactivation. The necessity of a photon for the reaction of $Fe(CO)_4$ and H_2 may have two origins. $Fe(CO)_4$ is thought to have a triplet ground state in argon [54] and there is the need for a spin conversion. Secondly, the orientation of the $Fe(CO)_4$ may be unsuitable for reaction. It has been shown in several elegant studies that photons can cause a coordinatively unsaturated complex to pseudorotate, the result of which may be a more-favorable orientation of the metal complex with respect to the entering group [55, 56].

2.1.5. Structures of Photoproducts

It has been possible to assign a geometry to the coordinatively unsaturated hydrides that are formed by photolysis. Often, the geometry is such that the vacancy can be considered a space-occupying ligand. Thus, $HMn(CO)_4$ adopts two geometries, one being of C_s symmetry with the vacancy cis to the hydride ligand and the other being of C_{4v} symmetry with the vacancy trans to the hydride [4]. The geometry of $HMn(CO)_4$ had earlier been claimed to be of C_{3v} symmetry [57]; this determination ranks as one of the very first reports of a coordinatively unsaturated metal carbonyl complex in the literature. Two useful observations were instrumental in making the revised assignments [4]. A factored force field was fit to the spectra of all ^{13}CO and ^{12}CO isotopomers from which intensities were calculated that agreed with the observed spectrum. Secondly, it was noted that there was a strong matrix dependence on the visible absorptions of $HMn(CO)_4$. This latter observation suggests that there is not so much a vacancy as there is a host molecule occupying a position in the inner coordination sphere, so that $HMn(CO)_4$ might still be considered a six-coordinate complex. This is a limitation in matrix experiments that only now is being fully appreciated; depending on the energy of the interaction of host with the coordinatively unsaturated metal, the structure that is observed in the matrix may not be what would be observed in the gas phase.

Although little can be learned from the spectrum of $HCo(CO)_3$ because of low intensities [34, 38, 39], a fruitful comparison was made between the spectra observed for the two isomers of $CH_3Co(CO)_3$ and what is known about the spectrum of $HCo(CO)_3$ [11]. Again, the geometries of the product isomers are derived from the geometry of the parent molecule with the vacancies taking the place of a coordinated atom. The specific interactions between the matrix atom and the metal may stabilize an isomer that would not appear in the gas phase, although calculations suggest that the energy difference between the two isomers is not large [58]. Only one isomer of $HMCp(CO)_2$ has been identified in the spectrum of photolyzed $HMCp(CO)_3$, for M = Mo and W. In fitting the spectrum of $HMCp(CO)_2$, both carbonyl ligands were considered equivalent, an assumption that cannot be challenged by the infrared spectral data [35]. A force field that allowed the two carbonyls to be different did not produce a dramatically better fit [59]. In order to determine whether there is a sterically significant vacancy in $HMCp(CO)_2$, one would have to observe the behavior of the molecule in a variety of matrices, looking for strongly matrix-dependent frequency shifts in the visible–ultraviolet spectrum, as was done for $HMn(CO)_4$ [4].

2.1.6. Charge Distribution in Hydrides

As a result of matrix isolation studies, a number of radical species formed from H—M bond homolysis have been characterized and one can, at least in thinking, observe the spectral changes for the reaction of $H + ML_n \rightarrow HML_n$. If a carbonyl ligand is present in the coordination sphere of the metal, then the frequency of the CO stretch and its associated force constant become larger as the hydrogen atom is added [60]. This behavior is consistent with the assertion that the hydrogen atom should be considered an electron-withdrawing ligand. Because hydrogen starts as zero-charged, it becomes negatively charged in the process. To what degree hydrogen can accept charge from the metal cannot be known from spectra, but it seems unlikely that hydrogen will become as negatively charged as has been claimed from ESCA data [61]. The hydrogen atom is probably more electronegative than the metal atoms of the organometallic radicals, but that electronegativity advantage drops drastically as electron density is transferred to hydrogen. In other words, the charge capacity of hydrogen is smaller than for any other atom [62].

If the hydride ligand is partially negatively charged, then one would not expect it to hydrogen-bond to polar molecules such as water. A molecule as acidic as $HCo(CO)_4$ has been deposited in matrices composed totally of water [63]. $HCo(CO)_4$ is a strong acid in aqueous solutions [64], but the mixtures are quenched so fast during the deposit that proton transfer is not observed. In such a polar environment, the infrared spectrum of $HCo(CO)_4$ as shown in Figure 2.3 is broadened and perturbed by the high degree of scattering, but the positions of the carbonyl bands and the Co—H deformation mode are not shifted as one might expect if there was a significant interaction with water. Typically, hydrogen-bonding causes the A—H stretching frequency to decrease and the intensity of the mode becomes greater [65]. The unperturbed Co—H stretching vibration is extremely weak and is rarely observed in infrared experiments. In solid water, nothing is observed that can be assigned to the Co—H stretching vibration. A—H deformation vibrations usually are shifted to higher frequency by hydrogen-bonding. This mode is easily observed in the spectrum of $HCo(CO)_4$ and appears nearly unshifted

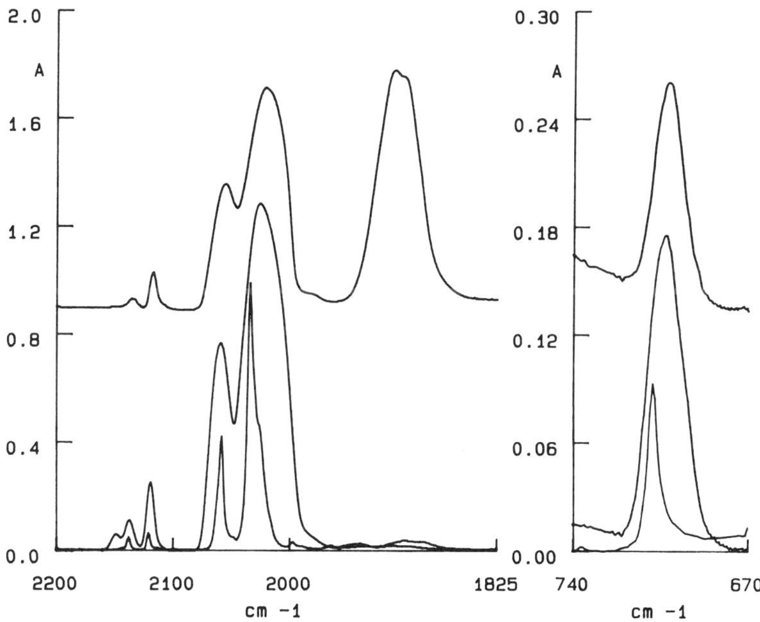

FIGURE 2.3
Spectra of $HCo(CO)_4$ in a water matrix superimposed on a much more narrow spectrum of $HCo(CO_4)$ in argon. The upper trace shows the transformation that occurs when the matrix is warmed to 120 K.

in solid water [63]. Finally, if hydrogen bonding were to occur, one might expect some charge to be drained off $HCo(CO)_4$, which could be detected by blue shifts of the carbonyl modes. However, they are largely unperturbed by water. Interestingly, when these H_2O–water matrices are warmed to 110 K, dramatic changes do occur, consistent with proton transfer.

2.1.7. Characterization of the Ultraviolet–Visible Spectrum

Ultraviolet spectra made in low-temperature matrices are often of high quality. The same matrix can be analyzed by both infrared and optical spectroscopy so that the purity of the matrix can be ascertained. For instance, matrices have been prepared of $HCo(CO)_4$ that are free of $Co_2(CO)_8$ impurities as judged by the infrared spectrum. The ultraviolet spectrum of $HCo(CO)_4$ shows no features that could be assigned to the decomposition products that are often encountered in fluid media at ambient temperatures [38]. Similarly, the spectrum of $H_2Fe(CO)_4$ that is shown in Figure 2.4 has

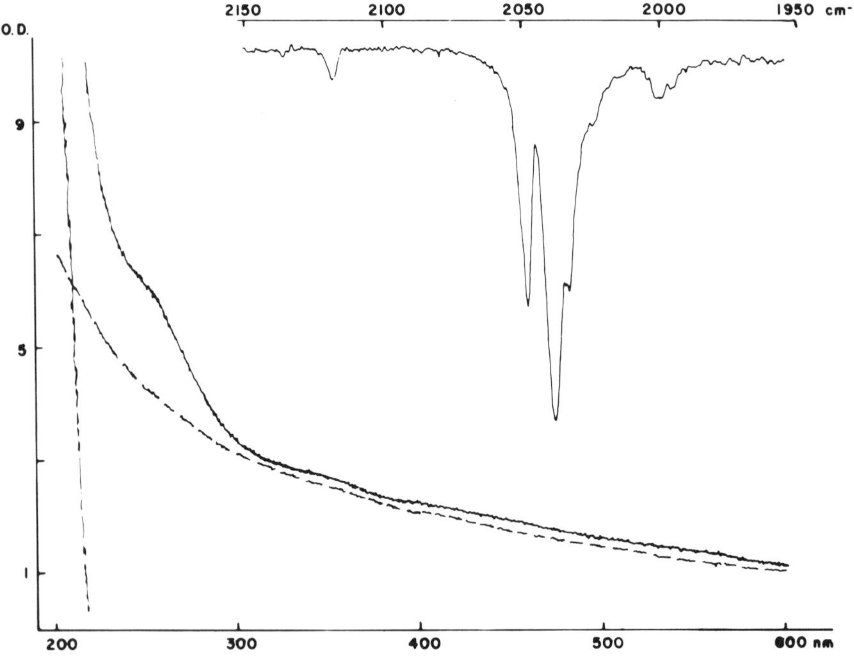

FIGURE 2.4
Optical and infrared spectrum of $H_2Fe(CO)_4$ is an argon matrix. Dashed line trace in the optical spectrum is of the NaCl windows prior to the deposit [66].

been measured of a matrix that contained virtually no $Fe(CO)_5$ [66]. These observations have been of use in understanding the photochemistry of hydrides from a theoretical standpoint [67]. Not only can spectra be measured that are free from impurity bands, but spectra made in matrices are usually sharper than those observed at higher temperature in fluid media [68].

Previously, I stated that the necessity of vaporizing a hydride limited the utility of matrix isolation for studies of hydride precursors. In several studies, it has been possible to characterize the decomposition products that form as the sample is vaporized. For example, ScH_2 and YH_2 were formed by the pyrolysis of the solid, binary hydrides, YH_3, and a variety of nonstoichiometric hydrides of scandium. The hydrides were characterized by ESR [69]. It was impossible to deposit $HCo(CO)_4$ without $Co(CO)_4$ being present, as detected by ESR [38]. The decomposition was catalyzed by metallic surfaces, which could result in the complete loss of the hydride unless the surfaces were cooled [38, 39].

2.2. HYDRIDES FORMED IN SITU

In addition to the hydrides formed by photolysis of hydride precursors, there have been a variety of interesting and unusual hydride-containing molecules that have formed as a result of intermolecular or intramolecular reactions with an array of hydrogen donors in the matrix. The metal-containing precursor in these reactions can be as simple as a single atom or it can be a stable organometallic complex, a coordination site of which has been vacated as a result of photolysis. The simplest reactions leading to hydrides start with hydrogen atoms; experimentally, these reactions are most difficult to observe because of the difficulty of obtaining large quantities of hydrogen atoms in matrices. There also has been a great deal of attention paid to reactions of dihydrogen with both metal atoms, as well as with coordinatively unsaturated metal complexes. Hydrides also have been observed to form from reactions of a variety of donors, reactions that can be classed as oxidative addition reactions analogous to the oxidative addition of H_2 to a metal center. Finally, hydrogen has been observed to transfer to the metal from coordinated ligands in processes that are similar to reactions at ambient temperatures.

2.2.1. Hydrides Formed from Metal Atoms and Hydrogen

Naked metal atoms have exhibited a wide range of reactions leading to hydrides; all of them can be classified as oxidative addition. They are summarized in Scheme 2.2. It should not be surprising that hydrogen atoms will form hydrides in reactions with metal atoms. A variety of hydrides were formed by cocondensing metal atoms with hydrogen atoms that were gener-

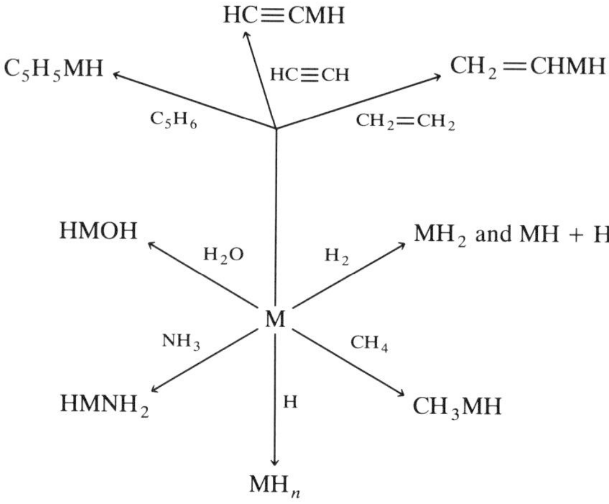

Scheme 2.2
Different means of forming metal hydrides, starting with metal atoms.

ated by heating H_2 to 2600 K in a tungsten oven. By these methods, MH, MH_2, and possibly MH_3 were synthesized by using Mn and Cr [70, 71]. The infrared, electronic, and electron-spin spectra were observed. Initially, it was argued that MnH_2 was bent, based on the band positions of the hydrogen isotopomers. A bond angle of 117° was calculated even though the totally symmetric M—H stretching vibration was not observed [70]. In a subsequent publication, Van Zee et al. [71] conceded that the bond angle might well be 180°.

One can conceive of a variety of reactions between naked metal atoms and dihydrogen. Unlike reactions with atomic hydrogen, these reactions may require activation. A number of theoretical treatments of the reaction have been made [72]. Dissociative chemisorption occurs without much activation on metal surfaces with atoms with incompletely filled *d* orbitals [73]. Nevertheless, in most instances, some photoexcitation is required in order to induce hydrogen to add oxidatively to a single metal center. A key question that can be addressed is one concerning the role of the nuclearity of the metal cluster in the oxidative addition. When oxidative addition does not occur, it may still be possible for the hydrogen to form a simple complex with a metal atom. Scheme 2.3 shows the conceivable reactions of hydrogen with single metal atoms. The heterolysis of the H—H bond is an unlikely event without the stabilization of H(I) by a polar solvent or an ionizable substituent.

The reaction of Fe + H_2 has been examined by two groups and there is disagreement with regards to some of the details [74, 75]. Prior to photoexci-

$$M + H_2(\text{matrix}) \longrightarrow MH_2$$
$$\xrightarrow{h\nu} MH_2$$
$$\longrightarrow MH + H \text{ (not favorable)}$$
$$\xrightarrow{h\nu} MH + H$$
$$\longrightarrow M(H_2)$$

Scheme 2.3
Modes of reaction between metal atoms and dihydrogen.

tation there is no evidence of an interaction between single iron atoms and H_2. Using 307-nm radiation, Ozin and McCaffrey [74] presumed that oxidative addition occurred via the $^5P_3$ excited state ($3d^64s^14p^1$), whereas Rubinovitz and Nixon [75] used considerably longer-wavelength radiation. In either study, the rate of reaction with diprotium is from five to seven times greater than that of dideuterium, indicating that the reaction of the excited state of iron with hydrogen is still an activated process. The product FeH_2 has been observed in Ar, Kr, and Xe. Ozin and McCaffrey [74] have argued that the molecule is bent in xenon. Key to concluding that FeH_2 is bent is the observation of *two* absorptions in the Fe—H stretching region, both of which must be assigned to fundamental vibrations of the same molecule. The two correspond to the symmetric and antisymmetric combination of the Fe—H stretching motion. They are separated by 24 cm^{-1} in solid xenon. Superficially, it might be argued that the two absorptions are really due to different molecules, both FeH_2 but occupying different sites in the lattice. The spectrum of FeHD is key to differentiating between the two possibilities. Two absorptions are expected for FeHD: One is essentially the Fe—H stretch and the other is essentially the Fe—D stretch. The two absorptions will be present no matter what the geometry of FeH_2 is. If, in addition to the two features, there are secondary features, they can only be due to a second trapping site. There was no evidence of a second trapping site in the spectrum of FeHD [74]. Therefore, it was argued that FeH_2 is bent. However, Rubinovitz and Nixon [75] could not find the symmetric stretch in argon or krypton matrices and concluded that the molecule was nearly linear. To accommodate the two reports, one can conclude that either xenon enforces a bent geometry on FeH_2 or that the weak absorption that was assigned to the symmetric stretching vibration in the Ozin experiment is really due to a second trapping site, in spite of the failure to show a second trapping site in the HD experiments.

A separation as large as 24 cm^{-1} should not be considered an incredibly large separation for absorptions due to two different sites. In a similar study of MnH_2, secondary features were observed in the spectrum of MnHD, separated by 26 cm^{-1} [76]. Moreover, a splitting of 24 cm^{-1} is not incredibly small for the separation between the symmetric and antisymmetric M—H stretching vibration. In complex hydrides it is never very large and is occasionally not resolved at all. For example, only a single broad absorption is observed in the spectrum of $H_2Os(CO)_4$ [77]. The splitting is not resolved in $H_2Fe(PF_3)_4$, either [78]. On the other hand, a splitting of 45 cm^{-1} was observed for two *cis*-H_2RuL_4 complexes [79]. If one assumes that FeH_2 is indeed bent, a bond angle of 120° is calculated from the intensity ratio of the two absorptions that are observed in solid xenon [74]. If, on the other hand, the second absorption is due to a different site, then Rubinovitz and Nixon estimated that the bond angle could be no less than 172° [75].

The reactions of Fe and Mn with H_2 can be reversed by light the frequency of which matches the absorption spectrum of MH_2 [74–76]. There is no evidence of radical products; thus, it appears that the reverse is a concerted process just as is the original oxidative addition. The yield of Fe + H_2 is quantitative whereas 50% of the starting manganese could be retrieved after irradiation as judged by the atomic spectrum of manganese [80]. The authors thought that the missing atoms may have become aggregated, although they could not detect evidence for several known small clusters. It may also be that a complex of H_2 was formed; the vibrational spectrum of $Mn(H_2)$ may be too weak to detect (*vide infra*), leaving only the electronic spectrum for detection. Typically, electronic spectra of molecules are much broader than atomic spectra, so that detection of minor molecular fragments may be problematic as well.

Margrave's group has explored matrices that are more concentrated in iron and found evidence for metal-cluster hydrides without the necessity of photolysis [81]. The implication from these experiments is that hydrogen will oxidatively add to metal clusters without photoactivation whereas hydrogen awaits photoassistance before it oxidatively adds at a single metal center. There are several reasons to be cautious in making such a conclusion. The furnaces used in producing metal atoms emit a substantial amount of visible radiation in addition to the infrared [82]. These photons cannot be screened from the developing matrix without also interrupting the metal atom flux itself. Although it might be supposed that the photon flux from such a source is inconsequential, it is at least the equivalent of a Nernst glower, photons from which have been very useful in inducing $M(CO)_{n-1}$ + CO reactions. It was reported that FeH_2 was destroyed by irradiation with light of 440 nm [74]. It is conceivable that FeH_2 forms in the developing deposit, but is destroyed or kept at negligibly low concentrations by the light from the furnace. If this were true, then the distinction between the FeH_2 system and the Fe_xH_2 system revolves around the inability of photoejecting dihydrogen from the cluster. This is not an unreasonable distinction because the hydrogen atoms of the cluster can presumably migrate to remote locations making it unlikely that they will recombine upon photoexcitation.

Without the necessity of photolysis, deposits of Ni atoms and H_2 yielded two bands, which were assigned to NiH_2 [83]. It was presumed that NiH_2 was bent, due to the observation of an absorption assigned to ν_s, the symmetric combination of M—H stretches. The bands assigned to ν_s and ν_a are separated by 38 cm^{-1}, which, once again, raises the question of a second trapping site. It should be pointed out that calculations have shown NiH_2 to be linear [84]. In contrast with the above chemistry, irradiation of matrices containing copper atoms and H_2 yields CuH and H atoms, the detection of which was done by using infrared and ESR spectroscopy. Silver atoms behave similarly [85]. Upon heating to 18 K, the reaction is reversed.

$$CuH + H \xrightarrow{18\ K} Cu + H_2$$

The rates of reaction between CuH and AgH and H are independent of the mass of hydrogen, suggesting that the reaction occurs by hydrogen-atom abstraction, rather than via the formation of CuH_2.

Returning to the studies of matrices containing nickel atoms, in addition to the bands that are assigned to NiH_2, other peaks were observed that were assigned to complexes of hydrogen, exhibiting both end-on (η^1) and side-on (η^2) coordination [83]. There were features at 2556 and 2314 cm^{-1} that were assigned to the H—H stretching vibration. Absorptions were also assigned to the M—H_2 stretching vibration between 800 and 700 cm^{-1}. The absorptions assigned to the H—H stretching vibration are located at considerably lower wave numbers than the 3210 cm^{-1} that was observed for H_2 on a nickel surface as determined by (EELS) [86]. In light of the near-record low frequency for the H—H stretching vibration, it is surprising that the M—H_2 stretching vibration is also lower than what is found in other H_2 complexes that have been vibrationally characterized. The low H—H stretch would indicate a strong interaction between Ni and H_2 whereas the low Ni—H_2 stretch would seem to indicate the contrary [20]. The formulation as a complex of hydrogen probably should be viewed with some skepticism.

Complexes of H_2 also have been claimed in the reaction of H_2 with atomic palladium [22]. In xenon, the product was thought to be Pd(η^2-H_2) whereas in krypton additional absorptions were noted that were assigned to Pd(η^1-H_2). More recently, a complex cation PdH_2^+ was characterized by ESR that was also thought to be an η^2 complex, because so little spin density was found on the hydrogen atoms [87]. Not all the vibrational coordinates of PdH_2 were observed; in particular, the H—H stretching vibration was not observed, even for the η^1-H_2 complex. The features that were observed were assigned to the Pd—H_2 stretch. For the η^2 complex in solid Xe, the absorption assigned to the Pd—H_2 stretch appears at 890 cm^{-1} with a splitting of 9 cm^{-1}. The splitting was ascribed to the librational motion of the hydrogen. In illustrating the motion, a rotation is suggested, but this is unlikely for reasons that have already been stated. A number of features in the spectra are remarkable. It is surprising that the features assigned to the

M—H_2 stretch of Pd(η^2-H_2) in Xe appear at 890 cm^{-1}, albeit split into a doublet, whereas the mode is found at 954 cm^{-1} in Kr. A matrix-induced shift of a vibrational band of this magnitude is unprecedented. It is also surprising that so much intensity is observed for the M—H_2 stretching vibration in light of the extreme difficulty in seeing the same mode in the spectra of organometallic dihydrogen complexes in matrices and liquid xenon [88] (*vide infra*). Pd(η^1-H_2) is reminiscent of recently characterized complexes of H_2 and bases such as $N(CH_3)_3$ [89]. In those complexes, the H—H stretch is readily observed, although it is only shifted about 20 cm^{-1} from the position at which p-H_2 absorbs in matrices. The infrared intensity is associated with rotating hydrogen molecules in association with base. Should the rotation be quenched, then it is conceivable that the mode would escape detection. The system Pd(H_2) has been the subject of numerous quantum-mechanical calculations. The η^2 complex is quite stable by about 10 kcal mol^{-1} [90, 91]. The η^1 geometry has only been reported in one communication, it too being stable by about 10 kcal mol^{-1} [91]. Thus, there is good theoretical underpinning for the experimental results, the aforementioned anomalies notwithstanding [92]. In attempting to evaluate the results, it is noteworthy that Pd atoms can be present in matrices in either a $^1S_0$ or a $^3D_3$ state [93]. The two states react differently with N_2, the latter state interacting so weakly that the electronic transitions of the metal are hardly perturbed. The infrared spectrum of what is presumably the $^1\Sigma^+$ PdN_2 shows an N—N stretch at 2215 cm^{-1} [94]. Morris et al. [95] have attempted to correlate the N—N stretch of organometallic complexes with the stabilities of H_2 complexes that are formed by replacing the N_2 ligand with H_2. An N—N stretch at 2215 cm^{-1} would lead one to expect the existence of an η^2-H_2 with a stability akin to that of $W(CO)_5(H_2)$ [95]. This type of correlation has never been tested for simple metal-atom–H_2 systems, however.

2.2.2. Hydrides Formed from Metal Atoms and Other Hydrogen Donors

In addition to reactions of naked metal atoms with H_2, a variety of other hydrogen-containing molecules react with metal atoms. This area has been reviewed recently by Margrave [96]. One might anticipate reaction to begin with some sort of coordination that is followed by insertion of the metal into the H—X bond analogous to current views on reactions of H_2 [97]. Other molecules will differ from H_2 by coordinating with either lone pairs or with π electrons in preference to the bonding electrons of the H—X bond. This interaction may prevent the metal from approaching a sigma bond prior to oxidatively adding. For example, Fe(η^2-C_2H_4)$_2$ does not lead directly to insertion products, presumably because the metal is associated with the π electrons of the carbon–carbon double bond [98]. In this respect, the complex mimics the behavior exhibited by CpIrCO(C_2H_4) in fluid media [99]. Interestingly, the 1 : 1 complex Fe(C_2H_4) is able to add ethylene oxidatively with the assistance of ultraviolet photons. The starting complex can be regenerated by using light with $\lambda > 400$ nm.

$$\text{Fe}(\text{H}_2\text{C}{-}\text{CH}_2) \underset{\text{visible light}}{\overset{\text{UV}}{\rightleftharpoons}} \text{HFe}{-}\text{CH}{=}\text{CH}_2$$

The infrared spectrum of $Fe(C_2H_4)$ was enough different from that of $Fe(C_2H_4)_2$ that it was suspected that the ethylene was not bound in the usual fashion as found in Zeise's salt. Rather, it was argued that the complex was bound to the ethylene via bridging hydrogens. One possible structure is shown in the preceding reaction. In this orientation, the σ bonding orbitals of ethylene can act as donor orbitals whereas the π^* orbital of the C—C bond can still act as an acceptor. In such an orientation, the metal can easily insert into the C—H bond.

Analogously to $Fe(C_2H_4)$, donor–acceptor complexes of metal atoms with HX have been observed for HX = CH_4 [100], NH_3 [101], H_2O [102], CH_2 [103], HCCH [104], and HF [105]. Except for CH_4 and HCCH, the interaction between the metal and HX apparently does not involve any aspect of the H—X bond. Acetylene forms η^2 complexes with Ni, Cu, and Au atoms [106] and does not give insertion products upon irradiation. Iron is dissimilar. Upon coordination to iron, the H—C asymmetric stretching vibration shifts to lower wave number by 18 cm^{-1} and the C—C stretch is apparently too weak to be observed, whereas it has been observed in the aforementioned examples [104]. The structural model that is most in accord with observations is

$$\text{Fe}\cdots\text{H}{-}\text{C}{\equiv}\text{C}{-}\text{H}$$

Excepting η^2-acetylene and η^2-ethylene, irradiation of complexes of metal atoms and HX yields insertion products. For CH_4 and CH_2, the insertion reaction could be reversed [99, 102].

In a few instances, insertion occurs without the need for photoassistance. The formation of MH(OH) requires no separate excitation for the early transition metals, Sc, Ti, and V [82]. Also, iron adds HCl, HBr, and HI without any need for separate photoactivation [105]. Cyclopentadiene reacts with iron to give η^5-C_5H_5FeH with no evidence of any intermediate stages in the reaction [107].

Finally, the oxidative addition of CH_4 on several metals has been noted after photolysis. In spite of having the metal atom present in a methane matrix prior to photolysis, the evidence of a complex formation has not always been available. In pure methane matrices the infrared bands of a complex are obscured by the intense absorptions of the host molecules. The observation of a complex with iron was possible in an argon matrix with 0.5 mol % CH_4 but not enough data are in hand to define the structure of the adduct [100]. The interaction between the iron and the methane is not sufficient to broaden or otherwise transform the optical spectrum [80]. Following photolysis, the optical spectrum of "isolated" Fe is attenuated dramatically as a very broad, weak feature grows in. The new feature is assigned to CH_3FeH, based on simultaneous observations of the infrared spectrum. The reaction can be reversed by irradiating with light of 420 nm. Very similar behavior is observed for Mn [80, 108], Co, Ag [108], and, to a degree, Cu atoms [108, 109]. With Cu atoms, CH_3CuH is formed upon irradiating CH_4 matrices of copper atoms, but the complex can be annihilated to produce $CuH + CH_3$, $CuCH_3 + H$, as well as $Cu + CH_4$. The particular mix of products depends on the wavelength of radiation. It should be recalled that Cu is incapable of forming CuH_2 from Cu and H_2, but rather forms CuH and H atoms [85]. At present there is no clear reason for the dissimilarity in behaviors.

2.2.3. Hydrides Formed from Complexed Metal Atoms with Hydrogen

Over the last 20 years, a number of coordinatively unsaturated organometallic complexes have been characterized in matrices [110]. These same complexes can be formed in dihydrogen-containing matrices with the additional possibility of forming either classical hydrides or dihydrogen complexes.

$$ML_n \xrightarrow{h\nu} ML_{n-1} + L$$
$$ML_{n-1} \xrightarrow{H_2} ML_{n-1}(H)_2 \quad \text{or} \quad ML_{n-1}(H_2)$$

These same syntheses could be accomplished at higher temperatures in fluid media, but the matrix synthesis offers considerable advantages. Because of the rigidity of the matrix, the coordinatively unsaturated complex cannot move; reactions that form binuclear complexes are therefore improbable in dilute matrices. A coordinatively unsaturated complex can only react with H_2 or with the photoejected L, if it reacts at all. Because matrices can be doped with as much as 25 mol % of H_2, hydrogen can effectively compete with L even if it binds the metal less tightly [111]. At such doping levels, more than one hydrogen molecule can be expected to occupy positions in the cage of the newly formed fragment. With such an abundance of hydrogen, it has been possible to characterize what are undoubtedly *bis*-dihydrogen complexes such as $Cr(CO)_4(H_2)_2$ [10] and possibly $HCo(CO)_2(H_2)_2$ [11].

When coordinatively unsaturated complexes are produced in hydrogen-containing matrices by photolysis, the detection of new compounds is facilitated if a carbonyl ligand is present in the coordination sphere. Although it is relatively easy to determine that new compounds have been synthesized that require the presence of hydrogen, it is not always clear whether hydrogen has been oxidatively added or coordinated. The hydrogen vibrational coordinates, which might be most diagnostic of the nature of the $M-H_2$ interaction, usually give rise to low intensity. Two criteria have been used to differentiate between oxidative addition and coordination by hydrogen. As noted earlier, when hydrogen oxidatively adds, electron density is removed from the metal [60]. As this happens, the metal is less effective as a π donor relative to the remaining carbonyl ligands. For example, when hydrogen oxidatively adds to $Fe(CO)_4$, the totally symmetric breathing mode of the carbonyls shifts 42 cm^{-1} to higher wavenumber [29]. Many of the complexes of molecular hydrogen that have been identified in matrices exhibit much more subtle changes as hydrogen adds to a previously coordinatively unsaturated metal center. By contrast to the preceding example, the totally symmetric breathing mode of $Cr(CO)_5$ only shifts by 3 cm^{-1} as hydrogen coordinates [10, 112].

In addition to using the carbonyl ligand spectrum as a diagnostic to determine whether hydrogen has coordinated or oxidatively added, it is also possible to determine that a hydride has formed by observing the metal–hydrogen deformation modes. Deformation modes, in contrast to the stretching modes, can be quite intense. For instance, the Co—H deformation mode of $HCo(CO)_4$ is nearly 10% of the intensity of the strongest carbonyl mode [66]. Typically, the δ_{MH} modes appear in the region between 950 and 700 cm^{-1} [7]. The appearance of intense features in this region can be taken as strong evidence that a hydride has formed, that is, that hydrogen has oxidatively added.

Neither of these criteria for the nature of $M—H_2$ bonding is without ambiguities. As the H_2 coordinates most strongly, the carbonyl ligands will become most perturbed and the modes and their associated force constants may shift to as high or higher wavenumber than that of the parent metal–carbonyl. For example, the spectrum of $CpW(CO)_2H(H_2)$ is interwoven with that of the parent, $CpW(CO)_3H$ [59]. The frequencies are considerably blue-shifted from the positions of the unsaturated $CpW(CO)_2H$. In this instance, it was not possible to observe the growth of any absorptions that could be assigned to the deformation coordinates, so it was concluded that the complex was $CpW(CO)_2(\eta^2\text{-}H_2)H$. Unfortunately, the absence of a mode assignable to the δ_{MH} vibration is not a dependable criterion for the absence of a hydride ligand; it is not always intense. For instance, the mode has not been observed in the spectrum of $HWCp(CO)_3$ [13]. To complicate analyses further, the region in which one expects the M—H deformation modes is also the region of the $M—H_2$ stretch of a coordinated H_2 ligand [14]. The behavior of the two modes may resemble each other upon isotopic substitution, especially for triatomics.

The dihydrogen complexes that have been prepared in matrices are listed in Table 2.2 [113]. In nearly every instance, the quantity of material synthesized was insufficient to allow the observation of the H—H stretching vibration or the other hydrogen vibrational coordinates. Typically, the dihydrogen ligand is photolabile. As increasing quantities of the dihydrogen complex build up in the matrix, it becomes more probable that the hydrogen

Table 2.2
Summary of Dihydrogen Complexes Formed in Matrix Experiments

Complex	Ref.
$M(CO)_5(H_2)$	[10, 21]
$M(CO)_4(H_2)_2$	[10, 21]
for M = Cr, Mo, W	
$HMCp(CO)_2(H_2)$	[59]
for M = Mo, W	
and Cp = η^5-C_5H_5 and η^5-$C_5(CH_3)_5$	
$RCo(CO)_3(H_2)$	[11, 114]
$HCo(CO)_2(H_2)_2$	[11]
$Ni(CO)_3(H_2)$	[115]
$IrCp(CO)(H_2)$	[30]

ligand is knocked off the metal. In several instances, the dihydrogen complex is lost due to other photoreactions. For example, $CH_3Co(CO)_3(H_2)$ loses CH_4 [11] and $HWCp(CO)_2(H_2)$ loses CO as hydrogen oxidatively adds [59].

$$CH_3Co(CO)_3(\eta^2\text{-}H_2) \xrightarrow{\lambda < 350\text{ nm}} HCo(CO)_3 + CH_4$$

$$HCo(CO)_3 \xrightarrow{CO} HCo(CO)_4$$

$$HWCp(CO)_2(\eta^2\text{-}H_2) \xrightarrow{h\nu} H_3WCpCO + CO$$

The use of ultraviolet photons to prepare dihydrogen complexes in matrices is required because of the necessity of ejecting a carbonyl group from a coordinatively saturated precursor, which is nearly always colorless. In one instance it was possible to matrix-isolate a colored precursor, $M(CO)_5NH_3$, where M = Cr and W, the ammonia of which could be ejected by using photons that are not absorbed by the product dihydrogen complex [21].

$$Cr(CO)_5NH_3 \xrightarrow[\lambda > 400\text{nm}]{H_2\text{–}Ar} Cr(CO)_5(H_2) + NH_3$$

The success of this stratagem is due in large part to the π-acceptor ability of H_2. The d orbitals split into a d_π and a d_{σ^*} set in C_{4v} symmetry [116]. The lowest unoccupied molecular orbital (LUMO) in these complexes is probably the d_{z^2} orbital. The HOMO–LUMO gap for Cr(0) is determined by the σ-donor and π-acceptor properties of the ligands. For $Cr(CO)_5NH_3$, the HOMO–LUMO transition is observed at 425 nm [117]. The same transition for $Cr(CO)_5(H_2)$ is at 367 nm. If the geometry of the $Cr(CO)_5$ moiety is not greatly changed in the two complexes, then either the H_2 ligand must be a better σ donor than NH_3, or H_2 must act as a π acceptor in order for the transition of the dihydrogen complex to appear to the blue of that of the ammine complex. The first of these two explanations is clearly unacceptable, meaning that there must be a π interaction sufficiently strong to compensate for the probability that the d_{z^2} orbital is not as destabilized in the H_2 complex as it is in the ammine complex. Although these data cannot serve to quantitate the amount of electron transfer from the metal to hydrogen, the results of several calculations have indicated that hydrogen functions primarily as a σ donor in similar complexes [118]. It is interesting to note that the transition in $Cr(CO)_5N_2$ complex is also at 367 nm; π-bonding by N_2 is thought to be a significant factor in interactions between N_2 and metals [117].

Sufficient quantities of $Cr(CO)_5(H_2)$ were obtained from matrices of H_2 and $Cr(CO)_5NH_3$ so that nearly all the vibrational coordinates of the hydrogen could be observed in the region of 4300–450 cm^{-1}. There are three absorptions assigned to the H—H stretching vibration, which are found at 3087, 3027, and 2997 cm^{-1} in argon. In liquid Xe, only one broad absorption is observed at 3030 cm^{-1} [112]. These features are exhibited in Figure 2.5. Although it is reasonable to assume that the splitting is induced by the matrix, the behavior is unprecedented in many respects. Neither the H—D nor D—D stretching vibration of the heavier isotopomers is split in an analogous fashion. The same absorptions are observed in Ar, Ar–NH_3 mixtures, and Kr, although the relative intensities of the three peaks do depend on the nature of the matrix. For only diprotium to be affected, the explanation for the phenomenon must involve the presumably hindered rotation of the hydrogen around the Cr–H_2 axis. It is unclear what the barrier to rotation for the hydrogen is; it is undoubtedly smaller than the barrier of 760 cm^{-1} that is exhibited by $H_2W(CO)_3(PR_3)_2$ [20].

The experiment that has yielded the most information concerning the vibrational spectrum of coordinated dihydrogen is similar to the matrix experiment in many respects. Turner and others have explored the use of liquid xenon as a "fluid matrix" [119–122]. A number of complexes of H_2 have been synthesized in liquid xenon, the H—H stretch of which has generally been characterized. It is possible to observe extremely thick solutions of liquid xenon in order to obtain adequate intensity in weak absorptions. Such thick solutions are unusual in infrared spectroscopy; they are useful because of the absence of any solvent absorptions in the infrared region. Although thick rare-gas matrices also can be made, photons that are used to produce the dihydrogen complexes may not penetrate the inner reaches of the matrix, whereas stirring of the liquid xenon constantly renews the solution that is most exposed to the ultraviolet light. Moreover, thick matrices may be seriously affected by light scattering caused by crystalline defects. Whether the use of liquid xenon will eliminate matrix isolation as a method of choice for preparing unstable dihydrogen complexes will depend on whether enough precursors can be dissolved in liquid xenon. The matrix can be considered a metastable solution. The success of forming this unstable solution depends on the energy required to volatilize the precursor complex. On the other hand, liquid xenon must form a stable solution. Moreover,

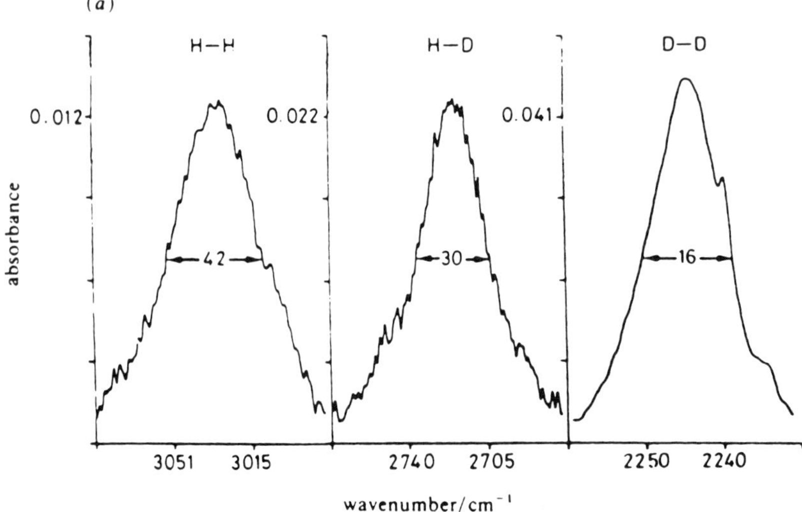

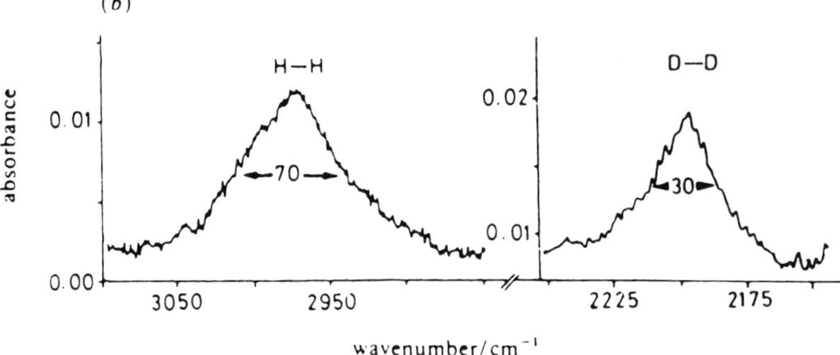

FIGURE 2.5
(a) Spectra of three isotopomers of $H_2Cr(CO)_5$ in liquid xenon.
(b) Spectra of two isotopomers of $H_2Fe(CO)(NO)_2$ in liquid xenon.
(Reprinted with permission from *Faraday Disc. Chem. Soc.* **1988**, 271 – 285. Copyright 1989, Royal Society of Chemistry.)

insoluble photo products will conceivably precipitate from the solution, complicating the spectroscopic characterization of the system.

In an early report, it was noted that the widths of the H—H stretching vibration (in liquid xenon, only a single feature) were dependent on the mass of the hydrogen [11]. Thus, the width for diprotium was 38 cm^{-1}, whereas the corresponding features were 30 and 16 cm^{-1} for HD and D_2, respectively. A similar mass dependence is observed in argon matrices, assuming that the feature at 3027 cm^{-1} in solid argon corresponds to the band at 3030 cm^{-1} in liquid xenon [21]. The mass-dependent width may result from the wide-amplitude torsional vibration. If the potential function for torsion were

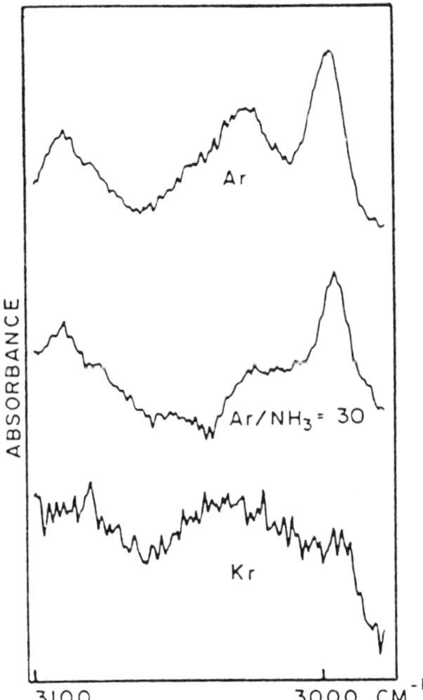

FIGURE 2.6
Spectrum of $H_2Cr(CO)_5$ in the H—H stretching region in various matrices. (Reprinted with permission from the *J. Am. Chem. Soc.* **1989**, *111*, 3577–3583. Copyright 1989, American Chemical Society.)

harmonic, then the torsional amplitude would vary with the fourth root of the reduced mass. The excursion of H_2 is expected to be 20% greater than D_2. The splitting that was exhibited by the H—H stretching vibration of $Cr(CO)_5(H_2)$ has not been observed in liquid xenon. The only analogous occurrence was the appearance of two absorptions in the spectrum of $Co(CO)_3NO(H_2)$ [23]. In this instance, the splitting was attributed to possible Fermi resonance. It should be noted that in matrices of $Cr(CO)_5(H_2)$, the Cr—H_2 stretching vibration is also split into two absorptions for diprotium [21]. There is a smaller splitting for the Cr—HD stretch and none is resolved for the Cr—D_2 stretch.

2.2.4. Hydrides Formed from Complexed Metal Atoms and Other Hydrogen Donors

There are far fewer examples of the oxidative addition of RH to coordinatively unsaturated metal complexes than to naked metal atoms. (R, in this context, is alkyl, aryl, NH_2, OH, or halide. See for example, Section 2.2.1.) Oxidative addition of methane and other polyatomic molecules may require considerable deformations of the metal coordination sphere as well as the oxidant in order for a sufficiently strong interaction to develop that can lead to the cleavage of the R—H bond. Aside from these kinetic barriers, the enthalpy change may not favor the oxidative addition of the R—H bond [123, 124]. In instances in which oxidative addition is thermodynamically disfavored, RH may still coordinate the unsaturated metal atom. In the earliest studies of CO-loss processes in methane matrices, it was found that yields of the coordinatively unsaturated complexes were greater in methane than in the rare-gas matrices. Moreover, the methane causes large perturbations on the electronic spectrum of the CO-deficient species [117]. In the light of recent developments, it is reasonable to formulate such complexes as adducts of methane in which the methane can be interacting as an η^1, η^2, or η^3 fashion. Oxidative addition of methane and other molecules with strong C—H bonds is not expected to occur except as the M—CH_3 and M—H bond energies are sufficiently strong to compensate for the cleavage of the CH_3—H bond. M—CH_3 bond energies are thought to be similar to M—H bond energies on naked metal centers, whereas the M—CH_3 bonds of complexed metal atoms are considerably weaker than M—H bonds [124, 125]. Therefore, one should not expect the oxidative addition of strong C—H bonds to a first-row transition metal except as it is a naked metal atom. Secondly, it is unlikely that a C—H bond will oxidatively add to a metal when H_2 does not oxidatively add.

Methane has been shown to add oxidatively to CpIr(CO) and CpRh(CO), where Cp is either the η^5-cyclopentadienyl or η^5-pentamethylcyclopentadienyl ligand [30, 126–129]. In this behavior, CpIr(CO) mimics the behavior observed in liquid xenon and in other fluid media [121, 122]. The CpM(CO) has been generated in a variety of ways. The photolysis of $CpM(CO)_2$ in argon matrices gives very little evidence of CO loss except as the CpM(CO) is scavenged by a reactant molecule such as CH_4 or ^{13}CO [126, 127]. In fact, one band was tentatively assigned to a simple complex of methane and CpIr(CO), a claim that is disputed by Perutz in a subsequent publication [128]. CpRh(CO) has also been generated from an ethylene complex that also activates the C—H bond of methane. In argon matrices, either CO or ethylene is lost, both of which return upon long-wavelength irradiation. The C—H bonds of the ethylene ligand appear to be immune to activation. The *bis*-complex, $CpRh(C_2H_4)_2$, undergoes ethylene loss upon ultraviolet irradiation forming $CpRh(C_2H_4)$. The coordinatively unsaturated complex will react with simple donors such as CO or N_2, or will recombine with the ethylene, but it will not oxidatively add either CH_4 or ethylene. This contrasts with the insertion of iridium into H_3C_2—H bonds when $CpIr(C_2H_4)_2$ is irradiated [129] (*vide infra*). Finally, CpIr(CO) has been prepared from $CpIr(CO)(H)_2$ by photolysis [30]. In methane, the activation of the C—H bond is observed once the H_2 is eliminated. In an argon matrix, observable quantities of CpIr(CO) can accumulate; however, the conversion to product is never complete. Long-wavelength irradiation can reverse the H_2

loss process, so it is reasonable that a photostationary state is achieved which accounts for the incomplete conversion of $CpIr(CO)(H)_2$ into $CpIr(CO)$. A minor absorption band appears in the spectrum of photolyzed matrices at 2002 cm^{-1}, especially when high-energy photons are used, along with that of $CpIr(CO)$. The band becomes attenuated at the same time as when H_2 oxidatively adds to $CpIr(CO)$, although no additional products are noted. It was suggested that it may be due to simple adduct of H_2 and $CpIr(CO)$ although it is not clear why it is formed only after photolyses with especially high frequencies, nor is it clear why more of the hydrogen does not associate with the iridium. A diene formulation was also offered for the complex:

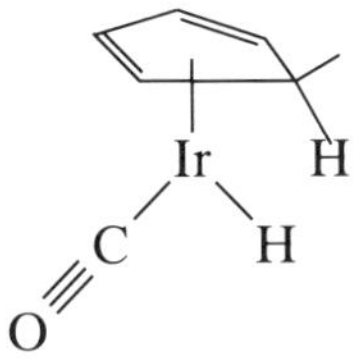

Ir H
C H
O

Such as proposal can be tested by looking for deuterium incorporation into the cyclopentadienyl group when $CpIr(CO)(D)_2$ is photolyzed. The extent of incorporation should be augmented by doping the matrix with additional D_2.

In a somewhat related work [130], it has been shown that $R_3SiCo(CO)_4$ undergoes CO loss upon photolysis in methylcyclohexane matrices at 77 K. In the presence of R_3SiH, $R_3SiCo(CO)_3$ is still the sole product, although a new product forms as soon as the matrix warms. The product was presumed to be the six-coordinate $HCo(SiR_3)_2(CO)_3$. For $R = C_2H_5$, the product is stable at mild temperatures. At room temperature, the starting material is regenerated.

2.2.5. Hydrides Formed by Intramolecular Rearrangements

There are several examples of intramolecular rearrangements of organometallic complexes that have resulted in hydride formation. The irradiation of $CH_3CrCp(CO)_3$ yields $CH_2CrCp(CO)_2H$ via an α elimination [131]. The starting material can be regenerated by annealing the matrix. Interestingly, the heavier congeners of chromium fail to give analogous compounds; rather, they produce the coordinatively unsaturated $CH_3MCp(CO)_2$ [5, 41, 132].

$$CH_3CrCp(CO)_3 \underset{\text{annealing}}{\overset{\text{UV}}{\rightleftharpoons}} CH_2Cr(H)Cp(CO)_2 + CO$$

$$CH_3MCp(CO)_3 \underset{\text{visible}}{\overset{\text{UV}}{\rightleftharpoons}} CH_3MCp(CO)_2 + CO \quad M = Mo, W$$

There is some evidence to suggest that the methyl group is interacting with the vacant coordination site in the unsaturated molybdenum and tungsten complexes [133]. Still it is surprising that the chromium complex undergoes α elimination whereas the tungsten complex does not, given what is generally conceded for M—X bond energies. It may be that α elimination is endergonic for all three congeners. The reversal of the elimination would then have to be kinetically impossible at such low temperatures for the chromium complex whereas the reversal would occur rapidly for the heavier congeners. In fluid media, M—CH_3 homolysis is believed to occur, but trapping of radical products is problematic except for reactions that lead to the hydrogen atom [46] (*vide supra*). It is interesting to speculate on the role of radicals in the unusual chemistry exhibited by the chromium complex. In a related reaction, $CpM(CO)_3C_2H_5$ undergoes CO loss upon irradiation in a paraffin matrix, followed by the formation of what is presumably an agostically stabilized intermediate [5, 132]. This, in turn, undergoes β elimination to yield an ethylene–hydride complex upon warming to 195 K. The results are reproduced in rare-gas matrices except for the detection of the agostic intermediate. However, it was noted that CO would not return upon long-wavelength illumination as is typical in matrix experiments. Rather, the complex undergoes β elimination. Interestingly, the β elimination can even be effected in a CO and N_2 matrix.

In paraffins, the yield of the ethylene–hydride complex is quantitative for $CpW(CO)_2C_2H_5$ whereas β elimination does not go to completion for the molybdenum complex.

$Cp(CO)_3M—CH_2—CH_3$ (in paraffin at 77 K)

UV light ↓

$Cp(CO)_2M—CH_2—CH_3 + CO$ (λ_{max} = 565 nm)

↓ irradiation with 633-nm light

$Cp(CO)_2M—CH_2$ (λ_{max} = 405 nm)
$\quad\vdots\qquad|$
$\quad H—CH_2$

heat ↓

$Cp(CO)_2M—\|$ ($CH_2=CH_2$)
$\quad|$
$\quad H$

Assuming that $CpMo(CO)_2(\eta^2\text{-}C_2H_5)$ is in equilibrium with $CpMo(CO)_2H(C_2H_4)$, one must conclude that the energies of the two complexes are nearly equal. Apparently, the legitimate 16-electron complex is not interacting with the hydrocarbon matrix in an agostic fashions, as judged by the ultraviolet transition. The matrix may not be flexible enough to establish the interaction once the coordination site is vacated.

Ethylene molecules coordinated to iridium(I) have undergone insertion reactions upon ultraviolet irradiation [129]. The isomerization is not apparently preceded by ethylene loss because irradiation of $CpIr(C_2H_4)_2$ in a CO matrix gives the same insertion products as are observed to form in argon, as well as some $CpIrCO(C_2H_4)$. This latter molecule can give the vinylidene hydride, $CpIrCO(H)C_2H_3$.

$$\mathrm{CpIr}(\mathrm{C_2H_4})_2 \xrightarrow[\mathrm{Ar}]{\lambda > 290\ \mathrm{nm}} \mathrm{CpIrH} \xrightarrow{h\nu} \mathrm{CpIrH_2}(=\mathrm{C}=\mathrm{CH_2})$$

A novel rearrangement of $(\eta^6\text{-}C_7H_8)Mo(CO)_3$ occurs on heating, the products of which can be detected by quenching the sublimate in a cold matrix [134]. New bands were present in the spectrum that correlated with the temperature at which the starting heptatriene sublimed. The new product was formulated as $(\eta^5\text{-}C_7H_7)Mo(CO)_3H$. It could be obtained in a variety of gas matrices including CO, suggesting that the metal was still coordinatively saturated.

2.3. CONCLUSION

At the beginning, I commented that matrix isolation studies of transition metal hydrides were probably more successful than matrix studies of any other major category of compounds. The validity of this statement depends a great deal on the nature of the hydrogen atom, including its size, mass, and limited, albeit strong, bonding capabilities. Its small size and mass has meant that it could separate from a metal in photochemically induced homolyses in spite of the rigidity of the matrix. Its limited bonding capability usually forces the hydrogen atom to reside on the periphery of molecules, where it is approachable by metal atoms. In the limit, the hydrogen molecule ranks as the most sterically unencumbered molecule. In particular, the sigma bond of the hydrogen molecule is more exposed to attack by electrophiles than any other sigma bond. These features are important for matrix work because it is more probable that reactions will have low activation barriers because there is no necessity for major distortions.

In looking to the future, there are a number of issues that can and should be investigated by the matrix isolation technique.

1. There are two reports of complexes of dihydrogen and naked metal atoms, the spectral characteristics of which differ from those of $(H_2)ML_n$. If the spectra of the metal dihydrogen complexes are correctly interpreted, an attempt must be made to understand the origin of the difference.

2. Mostly, the hydrogen-containing vibrational features of $(H_2)ML_n$ have not been observed. When they were seen, they were found to be unexpectedly complex. It will be important to gain a wider experience of these vibrational coordinates in order to achieve an understanding of the basis for the complexity.
3. As quantum chemists become increasingly able to model both the ground- and excited-state electronic structure and the total energy of transition metal compounds, it will be important to make available reliable optical spectra of transition metal hydrides, many of which are unstable at room temperature, especially those that are most amenable to calculation.
4. Also, it is still unclear to what extent metal–hydrogen homolyses experience a cage effect. Only by settling this question can it be determined what the quantum yields of the homolysis of the M—H bond are, relative to other ligand dissociations.
5. Attempts should be made to characterize associations between dihydrogen and d^0 transition metal and main-group acids. A four-center transition state has been proposed to explain the reactivity of hydrogen with these materials. It may be possible to observe such an assembly at low temperature.
6. For the syntheses of new, unstable η^2-dihydrogen complexes, liquid Xe offers several advantages over the classical matrix isolation technique. Both techniques should be exploited, especially as they may yield evidence of polyhydrogen complexes [23, 135].

REFERENCES

1. Whittle, E.; Dows, D. A.; Pimentel, G. C. *J. Chem. Phys.* **1954**, *22*, 1943.
2. Perutz, R. N. Inorganic and Organometallic Photochemistry in Matrices. In *Chemistry and Physics of Matrix-Isolated Species*; Andrews, L., Moskovits, M., Eds.; Elsevier Science: New York, 1989.
3. Buckingham, A. D. *Proc. Roy. Soc. London Ser. A* **1958**, *248*, 169; *Proc. Roy. Soc. London Ser. A* **1960**, *255*, 32; *Trans. Faraday Soc.* **1960**, *56*, 753 (cited by Hanlan, L. A.; Huber, H.; Kündig, E. P.; McGarvey, B. R.; Ozin, G. A. *J. Am. Chem. Soc.* **1975**, *97*, 7054–7068).
4. Church, S. P.; Poliakoff, M.; Timney, J. A.; Turner, J. J. *Inorg. Chem.* **1983**, *22*, 3259–3266.
5. Kazlauskas, R. J.; Wrighton, M. S. *J. Am. Chem. Soc.* **1982**, *104*, 6005–6015.
6. Hooker, R. H.; Rest, A. J. *J. Chem. Soc., Dalton Trans.* **1984**, 761–770.
7. Jesson, J. P. Stereochemistry and Stereochemical Nonrigidity in Transition Metal Hydrides. In *Transition Metal Hydrides*; Meutterties, E. L., Ed. Marcel Dekker: New York, 1971; pp. 75–201.
8. Gordan, A. J.; Ford, R. A. *The Chemist's Companion: A Handbook of Practical Data, Techniques, and References*; Wiley-Interscience: New York, 1972. See also, references 14 and 23.
9. Braterman, P. S.; Harrill, R. W.; Kaesz, H. D. *J. Am. Chem. Soc.* **1967**, *89*, 2851–2855.

10. Sweany, R. L. *J. Am. Chem. Soc.* **1985**, *107*, 2374–2379.
11. Sweany, R. L.; Russell, F. N. *Organometallics* **1988**, *7*, 719–727.
12. Girling, R. B.; Grebenik, P.; Perutz, R. N. *Inorg. Chem.* **1986**, *25*, 31–36.
13. Davidson, G.; Duce, D. A. *J. Organomet. Chem.* **1976**, *120*, 229–237. See also reference 59.
14. Kubas, G. J. *Acc. Chem. Res.* **1988**, *21*, 120–128.
15. Kubas, G. J.; Ryan, R. R.; Swanson, B. I.; Vergamini, P. J.; Wasserman, H. J. *J. Am. Chem. Soc.* **1984**, *106*, 451–452.
16. Aresta, M.; Giannoccaro, P.; Rossi, M.; Sacco, A. *Inorg. Chim. Acta* **1971**, *5*, 115 (cited by Hamilton, D. G.; Crabtree, R. H. *J. Am. Chem. Soc.* **1988**, *110*, 4126–4133).
17. Harmon, W. D.; Taube, H. *J. Am. Chem. Soc.* **1990**, *112*, 2261–2263.
18. Koehler, J. S.; Dennison, D. M. *Phys. Rev.* **1940**, *57*, 1006–1021. Lee, R. G.; Hunt, R. H.; Plyler, E. K.; Dennison, D. M. *J. Mol. Spectrosc.* **1975**, 138–154.
19. Herzberg. G. *Molecular Spectra and Molecular Structure II. Infrared and Raman Spectra of Polyatomic Molecules*; Van Nostrand: New York, 1945; pp. 220–227. Pauling, L. *Phys. Rev.* **1930**, *36*, 430–443.
20. Eckert, J.; Kubas, G. J.; Hall, J. H.; Hay, P. J.; Boyle, C. M. *J. Am. Chem. Soc*, **1990**, *112*, 2324–2332.
21. Sweany, R. L.; Moroz, A. *J. Am. Chem. Soc.* **1989**, *111*, 3577–3583.
22. Ozin, G. A.; García-Prieto, J. *J. Am. Chem. Soc.* **1986**, *108*, 3099–3100.
23. The splitting of the H—H stretching vibration was attributed to Fermi resonance with overtones and combination bands. In light of the weak coupling that has been observed in other molecules between the motions of the hydrogen atoms and other modes of the molecule this explanation should be treated with skepticism. Gadd, G. E.; Upmacis, R. K.; Poliakoff, M.; Turner, J. J. *J. Am. Chem. Soc.* **1986**, *108*, 2547–2552.
24. The interchange of protium nuclei via a quantum-mechanical tunnelling mechanism can cause anomalous couplings in the NMR spectrum. Heinekey, D. M.; Millar, J. M.; Koetzle, T. F.; Payne, N. G.; Zilm, K. W. *J. Am. Chem. Soc.* **1990**, *112*, 909–919. The theoretical treatment of such a phenomenon is very similar to that used to treat the quantum-mechanical doubling in infrared spectra. Zilm, K. W.; Heinekey, D. M.; Millar, J. M.; Payne, N. G.; Neshyba, S. P.; Duchamp, J. C.; Szczyrba, J. *J. Am. Chem. Soc.* **1990**, *112*, 920–929.
25. Symons, M. C. R.; Aly, M. M.; West, D. X. *J. Chem. Soc. Dalton Trans.* **1979**, 1744–1748.
26. Symons, M. C. R.; Aly, M. M.; Wyatt, J. L. *J. Chem. Soc., Chem. Commun.* **1981**, 176–178.
27. Knight, L. B., Jr.; Weltner, W., Jr. *J. Mol. Spectrosc.* **1971**, *40*, 317–327.
28. Geoffroy, G. L.; Bradley, M. G.; Pierrantozzi, R. Photochemistry of Transition Metal Hydride Complexes. In *Transition Metal Hydrides*; Bau, R., Ed.; American Chemical Society: Washington, DC, 1978; pp. 181–200. Geoffroy, G. L.; Wrighton, M. S. *Organometallic Photochemistry*; Academic: New York, 1979.
29. Sweany, R. L. *J. Am. Chem. Soc.* **1981**, *103*, 2410–2412.
30. Bloyce, P. E.; Rest, A. J.; Whitwell, I.; Graham, W. A. G.; Holmes-Smith, R. *J. Chem. Soc., Chem. Commmun.* **1988**, 846–848. Bloyce, P. E.; Rest, A. J.; Whitwell, I. *J. Chem. Soc., Dalton Trans.* **1990**, 813–821.
31. Chetwynd-Talbot, J.; Grebenik, P.; Perutz, R. N.; Powell, M. H. A. *Inorg. Chem.* **1983**, *22*, 1675–1684.
32. The irradiation of $HMCp(CO)_3$ in CO matrices also can cause the cyclopentadienyl group to be replaced by carbonyl groups (see reference 35).
33. Church, S. P.; Poliakoff, M.; Timney, J. A.; Turner, J. J. *J. Am. Chem. Soc.* **1981**, *103*, 7515–7520.
34. Sweany, R. L. *Inorg. Chem.* **1982**, *21*, 752–756.

35. Mahmoud, K. A.; Rest, A. J.; Alt, H. *J. Chem. Soc., Dalton Trans.* **1984**, 187–197.
36. Baynham, R. F. G.; Chetwynd-Talbot, J.; Grebenik, P.; Perutz, R. N.; Powell, M. H. A. *J. Organomet. Chem.* **1985**, *284*, 229–242.
37. Chetwynd-Talbot, J.; Grebenik, P.; Perutz, R. N. *Inorg. Chem.* **1982**, *21*, 3647–3657.
38. Sweany, R. L. *Inorg. Chem.* **1980**, *19*, 3512–3516.
39. Wermer, P.; Ault, B. S.; Orchin, M. *J. Organomet. Chem.* **1978**, *162*, 189–194.
40. In several papers, Rest and co-workers have proposed that homolysis of M—H and M—CH_3 groups would only occur from the coordinatively unsaturated $RMCp(CO)_2$ [35, 41]. This idea was heavily influenced by the work of Severson and Wojcicki [45], which reported that the homolysis of the CH_3—M bond in $CH_3MCp(CO)_3$ was inhibited by CO. Subsequently, it was argued that, because homolysis could not be entirely shut off by high pressures of CO, homolysis could still be considered a minor path for the photodecomposition of $CH_3MCp(CO)_3$ [46].
41. Mahmoud, K. A.; Narayanaswamy, R.; Rest, A. J. *J. Chem. Soc., Dalton Trans.* **1981**, 2199–2204.
42. An early report that CH_3—M homolyses become much more facile in polymer matrices [43] was withdrawn when the infrared features that originally were assigned to $MoCp(CO)_3$ were reassigned [44].
43. Hitam, R. B.; Hooker, R. H.; Mahmoud, K. A.; Narayanaswamy, R.; Rest, A J. *J. Organomet. Chem.* **1981**, *222*, C9.
44. Hooker, R. H.; Rest, A. J. *J. Chem. Soc., Dalton Trans.* **1984**, 761–770.
45. Severson, R. G.; Wojcicki, A. *J. Organomet. Chem.* **1978**, *157*, 173–185.
46. See, for example, Goldman, A. S.; Tyler, D. R. *J. Am. Chem. Soc.* **1986**, *108*, 89–94.
47. Symons, M. C. R.; Sweany, R. L. *Organometallics* **1982**, *1*, 834–836.
48. $D_{H—CO}$ is 18 kcal mol^{-1}; Kerr, J. A. *Chem. Rev.* **1966**, *70*, 465–500; Marling, J. *J. Chem. Phys.* **1977**, *66*, 4200–4225. This relatively weak bond contrasts with M—H bond energies in excess of 50 kcal mol^{-1} [49].
49. Tilset, M.; Parker, V. D. *J. Am. Chem. Soc.* **1989**, *111*, 6711–6717 (erratum: *J. Am. Chem. Soc.* **1990**, *112*, 2843).
50. Crichton, O.; Poliakoff, M.; Rest, A. J.; Turner, J. J. *J. Chem. Soc., Dalton Trans.* **1973**, 1321–1328.
51. Belt, S. T.; Grevels, F.-W.; Klotzbücher, W. E.; McCamley, A.; Perutz, R. N. *J. Am. Chem. Soc.* **1989**, *111*, 8373–8382.
52. Poliakoff, M.; Turner, J. J. *J. Chem. Soc., Dalton Trans.* **1974**, 2276–2285.
53. McNeill, E. A.; Scholer, F. R. *J. Am. Chem. Soc.* **1977**, *99*, 6243–6249.
54. Barton, T. J.; Grinter, R.; Thomson, A. J.; Davies, B.; Poliakoff, M. *J. Chem. Soc., Chem. Commun.* **1977**, 841–842.
55. Burdett, J. K.; Grzybowski, J. M.; Perutz, R. N.; Poliakoff, M.; Turner, J. J.; Turner, R. F. *Inorg. Chem.* **1978**, *17*, 147–154.
56. Davies, B.; McNeish, A.; Poliakoff, M.; Turner, J. J. *J. Am. Chem. Soc.* **1977**, *99*, 7573–7579.
57. Rest, A. J.; Turner, J. J. *J. Chem. Soc., Chem. Commun.* **1969**, 375–376.
58. Versluis, L.; Ziegler, T.; Baerends, E. J.; Ravenick, W. *J. A. Chem. Soc.* **1989**, *111*, 2018–2025 and references cited therein.
59. Sweany, R. L. *J. Am. Chem. Soc.* **1986**, *108*, 6986–6991.
60. Sweany, R. L.; Owens, J. W. *J. Organomet. Chem.* **1983**, *255*, 327–334.
61. Chen, H. W.; Jolly, W. L.; Kopf, J.; Lee, T. H. *J. Am. Chem. Soc.* **1977**, *99*, 2607–2610.
62. Huheey, J. E. *Inorganic Chemistry: Principles of Structure and Reactivity*; 3rd ed.; Harper & Row: New York, 1983; pp. 158–160.

63. "On the Charge Distribution in Hydridometal Carbonyls and Its Impact on Reactivity." Paper delivered at the Symposium on Organometallic Reaction Mechanisms, Southwest Regional Meeting, Lubbuck, TX, December 1984; Kristjánsdóttir, S. S.; Norton, J. R.; Moroz, A.; Sweany, R. L.; Whittenburg, S. L. "The Absence of Hydrogen Bonding between $HCo(CO)_4$ and Nitrogen or Oxygen Bases. An IR and Raman Study. *Organometallics*, in press.
64. Collman, J. P.; Hegedus, L. S. *Principles and Applications of Organotransition Metal Chemistry*, University Science Books: Mill Valley, CA, 1980.
65. Bellamy, L. J. *Infrared Spectra of Complex Molecules*, 2nd ed.; Chapman and Hall: New York, 1980; Vol. 2, pp. 240–248.
66. Sweany, R. L. Unpublished data.
67. Veillard, A.; Dedieu, A. *Theor. Chim. Acta* **1983**, *63*, 339–348.
68. Hermann, T. S. *Appl. Spectrosc.* **1969**, *23*, 461–472.
69. Knight, L. B., Jr.; Wise, M. B.; Fisher, T. A.; Steadman, J. *J. Chem. Phys.* **1981**, *74*, 6636–6643.
70. Van Zee, R. J.; DeVore, T. C.; Wilkerson, J. L.; Weltner, W., Jr. *J. Chem. Phys.* **1978**, *69*, 1869–1875.
71. Van Zee, R. J.; DeVore, T. C.; Wilkerson, J. L.; Weltner, W., Jr. *J. Chem. Phys.* **1979**, *70*, 2051–2056.
72. Li, J.; Balasubramanian, K. *J. Phys. Chem.* **1990**, *94*, 545–552 and references cited therein.
73. Clark, A. *The Chemisorptive Bond*; Academic: New York, 1974.
74. Ozin, G. A.; McCaffrey, J. G. *J. Phys. Chem.* **1984**, *88*, 645–648.
75. Rubinovitz, R. L.; Nixon, E. R. *J. Phys. Chem.* **1986**, *90*, 1940–1944.
76. Ozin, G. A.; McCaffrey, J. G. *J. Am. Chem. Soc.* **1984**, *106*, 807–809.
77. L'Eplattenier, F.; Calderazzo, F. *Inorg. Chem.* **1967**, *6*, 2092–2097.
78. Kruck, T.; Prasch, A. *Z. Anorg. Allg. Chem.* **1967**, *371*, 1–22.
79. Osborn, J. A.; Jardine, F. H.; Young, J. F.; Wilkinson, G. *J. Chem. Soc. A* **1966**, 1711–1736. L'Eplattenier, F.; Calderazzo, F. *Inorg. Chem.* **1968**, *7*, 1290–1293. Dewhirst, K. C.; Keim, W.; Reilley, C. A. *Inorg. Chem.* **1968**, *7*, 546–551.
80. Ozin, G. A.; McCaffrey, J. G.; McIntosh, D. F. *Pure Appl. Chem.* **1984**, *56*, 111–128.
81. Hauge, R. H.; Kafafi, Z. H.; Margrave, J. L. In *Physics and Chemistry of Small Clusters*; Jena, P.; Rao, B. K.; Khanna, S. N., Eds.; Plenum: New York, 1987.
82. Kauffman, J. W.; Hauge, R. H.; Margrave, J. L. *J. Phys. Chem.* **1985**, *89*, 3547–3552.
83. Park, M. A. Ph.D. Dissertation, Rice University, 1988.
84. Blomberg, M. R. A.; Siegbahn, P. E. M. *J. Chem. Phys.* **1983**, *78*, 5682–5692.
85. Ozin, G. A.; Gracie, C. *J. Phys. Chem.* **1984**, *88*, 643–645.
86. Mårtensson, A.-S.; Nyberg, C.; Andersson, S. *Phys. Rev. Lett.* **1986**, *57*, 2045–2048.
87. Knight, L. B., Jr.; Cobranchi, S. K.; Herlong, J.; Kirk, T.; Balasubramanian, K.; Das, K. K. *J. Chem. Phys.* **1990**, *92*, 2721–2732.
88. To support the claim that the features assigned to PdH_2 are intense, it should be noted that the pure rotational transition of o-H_2 is absent from the spectrum in the region of 580 cm^{-1}. Either the hydrogen has been relaxed to the para form or the scale amplification is not sensitive enough to see the rotational feature. Also, it should be noted that the Xe–Pd ratio is 10,000.
89. Moroz, A.; Sweany, R. L.; Whittenburg, S. L. *J. Phys. Chem.* **1990**, *94*, 1352–1357.
90. Balasubramanian, K.; Feng, P. Y.; Liao, M. Z. *J. Chem. Phys.* **1988**, *88*, 6955–6961 and references cited therein.
91. Novaro, O.; García-Prieto, J.; Poulain, E.; Ruiz, M. E. *J. Mol. Struct. (Theochem)* **1986**, *135*, 79–91.

92. An alternative assignment of the chemistry should be seriously considered. If the absorption that was observed was, in reality, a bending vibration of $Pd(H)_2$, then the same isotopic pattern would be observed in xenon. Moreover, if $Pd(H)_2$ were nonlinear, then there would be an opportunity for it to undergo a quantum-mechanical pairwise exchange of the hydrogen atoms, associated with the bending motion. The bands at 890 cm^{-1} are in the region where one expects deformation modes of classical hydrides. Balasubramanian et al. [90] have determined from theory that a classical hydride is stable with a HPdH bond angle of 62°. The mechanism for interchange of the hydrogen atoms might be viewed as an angle compression forming an η^2 complex, followed by a rotation that interchanges the hydrogen atoms, followed by the relaxation of the HPdH angle. In fairness to the authors, nothing was observed in the regions in which the Pd—H stretching modes would be observed. It is conceivable that they are as weak as is sometimes found to be the case of organometallic hydrides. Also, the bending modes for other binary hydrides that have been characterized in matrices characteristically fall to the red of the positions reported in this study, at least for the first-row transition elements.
93. Ozin, G. A.; García-Prieto, J. *J. Phys. Chem.* **1988**, *92*, 318–324.
94. Ozin. G. A.; García-Prieto, J. *J. Phys. Chem.* **1988**, *92*, 325–337.
95. Morris, R. H.; Earl, K. A.; Luck, R. L.; Lazarowych, N. J.; Sella, A. *Inorg. Chem.*, **1987**, *26*, 2674–2683.
96. Hauge, R. H.; Margrave, J. L.; Kafafi, Z. H. Reactions of First-row Atoms and Small Clusters in Matrices. In *Chemistry and Physics of Matrix-Isolated Species*; Andrews, L., Moskovits, M., eds.; Elsevier: New York, 1989; pp. 277–302.
97. Saillard, J.-Y.; Hoffmann, R. *J. Am. Chem. Soc.* **1984**, *106*, 2006–2026.
98. Kafafi, Z. H.; Hauge, R. H.; Margrave, J. L. *J. Am. Chem. Soc.* **1985**, *107*, 7550–7559.
99. Stoutland, P. O.; Bergman, R. G. *J. Am. Chem. Soc.* **1988**, *110*, 5732–5744.
100. Kafafi, Z. H.; Hauge, R. H.; Margrave, J. L. *J. Am. Chem. Soc.* **1985**, *107*, 5134–6135.
101. Ball, D. W.; Hauge, R. H.; Margrave, J. L. *Inorg. Chem.* **1989**, *28*, 1599–1601.
102. Kauffman, J. W.; Hauge, R. H.; Margrave, J. L. *J. Phys. Chem.* **1985**, *89*, 3541–3547.
103. Chang, S.-C.; Hauge, R. H.; Kafafi, Z. H.; Margrave, J. L.; Billups, W. E. *J. Am. Chem. Soc.* **1988**, *110*, 7975–7980.
104. Kline, E. S.; Kafafi, Z. H.; Hauge, R. H.; Margrave, J. L. *J. Am. Chem. Soc.* **1985**, *107*, 7559–7562.
105. Parker, S. F.; Peden, C. H. F.; Barrett, P. H.; Pearson, R. G. *J. Am. Chem. Soc.* **1984**, *106*, 1304–1308.
106. Kline, E. S.; Kafafi, Z. H.; Hauge, R. H.; Margrave, J. L. *J. Am. Chem. Soc.* **1987**, *109*, 2402–2409 and references cited therein.
107. Ball, D. W.; Kafafi, Z. H.; Hauge, R. H.; Margrave, J. L. *Inorg. Chem.* **1985**, *24*, 3708–3710.
108. Billups, W. E.; Konarski, M. M.; Hauge, R. H.; Margrave, J. L. *J. Am. Chem. Soc.* **1980**, *102*, 7393–7394.
109. Parnis, J. M.; Ozin, G. A. *J. Phys. Chem.* **1989**, *93*, 4023–4029.
110. Burdett, J. K.; Turner, J. J. In *Cryochemistry*; Muskovits, M.; Ozin, G. A., Eds.; Wiley: New York, 1976; pp. 493–525.
111. Because of the volatility of H_2, it is probable that the nominal concentration of the gas mixture is only the upper limit of H_2 in the matrix.
112. Upmacis, R. K.; Gadd, G. E.; Poliakoff, M.; Simpson, M. B.; Turner, J. J.; Whyman, R.; Simpson, A. F. *J. Chem. Soc., Chem. Commun.* **1985**, 27–30. Upmacis, R. K.; Poliakoff, M.; Turner, J. J. *J. Am. Chem. Soc.* **1986**, *108*, 3645–3651.

113. There are also reports of H_2 complexes of Cp_2Ni and $ArMo(CO)_3$ that are formed merely by codeposition at low temperature. Greenvald, I. I.; Lokshin, B. V.; Rudnieski, N. K.; Marin, V. P. *Dokl. Akad. Nauk CCCP* **1988**, *298*, 1142–1144.
114. An earlier report of this synthesis assigned the features of $HCo(CO)_3(H_2)$ to $(H)_3Co(CO)_3$. Sweany, R. L. *J. Am. Chem. Soc.* **1982**, *104*, 3739–3940.
115. Sweany, R. L.; Polito, M. A.; Moroz, A. *Organometallics* **1989**, *8*, 2305–2308.
116. Pacchioni, G. *J. Am. Chem. Soc.* **1990**, *112*, 80–85 and references cited therein.
117. Perutz, R. N.; Turner, J. J. *J. Am. Chem. Soc.* **1975**, *97*, 4791–4800.
118. Hay, P. J. *J. Am. Chem. Soc.* **1987**, *109*, 705–710. Jean, Y.; Eisenstein, O.; Volatron, F.; Maouche, B.; Sefta, F. *J. Am. Chem. Soc.* **1986**, *108*, 6587–6592.
119. Turner, J. J.; Poliakoff, M.; Howdle, S. M.; Jackson, S. A.; McLaughlin, J. G. *Faraday Disc. Chem. Soc.* **1988**, 271–285.
120. Howdle, S. M.; Poliakoff, M. *J. Chem. Soc., Chem. Commun.* **1989**, 1099–1101. See also references 23 and 108.
121. Weiller, B. H.; Wasserman, E. P.; Bergman, R. G.; Moore, C. B.; Pimentel, G. C. *J. Am. Chem. Soc.* **1989**, *111*, 8288–8290.
122. Sponsler, M. B.; Weiller, B. H.; Stoutland, P. O.; Bergman. R. G. *J. Am. Chem. Soc.* **1989**, *111*, 6841–6843.
123. Abis, L.; Sen, A.; Halpern, J. *J. Am. Chem. Soc.* **1978**, *100*, 2915–2916.
124. Halpern, J. *Inorg. Chim. Acta* **1985**, *100*, 41–48.
125. Georgiadis, R.; Fisher, E. R.; Armentrout, P. B. *J. Am. Chem. Soc.* **1989**, *111*, 4251–4262. Ziegler, T.; Tschinke, V.; Becke, A. *J. Am. Chem. Soc.* **1987**, *109*, 1351–1358 and references cited therein.
126. Rest, A .J.; Whitwell, A J.; Graham, W. A.; Hoyano, J. K.; McMaster, A. D. *J. Am. Chem. Soc., Chem. Commun.* **1984**, 624–626.
127. Rest, A. J.; Whitwell, I.; Graham, W. A. G.; Hoyano, J. K.; McMaster, A. D. *J. Chem. Soc., Dalton Trans.* **1987**, 1181–1190.
128. Haddleton, D. M.; McCamley, A.; Perutz, R. N. *J. Am. Chem. Soc.* **1988**, *110*, 1810–1817.
129. Haddleton, D. M.; Perutz, R. N. *J. Chem. Soc., Chem. Commun.* **1986**, 1734–1736. Duckett, S. B.; Haddleton, D. M.; Jackson, S. A.; Perutz, R. N.; Poliakoff, M.; Upmacis, R. K. *Organometallics* **1988**, *7*, 1526–1532.
130. Anderson, F. R.; Wrighton, M. S. *J. Am. Chem. Soc.* **1984**, *106*, 995–999.
131. Mahmoud, K. A.; Rest, A. J.; Alt, H. *J. Chem. Soc., Chem. Commun.* **1983**, 1011–1013.
132. Mahmoud, K. A.; Rest, A. J.; Alt, H. G.; Eichner, M. E.; Jansen, B. M. *J. Chem. Soc., Dalton Trans.* **1984**, 175–186.
133. Sweany, R. L.; Flavier, A. Unpublished data.
134. Hooker, R. H.; Rest, A. J. *J. Organomet. Chem.* **1982**, *234*, C23–C27.
135. Burdett, J. K.; Pourian, M. R. *Organometallics* **1987**, *6*, 1684–1691.

CHAPTER 3

Transition Metal Hydrides: Structure and Bonding

CHARLES W. BAUSCHLICHER, JR., and STEPHEN R. LANGHOFF

NASA Ames Research Center, Moffett Field, CA 94035

3.1. INTRODUCTION

The strength of bonds between hydrogen and systems containing transition metals is of fundamental importance for gaining a fundamental understanding of catalytic processes. Most practical applications of transition-metal–hydrogen chemistry involve either heterogeneous catalytic processes at metal surfaces or homogeneous catalysis involving organometallic compounds where the metal reactive center is surrounded by ligands. A particularly active area of research is the search for homogeneous catalysts that can activate chemically inert hydrocarbons. Experiments designed to explain homogeneous and heterogeneous catalysis have not been completely successful, so considerable theoretical and experimental effort has been directed toward understanding simpler model systems of chemisorption or catalysis involving transition metal systems.

In this chapter, we consider the transition metal monohydrides and dihydrides and their positive ions. We focus on the computational techniques that are required for accurate determination of their spectroscopic constants (r_e, ω_e), separation between states (T_e), and dissociation energies (D_e). The emphasis will be placed on obtaining a fundamental understanding of the bonding in these relatively simple molecules to develop a basis for understanding the bonding in more complex systems. The positive ions are considered because of the availability of accurate dissociation energies from the guided ion beam approach. This provides a calibration of the theoretical D_e values for the neutral diatomics, where little experimental data is presently available. We discuss the dipole moment of NiH in some detail, because an accurate experimental value is available for comparison and because the dipole moment is shown to provide a sensitive diagnostic of the quality of the wave function. Finally, we consider the effect of ligands on the metal–hydrogen bond strength. Theoretical calculations indicate that substantial stabilization or destabilization of the bond strength can be achieved by selecting the appropriate ligands.

3.2. QUALITATIVE FEATURES OF TRANSITION METAL BONDING

The bonding in transition metal hydrides depends to a large extent on the relative ordering of the atomic states of the metal [1]. Because there is typically a large number of low-lying atomic states, differing in the occupation of the *ns*, *np*, and $(n - 1)d$ valence orbitals, there are many low-lying molecular states in the transition metal hydrides. This results in very complex spectra involving many overlapping bands, especially in view of the high multiplicity of many of the states. The complexity of the spectra are further complicated by large spin–orbit effects and large hyperfine splittings due to the large magnetic moments of many of the transition metals [2]. The chemistry of transition metals can also be quite different from first-row chemistry, because additional mechanisms become viable when the *d* orbitals of the metal are occupied. For example, reactions often can proceed with smaller barriers to reaction by using the *d* orbitals to make a reaction a series of one-electron jumps [3]. Similar reactions that involve two-electron jumps (Woodward–Hoffmann forbidden) generally have larger reaction barriers—see later discussion of the Ni + H_2 reaction.

In this section, we discuss various aspects of the bonding in relation to the first-row transition metal neutrals, but most of the concepts are equally valid for the second and third transition rows, as well as for the positive ions of these molecules. The bonding can be quite diverse, due to the similar energy and radial extent of the $3d$, $4s$, and $4p$ orbitals. For example, the $3d^n4s^2$ state, which can form either one or two bonds by *sp* hybridization, can also form *sd* hybrids. This latter hybridization is particularly important on the left side of the row, where the $3d$ and $4s$ orbitals are more comparable in radial extent. The repulsion between the $4s$ orbital and a datively bound ligand is reduced when an electron is promoted to the $3d$ shell. For the $3d^{n+1}4s^1$ occupation of the metal, two bonds can be formed by *sd* hybridization. As for the $3d^n4s^2$ occupation, the open-shell $3d$ orbitals can form additional bonds. Bonding from the $3d^{n+2}$ occupation minimizes the repulsion between the metal and dative ligands and requires direct *d* involvement in the bonds.

Another consideration in determining optimal bonding mechanisms and bond strengths is the loss of atomic exchange energy. For the transition metal atoms this can be quite large, especially the loss of $3d$–$3d$ exchange energy for the first-row transition metals. This tends to lead to weaker bonds for those systems in the middle of the row that have high multiplicities.

A unique feature of transition metal compounds is that the bonding frequently arises from a mixture of several atomic states with different d occupations. Because the d orbitals are rather compact, there is a large differential correlation energy between these states, so that high levels of theory are required to describe this mixing accurately. This complicates the calculation of accurate dipole moments and the identification of the correct ground state in molecules like FeH, where the $X\,^4\Delta$ state is a mixture of $3d^64s^2$ and $3d^74s^1$, whereas the low-lying $a\,^6\Delta$ state is of nearly pure $3d^64s^2$ character.

3.3. METHODS

3.3.1. One-Particle Basis Sets

Large one-particle basis sets and an extensive treatment of correlation are required for a quantitative treatment of transition metal systems. For very accurate calculations on small first-row transition metal systems, we use the large $(20s12p9d)$ primitive basis sets optimized by Partridge [4]. These primitive sets are supplemented with a diffuse d function to improve the description of the $3d^{n+1}4s^1$ and $3d^{n+2}$ occupations, three p functions to describe the $4p$ orbital, and six $4f$ (and in some cases four $5g$) polarization functions. These large primitive sets are contracted to manageable size using the atomic natural orbit (ANO) procedure [5]. These large basis sets are of approximately the same accuracy as the $(13s8p6d4f)$ basis sets for the first-row atoms, which yield accurate results for most properties. These very large basis sets can be used to perform benchmark-quality calculations.

The basis sets most extensively used for first-row transition metals start with the $(14s9p5d)$ primitive sets of Wachters [6]; they are then contracted to $[8s4p3d]$ and supplemented with a diffuse d [7] and two p functions. Whenever possible, we also add an f function that is optimized for atomic correlation. This f function is primarily a d correlating function, which improves the metal atomic separations and helps avoid basis set superposition errors (BSSEs). Accurate atomic separations are important to ensure the correct mixing of atomic states in the molecule.

It is often possible to obtain qualitative accuracy for first-row molecules using small basis sets such as STO-3G. For transition metals, a minimal STO basis set results in $3d$ orbital energies that are in error by as much as 27 eV for Cu. However, Huzinaga and co-workers [8] showed that if the exponents and contraction coefficients of a three-term Gaussian function were optimized, the resulting orbital energies were in good agreement with the numerical Hartree–Fock (HF) values. If the outermost s and d functions are uncontracted and a diffuse d and two p functions are added, a reasonably accurate valence double-zeta basis set is produced. The accuracy of this basis set can be improved further by freezing the $1s$–$3p$ orbitals in their atomic form [9]. This compact basis set yields results in good agreement with the larger Wachters-derived sets. This basis set could be used in studies of quite large systems containing transition metals.

The size of the basis set also can be reduced by using effective core potentials (ECPs). The two formulations [10, 11] most frequently used are the Huzinaga type and the Kahn type. We concentrate on the Kahn-type formalism, because most of our experience is with this formulation. One advantage of the Kahn type is that it is easy to include the Darwin and mass–velocity relativistic contributions [12] to the ECP, commonly then referred to as an RECP. One disadvantage of ECPs is that the valence orbitals may be nodeless. For example, the $4s$ orbital is nodeless if the $3s$ and $3p$ orbitals are included in the ECP. This can lead to an overestimation of the correlation energy when the $3s$ and $3p$ electrons are correlated [13]. In addition, problems can occur even at the self-consistent field (SCF) level on the left side of the row if the $3s$ and $3p$ orbitals are included in the core. The problem is particularly severe for ionic systems where the valence electrons are donated to ligands. In the worst cases the molecule actually "collapses" into an absurdly deep well, because the Pauli repulsion between the $3s$ and $3p$ electrons and the ligand electrons is not accurately represented by the core potential. For example, the Ti—F bond length in TiF_4 is 0.08 Å too short and the vibrational frequencies are in error by up to 150 cm^{-1} compared with the experiment and all-electron calculations [14]. Inclusion of the $3s$ and $3p$ electrons into the valence space eliminates the problems with the ECP. The $4s$ then has a node and the $4s$ and $3d$ electrons can be correlated. Further, collapse is avoided in ionic systems. Thus we recommended that only the ECPs with the semicore electrons in the valence space be used. This allows the study of systems with both ionic and covalent ligands and allows the inclusion of correlation.

For the second-row transition metals we use the RECPs developed by Wadt and Hay [11]. We have published [15] additional valence basis functions to supplement those of Wadt and Hay. These RECPs in conjunction with our valence basis sets produce results of about equivalent accuracy to those obtained from the Wachters-derived sets for the first-row transition metal systems. We stress that we only use the RECP with the $4s$ and $4p$ orbitals in the valence space. We have calibrated the RECP treatment by performing benchmark all-electron calculations for selected second-row systems using large primitive sets and flexible contractions [15]. In order to compare with experiment and the RECP, the mass–velocity and Darwin terms have been included in the all-electron calculations using first-order perturbation theory [16]. It should be noted that, unlike the first transition row, where relativistic effects make only a small change in the computed results, relativistic effects are much more important for the second transition row. For example, only the weakly bound Pd—H_2 complex is found in treatments that do not include relativistic effects, as a result of the very poor ordering of atomic states at the nonrelativistic level [17]. The all-electron and RECP results are in excellent agreement when the $4d$ and $5s$ electrons are correlated [15]. Furthermore, in the all-electron calculations the $4s$ and $4p$ electrons can be correlated, because the orbitals have the correct nodal structure. However, benchmark all-electron calculations indicate only a small bond contraction and change in D_e when these semicore electrons are correlated. Thus the RECP treatment should be adequate, except where very high accuracy is required.

RECPs have also been developed for the third-row transition metals [11]. Although these RECPs account for the relativistic effects in the core electrons, spin–orbit effects are quite large, even for the valence electrons in the third row. For example, the $^3D_3$ and $^3D_1$ states of the $5d^96s^1$ occupation of Pt are split by 10,132 cm^{-1} [18]. Calculations [19] on PtH including spin–orbit effects show that the ground state is probably $^2\Delta_{5/2}$, whereas a $^2\Sigma^+_{1/2}$ state is obtained if spin–orbit effects are not included. Thus, when valence spin–orbit effects are large, the substantial interaction between states of the same total angular momentum can significantly affect both the energies and transition probabilities. Therefore, both spin–orbit and correlation effects must be included for accurate results for third-row transition metal systems.

3.3.2. Treatment of Electron Correlation

The large differential correlation effects require that high levels of correlation treatment be used, especially when states with different d occupations are studied. One correlation method that has been highly successful is the complete-active-space self-consistent field (CASSCF)–multireference-configuration-interaction (MRCI) approach, where the metal near-degeneracies are accounted for with the CASSCF calculation and more-extensive correlation is then added in the MRCI treatment. For first- and second-row chemistry, this level of treatment has been shown to reproduce the full CI (FCI) accurately in small basis sets and agrees well with experiment in large ANO basis sets [20]. However, even this level of theory does not always give accurate dipole moments for transition metal systems. Only the most important near degeneracies can be accounted for in the CASSCF, because of the factorial growth of the CASSCF CI expansion with the number of active orbitals and electrons. This may introduce an orbital bias that the MRCI treatment cannot fully overcome, despite the high level of correlation treatment. However, the iterative natural orbital (INO) procedure has been found to be effective in reducing the orbital bias, as the orbital determination now incorporates more of the effect of electron correlation [21]. In addition, recent work has shown that the average coupled-pair functional (ACPF) method [22] can, in many cases, accurately account for the effect of higher excitations and thus reduce the orbital bias. These considerations are discussed in more detail later.

Although the CASSCF–MRCI and CASSCF–ACPF methods have yielded accurate properties of transition metal systems, these approaches are expensive and therefore cannot be applied routinely to larger systems. Because single-reference-based techniques are much more economical, they have been extensively tested for transition metal systems. The SCF-based modified coupled-pair functional (MCPF) method [23] has proven to be reliable in many cases and has the advantage that the quality of the wave function can be judged by the percentage of the wave function composed by the reference configuration. The CPF method itself is much less stable to near degeneracies and is therefore less useful in transition metal chemistry, whereas for the first two rows of the Periodic Table the MCPF and CPF commonly agree. For those metal-containing systems where the two approaches agree, the wave function is expected to be highly accurate. The success of the MCPF method is much greater than that of the singles and doubles CI (SDCI) treatment because the large correlation energies result in important contributions from the higher than double excitations. We note the quadratic configuration-interaction (QCISD) method [and presumably the coupled-cluster singles plus doubles (CCSD) method], especially when an estimate for triple excitations is included, appear to be promising alternatives to the MCPF method [24].

Single-reference-based methods tend to be most applicable to transition metals on the far right or left of the row. Those metals in the middle of the row have many open-shell d electrons resulting in a high density of low-lying states. This causes a greater degree of mixing between states, resulting in wave functions that are multireference in character. For example, the MCPF method yields better results for $Fe(CO)_5$ and $Fe(CO)_4$, where the Fe atom is $3d^8$, than for $Fe(CO)_3$, $Fe(CO)_2$, and FeCO, where the Fe atom has a rather mixed occupation and there are more low-lying states [25]. The near-degeneracy effects caused by many low-lying states may also explain why the MP2 approach gives poor atomic separations for the transition metal atoms [26], whereas it gives reliable results for many first- and second-row molecules that are well-described by a single reference. Single-reference-based methods are, therefore, expected to work better for saturated transition metal systems, where all of the d electrons are involved in bonds.

3.4. RESULTS AND DISCUSSION

Because the bonding in transition metal systems commonly involves mixtures of several atomic states, it is important to account accurately for the atomic separations. Atomic calculations have been discussed in detail previously [24, 27, 28]. Here we simply note that in a DZ or TZ basis set with at least one polarization function, the CASSCF–MRCI, MCPF, and QCISD(T) methods all yield reasonably accurate separations between the $d^n s^2$ and $d^{n+1} s^1$ occupations. To obtain very accurate separations, larger polarization

Table 3.1
Experimental Atom and Ion Excitation Energies (in Electron Volts) for the First- and Second-Row Transition Metal Atoms[a]

	Lower State	Upper State	Expt.	Lower State	Upper State	Expt.
	Atom			Ion		
	First-row transition metal atoms					
Sc	$3d^14s^2(^2D)$	$3d^24s^1(^4F)$	1.43	$3d^14s^1(^3D)$	$3d^2(^3F)$	0.60
Ti	$3d^24s^2(^3F)$	$3d^34s^1(^5F)$	0.81	$3d^24s^1(^4F)$	$3d^3(^4F)$	0.10
V	$3d^34s^2(^4F)$	$3d^44s^1(^6D)$	0.25	$3d^4(^5D)$	$3d^34s^1(^5F)$	0.33
Cr	$3d^54s^1(^7S)$	$3d^44s^2(^5D)$	1.00	$3d^5(^6S)$	$3d^44s^1(^6D)$	1.52
Mn	$3d^54s^2(^6S)$	$3d^64s^1(^6D)$	2.14	$3d^54s^1(^7S)$	$3d^6(^5D)$	1.81
Fe	$3d^64s^2(^5D)$	$3d^74s^1(^5F)$	0.87	$3d^64s^1(^6D)$	$3d^7(^4F)$	0.25
Co	$3d^74s^2(^4F)$	$3d^84s^1(^4F)$	0.42	$3d^8(^3F)$	$3d^74s^1(^5F)$	0.43
Ni	$3d^94s^1(^3D)$	$3d^84s^2(^3F)$	0.03	$3d^9(^2D)$	$3d^84s^1(^4F)$	1.08
Cu	$3d^{10}4s^1(^2S)$	$3d^94s^2(^2D)$	1.49	$3d^{10}(^1S)$	$3d^94s^1(^3D)$	2.81
	Second-row transition-metal atoms					
Y	$4d^15s^2(^2D)$	$4d^25s^1(^4F)$	1.36	$5s^2(^1S)$	$4d^15s^1(^3D)$	0.15
Zr	$4d^25s^2(^3F)$	$4d^35s^1(^5F)$	0.59	$4d^25s^1(^4F)$	$4d^3(^4F)$	0.32
Nb	$4d^45s^1(^6D)$	$4d^35s^2(^4F)$	0.18	$4d^4(^5D)$	$4d^35s^1(^5F)$	0.33
Mo	$4d^55s^1(^7S)$	$4d^45s^2(^5D)$	1.47	$4d^5(^6S)$	$4d^45s^1(^6D)$	1.59
Tc	$4d^55s^2(^6S)$	$4d^65s^1(^6D)$	0.41	$4d^55s^1(^7S)$	$4d^6(^5D)$	0.52
Ru	$4d^75s^1(^5F)$	$4d^65s^2(^5D)$	0.87	$4d^7(^4F)$	$4d^65s^1(^6D)$	1.09
Rh	$4d^85s^1(^4F)$	$4d^9(^2D)$	0.34	$4d^8(^3F)$	$4d^75s^1(^5F)$	2.13
Pd	$4d^{10}(^1S)$	$4d^95s^1(^3D)$	0.95	$4d^9(^2D)$	$4d^85s^1(^4F)$	3.19
Ag	$4d^{10}5s^1(^2S)$	$4d^95s^2(^2D)$	3.97	$4d^{10}(^1S)$	$4d^95s^1(^3D)$	5.04

[a] The results are the *j*-averaged values reported in reference 18.

sets are required. Because we frequently refer to the atomic separations, we summarize the experimental values in Table 3.1.

3.4.1. Transition Metal Hydride Positive Ions

There have been many studies of the transition metal hydride positive ions [29–35]. Overall, there is good agreement between the different theoretical treatments. Most have been directed at determining the identity of the ground state and computing the dissociation energies for comparison with experiment. Although there are several studies on the low-lying excited states of the ions [34, 35], no experimental data are available for comparison. Therefore, we defer the discussion of spectroscopy to the neutral hydrides (see Section 3.4.5).

Very accurate dissociation energies have been obtained for the transition metal monohydride positive ions using the guided ion beam approach [36].

This experimental technique relies on the fact that an ion can be accelerated to a specific energy, so that an endothermic reaction such as

$$M^{+} + H_2 \rightarrow MH^{+} + H \tag{3.1}$$

can be studied as a function of the kinetic energy of M^{+}. By combining the reaction threshold with the known H_2 dissociation energy, an accurate D_0 for MH^{+} can be determined. This analysis assumes that there is no barrier to reaction, which is reasonable considering that the electrostatic charge-induced dipole interaction is attractive at large r values. It should be noted that the experimental analysis can be complicated by excited ions in the beam if they are more reactive than the ground state. This can result in the determination of incorrect thresholds and thus D_0 values. However, when these effects are properly accounted for, very accurate D_0 values can be derived from experiment.

Although reaction (3.1) is endothermic, the reaction

$$M^{+} + H_2 \rightarrow MH_2^{+} \tag{3.2}$$

is much less so or can even be exothermic. However, MH_2^{+} is not always observed experimentally. Rappé and Upton [37] considered both reactions (3.1) and (3.2) for Sc^{+}. They found that the abstraction reaction (3.1) proceeds with essentially no barrier for the ^3D ground state of Sc^{+} from any interaction geometry, thereby confirming the assumption of no barrier in the experimental analysis. Although four of the five components of the ^1D state reacted to form $ScH^{+} + H$, they found that one component of the ^1D state correlated with ground state ScH_2^{+}. This reaction had a 13-kcal-mol^{-1} barrier and a small exothermicity of 4.4 kcal-mol^{-1}, as compared with a 55.5-kcal-mol^{-1} endothermicity of reaction (3.1). They further noted that one of the fifteen spin–orbit components of the ^3D state of Sc^{+} correlates with the ground state of ScH_2^{+}. The barrier for this component of the ^3D state is computed to be 19 kcal-mol^{-1}. Thus ScH_2^{+} can be formed either directly from the ^1D state of Sc^{+}, because some ^1D is present in the beam, or from the ^3D state if the spin–orbit coupling is sufficient to promote crossing to the singlet surface. The failure to observe ScH_2^{+} experimentally is attributed to the fact that the products cannot accommodate the excess energy (caused by the barrier) under the experimental conditions [38].

Considering that a nearly complete set of accurate experimental D_0 values is available for MH^{+} [36], several theoretical calculations have been performed, partially for calibration purposes. Some of these studies treated all of the first- and second-row transition metal hydride ions at the same level of theory, so that trends in the bonding and D_0 values could be observed. The systematic studies allow a critical evaluation of the experimental data. We compare our MCPF D_0 values [29] for the first- and second-row MH^{+}

Table 3.2
MCPF Spectroscopic Constants (in a_0 and kcal mol^{-1}) and Metal d and Net Populations for the MH^+ Molecules

	State	r_e	d Population	Net	D_0	D_0 (Expt.[a])
ScH^+	$^2\Delta$	3.457	1.29	1.23	56.0	55.4(2.1)[b]
TiH^+	$^3\Phi$	3.289	2.34	1.11	53.3	53.3(2.5)
VH^+	$^4\Delta$	3.143	3.37	1.01	48.6	47.3(1.4)
CrH^+	$^5\Sigma^+$	3.031	4.45	0.91	27.7	31.6(2.0)
MnH^+	$^6\Sigma^+$	3.122	5.10	0.85	43.7	47.5(3.4)
FeH^+	$^5\Delta$	3.029	6.19	0.78	52.3	48.9(1.4)
CoH^+	$^4\Phi$	2.923	7.22	0.74	44.5	45.7(1.4)
NiH^+	$^3\Delta$	2.810	8.26	0.73	41.0	38.6(1.8)
CuH^+	$^2\Sigma^+$	2.730	9.45	0.48	18.5	21.2(3.0)
YH^+	$^2\Sigma^+$	3.576	0.80	1.00	59.4	60.2(1.4)[b]
ZrH^+	$^3\Delta$	3.459	1.97	1.10	56.7	54(3)
NbH^+	$^4\Delta$	3.295	3.57	1.09	52.5	53(3)
MoH^+	$^5\Sigma^+$	3.174	4.72	0.95	35.3	41(3)
TcH^+	$^6\Sigma^+$	3.198	5.18	1.00	50.7	
RuH^+	$^5\Delta$	3.158	6.35	0.95	37.8	40(3)
RhH^+	$^2\Delta$	2.864	7.99	0.77	41.5	35(3)
PdH^+	$^1\Sigma^+$	2.817	9.03	0.69	49.0	46(3)
AgH^+	$^2\Sigma^+$	3.442	9.83	0.88	10.6	15(3)

[a]Taken from the recent review of the literature by Armentrout and Georgiadis [36]; the values (with uncertainties in parentheses) have been converted from 300 K to 0 K.

[b]Revised values (Armentrout, private communication).

with experiment in Table 3.2. In this study several low-lying states were studied to ensure a correct determination of the ground state.

Except possibly for MoH^+, RhH^+, and AgH^+, the MCPF and experimental D_0 values agree to within their combined uncertainty, thereby confirming that the experimental values have been adjusted correctly for excited-state reactions. This also gives us considerable confidence in the comparable quality MCPF D_0 values for the neutral MH molecules.

The d populations in Table 3.2 illustrate that both the d^ns^1 and d^{n+1} atomic states potentially contribute to the ground-state wave functions of MH^+. The d population is less than 1 for the $^2\Sigma^+$ state of YH^+, because the lowest two atomic states of Y^+ are $^1S(5s^2)$ and $^3D(4d^15s^1)$. For Mn^+, Rh^+, Pd^+, and Ag^+, the large separation between the d^ns^1 and d^{n+1} atomic states results in relatively little mixing. The bonding orbital is also a mixture of different angular momentum contributions. For ScH^+ through CrH^+, the average composition of the bonding orbital is 46% $4s$, 16% $4p$, and 38% $3d\sigma$. The composition of the bonding orbital for the second half of the row is essentially the same, 51% $4s$, 13% $4p$, and 36% $3d\sigma$, at the MCPF level. Note, however, that a much smaller average d population of 14% is obtained at the generalized valence bond (GVB) level for the first transition row [31]. This difference between the MCPF and GVB results reflects the importance of including d–d correlation on the right side of the row. Although the computed D_0 values are not very sensitive to this difference, other properties such as dipole moments can change substantially. The d populations in the second row are considerably larger, because the d^{n+1} occupation is stabilized relative to the $d^n s^1$ occupation. In addition, there is a reduction of the $\langle r \rangle_s / \langle r \rangle_d$ ratio for the second row. Both of these factors favor a larger d contribution to the bonding orbital. The MCPF populations reflect this: The bonding orbital is 56% $4d\sigma$ for YH^+ through MoH^+ and 78% $4d\sigma$ for TcH^+ through PdH^+.

The mixing of atomic states is an important factor in determining the ground state [29]. The identity of this state usually can be predicted by considering the molecular states that arise from the low-lying metal asymptotes. Any state that arises from only one asymptote usually can be excluded from consideration. Also, the difference in ground states between the corresponding first- and second-row transition metal systems can generally be understood in terms of a differing order of atomic states. For example, the $X\,^2\Delta$ state of ScH^+ arises from a mixture of the $3d\delta^1 4s^1$ component of the $^3D(3d^1 4s^1)$ state and the $3d\sigma^1 3d\delta^1$ component of the $^3F(3d^2)$ state. This results in a bonding orbital that is a mixture of the $4s$ and $3d\sigma$ orbitals, which is reflected in the d population in Table 3.2. The strongest bond is thus formed by mixing an optimal amount of s and d character. For YH^+, although the bonding orbital is again a mixture of s and d, the ground state is $X\,^2\Sigma^+$, because the mixing is determined by the $^1S(5s^2)$ and $^3D(4d^1 5s^1)$ states of Y^+. The difference in the ground state for Co versus Rh and Ni versus Pd is due to the enhanced stability of the d^{n+1} states of Rh and Pd.

The reason that TiH^+ has a $^3\Phi$ ground state whereas ZrH^+ has a $^3\Delta$ ground state, even though the metal ions have the same ordering of atomic states, is more subtle. The preference for a $^3\Delta$ ground state for ZrH^+ is due to the greater ease of forming the sd hybrid orbitals in the second row. For example, an sd hybrid can be formed by using the $4d\sigma^1 4d\delta^1 5s^1$ occupation derived from the ^4F state of Zr^+. For Zr^+, the slightly larger $d^n s^1$–d^{n+1} separation, coupled with the greater size of the $4d$ orbital leads to a preference for this bonding mechanism, thereby resulting in a $^3\Delta$ ground state. However, the $^3\Delta$–$^3\Phi$ separation is small for both TiH^+ and ZrH^+.

Although both CuH^+ and AgH^+ have $^2\Sigma^+$ ground states, the nature of the bonding is considerably different. The bonding in CuH^+ involves a mixture of the $3d^{10}$ and $3d^9 4s^1$ occupations of Cu^+. That is, Cu^+ is promoted to the $3d^9 4s^1$ occupation to form a two-electron bond. For Ag^+, the promotion energy is so large that the bonding arises predominantly from the $4d^{10}$ occupation. Thus only a one-electron bond is formed by H donating charge into the empty Ag $5s$ orbital. This results in a D_0 value that is significantly smaller than for any other MH^+. The D_0 of CuH^+ is also relatively small even though it forms a two-electron bond, because of the large (2.81 eV) promotion energy [18].

The net population shows that, in spite of its positive charge, the metals on the left side of the row still donate electrons to the hydrogen. However, as the ionization potential (IP) of the metal increases with increasing Z, the metal donation decreases and eventually H donates to the metal. The computed dipole moments given in reference 29 support this trend, showing it is not a population artifact.

With a few exceptions, the bond lengths decrease with increasing Z. One exception is Pd–Ag, where the bonding changes from being two-electron to one-electron in character. The other exceptions are Cr–Mn and Mo–Tc, where a $d\sigma$ electron is added, increasing the repulsion.

Although there are trends in the r_e values and populations with increasing Z, the D_0 values have an irregular pattern. This is related to the differing atomic separations for the ions and the fact that the loss of exchange energy with bond formation depends on the number of open shells. Elkind and Armentrout [39] accounted for this in their analysis by referring all of the D_0 values to the d^ns^1 asymptote. After accounting for the loss of exchange energy, they observed a trend in the D_0 values, except for the Rh–Pd metals in the second row. Our calculations show that Rh–Pd metals are exceptions: They have little contribution from the d^ns^1 state, but rather bond primarily from the d^{n+1} ground states by formation of a $d\sigma$-H bond.

3.4.2. Transition Metal Hydrides

Previous theoretical studies [40–49] of the neutral transition metal hydrides have focussed on determining the identity and spectroscopic constants (r_e, ω_e, D_0) of the ground state, as well as gaining an understanding of the nature of the bonding. A few studies also have been directed toward elucidating the spectroscopy of these molecules.

Comparable MCPF calculations have been carried out for both the first- and second-row transition metal hydride neutrals and ions [15, 29, 40]. The Mulliken d populations indicate considerable mixing of atomic asymptotes for both the neutrals and ions. However, there are again many differences between the corresponding first- and second-row systems (Table 3.3). For example, TiH and ZrH have different ground states, as do TiH^+ and ZrH^+. The ZrH bond arises from the $^3F(4d^25s^2)$ state by the formation of sd hybrids, whereas the bonding in TiH involves a mixture of the $^3F(3d^24s^2)$ and $^5F(3d^34s^1)$ atomic states. The different bonding mechanisms are reflected in the d populations (2.3 for TiH and 1.90 for ZrH). Even though the separation between atomic states is smaller in Zr, the greater extent of the $4d$ orbital in Zr leads preferentially to sd hybridization. In contrast, both ScH and YH involve the same bonding mechanism, namely sd hybridization leading to $^1\Sigma^+$ ground states.

Table 3.3
Summary of Theoretical Spectroscopic Constants for the Ground States of the First- and Second-Row Transition Metal Hydrides

Molecule	State	r_e (a_0)	ω_e (cm^{-1})	D_e (eV)	μ (D)	nd
ScH	$^1\Sigma^+$	3.390	1587	2.27	1.37	0.84
TiH	$^4\Phi$	3.440	1548	2.06	2.19	2.30
VH	$^5\Delta$	3.249	1635	2.33	2.02	3.40
CrH	$^6\Sigma^+$	3.201	1647	2.13	3.81	4.83
MnH	$^7\Sigma^+$	3.313	1530	1.67	1.24	5.05
FeH	$^4\Delta$	2.973	1915	1.67	2.90	6.52
CoH	$^3\Phi$	2.895	1842	1.94	2.74	7.60
NiH	$^2\Delta$	2.807	1987	2.69	2.56	8.65
CuH	$^1\Sigma^+$	2.851	1852	2.63	2.95	9.80
YH	$^1\Sigma^+$	3.706	1559	2.95	1.54	0.74
ZrH	$^2\Delta$	3.509	1483	2.45	1.23	1.90
NbH	$^5\Delta$	3.384	1583	2.60	2.45	3.69
MoH	$^6\Sigma^+$	3.300	1642	2.19	3.03	4.89
TcH	$^5\Sigma^+$	3.158	1797	1.95	2.15	5.35
RuH	$^4\Phi$	3.114	1801	2.34	2.70	7.00
RhH	$^3\Delta$	2.976	2057	2.81	2.24	8.10
PdH	$^2\Sigma^+$	2.911	1958	2.22	2.03	9.24
AgH	$^1\Sigma^+$	3.130	1703	2.22	2.95	9.80

The MnH and TcH molecules also have different ground states. The MnH molecule has a $^7\Sigma^+$ ground state derived from the ^6S state of Mn. The $^5D(3d^6)$ state of Mn is too high in energy to contribute significantly to the wave function. However, the ground state of TcH is $^5\Sigma^+$, because the ^5D state of Tc is much lower lying and mixes significantly into the wave function. For Fe–Ru to Ni–Pd, the difference in the hydride ground state is due to the different order of atomic states. For example, the bonding in NiH is a mixture of $^3D(3d^94s^1)$ and $^3F(3d^84s^2)$, whereas PdH is a mixture of $^1S(4d^{10})$ and $^3D(3d^94s^1)$. Thus the ground state of NiH is $^2\Delta$, whereas PdH has a $^2\Sigma^+$ ground state.

As for the MH^+ systems, the MH bond lengths generally decrease with increasing Z. There is an exception at TiH, because the $X\,^1\Sigma^+$ state of ScH has an unusually short bond length, as a result of the *sd* hybridization on Sc. The bond length also increases at MnH, CuH, and AgH. This is a consequence of the large atomic separations in Mn, Cu, and Ag, which results in less mixing. Note that the bond length does not increase at TcH, because the atomic separation in Tc is relatively small.

The D_e values, when referred to a common asymptote, indicate that the bonds are stronger for the second row and that the bond strength decreases with increasing Z. These trends are similar to those for the positive ions, but there is no systematic experimental study of all the neutrals, so that the theoretical D_e values are the most consistent set available.

3.4.3. Dipole Moments

The electric dipole moment provides a measure of the quantity of the wave function, because it is a sensitive measure of the relative mixings of the atomic asymptotes [42]. This is because bonding from the $d^{n+1}s^1$ metal asymptote involves a metal s-H s bond that is polarized toward the hydrogen. This leads to a large dipole moment of M^+H^- polarity in molecules such as CrH, which involves almost exclusively the $3d^54s^1$ asymptote. On the other hand, the d^ns^2 asymptote bonds by the formation of sp hybrids. Although the bonding hybrid is polarized toward H, this is balanced by the polarization of the nonbonding hybrid orbital away from hydrogen. This results in a relatively small dipole moment for MnH, for example, which bonds from the $3d^54s^2$ atomic state.

In the remainder of this subsection, we discuss the dipole moment for the $X\,^2\Delta$ ground state of NiH, which is very sensitive to the relative contributions of the nearly degenerate $^3F(3d^84s^2)$ and $^3D(3d^94s^1)$ atomic states of Ni. Because of this and the fact that the dipole moment of NiH has been measured [50] to be 2.4 ± 0.1 D, correctly computing this property has been the goal of many calculations.

The problem of obtaining an accurate dipole moment for the NiH ground state is related to that of accurately computing the atomic separations in the Ni atom [27–29]. At the SCF level the ^{3}F state is greatly favored and it is only with the inclusion of extensive correlation that an accurate ^3F–^3D atomic separation is obtained. Thus the SCF treatment of NiH, which favors the ^3F state, might be expected to produce a small dipole moment. However, because of lack of flexibility, the SCF wave function can only describe a metal s-H s bond, which results in a large dipole moment (around 4.0 D). If the flexibility of the wave function is increased by performing a small CASSCF calculation, the ^3F state is now favored, because the CASSCF does not include extensive correlation. This results in a dipole moment that is smaller than experiment.

In Table 3.4 we summarize the spectroscopic constants for NiH for various levels of correlation treatment in a large ANO basis set [51]. All CASSCF treatments have one active orbital on the hydrogen and thus differ only in their treatment of Ni. The MRCI reference wave function includes all configurations in the CASSCF wave function with coefficients greater than 0.05. The ACPF reference wave function includes, in addition, all configurations that appear in the ACPF wave function with a coefficient greater than 0.05. When the CASSCF active space includes only the Ni $3d\sigma$, $3d\delta$, and $4s$ orbitals, the spectroscopic constants are poor, because the $3d\sigma$ and bonding orbitals share one correlating orbital. Adding the $4p\sigma$ orbital to the CASSCF active space eliminates this problem, and the spectroscopic constants, exclud-

Table 3.4
Spectroscopic Constants for the $X^2\Delta$ state of NiH

Calculation	r_e (a_0)	ω_e (cm^{-1})	μ (D)
CASSCF	2.975	1580	1.544
MRCI	2.663	3096	2.241
ACPF	2.707	2603	2.493
CASSCF($4p\sigma$)	3.036	1508	0.876
MRCI($4p\sigma$)	2.659	2016	1.861
ACPF($4p\sigma$)	2.721	2107	2.559
CASSCF($4p$)	2.997	1460	1.024
MRCI($4p$)	2.677	2022	1.830
ACPF($4p$)	2.727	2060	2.515
CASSCF($3d$–$3d'$)	2.912	1664	3.310
MRCI($3d$–$3d'$)	2.778	1894	3.056
ACPF($3d$–$3d'$)	2.780	1918	2.355
Experiment	2.76[a]	2003[b]	2.4 ± 0.1[c]

[a] Reference 44.
[b] Reference 52.
[c] Reference 50.

ing the dipole moment, are much improved. The dipole moment is much too small at the CASSCF level and, although correlation improves it, it is still more than 0.6 D too small at the MRCI level. The ACPF dipole moment is just slightly too large based on this CASSCF space. Adding the $4p\pi$ orbital to the CASSCF active space to account for the angular correlation of the bonding orbital improves the CASSCF results, but the MRCI and ACPF results are not significantly changed. However, including $3d$–$3d$ correlation in the CASSCF calculation improves the results significantly. The MRCI dipole moment in this active space is too large, because the treatment now favors the $3d^9 4s^1$ occupation. However, the ACPF results are changed by only about 0.2 D from those in the smaller active spaces.

The theoretical calculations [51] show that a substantial orbital bias is introduced even at the CASSCF level, which is difficult to overcome in the correlation treatment. Another method to overcome this orbital bias, apart from the ACPF, is the iterative natural orbital (INO) procedure off the MRCI wave function. The INO approach gave good agreement with the full configuration interaction (FCI) calculations for the dipole moments of the $^4\Phi$ and $^2\Delta$ states of TiH [21]. However, the INO approach can be time-consuming, as five or more iterations are commonly required to reach convergence. The NO-MRCI dipole moment for NiH at $r = 2.7\ a_0$ based on the CASSCF($4p$) treatment is 2.69 D, which is in better agreement with experiment than the MRCI expectation value of 1.83 D. Note that the converged NO-ACPF value of 2.41 D is in excellent agreement with the ACPF expectation value of 2.49 D at this r value. The finite-field approach yields 3.10 D and 3.05 D at the MRCI and MRCI + Q levels. This error is probably a result of the MRCI treatment having an incorrect balance of the $3d^8 4s^2$ and $3d^9 4s^1$ atomic states, so that the field response is incorrect. This is supported by the MRCI and MRCI + Q results of 2.84 and 2.68 D when the superior CASSCF($3d$–$3d'$) treatment is used. However, the ACPF field value of 2.66 D is in good agreement with the ACPF expectation value. Thus it is clear that the INO and ACPF methods are the most promising for the computation of accurate dipole moments. We recommend at least one NO iteration be performed to assess the convergence of an MRCI value.

3.4.4. $X\,^4\Delta$–$a\,^6\Delta$ State Separation in FeH

Ab-initio calculations are now capable of determining energy separations between low-lying states of first- and second-row molecules to within a few hundred wave numbers [20]. This has allowed the determination of the ground states of the Al_2 [53] and Si_2 [54] molecules, even though the separation between the lowest two states in each case is very small. However, the accurate determination of energy separations for transition metal systems is more difficult. The worst case is when the states are of different multiplicities, because high-spin states generally arise from one atomic state, whereas the states of lower spin involve a mixture of several atomic states and have larger correlation energies. These concepts are well-illustrated by calculations [41] on the FeH molecule.

The ground state of FeH is known to be $^4\Delta$ from magnetic deflection experiments [55]. The photodetachment experiments of Stevens, Feigerle, and Lineberger [56] place the $a\,^6\Delta$ state at 0.24 eV. However, most theoretical calculations place the $^6\Delta$ state lower in energy.

The $^6\Delta$ state is derived from the $^5D(3d^6 4s^2)$ ground state of the Fe atom: The Fe atom undergoes sp hybridization and one hybrid bonds with the hydrogen. In the $^4\Delta$ state, the electron in the nonbonding sp hybrid is low-spin coupled to the $3d$ shell, allowing the $^5F(3d^7 4s^1)$ state to contribute to the wave function. The bonding then involves some d character, which enhances it sufficiently to compensate for the promotion energy and loss of sd exchange energy. This illustrates the importance of the mixing of atomic states, because a $^6\Delta$ ground state is predicted from Hund's rules if only the contribution from the lower atomic state is considered.

We now consider what level of theory is required to predict the $X\,^4\Delta$–$a\,^6\Delta$ separation correctly. The correct positioning of these two spin manifolds is important for an understanding of the spectroscopy of FeH (see Section 3.4.5). An accurate determination of the $X\,^4\Delta$–$a\,^6\Delta$ separation in FeH, requires that the corresponding atomic separations can be accurately determined. Benchmark calculations indicate that this requires a large one-particle basis set. Further, the benchmark calculations [57] show that $3s3p$ correlation affects the atomic separation. Thus we have performed a series of calculations in a large $(20s15p10d6f4g)/[7s6p4d3f2g]$ Fe and $(8s6p4d)/[4s3p2d]$ H ANO basis set. The MCPF separation is -0.01 eV (where the minus sign indicates that the $^6\Delta$ is lower) in this basis, as compared with -0.10 eV obtained in the Wachters-derived basis set. This is a dramatic improvement over the SCF separation of -2.24 eV. A CASSCF calculation including the Fe $3d$, $4s$, and $4p$ and the H $1s$ in the active space gives a separation of -0.62 eV, which is also a substantial improvement over SCF. The CASSCF–MRCI separation in the large ANO basis set is -0.09 eV, which is not quite as good as the MCPF value. The problem is that the CASSCF does not include any $3d$–$3d$ correlation and is therefore biased toward the $3d^64s^2$ occupation. Inclusion of the multireference Davidson correction ($+$Q) for higher excitations helps account for this orbital bias; the MRCI + Q separation is 0.08 eV. Unfortunately, it is prohibitively expensive to include $3s3p$ correlation at the MRCI level. However, because the MRCI and MCPF separations agree rather well, we investigated the importance of $3s3p$ correlation at the MCPF level. We find that it favors the $^4\Delta$ state by 0.10 eV. Relativistic effects were found to have only a small effect on the separation (0.02 eV, favoring the $^6\Delta$ state). Adding the MCPF estimate of the $3s3p$ effect to the MRCI + Q separation and including the estimate of relativistic effects leads to a separation of 0.15 eV, which is 0.1 eV smaller than the experimental separation. Thus the calculations on FeH indicate that very high levels of theory are required to obtain accurate state separations between states of different spin. In contrast, the $^3B_1$–$^1A_1$ separation in CH_2 can be computed to within 100 cm^{-1} at a fraction of the cost [58].

3.4.5. Spectroscopy of the Transition Metal Hydrides

The spectra of the transition metal hydrides tend to be very complex as a result of the large number of low-lying states; even ScH, which can have no more than two electrons in molecular orbitals derived from the atomic d shell, has more than 30 states below 3 eV [47]. The transition metal hydride spectra are further complicated by large spin–orbit and hyperfine interactions, although a few electronic transitions have been rotationally analyzed. In this subsection we illustrate the spectroscopy of first-row transition metal systems by considering ScH and FeH.

The singlet and triplet states of ScH up to about 4 eV in energy have been investigated by Anglada, Bruna, and Peyerimhoff [47]. They assign an experimental absorption band at 18,000 cm^{-1} to the transitions $X^1\Sigma^+$–$(2)^2\Sigma^+$ and $X^1\Sigma^+$–$(2)^1\Pi$. Two emission regions (12,000 and 18,000 cm^{-1}) are assigned to several possible combinations between higher-lying upper states and all triplet and singlet states from the first dissociation limit. The reason that more definitive assignments cannot be made is that there are a large number of dipole-allowed transitions carrying substantial oscillator strength occurring in a narrow energy range. This is a consequence of the large number of ScH states that can be derived from the atomic states below 2.5 eV (see Table III of reference 47). Some insight into these complex spectra is obtained by considering the hydride to be a metal atom perturbed by the H atom. Thus the molecular states can be associated with well-defined atomic states, particularly for the high-spin states where the metal retains its atomic coupling. This concept is very useful for explaining the sextet potentials of FeH [59] and the septet potentials of MnH [60], for example.

The diversity of bonding in the low-spin states requires accurate calculations to explain the spectra. Consider, for example, the low-lying states of ScH. The $^2D(3d^14s^2)$ state of Sc can bond by either *sd* or *sp* hybridization. The most favorable mechanism is *sd* hybridization, which leads to the $X\,^1\Sigma^+$ ground state, because this avoids the $4s$ to $4p$ promotion energy required to form *sp* hybrids. When *sp* hybridization occurs, one hybrid orbital bonds with H, while the nonbonding orbital contains a single electron. The electron in the nonbonding *sp* orbital can be either singlet or triplet coupled to the spectator $3d$ electron, leading to the $^{1,3}\Delta$, $^{1,3}\Pi$, and $^{1,3}\Sigma^+$ states. The exchange energy is small because of the very different spatial extent of the $3d$ and *sp* orbitals. Furthermore, because the $3d$ orbital is nonbonding, the states with different angular momentum are close in energy. The energy separation between the singlet and triplet Π and Δ states is less than 0.4 eV, and excluding the $(2)\,^1\Sigma^+$ state only 0.69 eV separates the other five states. The $(2)\,^1\Sigma^+$ state is more than an additional electron volt higher in energy, because the $X\,^1\Sigma^+$ ground state mixes in a small amount of *sp* hybridization, thus pushing the $(2)\,^1\Sigma^+$ state up in energy. This places the $(2)\,^1\Sigma^+$ state derived from the $^2D(3d^14s^2)$ atomic state in the same energy region as the manifold of states derived from the ^4F and ^2F atomic states with a $3d^24s^1$ occupation and the ^4F, ^4D, and ^2D atomic states with a $3d^14s^14p^1$ occupation. These large shifts in energy that arise from the mixing of asymptotes complicate the identification of the origin of the states, and therefore calculations can serve a unique role in helping to interpret spectra.

The spectra of the transition metal hydrides in the middle of the row, such as FeH, can also be very complex [59], due to the large number of open-shell *d* electrons. This is further complicated by the large differential correlation contributions from states derived from the $3d^64s^2$ and $3d^74s^1$ atomic states. This does not present a problem in the sextet manifold, because all of the states below about 25,000 cm^{-1} are derived from either the $3d^64s^2$ or $3d^64s^14p^1$ occupations. Because all the states have six $3d$ electrons, the differential correlation effects are small, and even the CASSCF T_e values are reasonably accurate. However, the quartet states are much more difficult to treat accurately, because there are states derived from both d^6s^2 and d^7s^1 parentage. For example, the three lowest quartet states are derived principally from the ^5D state of Fe, whereas the next five states are derived from the ^5F state. At the SA-CASSCF level, the lowest state derived from the ^5F asymptote is more than 18,000 cm^{-1} above the ground state but once correlation is included, the $B\,^4\Sigma^-$ and $C\,^4\Phi$ states derived from the ^5F state lie below the $D\,^4\Sigma^+$ state derived from the ^5D state. Thus there is over 10,000 cm^{-1} of differential correlation between the states derived from d^6s^2 and d^7s^1.

Experimentally, only one transition, the so-called infrared system, has been rotationally analyzed for FeH [61, 62]. Theoretical calculations confirm the experimental assignment of this band system to a $^4\Delta$–$^4\Delta$ transition. The visible region of FeH is dominated by two regions of complex structure in the green and blue centered at 5320 and 4920 Å, respectively [63, 64]. As was the case for ScH, there are several dipole-allowed transitions in this energy region, so that a definitive assignment of these transitions is not possible. However, the theoretical calculations present several well-defined alternatives, which may aid spectroscopists in assigning these band systems.

3.4.6. Transition Metal Dihydrides

In this section we consider the nature of the bonding in the transition metal dihydrides [3, 65–71] and their positive ions [72–74], as well as the insertion reaction of a transition metal into the H_2 molecule. Although these systems have not been studied as systematically as the monohydrides, these systems are interesting as models for the effect of ligands.

As for the transition metal monohydrides, the bonding in the dihydrides is quite diverse. The d^ns^2 occupation of the metal bonds to a hydrogen atom by the formation of *sp* hybrid orbitals, leading to a linear molecule with two metal–H bonds. This mechanism has a favorable exchange energy, as the metal *d* electrons retain their high-spin atomic coupling. Thus, linear ground states are observed for the ScH_2 [66], FeH_2 [65], CoH_2 [65], NiH_2 [3], and TcH_2 [71] molecules, because these atoms have low-lying d^ns^2 states. However, there is still some *d* character in the bonding orbital. This results, for example, in a $^2\Sigma_g^+$ ground state for ScH_2, whereas a $^2\Delta_g$ state would occur if the $3d$ electron was strictly a spectator.

Because the ground state of Y is $^2D(4d^15s^2)$, YH_2 might be expected to be linear like ScH_2. However, the ground state structure is bent [67], because *sd* hybridization is more favorable for Y, due to the more comparable size of the *d* and *s* orbitals for Y than Sc. The optimal bond angle for *sd* hybrids is 90°, but because the bonding involves a mixture of $4d$, $5s$, and $5p$, the actual bond angle is 123°. The same is true of Fe versus Ru, where FeH_2 is linear and RuH_2 is bent [71].

As the $d^{n+1}s^1$ and d^{n+2} occupations become more stable, the bonds involve more *d* character. Thus the bond angle is reduced, becoming only 84° for RhH_2 [68]. A bond angle smaller than 90° indicates that the bond is more *d* than *s* in character. Recall that the optimal angle for two bonds arising from pure metal *d* orbitals is 60°. For Cu [65] and Ag [70], where there is a large promotion energy required to reach the d^9s^2 occupation, the MH_2 structure is higher in energy than $M + H_2$.

The positive ions of the monohydrides contain more *d* character in the bonds than the neutrals, because of the enhanced stabilization of the *d* orbitals in the metal positive ions. This same trend is observed for the dihydride positive ions and results in smaller bond angles. For example, the ground state of ScH_2^+ [30] has a bond angle of 107°, as compared with a linear structure for the neutral. This decrease even occurs for YH_2 where both Y and Y^+ have s^2 occupations in the ground state; YH_2^+ [76] has an angle that is 7° smaller than YH_2. For NbH_2 [69] and NbH_2^+ [74] where the bonding is *sd* in origin for both, there is still a large reduction from 127° to 105°.

In the case where the ground state of the ion is derived from the d^{n+1} occupation, it is possible to form two *d* bonds. However, this does not occur often, because the bonding is generally enhanced by mixing the d^ns^1 character. Schilling, Goddard, and Beauchamp [72] have found both *sd* and *dd* bonding in MoH_2^+. The bond angles for these two bonding mechanisms are quite different, 112° and 65°, even though the Mo—H bond lengths are essentially the same. Note that there is no chemical H—H bond even for the small-angle structure; the H—H bond distance is 1.82 Å. The two structures are of nearly equal stability, with only a small barrier to interconversion. Two distinct structures are not observed for CrH_2^+, but the bending potential is very flat, indicating that both bonding mechanisms are present, but that there is no barrier between them.

Although these calculations on selected systems have given insight into the bonding of the dihydrides, there has not been a study of all of the systems at the same level of treatment. However, some additional insight into the dihydride positive ions can be obtained from the dimethyl positive ions, because calculations have shown that the monohydride and monomethyl positive ions are quite similar [28]. In all of the dimethyl positive ions, the bonding is *sd* in origin, leading to a bent molecule [75]. The common bonding mechanism also results in the same ground states for the first- and second-row transition metal systems. The two metal–carbon bonds are of a_1 and b_2 symmetry. This leaves four nonbonding metal valence orbitals, $(1)a_1$, $(2)a_1$, b_1, and a_2. Because Sc^+ and Y^+ have only two valence electrons, both $Sc(CH_3)_2^+$ and $Y(CH_3)_2^+$ have singlet ground states with both metal electrons in the bonding orbitals. As more electrons are added, they fill the nonbonding valence orbitals in the order $(1)a_1 > a_2 > b_1 > (2)a_1$. For Ti^+ (Zr^+) to Mn^+ (Tc^+) the open-shell electrons are high-spin coupled. From Fe^+ (Ru^+) to Cu^+ (Ag^+), the added electrons then begin to pair in the same order as the first electron. From the small separations between some of the excited states, it is clear that the nonbonding orbitals are relatively close in energy. This bonding mechanism and resulting ground state is also consistent with the available data for the dihydrides and we expect it will be true for all of them. This is an indication of how the bonding becomes simpler with increasing number of metal ligand bonds.

The most important step in a catalytic process frequently involves a bond-cleavage reaction such as insertion of a metal into the H_2 bond. For Cu there is a large barrier to insertion, because the reaction is Woodward–Hoffmann forbidden [65]. This is also true for Ni(^3F) + H_2, if the reaction proceeds on the $^3A_1$ potential surface [3]; a barrier of 80 kcal mol^{-1} occurs, as the 3d electrons are essentially spectators. The barrier is reduced to about half this value for the $^3B_1$ potential, because the reaction proceeds not by a two-electron jump, but rather by a three-step process, where the electrons that are involved in the Ni—H bonds move in one-electron jumps. Although the 3d electrons are not directly involved in the bonding, they facilitate the reaction by changing their d occupation to make the process symmetry-allowed. Although this mechanism is potentially important for small metal clusters, it does not occur for saturated organometallic systems, because the d electrons are involved in bonds and cannot change their occupation. However, the Ni + H_2 reaction also can occur by another mechanism that retains a fixed d occupation. This alternative mechanism involves the excited $^1D(3d^94s^1)$ state of Ni. The d and s open-shell electrons are low-spin coupled, allowing the formation of sd hybrids, and therefore two bonds can form. Because there is no electron jump associated with this reaction, there is a low barrier. The $^1A_1$ product of this reaction is a bent molecule due to the sd hybridization of Ni. These two mechanisms for the H_2 insertion are not unique to Ni, but can also occur for Fe, Co, Ru, and Rh [65, 68, 71]. Note also that the low-spin mechanism can occur for saturated organometallic systems. Thus, studies of this reaction mechanism can give insight into the effect of ligands on reaction rates, as we show in the next subsection.

3.4.7. Metal Dihydrides Including Ligands

In this subsection we consider the effect of ligands on transition metal hydrogen bonding. By either stabilizing or destabilizing the M—H bond, ligands can significantly affect the exothermicity of chemical reactions. Low and Goddard [17] considered the reductive H–H coupling reactions,

$$MH_2 \rightarrow M + H_2 \tag{3.3}$$

and

$$M(PH_3)_2H_2 \rightarrow M(PH_3)_2 + H_2 \tag{3.4}$$

for the Pd and Pt metals. The bonding in both PtH_2 and PdH_2 was found to involve sd hybrids derived from the d^9s^1 occupation. Spin is conserved in both reactions so that the product metal atom in (3.3) is the $^1S(d^{10})$ state.

This favors the Pd atom, which has a $^1S(d^{10})$ ground state, over Pt, which has a $^3D(d^9s^1)$ ground state. Thus both reactions are exothermic for Pd, but endothermic for Pt. Further, Low and Goddard showed that the PH_3 groups preferentially stabilize the d^{10} occupation, because the PH_3 groups donate electrons into the s and p orbitals. Therefore, the $M(PH_3)_2$ product in (3.4) is substantially stabilized by the phosphine ligands; reaction (3.4) is 21 kcal mol^{-1} less endothermic than (3.3) for Pt. Thus, it is a useful concept to think of the effect of ligands in terms of their stabilizing effect on molecular states derived from different atomic occupations, in determining their effect on the energetics of the reductive coupling process.

Blomberg, Schüle, and Siegbahn [76] have also considered the effect of ligands on metal–hydrogen bond energies for Ni and Pd systems. They argued that there is a correlation between the amount of covalency in the metal-ligand bonding and the destabilization effect of the ligand on the metal–H bond strength. To facilitate reductive elimination, they find that strongly covalently bound ligands like olefins should be added. Their view that stabilization (destabilization) effects are related to the degree of covalency in the metal–ligand bonds is consistent with the view of Low and Goddard that these effects should be viewed as affecting the relative stability of the d^ns^1 and d^{n+1} derived states. For example, covalent character results in some sd character in the ligand–metal bond, which favors the d^9s^1 state. Thus, whether one considers the covalent character, which favors occupations with s electrons, or the dative character, which favors states without s electrons, the character of the bond is largely determined by how the ligands affect the relative contribution of the atomic states.

3.5. CONCLUSIONS

In this chapter we have shown that the bonding in the transition metal hydrides is largely dictated by the relative positioning of the atomic states of the metal, because this primarily determines the character of the bond, which ranges from almost pure sp hybrid orbitals to nearly pure d orbitals. The bond dissociation energies are determined by these considerations and the loss of atomic exchange energy. The excellent agreement between the theoretical values for the transition metal monohydride positive ions and the experimental values determined from the guided ion beam approach supports the accuracy of the theoretical calculations as well as the assumptions made in the experimental analysis.

Molecular properties such as dipole moments are also found to be very sensitive to the relative mixing of atomic states and thus the level of theoretical treatment. The ACPF and INO methods have proven to be the most reliable for the computation of dipole moments. In fact, we recommended at least one natural orbital iteration be performed to ensure that calculated properties are relatively invariant to the molecular orbital basis.

Another challenge for theory is the calculation of T_e values for transition metal systems. This proves to be most difficult when the states are of different multiplicities. This is a particular problem for FeH, where the $X\,^4\Delta$ and $a\,^6\Delta$ states are separated by only 0.24 eV.

Although very sophisticated theoretical calculations are required for quantitative results on many transition metal systems, especially for those transition metals in the middle of the row, larger saturated organometallic systems actually appear to be more tractable, because the metal *d* electrons are involved in the bonding. Thus less mixing of atomic states occurs and the wave functions are less multireference in character so that single-reference-based methods can often give quantitative results.

REFERENCES

1. Langhoff, S. R.; Bauschlicher, C. W. *Ann. Rev. Phys. Chem.* **1988** *39*, 181.
2. Merer, A. J. *Ann. Rev. Phys. Chem.* **1989**, *40*, 407.
3. Blomberg, M. R. A.; Siegbahn, P. E. M. *J. Chem. Phys.* **1983**, *78*, 5682.
4. Partridge, H. *J. Chem. Phys.* **1989**, *90*, 1043.
5. Almlöf, J.; Taylor, P. R. *J. Chem. Phys.* **1987**, *86*, 4070.
6. Wachters, A. J. H. *J. Chem. Phys.* **1970**, *52*, 1033.
7. Hay, P. J. *J. Chem. Phys.* **1977**, *66*, 4377.
8. Tavouktsoglou, A. N.; Huzinaga, S. *J. Chem. Phys.* **1980**, *72*, 1385. Tatewaki, H.; Huzinaga, S. *J. Chem. Phys.* **1979**, *71*, 4339.
9. Pettersson, L.; Wahlgren, U. *Chem. Phys.* **1982**, *69*, 185.
10. Krauss, M.; Stevens, W. J. *Ann. Rev. Phys. Chem.* **1984**, *35*, 357.
11. Hay, P. J.; Wadt, W. R. *J. Chem. Phys.* **1985**, *82*, 299.
12. Cowan, R. D.; Griffin, D. C. *J. Opt. Soc. Am.* **1976**, *66*, 1010.
13. Rohlfing, C. M.; Hay, P. J.; Martin, R. L. *J. Chem. Phys.* **1986**, *85*, 1447.
14. Bauschlicher, C. W.; Taylor, P. R.; Komornicki, A. *J. Chem. Phys.* **1990**, *92*, 3982.
15. Langhoff, S. R.; Pettersson, L. G. M.; Bauschlicher, C. W.; Partridge, H. *J. Chem. Phys.* **1987**, *86*, 268.
16. Martin, R. L. *J. Phys. Chem.* **1983**, *87*, 750.
17. Low, J. J.; Goddard, W. A. *Organometallics* **1986**, *5*, 609. Low, J. J.; Goddard, W. A. *J. Am. Chem. Soc.* **1986**, *108*, 6115.
18. Moore, C. E. *Atomic Energy Levels*; US National Bureau of Standards: Washington, DC; (US) Circ. no. 467.
19. Wang, S. W.; Pitzer, K. S. *J. Chem. Phys.* **1983**, *79*, 3851.
20. Bauschlicher, C. W.; Langhoff, S. R.; Taylor, P. R. *Adv. Chem. Phys.* **1990**, *77*, 103.
21. Bauschlicher, C. W. *J. Phys. Chem.* **1988**, *92*, 3020.
22. Gdanitz, R. J.; Ahlrichs, R. *Chem. Phys. Lett.* **1988**, *143*, 413.
23. Chong, D. P.; Langhoff, S. R. *J. Chem. Phys.* **1986**, *84*, 5606. See also Ahlrichs, R.; Scharf, P.; Ehrhardt, C. *J. Chem. Phys.* **1985**, *82*, 890.
24. Raghavachari, K.; Trucks, G. W. *J. Chem. Phys.* **1989**, *91*, 1062; *J. Chem. Phys.* **1989**, *91*, 2457.
25. Barnes, L. A.; Rosi, M.; Bauschlicher, C. W. *J. Chem. Phys.* **1991**, *94*, 2031.
26. Rohlfing, C. M.; Martin, R. L. *Chem. Phys. Lett.* **1985**, *115*, 104.
27. Bauschlicher, C. W.; Siegbahn, P.; Pettersson, L. G. M. *Theor. Chim. Acta* **1988**, *74*, 479.

28. Bauschlicher, C. W.; Langhoff, S. R.; Partridge, H.; Barnes, L. A. *J. Chem. Phys.* **1989**, *91*, 2399.
29. Pettersson, L. G. M.; Bauschlicher, C. W.; Langhoff, S. R.; Partridge, H. *J. Chem. Phys.* **1987**, *87*, 481.
30. Alvarado-Swaisgood, A. E.; Harrison, J. F. *J. Phys. Chem.* **1985**, *89*, 5198. Alvarado-Swaisgood, A. E.; Allison, J.; Harrison, J. F. *J. Phys. Chem.* **1985**, *89*, 2517.
31. Schilling, J. B.; Goddard, W. A., III; Beauchamp, J. L. *J. Am. Chem. Soc.* **1986**, *108*, 582; *J. Am. Chem. Soc.* **1987**, *109*, 5565.
32. Anglada, J.; Bruna, P. J.; Peyerimhoff, S. D.; Buenker, R. J. *J. Mol. Struct.* **1984**, *107*, 163.
33. Anglada, J.; Bruna, P. J.; Peyerimhoff, S. D. unpublished.
34. Anglada, J.; Bruna, P. J.; Peyerimhoff, S. D. unpublished.
35. Anglada, J.; Bruna, P. J.; Grein, F. *J. Chem. Phys.* **1990**, *92*, 6732.
36. Armentrout, P. B.; Georgiadis, R. *Polyhedron* **1988**, *7*, 1573, and references therein.
37. Rappé, A. K.; Upton, T. H. *J. Chem. Phys.* **1986**, *85*, 4400.
38. Tolbert, M.; Beauchamp, J. *J. Am. Chem. Soc.* **1984**, *106*, 8117.
39. Elkind, J. L.; Armentrout, P. B. *Inorg. Chem.* **1986**, *25*, 1078.
40. Chong, D. P.; Langhoff, S. R.; Bauschlicher, C. W.; Partridge, H. *J. Chem. Phys.* **1986**, *85*, 2850.
41. Bauschlicher, C. W.; Langhoff, S. R. *Chem. Phys. Lett.* **1988**, *145*, 205.
42. Walch, S. P., Bauschlicher, C. W.; Langhoff, S. R. *J. Chem. Phys.* **1985**, *83*, 5351.
43. Blomberg, M. R. A.; Siegbahn, P. E. M.; Roos, B. O. *Mol. Phys.* **1982**, *47*, 127.
44. Marian, C. M.; Blomberg, M. R. A.; Siegbahn, P. E. M. *J. Chem. Phys.* **1989**, *91*, 3589.
45. Bruna, P. J.; Anglada, J. *Quantum Chemistry: The Challenge of Transition Metals and Coordination Chemistry*; Veillard, A., Ed.; D. Reidel: Dordrecht, 1986.
46. Anglada, J.; Bruna, P. J.; Peyerimhoff, S. D.; Buenker, R. J. *J. Mol. Struct.* **1983**, *93*, 299.
47. Anglada, J.; Bruna, P. J.; Peyerimhoff, S. D. *Mol. Phys.* **1989**, *66*, 541.
48. Anglada, J.; Bruna, P. J.; Peyerimhoff, S. D. *Mol. Phys.* **1990**, *69*, 281.
49. Anglada, J.; Bruna, P. J.; Peyerimhoff, S. D. unpublished.
50. Gray, J. A.; Rice, S. F.; Field, R. W. *J. Chem. Phys.* **1985**, *82*, 4717.
51. Bauschlicher, C. W.; Langhoff, S. R.; Komornicki, A. *Theor. Chim. Acta* **1990**, *77*, 263.
52. Åslund, N.; Neuhaus, H.; Lagerqvist, A.; Andersen, E. *Ark. Fys.* **1964**, *28*, 271.
53. Bauschlicher, C. W.; Partridge, H.; Langhoff, S. R.; Taylor, P. R.; Walch, S. P. *J. Chem. Phys.* **1987**, *86*, 7007.
54. Bauschlicher, C. W.; Langhoff, S. R. *J. Chem. Phys.* **1987**, *87*, 2919.
55. Beaton, S. P.; Evenson, K. M.; Nelis, T.; Brown, J. M. *J. Chem. Phys.* **1988**, *89*, 4446.
56. Stevens, A. E.; Feigerle, C. S.; Lineberger, W. C. *J. Chem. Phys.* **1983**, *78*, 5420.
57. Bauschlicher, C. W. *J. Chem. Phys.* **1987**, *86*, 5591.
58. Bauschlicher, C. W.; Langhoff, S. R.; Taylor, P. R. *J. Chem. Phys.* **1987**, *87*, 387.
59. Langhoff, S. R.; Bauschlicher, C. W. *J. Mol. Spectrosc.* **1990**, *141*, 243.
60. Langhoff, S. R.; Bauschlicher, C. W.; Rendell, A. P. *J. Mol. Spectrosc.* **1989**, *138*, 108.
61. Phillips, J. G.; Davis, S. P.; Lindgren, B.; Balfour, W. J. unpublished.
62. Phillips, J. G.; Davis, S. P. unpublished.
63. Carroll, P. K.; McCormack, P.; O'Connor, S. *Astrophys. J.* **1976**, *208*, 903.
64. McCormack, P.; O'Connor, S. *Astron. Astrophys. Suppl.* **1976**, *26*, 373.
65. Siegbahn, P. E. M.; Blomberg, M. R. A.; Bauschlicher, C. W. *J. Chem. Phys.* **1984**, *81*, 1373.

66. Balasubramanian, K. *Chem. Phys. Lett.* **1987**, *135*, 288.
67. Balasubramanian, K.; Ravimohan, Ch. *Chem. Phys. Lett.* **1988**, *145*, 39.
68. Balasubramanian, K.; Liao, D. *J. Phys. Chem.* **1988**, *92*, 6259.
69. Balasubramanian, K.; Ravimohan, Ch. *J. Phys. Chem.* **1989**, *93*, 4490.
70. Balasubramanian, K.; Liao, M. Z. *J. Phys. Chem.* **1989**, *93*, 89.
71. Balasubramanian, K.; Wang, J. Z. *J. Chem. Phys.* **1989**, *91*, 7761.
72. Schilling, J. B.; Goddard, W. A., III; Beauchamp, J. L. *J. Phys. Chem.* **1987**, *91*, 4470.
73. Das, K. K.; Balasubramanian, K. *J. Chem. Phys.* **1989**, *91*, 2433.
74. Das, K. K.; Balasubramanian, K. *J. Chem. Phys.* **1989**, *91*, 6254.
75. Rosi, M.; Bauschlicher, C. W.; Langhoff, S. R.; Partridge, H. *J. Phys. Chem.* **1990**, *94*, 8656.
76. Blomberg, M. R. A.; Schüle, J.; Siegbahn, P. E. M. *J. Am. Chem. Soc.* **1989**, *111*, 6156.

CHAPTER 4

Theoretical Studies of Oxidative Addition and Reductive Elimination of Hydrogen and Alkanes

P. JEFFREY HAY

Theoretical Division, Mail Stop B268, Los Alamos National Laboratory, Los Alamos, New Mexico 87545

This chapter reviews theoretical calculations, involving both *ab initio* and semiempirical methods, of oxidative addition of hydrogen and alkanes to transition metal complexes. It also discusses the reverse reductive elimination process and compares theoretical results and experimental data. The cases discussed include reactions of two-coordinate d^{10} ML_2 complexes, four-coordinate square-planar d^8 ML_4 complexes, H_2 addition to bare metal M species, C—H activation by metal complexes, and oxidative addition of dihydrogen ligands.

4.1. INTRODUCTION

The oxidative addition and reductive elimination of ligands are two of the most fundamental steps in reactions occurring at metal centers in inorganic and organometallic complexes. In oxidative addition, insertion of the metal into a single X—Y bond typically proceeds with scission of the X—Y bond and the formation of two metal–ligand bonds

$$L_nM + \begin{matrix} A \\ | \\ B \end{matrix} \rightarrow L_nM\begin{matrix} \diagup A \\ \\ \diagdown B \end{matrix}$$

as exemplified by the reaction

$$IrCl(CO)(PR_3)_2 + H_2 \rightarrow IrCl(CO)(PR_3)_2(H)_2$$

In the reverse reductive elimination step, a new bond is formed as the two ligands depart,

$$L_nM\begin{matrix} \diagup A \\ \\ \diagdown B \end{matrix} \rightarrow L_nM + \begin{matrix} A \\ | \\ B \end{matrix}$$

as illustrated in the reaction

$$Pt(PR_3)_2(H)(CH_3) \rightarrow Pt(PR_3)_2 + CH_4$$

The stereochemistry and kinetics of experimental studies of oxidative addition reactions have been reviewed by several authors [1–4], and the qualitative features of the molecular orbitals involved in these symmetry-allowed processes have been developed in various early studies [5–7]. Only in recent years, however, is a detailed picture emerging of the nature of the potential energy surfaces involved in these reactions from theoretical studies [8]. In this review we shall attempt to summarize the finding of semiempirical, approximate, and *ab initio* calculations on the oxidative addition and reductive elimination of hydrogen and simple alkanes. Among the issues to be discussed are the role of the coordination number around the metal, the oxidation states of the metal, and the role of low-lying electronic states in the reaction. In addition, the role of electron-withdrawing and electron-donating ligands on both the activation barriers and the overall thermochemistry will be discussed.

4.2. OXIDATIVE ADDITION TO d^{10} ML_2 AND RELATED SYSTEMS

4.2.1. H_2 Addition

The oxidative addition of H_2 to ML_2 complexes has been observed for the case of PtP_2 complexes [9], where P represents phosphine ligands PR_3 with bulky alkyl groups

$$H_2 + PtP_2 \rightarrow PtP_2H_2$$

Almost all dihydrides assume a trans geometry except in cases where the ligands are chelating diphosphine groups. From orbital symmetry considerations, the lowest-energy pathway for oxidative addition is presumed to occur by initial cis addition of H_2 to the complex. By contrast, *trans* addition to ML_2 with d^{10} configuration is symmetry-forbidden.

For the case of trimethylphosphine (PMe_3) ligands, the presence of both *cis* and *trans* PtP_2H_2 species has been found in equilibrium in solution [10],

$$trans\text{-}PtP_2H_2 \rightarrow cis\text{-}PtP_2H_2$$

upon oxidative addition of H_2, and the kinetics of reductive elimination of H_2 from *cis*-$Pt(PMe_3)_2H_2$ have also been observed [11, 12].

Because of the much stronger M—H (60 kcal mol^{-1}) bond strengths compared to M—R bond strengths (30 kcal mol^{-1}), oxidative addition of alkanes to ML_2 complexes (and other complexes in general) usually is observed only in the case of strained systems, as in the case of PtP_2 inserting into cyclopropane. Reductive elimination is much more prevalent and the kinetics of alkane elimination has been studied for the cases of $PtP_2(H)(Me)$ [13], PdP_2R_2 [14], and AuR_3 [15] systems.

The oxidative addition of H_2 to d^{10} PtL_2 has been the subject of several studies [16–20] using *ab initio* theoretical techniques on the model reaction

$$Pt(PH_3)_2 + H_2 \rightarrow cis\text{-}Pt(PH_3)_2(H)_2$$

Calculations by Obara, Kitaura, and Morokuma (OKM) [19], Noell and Hay (NH) [17], and Low and Goddard (LG) [18] examined the reaction path and transition state for this process. A related model reaction,

$$Pd(H_2O)_2 + H_2 \rightarrow Pd(H_2O)_2(H)_2$$

was the subject of an *ab initio* study by Brandemark et al. [20].

In the studies of the PtL_2 system, a remarkably consistent overall picture emerged from the calculations despite the differences in the theoretical approaches. In each case the *cis*-PtP_2H_2 complex was found to be thermodynamically stable with respect to loss of H_2 and to have a square-planar-like geometry analogous to other d^8 ML_4 complexes. The predicted activation barriers ranged from 2 to 17 kcal mol^{-1} for the oxidative addition process, and the transition state resembled the reactants in that the H_2 bond was lengthened only slightly from the bond length in H_2 itself. OKM carried out a full vibrational analysis and established that the intermediate for *cis* addition was in fact a true transition state with only one imaginary frequency. The reaction path corresponding to this mode consisted mainly of H_2 relative motion and PPtP bending motion. OKM and NH examined the *cis* and *trans* PtP_2H_2 structures and found the *trans* form to be the more stable, but only by about 2 kcal mol^{-1}, indicating that there is no overwhelming driving force for stabilization of either form. Rather, the steric repulsion between bulky PR_3 groups primarily appears to account for the predominance of trans structures observed experimentally. Calculations for R = isopropyl in the work by OKM showed that steric repulsion energies of 10–20 kcal mol^{-1} arose for P—M—P bond angles smaller than 140°. Comparison of *cis* and

Table 4.1
Theoretical Studies of the $PtP_2 + H_2$ Reaction

	OKM[a]	NH[b]	LG[c]	H[d]
	Heat of reaction (kcal mol^{-1})			
SCF	36.9	6.7	16.2	6.3
CI	27.0	5.0	15.9	10.9
	Barrier to addition (kcal mol^{-1})			
SCF	5.2	17.4	4.0	13.9
CI	7.3	16.6	2.3	9.2
	Barrier to elimination (kcal mol^{-1})			
SCF	42.1	24.1	20.2	20.2
CI	34.3	21.6	18.2	20.1
	Geometrical parameters: PtP_2H_2			
R(Pt—H)	1.522	1.55	1.504	1.576
R(Pt—P)	2.194	[2.268][e]	2.447	2.526
θ(H—Pt—H)	81.5	80	79.5	81.9
θ(P—Pt—P)	104.4	100	99.9	99.4
	Geometrical parameters: Transition state			
R(Pt—H)	2.066	1.807	2.199	1.893
R(H—H)	0.766	0.90	0.75	0.806
R(Pt—P)	2.153	[2.268]	2.35	2.410
θ(P—Pt—P)	147.7	120	137.4	125.0

[a]Reference 19.
[b]Reference 17.
[c]Reference 18.
[d]Reference 22.
[e]Fixed.

trans forms of $Pt(PR_3)_2(H)_2$ for R = H and R = Me, neither of which is a bulky ligand, by Noell and Hay showed an increased preference for the trans structure in the case of Me compared to H.

The quantitative variations in the structures and energies from the different calculations in Table 4.1 arise from differences in basis sets, effective potentials, treatments of electron correlation effects, and computational procedures in the calculations. The calculations by OKM and LG used gradient optimization techniques to vary all geometrical parameters along each step of the reaction path, whereas only the P—Pt—P angle and Pt—H bond length were varied as a function of each H–H separation in the NH work. The effective core potential (ECP) used by OKM to represent Pt was somewhat deficient in representing the core orbitals and, as a result, apparently overestimated Pt—H bond strengths, whereas the ECP used by LG and NH did not have this problem. More recent ECPs, by requiring "norm-conserving" electron density in the valence region of the atom, have been found to provide more-realistic descriptions in molecular calculations. The results of a full optimization study of this reaction using norm-conserving ECPs for both P and Pt by this author are also shown in Table 4.1. In addition, the ECP for Pt also explicitly treats the outer core $5s$ and $5p$ electrons in the calculation. The latter calculation predicts consistently longer metal–ligand bond distances than the earlier studies, and the resultant barrier [14 kcal mol^{-1} for SCF, 9 kcal mol^{-1} for configuration interaction (CI)] is somewhat higher than previous calculations.

Although no examples of *cis* dihydride PtP_2H_2 complexes were known prior to the calculations for monodentate P ligands, subsequent experimental studies have observed *cis* and *trans* forms of PtP_2H_2 complexes for both P = PMe_3 and PEt_3 to be in equilibrium with the ratio of *cis* to *trans* varying from 0.03 to 2.2, depending on solvent, and a relative energy of 0.3–0.1 kcal mol^{-1} between the forms [10]. Further kinetic studies [11, 12] of the reductive elimination of H_2 from $Pt(PMe_3)_2H_2$ deduced a barrier of 20 kcal mol^{-1} in the limit of noncoordinating solvents, which may be compared to the values of 18–34 kcal mol^{-1} from the CI calculations of these authors.

In the calculations by Brandemark *et al.* on oxidative addition of H_2 to $Pd(H_2O)_2$, qualitatively similar results were obtained, although they did not investigate the reaction pathway to identify a transition state. The formation of the *cis* dihydride was found to be exothermic by 6.9 kcal mol^{-1}, with Pd—H bond lengths of 1.53 Å. This may be compared with a similar value of 5.5 kcal mol^{-1} for the oxidative addition to a Pd atom.

The works by Low and Goddard (LG) and by Brandemark *et al.* both emphasize the importance of the competition between the d^9s^1 and d^{10} states of the metal atom in influencing the course of the reaction. This point is extensively developed by LG in their analysis of the bonding in terms of the generalized valence bond (GVB) picture. In the ML_2 fragment, the metal can be described qualitatively by a d^{10} configuration with the bonding to the ligands primarily involving donation of the ligand lone pairs to s and p metal orbitals. In order to form two more bonds, the promotion of the metal to the d^9s^1 state is required. As many authors have pointed out, the term oxidative addition is a misnomer in the narrow classical sense of charge transfer. Relatively little charge is actually transferred to the hydrogen atoms during the process, so that the resulting dihydride exhibits relatively little "hydridic" character. This is particularly evident in the analysis of the GVB orbitals in $Pt(PH_3)_2H_2$, which consist of a covalent pairing involving a H $1s$ orbital singlet coupled to an sd hybrid orbital on the metal.

In this view, as the H_2 approaches the ML_2 fragment, the bending of the ligands away from the H_2 serves to decrease the separation between the d^{10} and d^9s^1 states to the point where the d^9s^1-like state becomes accessible and formation of two M—H bonds can follow. Comparison of the addition of H_2

to Pt and Pd, for example, is exothermic for the case of Pt (which has a d^9s^1 ground state) by 34 kcal mol^{-1} but is endothermic for the case of Pd (which has a d^{10} ground state) by 4 kcal mol^{-1}. These comparisons will be discussed further in the next section.

4.2.2. *Ab Initio* Studies of CH_4 and C_2H_6 Reductive Elimination

In addition to the studies of oxidative addition involving the H—H bond of H_2 with ML_2 d^{10} systems discussed previously, there have been several extensive series of calculations [19–26] on related reactions involving the C—H bond of CH_4 and the C—C bond of C_2H_6. Because the M—C bonds are inherently weaker than M—H bonds, the reductive elimination process has been observed more frequently in experiments as compared with oxidative addition, as shown in the following examples [13, 14]:

$$PtH(CH_3)(PPh_3)_2 \rightarrow Pt(PPH_3)_2 + CH_4$$

$$Pd(CH_3)_2(PPh_3)_2 \rightarrow Pd(PPh_3)_2 + C_2H_6$$

We shall adopt this perspective for purposes of discussion in this section.

The reductive elimination of CH_4 was investigated by Obara, Kitaura, and Morokuma (OKM) [19] for the model complex $Pt(H)(CH_3)(PH_3)_2$ analogous to their study of H_2 oxidative addition. The elimination process was calculated to be slightly exothermic at the highest level of theory (−5 kcal mol^{-1}) and to have a large barrier (30 kcal mol^{-1}). In the transition state, the CH bond was still considerably stretched (1.35 Å) compared to the equilibrium CH distance of 1.08 Å in CH_4. By contrast, for the transition state in H_2 elimination, the HH bond had shortened nearly to the distance in H_2 itself.

Low and Goddard [23] followed the reaction paths for elimination of H_2, CH_4, and C_2H_6 from Pd and Pt atoms as models for d^{10} complexes. Earlier work by Blomberg *et al.* [24] had investigated analogous reactions of methane and ethane with Ni. In a later work [25], studies of the energies and equilibrium geometries of more realistic Pt(II) complexes $Pt(CH_3)_2(PH_3)_2$, $Pt(H)(CH_3)(PH_3)_2$, and $PtH_2(PH_3)_2$ and Pt(IV) complex $Pt(CH_3)_2Cl_2(PH_3)_2$ were analyzed to provide a more comprehensive model of these processes. The results of these studies are summarized in Table 4.2. The differences observed between Pt and Pd can be traced primarily to the relative ordering of the d^{10} and d^9s^1 states, as discussed for the case of H_2. When the product of reductive elimination is a d^{10} complex, such as $M(PH_3)_2$, the reaction is

Table 4.2
Comparison of Reaction Energies and Activation Energies (kcal mol^{-1})

		Activation Energies		
	Reaction Energies[a]	Reductive Elimination	Oxidative Addition	Ref.
$PtP_2HMe \rightarrow PtP_2 + CH_4$	−1.2	34.1	35.3	[19][c]
	4.7	34.3	29.6	[19][d]
$NiH_2 \rightarrow Ni(^1D) + H_2$	9.0	11.0	2.0	[55]
$NiMeH \rightarrow Ni(^1D) + CH_4$	−20.7	33.4	54.1	[24]
$NiMe_2 \rightarrow Ni(^1D) + C_2H_6$	−5.3	37.0	42.3	
$NiMe_2 \rightarrow Ni(^3F) + C_2H_6$	−8.0	56.4	64.0	
$PtH_2 \rightarrow Pt + H_2$	33.6	33.6	0	[23]
$PtHMe \rightarrow Pt + CH_4$	16.1	29.0	12.9	
$PtMe_2 \rightarrow Pt + C_2H_6$	18.3	53.5	35.2	
$PdH_2 \rightarrow Pd + H_2$	−3.6	1.6	5.2	[23]
$PdHMe \rightarrow Pd + CH_4$	−20.1	10.4	30.3	
$PdMe_2 \rightarrow Pd + C_2H_6$	−26.0	22.6	48.6	
$PtP_2H_2 \rightarrow PtP_2 + H_2$	15.9	18.1	2.2	[23, 24]
$PtP_2HMe \rightarrow PtP_2 + CH_4$	−7.2	[17.3][b]	[25.5]	
$PtP_2Me_2 \rightarrow PtP_2 + C_2H_6$	−8.1	[40.3]	[48.4]	
$PtP_2Cl_2Me_2 \rightarrow PtP_2Cl_2 + C_2H_6$	−25.0	[31.8]	[56.8]	[24]

[a] *E*(products) − *E*(reactants)
[b] Quantities in brackets are estimates.
[c] SD-CI results.
[d] SD-CI plus corrections for quadruple excitations.

more exothermic in the case of Pd, where d^{10} is the ground state, than for Pt, which has a d^9s^1 ground state. This is consistent with the fact that reductive elimination of C_2H_6 from four-coordinate Pd(II) complexes is observed experimentally, but not in the case of Pt. Instead, Pt(II) dialkyls prefer beta-hydride elimination [27], leading to the formation of olefins and alkanes.

The trends in the activation barriers for reductive elimination generally follow the pattern $C_2H_6 > CH_4 > H_2$. This has been shown to derive from the capability of the spherical 1*s* orbital of hydrogen to form multicenter bonds and hence to maintain a high overlap between both the HH and MH bonds in the transition state. The more-directional sp^3 hybrid orbitals in alkyl ligands, by contrast, cannot simultaneously maintain such a high overlap between the bonds being broken and formed, and this results in a higher-energy transition state. The elimination of C_2H_6 from $Pt(CH_3)_2Cl_2(PH_3)_2$ was found to be twice as exothermic (−27 kcal mol^{-1}) than from $Pt(CH_3)_2(PH_3)_2$ (−11 kcal mol^{-1}). Comparisons with experiment for reductive elimination of alkanes from six-coordinate complexes is more difficult because five-coordinate intermediates are often involved.

The earlier studies [24] of elimination from Ni also found considerably higher activation barriers for CH_4 and C_2H_6 compared to H_2. The CH bond was considerably stretched (1.62 Å) compared to the CH_2 distance in the product CH_4 (1.09 Å), and similarly the CC bond in the transition state (2.07 Å) was intermediate between the complex (3.05 Å) and the product C_2H_6 (1.55 Å).

4.2.3. Reductive Elimination from Four-Coordinate Complexes

The reductive elimination (RE) of X_2 from four-coordinate d^8 complexes L_2MX_2 has been considered by numerous authors, using semiempirical or approximate methods [28–32]. Tatsumi *et al.* [28] employed extended Hückel analyses to study the factors influencing the ease of reductive elimination for the generic process

$$cis\text{-}L_2MX_2 \rightarrow L_2M + X_2$$

As noted by many authors, the overall reaction is symmetry-allowed, as shown in Figure 4.1. The basic correlation involved is the evolution of the a_1

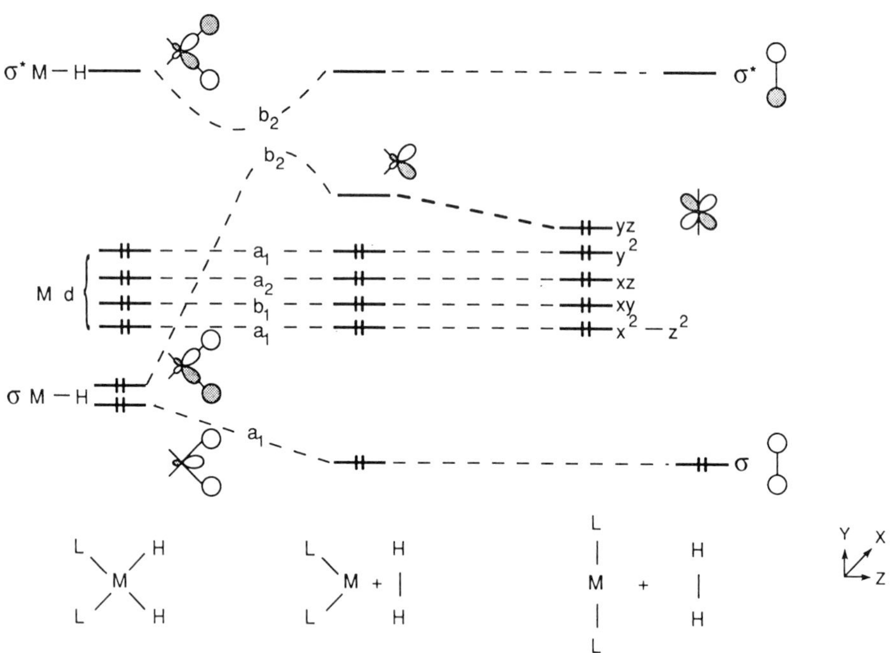

FIGURE 4.1
Orbital correlation diagram for the oxidative addition of H_2 to a d^{10} ML_2 complex.

metal–ligand bonding orbitals into the sigma-bonding orbital of the X_2 species and the b_2 metal–ligand bonding orbital into the d_{xy} orbital of the L_2M complex. The a_1 orbital remains bonding in character throughout the course of the elimination process and remains low in energy. By contrast, the b_2 orbital rises significantly as it is transformed from a bonding orbital into the nonbonding d orbital of the metal. The change in character can be ascribed to an avoided curve-crossing in the sense that it contains X_2 antibonding character and tends to correlate with the high-lying σ^* orbital of X_2. As the X–X distance becomes progressively smaller, the X–X antibonding character decreases and eventually the b_2 orbital assumes totally metal d character.

In the model studies, Tatsumi *et al.* ascribed many of the controlling features of the reaction to the b_2 orbital. They found that the reductive elimination would be faster if the b_2 energy in ML_2 lies lower in energy. if the metal d orbital lies lower in energy, for example, this would tend to lower the ML_2 b_2 energy. The analogy can be made with the analysis of Low and Goddard [23], who compared the relative energies of the s^1d^9 states with the d^{10} state of the metal and found reductive elimination was favored in the case of Pd, where the d^{10} state was lower in energy. Because a lower metal d orbital energy would favor the d^{10} state, the two arguments are similar in spirit. The relative energy of the s orbital does not map smoothly from the one picture to the other, however.

The other major conclusions reached by the model study by Tatsumi *et al.* were (a) the rate for RE should decrease as L is made more electron-withdrawing and (b) the rate should increase as X is made more electron-donating. A later analysis by Flores-Riveros and Novaro [29] examined RE of CH_4 from $Pt(PX_3)_2(H)(CH_3)$ and also emphasized the role of the b_2 orbital. They differed with regard to conclusion (a) in that, as X in the PX_3 ligand was made more electron-withdrawing, the barrier for elimination decreased and implied a faster overall rate. This was supported by experimental kinetic studies by Abis, Sen, and Halpern [13] on this system. Yet another analysis by Balazs, Johnson, and Whitesides [30] attempted to use $X\alpha$–scattered-wave molecular orbitals as a basis for understanding the observed trend in rate found for the metals Pt < Pd < Ni in RE from $L_2M(CH_3)_2$ and the previously mentioned trend in rates for L_2PtXY as XY varies from $H_2 > CH_4 > C_2H_6$, but no definitive quantitative correlation emerged from their analysis.

4.2.4. Three- and Five-Coordinate d^8 Species

Reductive elimination is also found to occur in three-coordinate species resulting from ligand loss from four-coordinated complexes [31]. The potential surface for the d^8 ML_3 species such as $Au(CH_3)_3$ has an interesting topology as a function of L—M—L angle (α) because the C_{3h} geometry (**1**) is unstable with respect to a Jahn–Teller distortion away from trigonal symmetry ($\alpha = 120°$) to either a T-shaped ($\alpha = 180°$) structure (**2**) or a

Y-shaped ($\alpha = 90°$) structure (**3**), with the T-shaped structure (**2**) corresponding to global minima on the surface.

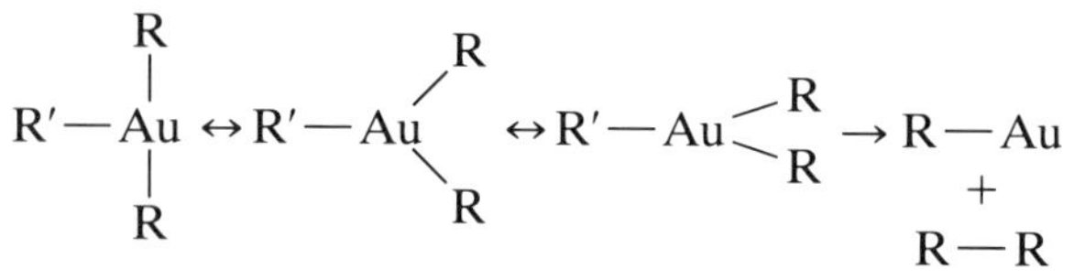

$$\underset{\mathbf{2}}{R'{-}Au(R)_2\ \text{(T-shaped)}} \leftrightarrow \underset{\mathbf{1}}{R'{-}Au(R)_2} \leftrightarrow \underset{\mathbf{3}}{R'{-}Au{<}^{R}_{R}} \rightarrow \underset{\mathbf{4}}{R{-}Au + R{-}R}$$

Reductive elimination is symmetry-allowed from either structure to form M—R + R—R (**4**), as discussed by Komiya et al. [31] and Tatsumi et al. [28]. In the latter paper, the barrier for RE from three-coordinate species was found to be comparable to or slightly smaller than from four-coordinate species, although quantitative predictions are more difficult using this method when comparing molecules with different coordination.

Tatsumi *et al.* [32] also considered associative mechanisms for assisting RE processes that occur when a fifth ligand coordinates to the metal,

$$ML_2X_2 + L \rightarrow ML_3X_2 \rightarrow ML_3 + X_2$$

as is observed when RE is accelerated by added phosphine ligands. The overall process was also found to be quite facile from their extended Hückel studies, but no single controlling orbital was found to be the crucial factor in these processes. Elimination of R_2 from either edge-shared square-planar positions (**5**) and trigonal bipyramidal positions (**6**) were identified as low-energy pathways.

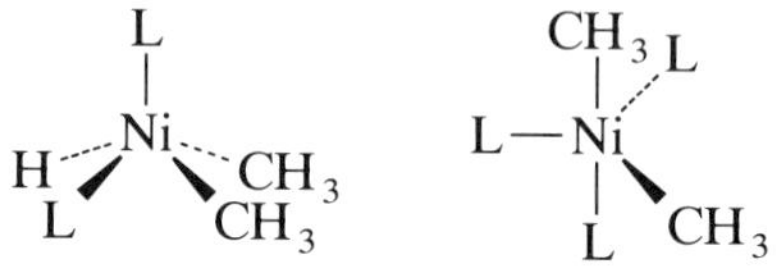

5: $Ni(L)_2(H)(CH_3)_2$ — L, H, L, CH_3, CH_3 on Ni

6: $Ni(L)_3(CH_3)_2$ — CH_3, L, L, L, CH_3 on Ni

4.3. ADDITION OF H_2 AND CH_4 TO d^8 ML_4 AND RELATED COMPLEXES

4.3.1. H_2 Addition

Among the most widely studied oxidative addition reaction is the hydrogenation of d^8 ML_4 complexes as in the case of Vaska's complex $IrCl(CO)(PPH_3)_2$ and Wilkinson's catalyst $RhCl(PPH_3)_3$. These are also symmetry-allowed processes (Figure 4.2) as the H_2 bonding orbital and d_{yz} orbital of the ML_4

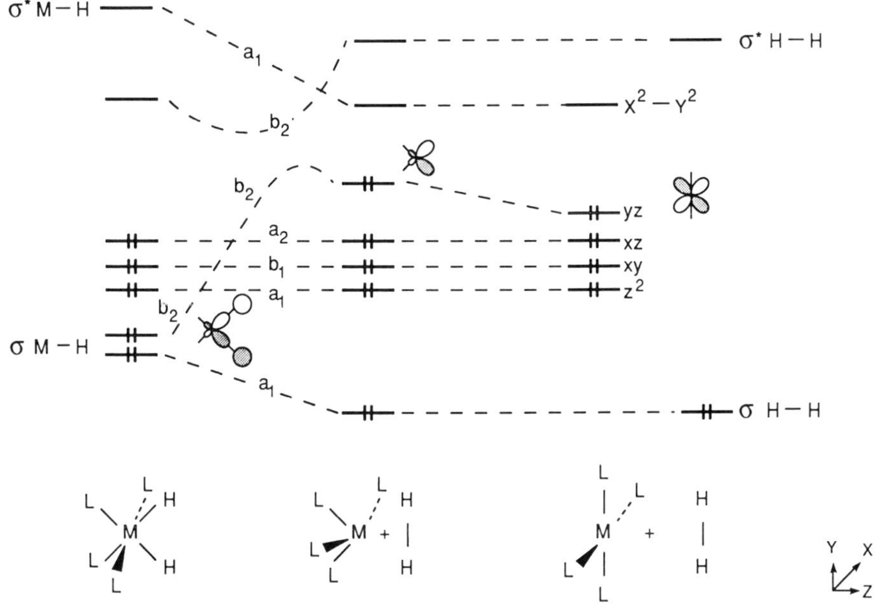

FIGURE 4.2
Orbital correlation diagram for the oxidative addition of H_2 to a d^8 ML_4 complex.

reactant transform into the two M—H bonding orbitals of the six-coordinate product. The four-electron repulsion arising from the sigma H_2 orbital interacting with the filled d_{z^2} orbital on the metal must be overcome while the other orbital changes are occurring. The bending of two of the ligands raises the energy of the d_{yz} orbital to enable the formation of two M—H bonds.

The numerous theoretical calculations on $H_2 + ML_4$ systems [33–37] include extended Hückel calculations by Sevin [35] and by Jean and Lledos [36]. In the latter study of H_2 addition to $Rh(PR_3)_2(CO)Cl$ and related species, the authors considered the issue of which products will be formed depending on whether the H_2 aligns with the CO–X plane or P–P plane to form the two possible products (**8** or **9**).

X—M(L)(L)—CO + H_2 → H—M(H)(L)(L)(X)—CO or X—M(H)(H)(L)(L)—CO

7 **8** **9**

The steric factors associated with bulky phosphine ligands complicate the analysis of Vaska's complex, but more recent kinetic studies [38, 39] involving the chemistry of *cis* bidentate phosphine ligands (dppe) offers the possibility of isolating the electronic factors operative in this reaction leading to **11** or **12**.

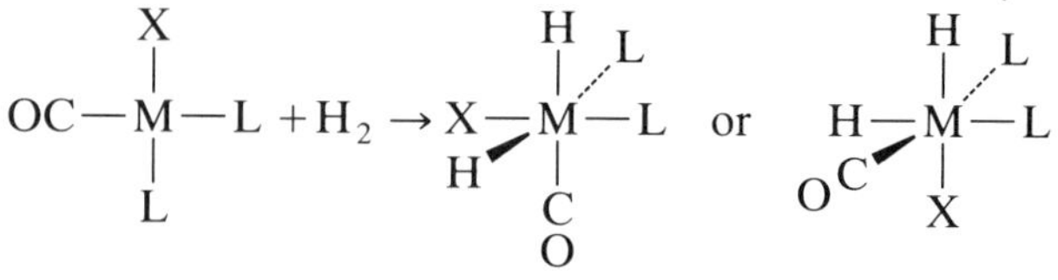

$$\text{OC—M(X)(L)—L} + H_2 \rightarrow \text{X—M(H)(H)(L)(CO)—L} \quad \text{or} \quad \text{H—M(H)(L)(CO)(X)—L}$$

10 **11** **12**

In this case, the H_2 could align either with the M–X or M–CO axes along the approach to oxidative addition. Experimentally, it was observed that in all cases (X = Cl, Br, I, H, PPH_3, and CN) the kinetic product of H_2 addition is the structure corresponding to H_2 aligned along the L–Ir–CO axis.

In order to investigate the orbital interactions involved in this reaction, Jean and Lledos [36] examined $Rh(Cl)(CO)(PH_3)_2$ and *trans*-$RhCl_2(CO)_2$ model complexes. In the latter case, where the analysis is simpler because of the higher symmetry, alignment of the H_2 along the CO axes requires bending of the CO ligands in forming the octahedral complex. Conversely, alignment of H_2 along the Cl axes would require bending of the Cl ligands in the plane. They found that the overriding factor leading to the lower transition state is the favorable interaction of the d_{z^2} orbital with the π^* CO orbitals upon bending compared with the unfavorable interaction of the d_{z^2} orbital with the filled π halogen orbitals upon bending. This results in the H_2 adding parallel to the ligand that is the strongest π acceptor (i.e., CO in this case), in agreement with the observed experimental studies.

In earlier *ab initio* studies, Dedieu and Strich [33] examined the relative thermodynamic stabilities of the products of the reaction of H_2 with the Wilkinson catalyst [34],

$$RhCl(PH_3)_3 + H \rightarrow RhCl(PH_3)_3H_2$$

by Hartree–Fock calculations using idealized geometries. The *cis* addition product (**13**), where the H_2 adds in the P–Rh–Cl plane was lower than the case (**14**), where it aligns with the P–Rh–P plane, and both *cis* products were more stable than the trans form (**15**).

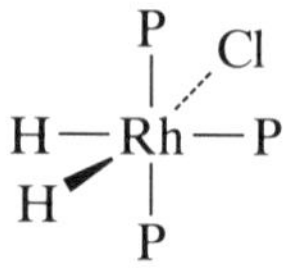

$$\text{H—Rh(P)(Cl)(H)(P)—P} \qquad \text{H—Rh(H)(Cl)(P)(P)—P} \qquad \text{P—Rh(H)(Cl)(P)(H)—P}$$

13 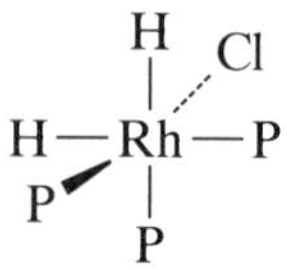 **14** 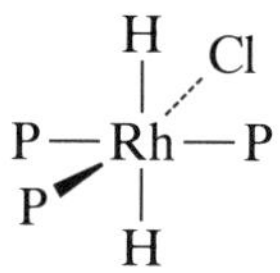 **15**

A more recent study by Daniel *et al.* [37] revisited the catalytic cycle of olefin hydrogenation by $RhCl(PR_3)_3$ originally studied by Dedieu [34]. In this account we shall concentrate solely on the oxidative addition step, as other aspects of such reactions are considered in more detail in separate chapters in this book. In their study the reaction was presumed to proceed by prior dissociation of the phosphine ligand as suggested by kinetic and spectroscopic studies,

$$RhClP_3 \rightarrow RhClP_2 + P$$

The three-coordinate intermediate then undergoes oxidative addition of H_2,

$$RhCl(PH_3)_2 + H_2 \rightarrow H_2RhCl(PH_3)_2$$

before the catalytic cycle porceeds with coordination of the olefin. According to the *ab initio* results at the Hartree–Fock level, there is no overall barrier for oxidative addition. A stable H_2 adduct (-20.1 kcal mol^{-1} relative to reactants) lies slightly only 0.6 kcal mol^{-1} below the transition state (which is itself -19.5 kcal mol^{-1} relative to the reactants) along the pathway to the five-coordinate cis dihydride (-26.4 kcal mol^{-1}).

The treatments to date have discussed at various levels some of the electronic factors affecting (a) the relative stabilities of transition states leading to kinetic products, as well as (b) relative energies of thermodynamic products. Full treatments using rigorous theoretical methods have yet to examine potential-energy surfaces for $H_2 + ML_4$ reactions in a systematic fashion so that the transition state and overall reaction can be discussed on a consistent level.

4.3.2. C—H Activation by ML_4 and CpML Species

The intermolecular activation of C—H bonds by oxidative addition was realized during the past decade by Janowicz and Bergman [40] and by Hoyano and Graham [41] according to the reaction

$$(C_5Me_5)IrL + CH_4 \rightarrow (C_5Me_5)(L)Ir(H)(CH_3)$$

where the coordinatively unsaturated Ir complex was achieved by photolysis of 18-electron species $(C_5Me_5)Ir(CO)_2$ or $(C_5Me_5)Ir(PMe_3)(H_2)$. A similar reaction involving oxidative addition to CH_4 to $(C_5Me_5)Rh(PMe_3)$ was observed by Jones and Feher [42]. Summaries of experimental progress in C—H activation during the intervening years have appeared recently [43, 44]. This topic has been the study of several theoretical studies. A comprehensive analysis of C—H activation by organometallic compounds and by metal surfaces was presented by Saillard and Hoffmann [45] on the basis of extended Hückel calculations. In this discussion we shall focus on the aspects treated by the authors involving interactions of CH_4 with the d^8 species: CpRh(CO) (**16**) as a model for the above Ir and Rh complexes given previously and $Rh(CO)_4{}^+$ (**17**) as a model ML_4 system.

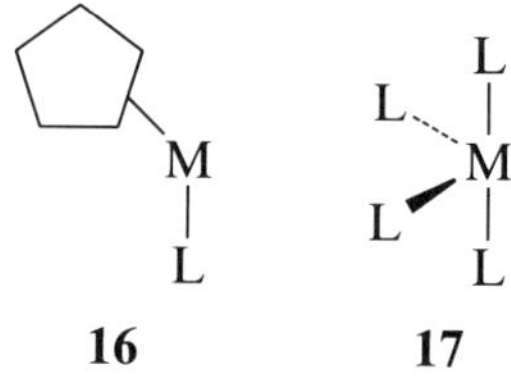

The orbital interactions along the reaction path for oxidative addition of CH_4 to CpRh(CO) and bent ML_4 are conceptually similar to the ones discussed previously in addition of H_2 to d^8 complexes. The controlling feature is the *yz* orbital on the metal, which is already relatively high in energy in the CpML moiety and which becomes destabilized as the two of the ML_4 ligands bend back. Eventually there is, in effect, an avoided curve crossing (although both orbitals are the same symmetry in this allowed reaction) with the C—H* orbital of CH_4 as it is gradually transformed into one of the two bonding orbitals of the adduct. The authors argue that the CpML framework is already well-adapted to facilitate oxidative addition by virtue of having the same orbital makeup (i.e., "isolobal") with the activated bent ML_4 structure.

This reaction was studied later by Ziegler et al. [46] using the first-principles density functional LCAO-HFS technique, which also incorporates relativistic effects. The results are summarized in Table 4.3. They find that the first stage of the reaction consists of formation of an adduct $CpML(H—CH_3)$ with the C—H bond elongated but intact. The binding energies of the CH_4 adduct are found to be about 6 kcal mol^{-1} for Rh and 15 kcal mol^{-1} for Ir complexes. There is then a modest barrier in the transition state region (8 kcal mol^{-1} for Rh and 2 kcal mol^{-1} for Ir) before the final exothermic stage of the reaction. There is, therefore, a slight barrier (2 kcal mol^{-1}) calculated for addition relative to CpML + CH_4 in the case of

Table 4.3
Calculated Energies for CH_4 Addition to CpML Species (kcal mol^{-1}) [46]

	Adduct Energy[a]	Reaction Energy[b]	Barrier (Relative to Adduct)
CpRh(CO)	−6.9	−15	8
$CpRh(PH_3)$	−5.7	−17	8
CpIr(CO)	−14	−33	2
$CpIr(PH_3)$	−12	−36	2

[a] $E[CpML—CH_4] - E[CpML + CH_4]$
[b] $E[CpML(H)(CH_3)] - E[CpML + CH_4]$
[c] $E[\text{transition state}] - E[CpML—CH_4]$

Rh but no net barrier in the case of Ir. Experimentally, estimates for the barrier range up to 5 and 10 kcal mol^{-1} for Rh and Ir, but the analysis is somewhat ambiguous. The estimated experimental reaction enthalpies are approximately -25 kcal mol^{-1} for Ir and -8 kcal mol^{-1} for Rh [44] as compared with -35 and -15 kcal mol^{-1}, respectively, for Ir and Rh for the HFS calculations.

4.4. SPECIAL TOPICS

4.4.1. Oxidative Addition of H_2 as Dihydrogen Ligand

In recent years the synthesis and isolation of the first metal complexes containing adducts of molecular hydrogen have been achieved [47]. In these species, the H—H bond remains intact as a "sideways-bonded" η^2 ligand rather than proceeding onward to form the classical dihydride product of classical oxidative addition. This subject will be reviewed elsewhere in this book. In this section we shall touch on the issue of the equilibrium between such dihydrogen species and the dihydride form corresponding to complete oxidative addition. In the first prototypical complex studied by Kubas [47], the stable form of $W(CO)_3(PR_3)_2(H_2)$ is the η^2-dihydrogen species. As studied by Jean *et al.* [48] in extended Hückel calculations and by Hay [49] in *ab initio* studies, the process of forming the seven-coordinate dihydride $W(CO)_3(PR_3)_2(H)_2$ is symmetry-allowed. In the *ab initio* studies using model PH_3 ligands, the dihydrogen form was lower in energy by 17 kcal mol^{-1} relative to separated $H_2 + W(CO)_3(PR_3)_2$ and by 10 kcal mol^{-1} relative to the dihydride species (Figure 4.3). When the pi-accepting CO ligands were replaced by better electron-donating PR_3 groups, the situation was reversed, with the dihydride more stable than the dihydrogen species by 14 kcal mol^{-1}. The relative energies in these studies were based on partial geometry optimizations. In the analysis by Jean *et al.*, they reached similar conclusions in noting that the dihydrogen species is favored by replacing sigma-donor ligands with pi-accepting ligands (i.e., PR_3 by CO) and by substituting a metal with higher-lying *d* orbitals for lower-lying *d* orbitals (i.e., W by Fe).

Experimentally the $W(CO)_3(PR_3)_2(H_2)$ complex has more recently been found to be in equilibrium with the dihydride form [50], with the latter form 1–2 kcal mol^{-1} higher in energy for R = *i*-Pr. Replacing alkyl groups on phosphines with more-electron-donating groups (i.e., phenyl by ethyl) in $Mo(CO)(P_2)_2(H_2)$ complexes, where $P_2 = R_2P—CH_2—CH_2—PR_2$, can alter the equilibrium from favoring the dihydrogen species (phenyl) to the dihydride (ethyl) [51]. Equilibrium between dihydrogen and dihydride forms has also been observed in $RuH_2(OCOCF_3)_2(PR_3)_2$ [52], in $[Cp^*Re(CO)(NO)H_2]^+$ [53], and in $[Ru(dmpe)CpH_2]^+$ [54]. Clearly the subtle balance between these forms at various stages along the reaction coordinate of oxidative addition can be tuned by varying the characteristics of the metal and ligands and will remain an active research area.

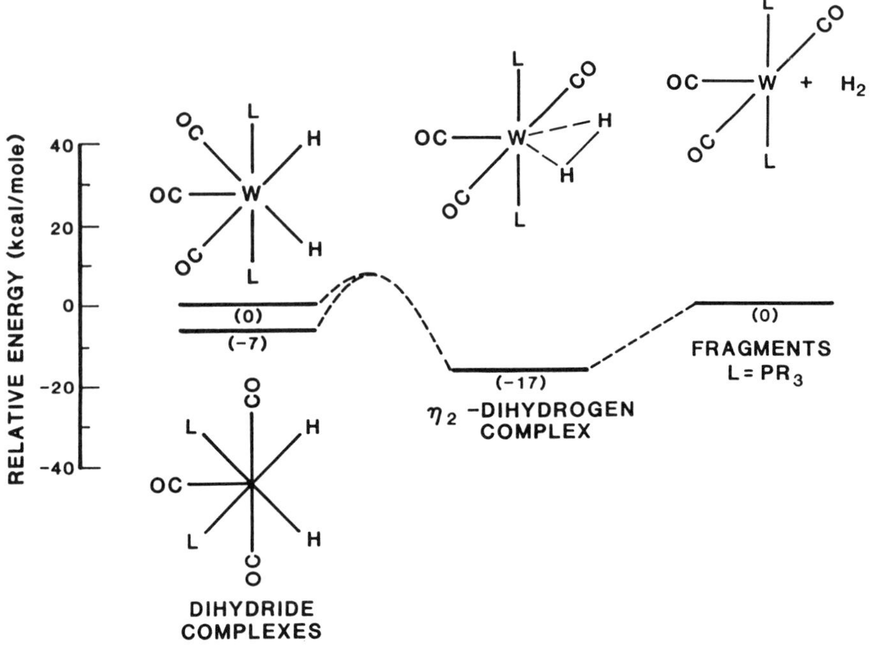

FIGURE 4.3
Relative energies of $ML_5 + H_2$ fragments, the $ML_5(H_2)$ dihydrogen complex, and the ML_5H_2 dihydride complex after oxidative addition has occurred as computed from *ab initio* calculations for the case $ML_5 = W(CO)_3(PH_3)_2$.

4.4.2. Addition of Hydrogen to Bare Metal Atoms

The addition of bare metal atoms to H_2 and other molecules has received increasing attention as the capabilities for experimental and theoretical studies of metal cluster chemistry have improved in recent years. In an earlier section, metal atoms were used as models for complexes involved in oxidative addition. In this section we shall focus on studies of MH_2 species as entities in their own right [55–63]. Results for selected theoretical studies of metal–H_2 addition reactions are summarized in Table 4.4. Of the calculations in the table, the studies by Balasubramanian and co-workers employ the most-sophisticated methodology on these systems. Typically, the potential-energy surfaces for MH_2 species are determined by full configuration interaction (CI) within a limited orbital subspace using the complex active space self-consistent field (CASSCF) treatment. This is then followed by a larger CI calculation based on the most important configurations from the first set of calculations.

For the series NiH_2, PdH_2, and PtH_2, each molecule has a stable singlet state ($^1A_1$ state in C_{2v} symmetry) with a bent geometry corresponding to

Table 4.4
Insertion of Bare Transition Metal Atoms into H_2

Reaction	M—H Distance (Å)	H—M—H Angle (degrees)	Reaction Barrier (kcal)	Reaction Energy (kcal)	Ref.
Ni(^1D) + H_2					
Bent ($^1A_1$)	1.43	49.4	3	−8	[60]
Linear ($^1\Sigma_g{}^+$)	1.53	180	27	15	
Ni (^3F) + H_2					
Linear ($^3\Delta_g$)	1.59	180	42	−0.4	[60]
Pd (^1S) + H_2					
$^1A_1$ outer min.	1.67	30	None	ca. 0	[58]
	1.83	25.9	None	−5	[23]
	1.7		None	−14	[61]
$^1A_1$ inner min.	1.50	62	Small	6	[58]
	1.51	73	1.6	3.6	[23]
Linear ($^1\Sigma_g{}^+$)	1.62	180	Large	48	[58]
Pt (^1S) + H_2					
Bent ($^1A_1$)	1.52	85	None	−34	[62]
Linear ($^1\Sigma_g{}^+$)	1.675	180	Large		
Pt (^3D) + H_2					
Linear ($^3\Delta_g$)	1.69	180	Large	57	[62]
Rh (^2F) + H_2					
Bent ($^2A_1$)	1.51	84	None		[61]
Rh (^4F) + H_2					
Linear ($^4\Phi_g$)	1.67	180	Large		[61]
Au (^2S) + H_2					
Linear ($^2\Sigma_g{}^+$)	1.66	180	Large		[60]
Au (^2P) + H_2					
Bent ($^1B_2$)	1.62	127	None		[60]

insertion of the metal into the H—H bond. The H—M—H bond angles increase in the order NiH_2 (49°), PdH_2 (73°), and PtH_2 (85°).

For Pd there is also a second bent minimum corresponding to a weakly bound dihydrogen complex [Pd—H_2] with a T-shaped geometry. This results from the unique situation for Pd, which has a closed-shell $d^{10}(^1S)$ ground state compared to d^9s^1 for Pt and nearly degenerate d^9s^1 and d^8s^2 ground states for Ni. The existence of such a weakly bound dihydrogen complex to Pd atoms was first noted in the calculations of Bagaturyants *et al.* [56]. It is interesting that these calculations were carried out several years before the experimental discovery of transition metal complexes involving molecular dihydrogen, discussed in the preceding section.

On the singlet potential-energy surfaces, insertion proceeds with little or no barrier. In addition, there is also a linear H—M—H structure ($^1\Sigma_g$ state) at much higher energy that is separated from the bent form by a significant barrier of approximately 50 kcal mol^{-1}. By contrast, insertion of triplet-state metal atoms occurs only by surmounting large (40 kcal mol^{-1}) activation barriers and leads ultimately to linear H—M—H structures ($^3\Sigma_g$ state). The same phenomenon is encountered in other cases of d^n metals, where low-spin species for bent structures with little activation barrier and high-spin species producing linear species but requiring enormous activation barriers.

4.4.3. Dinuclear Oxidative Addition

The entire review up to this point has focussed on processes involving the oxidative addition or reductive elimination from a single metal center. In this section the possibilities of simultaneous addition or elimination from two adjacent metal centers will be considered, as represented in the overall reaction

$$L_nM{-}ML_n + H_2 \rightarrow L_nM(H){-}M(H)L_n$$

for the case of addition to a metal–metal bond. In the case of bridging ligands between the metal centers, the process would be

$$L_nM{-}(\mu\text{-}Z)_2{-}ML_n + H_2 \rightarrow L_nM(H){-}(\mu\text{-}Z)_2{-}M(H)L_n$$

The latter process was investigated by Sevin, Hengtai, and Chaquin [64], employing extended Hückel calculations for the two cases of d^8–d^8 and d^7–d^7 dimers. The d^8–d^8 case is exemplified by Ir(I) complexes such as L_4—Ir—(μ-SR)$_2$—Ir—L_4, whereas the d^7–d^7 case is represented by L_3—Fe—(μ-SR)$_2$—Fe—L_3. An extensive experimental literature is cited by these authors in their study of oxidative addition to dinuclear complexes. Later we shall return to the issue of whether the process occurs by simultaneous addition to both centers or by other stepwise mechanisms.

The analysis of the molecular orbital correlation diagrams for these processes involving bridging ligands does not lead directly to predictions about likely barriers that may be encountered along the reaction path. This is primarily a consequence of two nearly degenerate molecular orbitals involving the d_{z^2} orbitals on each center that are involved in bond-breaking or bond-forming processes. In such processes there can be significant diradical character to the electronic states, which must be treated by configuration mixing between states involving both orbitals. In oxidative addition to d^8–d^8 complexes, a metal–metal bond is formed in the process, whereas in the case of d^7–d^7 complexes, which contain a formal metal–metal bond, a bond is broken during oxidative addition. Sevin et al. carried out extended Hückel studies on the reaction of H_2 with the model d^7–d^7 system $[H_3Fe(\mu\text{-}SH)]_2^{6-}$. From their analysis they concluded that addition to d^7–d^7 centers would occur with a small barrier (a few kilocalories per mole in the extended Hückel calculations). In this process, the metal–metal bond breaks while two metal–hydrogen bonds are formed on the same face of the dinuclear complex, while also preserving the bonds involving the bridging ligands [Fe—(SH)—Fe here]. Although they did not report results for the d^8–d^8 analogue, Sevin *et al.* argued that this process would proceed with much more difficulty, because H_2 addition would result initially in formation of a diradical that subsequently would rearrange before metal–metal bond formation could occur.

Reactions of hydrogen with binuclear complexes can occur either by simultaneous addition of H_2 to both metal centers, as discussed previously, or by a sequential process involving oxidative addition to one center,

$$L_nM\text{—}(\mu\text{-Z})_2\text{—}ML_n + H_2 \rightarrow L_nH_2M\text{—}(\mu\text{-Z})_2\text{—}ML_n$$

followed by hydrogen migration,

$$L_nH_2M\text{—}(\mu\text{-Z})_2\text{—}ML_n \rightarrow L_nHM\text{—}(\mu\text{-Z})_2\text{—}MHL_n$$

The ease of addition of H_2 to d^8 mononuclear complexes and the preceding analysis suggest this would be a more likely pathway for d^8–d^8 systems. Reference 64 gives an extensive experimental literature on this point. The details of the hydrogen migration reaction have been studied recently by Branchadell and Dedieu [65] for the model system

$$L_2H_2Rh\text{—}(\mu\text{-H})_2\text{—}RhL_2 \rightarrow L_2HRh\text{—}(\mu\text{-H})_2\text{—}RhHL_2$$

where L = PH_3, using *ab initio* SCF and CASSCF techniques. The original d^8–d^8 dinuclear complex formally becomes d^6–d^8 upon oxidative addition to one center and then becomes d^7–d^7 as the hydrogen migration proceeds. Their results indicated that hydrogen transfer to the d^8 center was unfavorable if the square-planar geometry of the H_2RhL_2 moiety was maintained. If, however, the migration was accompanied by a square-planar to trigonal-pyramid-like rearrangement (with the entering hydrogen forming the fifth vertex), a relatively facile transfer would occur, with a calculated activation barrier of 19 kcal mol^{-1}. Qualitatively, the reaction can be viewed as a hydride transfer, according to their analysis. These calculations lend support to the postulated mechanisms for addition to dinuclear complexes involving addition at one center followed by hydrogen migration.

Finally, the possibility of simultaneous addition of H_2 to unbridged dimeric metal complexes was studied by Trinquier and Hoffmann [66]. Among the motivations for the calculations were experimental observations that d^7–d^7 dimers such as $(CO)_4Os(R)\text{—}Os(H)(CO)_4$ do not readily eliminate RH. Their semiempirical analysis of model reactions, such as

$$(CO)_4Mn(H)\text{—}Mn(H)(CO)_4^{2-} \rightarrow (CO)_4Mn\text{—}Mn(CO)_4^{2-} + H_2$$

for elimination of H_2 from d^7–d^7 centers, showed them to be symmetry-forbidden reactions with large activation barriers for least-motion pathways. Although distortions away from least-motion pathways reduced the barriers slightly, Trinquier and Hoffmann concluded that these reactions normally would represent energetically unfavorable pathways that are unlikely to occur in the case of d^7–d^7 dinuclear complexes.

ACKNOWLEDGMENTS

This work has been carried out under the auspices of the US Department of Energy. The author wishes to thank Dr. Alain Dedieu for providing numerous literature references and Dr. Greg Kubas for helpful discussions.

REFERENCES

1. Collman, J. P. *Acc. Chem. Res.* **1968**, *1*, 169.
2. Vaska, L. *Acc. Chem. Res.* **1968**, *1*, 335.
3. Halperin, J. *Acc. Chem. Res.* **1970**, *3*, 386.
4. James, B. R. *Homogeneous Hydrogenation*; Wiley: New York, 1973.
5. Pearson, R. G. *Symmetry Rules for Chemical Reactions*, Wiley-Interscience: New York, 1976; pp. 286, 405.
6. Braterman, P. S.; Cross, R. J. *Chem. Soc. Rev.* **1973**, *2*, 271.
7. Akermark, B.; Ljungqvist, A. *J. Organomet. Chem.* **1979**, *182*, 59.
8. Dedieu, A. In *Topics in Physical Organometallic Chemistry*; Giclan, M., Ed.; Elsevier: London, 1985; pp. 1–145.
9. Yoshida, T.; Yamagata, T.; Tulip, T. H.; Ibers, J. A.; Otsuka, S. *J. Am. Chem. Soc.* **1978**, *100*, 2063.
10. Paonessa, R. S.; Trogler, W. C. *J. Am. Chem. Soc.* **1982**, *104*, 1138.
11. Packett, D. L.; Trogler, W. C. *J. Am. Chem. Soc.* **1986**, *108*, 5036.
12. Packett, D. L.; Jensen, C. M.; Cowan, R. L.; Strouse, C. E.; Trogler, W. C. *Inorg. Chem.* **1985**, *24*, 3578.
13. Abis, L.; Sen, A.; Halperin, J. *J. Am. Chem. Soc.* **1978**, *100*, 2915.
14. Gillie, A.; Stille, J. K. *J. Am. Chem. Soc.* **1980**, *102*, 4933.
15. Tomaki, A.; Magennis, S. A.; Kochi, J. K. *J. Am. Chem. Soc.* **1974**, *96*, 640.
16. Kitaura, K.; Obara, S.; Morokuma, K. *J. Am. Chem. Soc.* **1981**, *103*, 2891.
17. Noell, J. O.; Hay, P. J. *J. Am. Chem. Soc.* **1982**, *104*, 4578.
18. Low, J. J.; Goddard, W. A., III. *J. Am. Chem. Soc.* **1984**, *106*, 6928.
19. Obara, S.; Kitaura, K.; Morokuma, K. *J. Am. Chem. Soc.* **1984**, *106*, 7482.
20. Brandemark, U. B.; Blomberg, M. R. A.; Petterson, L. G. M.; Siegbahn, P. E. M. *J. Phys. Chem.* **1984**, *88*, 4617.
21. Hay, P. J.; Wadt, W. R. *J. Chem. Phys.* **1985**, *82*, 270; Wadt, W. R.; P. J. Hay. *J. Chem. Phys.* **1985**, *82*, 284; Hay, P. J.; Wadt, W. R. *J. Chem. Phys.* **1985**, *82*, 299.
22. Hay, P. J. *New J. of Chem.*, submitted for publication.
23. Low, J. J.; Goddard, W. A., III. *Organometallics* **1986**, *5*, 609.
24. Blomberg, M. R. A.; Brandemark, U.; Siegbahn, P. E. M. *J. Am. Chem. Soc.* **1983**, *105*, 5557.
25. Low, J. J.; Goddard, W. A., III. *J. Am. Chem. Soc.* **1986**, *108*, 6115.

26. Akermark, B.; Johansen, H.; Roos, B.; Wahlgren, U. *J. Am. Chem. Soc.* **1979**, *101*, 5876.
27. Young, G. B.; Whitesides, G. M. *J. Am. Chem. Soc.* **1978**, *100*, 5808 and references therein.
28. Tatsumi, K.; Hoffmann, R.; Yamamoto, A.; Stille, J. K. *Bull. Chem. Soc. Jpn.* **1981**, *54*, 1857.
29. Flores-Riveros, A.; Novaro, O. *J. Organomet. Chem.* **1982**, *235*, 383.
30. Balazs, A. C.; Johnson, K. H.; Whitesides, G. M. *Inorg. Chem.* **1982**, *21*, 2162.
31. Komiya, S.; Albright, T. A.; Hoffmann, R.; Kochi, J. K. *J. Am. Chem. Soc.* **1976**, *98*, 7255.
32. Tatsumi, K.; Nakamura, A.; Komiay, S.; Yamamoto, A.; Yamamoto, T. *J. Am. Chem. Soc.* **1984**, *106*, 8181.
33. Dedieu, A.; Strich, A. *Inorg. Chem.* **1979**, *18*, 2940.
34. Dedieu, A. *Inorg. Chem.* **1980**, *19*, 375.
35. Sevin, A. *Nouv. J. de Chimie* **1981**, *5*, 233.
36. Jean, Y.; Lledos, A. *Nouv. J. de Chimie* **1986**, *10*, 635.
37. Daniel, C.; Koga, N.; Han, J. Fu, X. Y.; Morokuma, K. *J. Am. Chem. Soc.* **1988**, *110*, 3773.
38. Johnson, C. E.; Fisher, B.J.; Eisenberg, R. *J. Am. Chem. Soc.* **1983**, *105*, 7772.
39. Johnson, C. E.; Eisenberg, R. *J. Am. Chem. Soc.* **1985**, *107*, 3148.
40. Janowicz, A. H.; Bergman, R. G. *J. Am. Chem. Soc.* **1982**, *104*, 352.
41. Hoyano, J. K.; Graham, W. A. G. *J. Am. Chem. Soc.* **1982**, *104*, 3723.
42. Jones, W. D.; Feher, F. J. *J. Am. Chem. Soc.* **1982**, *104*, 4240.
43. Crabtree, R. H. *Chem. Rev.* **1985**, *85*, 245.
44. Stoutland, P. O.; Bergman, R. G.; Nolan, S. P.; Hoff, C. D. *Polyhedron* **1988**, *7*, 1429.
45. Saillard, J. Y.; Hoffmann, R. *J. Am. Chem. Soc.* **1984**, *106*, 2006.
46. Ziegler, T.; Tschinka, V.; Fan, L.; Becke, A. D. *J. Am. Chem. Soc.* **1989**, *111*, 9177.
47. Kubas, G. J. *Acc. Chem. Res.* **1988**, *21*, 120.
48. Jean, Y.; Eisenstein, O.; Volatron, F.; Maouche, B.; Sefta, F. *J. Am. Chem. Soc.* **1986**, *108*, 6587.
49. Hay, P. J. *J. Am. Chem. Soc.* **1987**, *109*, 305.
50. Kubas, G. J.; Unkefer, C. J.; Swanson, B. I.; Fukishima, E. *J. Am. Chem. Soc.* **1986**, *108*, 7000.
51. Kubas, G. J.; Ryan, R. R.; Unkefer, C. J. *J. Am. Chem. Soc.* **1987**, *109*, 8113.
52. Arliguiz, T.; Chaudret, B. *Chem. Commun.* **1989**, 155.
53. Chinn, M. S.; Heinekey, D. M.; Payne, N. G.; Sofield, C. D. *Organometallics* **1989**, *8*, 1824.
54. Chinn, M. S.; Heinekey, D. M. *J. Am. Chem. Soc.* **1987**, *109*, 5865.
55. Blomberg, M. R. A.; Siegbahn, P. E. M. *J. Chem. Phys.* **1983**, *78*, 986; *J. Chem. Phys.* **1983**, *78*, 5682.
56. Bagaturyants, A. A.; Anikin, N. A.; Zhidomirov, G. M.; Kzaznskii, V. O. *Zh. Fiz. Khim.* **1981**, *58*, 2035.
57. Bagaturyants, A. A.; Anikin, N. A.; Zhidomirov, G. M.; Kzaznskii, V. O. *Zh. Fiz. Khim.* **1986**, *60*, 1310.
58. Balasubramanian, K.; Feng, P. Y.; Liao, M. Z. *J. Chem. Phys.* **1988**, *88*, 6955.
59. Sevin, A.; Chaquin, P. *Nouv. J. de Chimie* **1983**, *7*, 353.
60. Balasubramanian, K.; Liao, D. *J. Phys. Chem.* **1988**, *92*, 361.
61. Balasubramanian, K.; Liao, D. *J. Phys. Chem.* **1988**, *92*, 6259.
62. Balasubramanian, K. *J. Chem. Phys.* **1987**, *87*, 2800.
63. Balasubramanian, K.; Liao, M. Z. *J. Phys. Chem.* **1989**, *93*, 89.
64. Sevin, A.; Hengtai, Y.; Chaquin, P. *J. Organomet. Chem.* **1984**, *262*, 391.
65. Branchadell, V.; Dedieu, A. *New. J. of Chem.* **1988**, *12*, 443.
66. Trinquier, G.; Hoffmann, R. *Organometallics* **1984**, *3*, 370.

CHAPTER 5

Transition Metal Dihydrogen Complexes: Theoretical Studies

JEREMY K. BURDETT

Department of Chemistry and The James Franck Institute, The University of Chicago, Chicago, IL 60637

and

ODILE EISENSTEIN and SARAH A. JACKSON

Laboratoire de Chimie Théorique, Université de Paris Sud, 91405 Orsay, France

5.1. INTRODUCTION

The discovery [1] of η^2-dihydrogen complexes by Kubas and co-workers in 1984 generated a set of obvious questions: What are the factors influencing their stability? What are the electronic features that stabilize this nonclassical arrangement (**1**) relative to its classical analogue (**2**)?

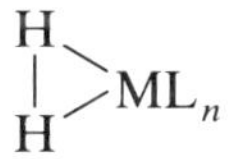

$$\underset{\mathbf{1}}{(\text{H–H})\text{ML}_n} \qquad \underset{\mathbf{2}}{(\text{H})_2\text{ML}_n}$$

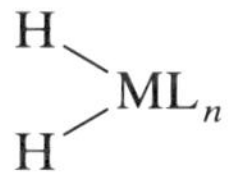

What are the possibilities for the observation of other H_n units (where $n > 2$) attached to a transition metal? In this chapter we study some aspects of this problem. Clearly, the numerical estimation of the energy difference between the dihydride and dihydrogen forms requires the use of accurate calculations, and, indeed, we review some of the calculations made to this end; however, the major thrust of this chapter is to present a picture based on rather simple orbital results predicted by symmetry and overlap arguments. We show how these η^2-dihydrogen complexes are, in orbital terms, a part of a much larger family of molecules, and we use the isolobal analogy to make comparisons between existing main group and transition metal complexes.

5.2. QUALITATIVE DESCRIPTION OF M–(H_2) BONDING

5.2.1. Synergic Interactions

The first [1] η^2-dihydrogen complex to be isolated was $M(CO)_3(PR_3)_2(H_2)$ (the Kubas complex), where M = Mo, W and R = Cy, *i*-Pr. The presence of the sideways-bound η^2-H_2 molecule [written in this chapter as (H_2)] was established by vibrational spectroscopy and by X-ray and neutron diffraction [2]. H–H and W–H distances of 0.82 and 1.89 Å, respectively, have been established by low-temperature neutron diffraction, for $W(CO)_3(P\textit{i}\text{-}Pr_3)_2(H_2)$. Two other complexes, $FeH(PPh_2CH_2CH_2PPh_2)_2(H_2)^+$ [3] and FeH_2-$(PEtPh_2)_3(H_2)$ [4a], were later characterized by the same methods and similar H–H distances were obtained (0.816 Å in the cationic Fe complex and 0.821 Å in the neutral one). Many other nonclassical H_2 complexes have been observed in various media (crystalline state, liquid Xe, gas phase, matrix isolation) and the existence of an H_2 ligand has been verified via several experimental techniques (IR, NMR, X-ray). Although these techniques give good evidence for the presence of a nonclassical H_2 ligand (especially when one or more different techniques give the same result), they do not provide more-precise structural details. Some of the recent key nonclassical (H_2) species include the following:

$M(CO)_5(H_2)$, where M = Cr, Mo, W [5]

$Fe(NO)_2(CO)(H_2)$ [6]

$IrH(H_2)(bq)(PPh_3)_2{}^+$, where bq = 7,8-benzoquinolinate [7]

$RuH_2(H_2)$(cyttp), where cyttp = $PhP(CH_2CH_2CH_2PCy_2)_2$ [8]
$CpRu(CO)(PCy_3)(H_2)^+$, where Cp = η^5-C_5H_5 [9]
(η^4-norbornadiene)$M(CO)_3(H_2)$, where M = Cr, Mo, W [10]
$IrH_2Cl(P(\textit{i}\text{-}Pr)_3)_2(H_2)$ [11]
(P-N)$Ru(H_2)(\mu\text{-}H)(\mu\text{-}Cl)_2Ru(PPh_3)_2H$, where P-N = Fe($\eta^5$-$C_5H_3$ ($CHMeNMe_2$)$P(\textit{i}\text{-}Pr)_2$(1,2)Cp) [12]
(P-P)$Ru(H_2)(\mu\text{-}Cl)_3RuCl$(P-P), where P-P = $Ph_2P(CH_2)_4PPh_2$ [13]
A metalloporphyrin, $Ru(H_2)$ [14a]
An organolanthanide, $(\eta^5\text{-}C_5Me_5)_2Eu(H_2)$ [14b]

Others are reported in reviews by Kubas and Crabtree [15a–15c].

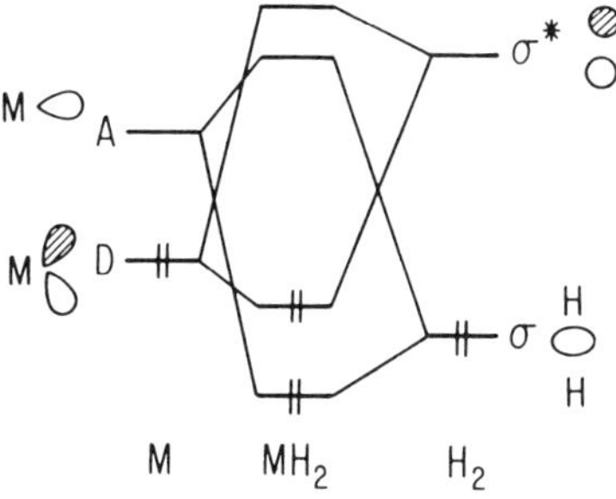

Scheme 5.1

The simplest model (Scheme 5.1) for the η^2 attachment of H_2 to a transition metal is analogous to that used to describe the bonding of olefins and CO to metals. There is an interaction between the filled σ molecular orbital of H_2 (σ_{H_2}) and an empty σ-type orbital on the metal, leading to the formation of a σ bond between the metal and H_2 and to electron donation from H_2 to the metal. There is a second interaction between a filled d orbital on the metal and the σ^* orbital on H_2 ($\sigma^*_{H_2}$). This creates a bond of π-type symmetry and leads to charge transfer in the opposite direction. What makes the H_2 ligand somewhat different from either CO or olefins is that there is no underlying supporting framework in the H_2 molecule analogous to the σ skeletons of olefins and CO. Thus, we can expect that too much back-donation will lead to fission of the H—H bond and the generation of the dihydride MH_2. This result suggests that, by ensuring that the metal is a poor electron donor, the $M(H_2)$ species may be stabilized. Most of the (H_2) compounds characterized so far satisfy this condition in some way. They are either associated with a positive-oxidation-state metal, carry a positive metal charge via protonation, or contain strong π-acceptor ligands on the metal, particularly in the position trans to (H_2). An increase in the oxidation state or a positive charge on the complex will result in a lowering of the metal-centered levels, thus reducing the interaction with $\sigma^*_{H_2}$. The same effect is obtained by moving up a column in the Periodic Table. For this reason, the group VIII complexes like $M(PR_3)_3H_2(H)_2$, where M = Fe, Ru, Os, change from a nonclassical formulation to a classical one going down the group [16a], as the metal-centered levels move up in energy. The kinetics and thermodynamics of the interconversion between dihydride and dihydrogen forms were obtained in the case of $W(CO)_3(PR_3)_3H_2$ [16b]. A strong π-acceptor ligand is particularly effective in favoring $M(H_2)$ [16c] because it reduces the effect of back-donation by diluting the d orbital character on the metal, leading to a commensurately smaller overlap with $\sigma^*_{H_2}$. Thus CO and NO groups play a major role. The results of calculations [17] on some 18-electron carbonyl η^2-(H_2) complexes and their substituted nitrosyl analogs are shown in Scheme 5.2. The population of $\sigma^*_{H_2}$ in the complex is reduced by the presence of the better π acceptor (NO) coordinated to the metal. This reduction in popula-

	H—H	H—H NO	H H	ON H H	H—H	H—H NO	H—H ON NO NO
bop	0.57	0.57	0.54	0.54	0.57	0.57	0.58
σ	1.85	1.79	1.81	1.81	1.85	1.84	1.82
σ^*	0.11	0.09	0.15	0.12	0.11	0.10	0.08

Scheme 5.2

Computed H—H bond overlap populations (bop) and extent of involvement of the H_2 σ and σ^* orbitals in metal–hydrogen bonding for some $M(H_2)$ complexes.

tion represents a strengthening of the H—H bond. Such considerations allow us to understand why $W(CO)_3(PR_3)_2(H_2)$ and $Mo(CO)_3(PR_3)(H_2)$ contain η^2-(H_2) whereas $Mo(PMe_3)_5H_2$ is a dihydride [15a]. In the latter the CO ligands are replaced by the weaker π-acid ligand PR_3. The stability of $Fe(NO)_2(CO)(H_2)$ originates from a different but related phenomenon [6].

Spectral data provides experimental evidence that such effects are important. Carbonyl vibrational frequencies in $M(CO)_5L$ molecules are quite sensitive to the nature of L [18], and the carbonyl stretching vibration bands of $Cr(CO)_5(H_2)$ and $Cr(CO)_5N_2$ are close in frequency [5], showing that (H_2) and N_2 have similar bonding properties and that H_2 is both a σ donor and a π acceptor. By contrast, the vibrational frequencies of carbonyls in dihydrides are intrinsically higher than the analogous nonclassical complexes [10, 19]. The dihydride $W(CO)_3(P\textit{i}\text{-}Pr_3)_2H_2$ has carbonyl bands that are 30–40 cm^{-1} higher than those of its nonclassical isomer [20], which illustrates well the decreased electron-donating ability of the metal that accompanies an increase in the oxidation state.

Morris et al. [21] found a general result concerning the correlation between the stability of **1** and **2** and the N—N stretching vibrational frequency in d^6 molecules where H_2 has been replaced by N_2. If the N—N stretching force constant is greater than 17.5 mdyn $Å^{-1}$ (corresponding to an N—N stretching frequency of 2060 cm^{-1}), then **1** is possible for the H_2 analogue. If these parameters are less than this critical value, then only the classical dihydride structure **2** will be found. Because lower N—N stretching frequencies are associated with increased π back-donation in this ligand, this result then quantifies the maximum extent of backbonding that may occur in an (H_2) complex before it breaks up into the dihydride.

5.2.2. Closed and Open Three-Center Interactions

One of the challenges associated with the description of the electronic structure of molecules and solids is the generation of a model that allows us

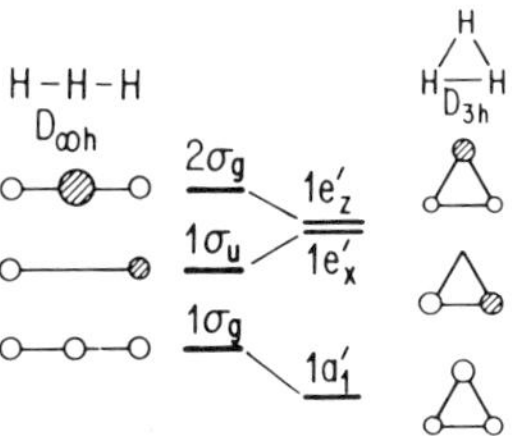

Scheme 5.3

to place the systems under consideration in a broader context. Here we do this by making analogies with main group and organic chemistry. The result will be a greater insight into the structures of these molecules along with some suggestions for further synthetic effort. Here we will start by using simple ideas associated with the stability of closed three-membered rings versus the open counterparts. The orbital correlation diagram (Scheme 5.3) links triangular and linear H_3 units using the Hückel model [22]. It is obvious that the cyclic form is stable for two electrons. The experimentally [23] determined structure for H_3^+ is indeed an equilateral triangle. This result is related to Hückel's $4n + 2$ rule and similar topological stability criteria have been developed recently using the method of moments [24]. In contrast, the cyclic form is Jahn–Teller unstable for the three- or four-electron case and a linear structure is favored because the corresponding Jahn–Teller active mode is of e' symmetry (Scheme 5.4) (one of its components leads to a lengthening of one side of the equilateral triangle to give the linear structure). This result is rather general. Recall that closed, three-center bonding often is associated with "electron-deficient" chemistry, in both main group polyhedral (e.g., the boranes) and transition metal clusters. The presence of extra electrons (either one or two in this case) induces an opening of the ring.

The natural extension of these arguments is to species of the type EH_2, where E is some main group or transition metal fragment. Now the D_{3h} structure has been replaced by one of C_{2v} symmetry. First-order Jahn–Teller arguments are now inappropriate due to the reduction in symmetry, but a second-order Jahn–Teller description of the electronic situation is very useful. A correlation diagram linking the energy levels of linear and bent units is shown in Scheme 5.5. Using this picture, it is easy to appreciate why the ground state of CH_2^+ [with the configuration $(1a_1)^2(1b_2)^2(2a_1)^1$, $^2A_1$] has a bond angle of around 140°. Let us ask what might happen to the geometry

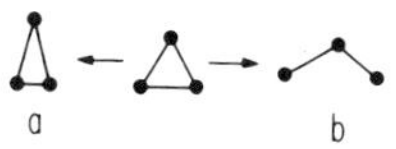

a b

Scheme 5.4

Scheme 5.5

of the molecule in the $^2B_2$ excited state derived from the configuration $(1a_1)^2(1b_2)^1(2a_1)^2$. The $1b_2$ orbital, which is destabilized on bending, is now only half-occupied and the $2a_1$ orbital, which is stabilized on bending via a strong second-order Jahn–Teller interaction, is now doubly occupied. The predicted result is that the HCH angle could well be very small. In fact, Schaefer et al. [25] concluded from ab-initio calculations that the structure of this $^2B_2$ state of CH_2^+ is described by r(C—H) = 1.41 Å, H—C—H angle = 35.2°, and r(H—H) = 0.84 Å. The H—H distance is strikingly similar to that in the coordinated H_2 unit in the nonclassical H_2 complexes. The electronic difference between the triangular–linear H_3 problem and the linear–sharply bent EX_2 problem is the relative energetic location of the orbitals that go up and down in energy on bending.

Using such three-center ideas, we can now examine the general requirements for a stable nonclassical H_2 species, **1**. Because the H_2 molecule contributes two electrons to the bonding picture, one criterion is that the fragment ML_x must have a vacant, energetically accessible orbital, geometrically located so as to provide a strong interaction between metal and H_2, that is, orbital A in Scheme 5.1. The isolobal analogy may be applied now. Instead of constructing molecular orbital diagrams for each case, we may simply use the frontier orbitals of relevant fragments. Using this strategy, many unsaturated transition metal organometallic fragments and main group units are candidates for H_2 coordination, as shown in **3–7**.

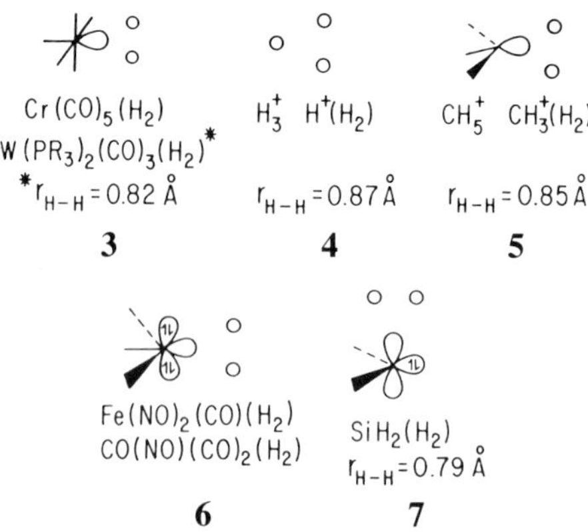

The species $CH_3(H_2)^+$ (**5**) and $SiH_2(H_2)$ (**7**) are good starting systems for the discussion, although here we have to rely on high-quality calculations and not on experimental facts. The species CH_5^+ has been observed [26], but only in a mass spectrometer, and **7** lies [27] in a local energy minimum between silane (the "dihydride") and separate SiH_2 and H_2 units. Minima of this type also have been calculated [28] in the reaction between H_2 (or HF) and the SiF_2 radical. Notably, no minimum is found [29] for similar reactions with CH_2, an important result we will return to later. The result for H_3^+ (**4**) comes from high-resolution gas-phase spectra [23]. There are numerous examples of suitable metal-containing fragments, falling into two groups represented in **3** and **6**. In one group, such as **3**, σ_{H_2} can overlap only with an empty orbital of the fragment to which it is bonded. In the other group, such as in **6**, σ_{H_2} can overlap in addition with occupied orbitals, which leads to some destabilization that might or might not be compensated by the influence of the empty orbital. Systems of the first group therefore are more stable in general.

A simple extension of these ideas leads to the anticipation of cases where there are two empty frontier orbitals and the possibility of coordinating a second H_2 molecule. There are two examples of *bis*-dihydrogen metal complexes: $IrH_2(H_2)_2(PR_3)_2]^+$ [7b, 7c] and $Cr(CO)_4(H_2)_2$ [5c] (**8**).

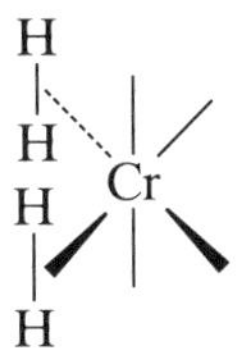

8 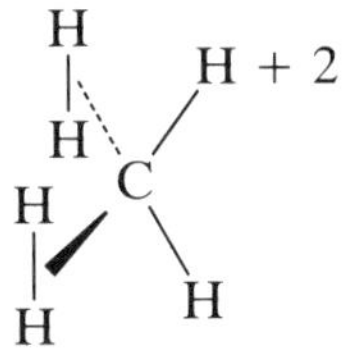 **9**

Ab-initio calculations on **8** recently confirmed the preference for the molecular H_2 complex over a polyhydride structure [30]. A main group analogue is Olah's predicted structure of CH_6^{2+} (**9**) [31], whose stability and geometry was confirmed by accurate ab-initio calculations. In terms of the isolobal analogy, the fragment $Cr(CO)_4$ is isolobal with CH_2^{2+}, just as the fragment $Cr(CO)_5$ is isolobal with CH_3^+.

5.2.3. Reactivity of H_2 toward Metal Centers

Such a model should be applicable to reactions of dihydrogen too. One example is the generation of a dihydride **2** in the oxidative addition reaction of H_2 to low-spin d^8 square-planar complexes. A typical orbital diagram for such ML_4 molecules is shown in Scheme 5.6. Bringing up an H_2 molecule along the z axis of such a system, parallel to the square plane, will lead initially to a strong interaction between the filled a_{1g} (z^2/s) hybrid orbital of the square-planar unit and σ_{H_2}. A two-orbital–four-electron destabilization results (Scheme 5.7), which can be diminished by approaching the H_2

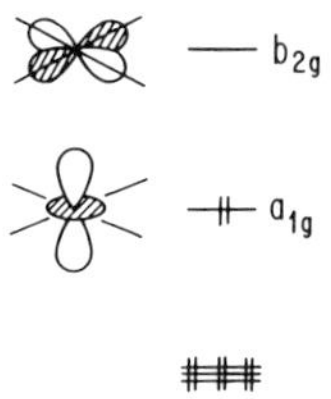

b_{2g}

a_{1g}

Scheme 5.6

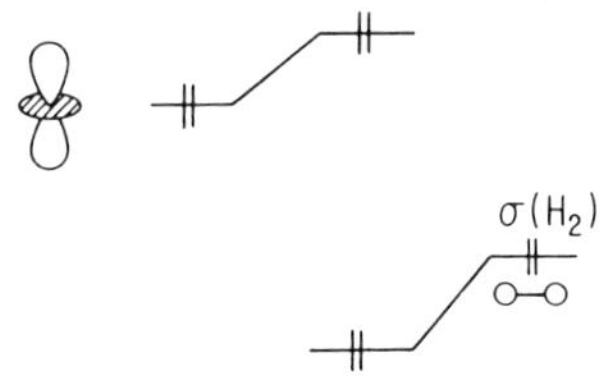

$\sigma(H_2)$

Scheme 5.7

molecule in an almost-η^1 manner, as has been shown by Saillard and Hoffmann [32] and by Sevin [33] by means of extended Hückel-type (EHT) calculations. Jean and Lledos [34] have shown that this four-electron repulsion is responsible for the stereoselectivity of the addition of H_2 to unsymmetrical square-planar complexes, in agreement with the experimental results of Eisenberg and co-workers [35]. By using the extended Hückel method, they have shown that this four-electron destabilization is minimized when the H_2 lies in the plane of stronger π-acceptor ligands (CO) because there is a delocalization of the filled a_{1g} into π^*_{CO}, which becomes possible when CO bends away from the incoming H_2 ligand. Notice that coordination of H_2 is not favored in this case because, in addition to the presence of the occupied metal orbital, the first empty orbital of the molecule (b_{2g}) is not only of higher energy but also is situated inconveniently in a geometric sense for good metal-hydrogen overlap (an incoming molecule must have a δ-type overlap with b_{2g}). In broader terms, because this three-orbital problem contains four electrons, the closed three-center arrangement is unstable and the eventual generation of a dihydride is encouraged.

We can also envisage coordination of H_2 in an end-on, or η^1, fashion. For stability of such a structure, the ideas developed here suggest that we need a filled fragment orbital for interaction with H_2 so as to produce a four-electron-three-orbital picture. A d^8 square-planar ML_4 fragment like that in Scheme 5.6 would fit this requirement. The resulting molecule would have a linear M—H—H unit (perpendicular to the ML_4 plane) and be isoelectronic with linear H_3^-. A related species has been made recently [36], consisting of a PtL_4 unit attached to an I_2 molecule such that the Pt—I—I fragment is linear. This molecule is isoelectronic with I_3^-. Using the same

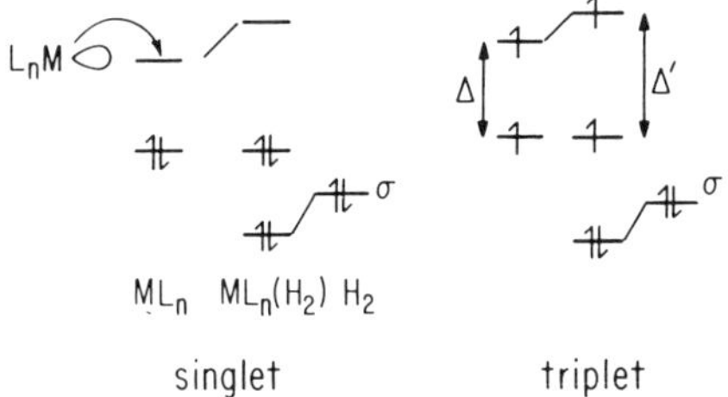

Scheme 5.8

arguments, a square-pyramidal d^8 $ML_4(\eta^1\text{-}H_2)$ species also might be possible.

Along similar lines, the availability of a low-lying triplet state may alter dramatically the propensity for dihydrogen complex formation. The one-electron energy levels and their occupancy for the singlet and triplet forms of an $ML_2(H_2)$ complex is shown in Scheme 5.8. Clearly, the singlet form could give rise to a dihydrogen complex, but the triplet form with electrons as in Scheme 5.8 most certainly will not. The isolobal pair $Fe(CO)_4$ and CH_2 are species where such arguments may be appropriate. For CH_2 the triplet state lies [37] lower in energy than the singlet by about 0.47 eV. $Fe(CO)_4$ is an unusual carbonyl complex because MCD studies have demonstrated that it exists as a triplet in its ground electronic state [38]. As we mentioned before, theoretical studies [29] on the reaction of H_2 with CH_2 show no evidence for a local minimum at a species akin to $SiH_2(H_2)$ (**7**). This is a result that may be understood by recognizing the prevalence of the triplet state of CH_2 as H_2 approaches. In contrast, we note that the singlet state of SiH_2 lies at lower energy than the triplet by about 0.8 eV [37]. Experimental studies on the addition of H_2 to $Fe(CO)_4$ in low-temperature matrices [39] show only the formation of the dihydride [the well-known molecule $Fe(CO)_4H_2$] and no evidence for the formation of a dihydrogen complex. Because these experiments are carried out at extremely low temperature, any local well at the dihydrogen complex geometry must be either tiny or nonexistent. The three types of potential surface for the reaction of H_2 with a suitable fragment are shown in **10–12**.

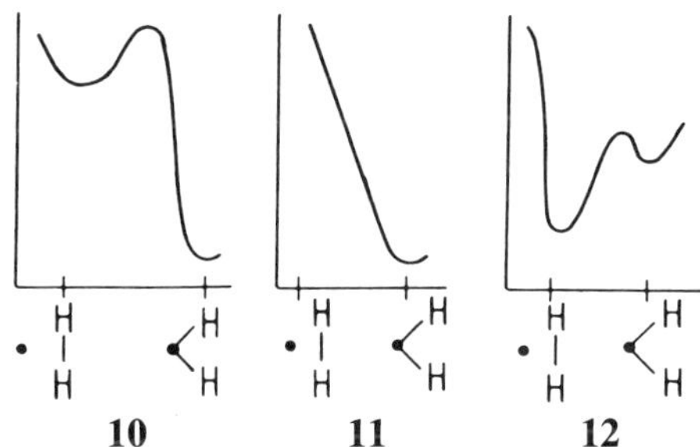

When the dihydrogen complex exists as a local minimum, for example,

$SiH_2(H_2)$ (7), the surface is described by **10**. However, even though it is predicted by high-level theory, the species $SiH_2(H_2)$ was not detected in a matrix experiment [40a]. Photophysical quenching of alkali metals by small molecules such as HX, N_2, and C_2H_2 has been explained using related phenomena [40b]. The surface in **11** represents the barrierless case as found for CH_2 and H_2 and perhaps for $Fe(CO)_4$ and H_2. Finally, **22** corresponds to the situation where the dihydrogen complex lies lowest in energy.

Many structural problems in chemistry are controlled by the balance between one- and two-electron terms in the energy. Perhaps the best known are the factors that control the high-spin–low-spin question in transition metal complexes. A similar balance exists here. We know that Δ in Scheme 5.8 is smaller than the pairing energy in the $Fe(CO)_4$ molecule because the triplet is the lower-energy structure, but whether Δ' is large enough to stabilize a dihydrogen complex in a thermodynamic sense is a question that is extremely difficult to answer. There is one piece of information that might help us here. Reaction of triplet $Fe(CO)_4$ with X = CH_4, Xe, N_2, or CO in matrices leads to formation of singlet $Fe(CO)_4X$ species [41]. If H_2 is a better σ donor than X then nonclassical $Fe(CO)_4(H_2)$ is a possibility. Naturally this does not exclude the existence of the dihydride as a lower-energy species [as in the case of $SiH_2(H_2)$].

5.3. MORE-DETAILED FORM OF THE ORBITALS

5.3.1. Rotation Barrier of H_2 in d^6 ML_5 Dihydrogen Complexes

Neutron diffraction studies of some d^6 ML_5 dihydrogen complexes have revealed that H_2 has a preferred orientation at low temperature. However, the rotation barrier of the weakly bonded ligand [42a] H_2 ligand about the metal H_2 axis is extremely small. Indeed inelastic neutron scattering (INS) measurements have estimated it to be between 1 and 2.5 kcal mol^{-1} [4a, 42b, 42c]. The preferential orientation of the H_2 is given in **13–15**.

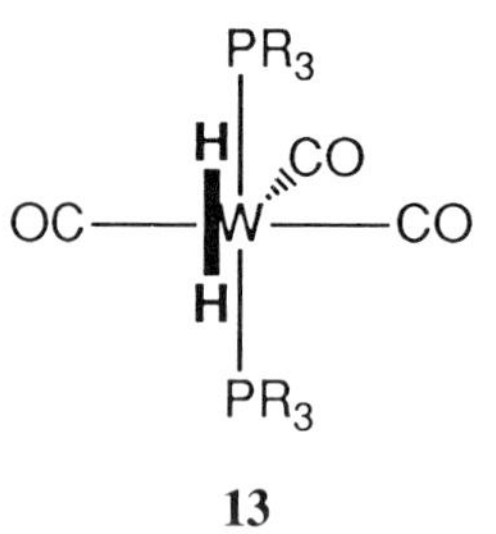

13 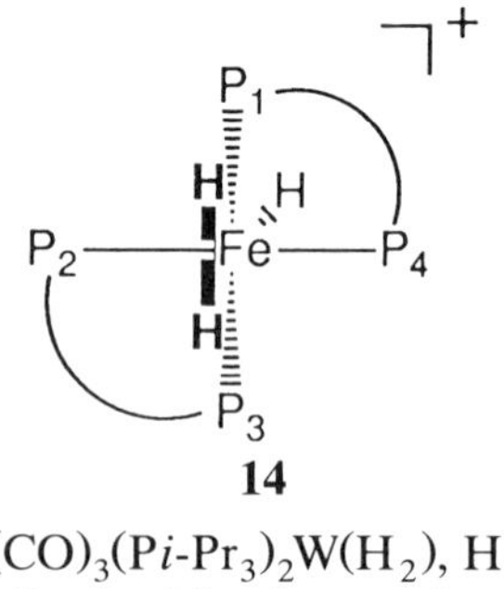 **14** 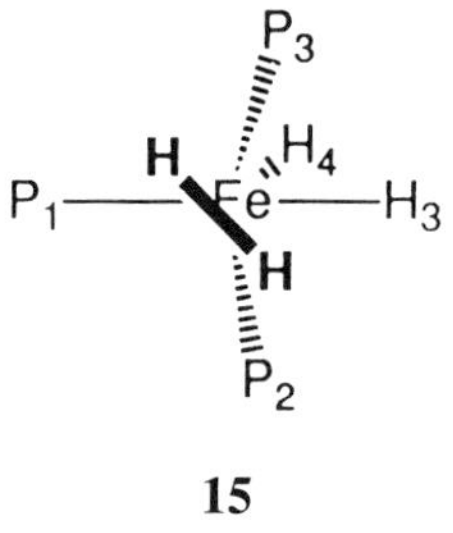 **15**

In the Kubas complex **13**, $(CO)_3(Pi\text{-}Pr_3)_2W(H_2)$, H_2 is found to lie parallel to the W–P direction [2, 15a]. In the Morris complex **14**, $H(P\text{-}P)_2Fe(H_2)$, where P-P = $P(Ph)_2CH_2CH_2P(Ph)_2$, H_2 is cis with respect to all four phosphines

[3]. However, the environment is not isotropic and H_2 eclipses a specific P—Fe—P vector. The octahedron is distorted and the H_2 lies in the plane of the two trans phosphine ligands, which are tilted away from H_2 (P_1 and P_3). In the Caulton complex **15**, $(PEtPh_2)_3(H)_2Fe(H_2)$, the orientation is different [4a]. In this complex the dihydrogen does not eclipse a metal–ligand bond but is found staggered in between the *cis*-Fe—P and *cis*-Fe—H axes. This orientation is not only unique among structurally characterized octahedral H_2 complexes but is also unprecented for other dihapto ligands such as olefins bonded to a d^6 ML_5 fragment.

The preferential orientation has been correctly reproduced by EHT [43] and ab-initio [42b, 44] calculations for **13**, and by EHT calculations for the two Fe(II) complexes **14** and **15** [4a]. The rotational barrier has been calculated to be between 0.3 and 4 kcal mol^{-1} depending on the method used. The preferential orientation of the H_2 ligand can be explained easily by using a fragment molecular orbital analysis. Here one looks for the best interaction between the square-pyramidal ML_5 fragment and a dihapto H_2 ligand (Figure 5.1). The frontier orbitals of ML_5 are shown on the left-hand side of Figure 5.1. The lowest set of orbitals are *xy*, *yz*, *xz* in the present coordinate system. They are σ nonbonding and therefore are degenerate if L is purely a σ donor. Higher in energy, one finds an orbital made of z^2, z, and s metal character, which is σ antibonding with all ligands and especially the one opposite the empty site in ML_5. This metal orbital is directed toward the

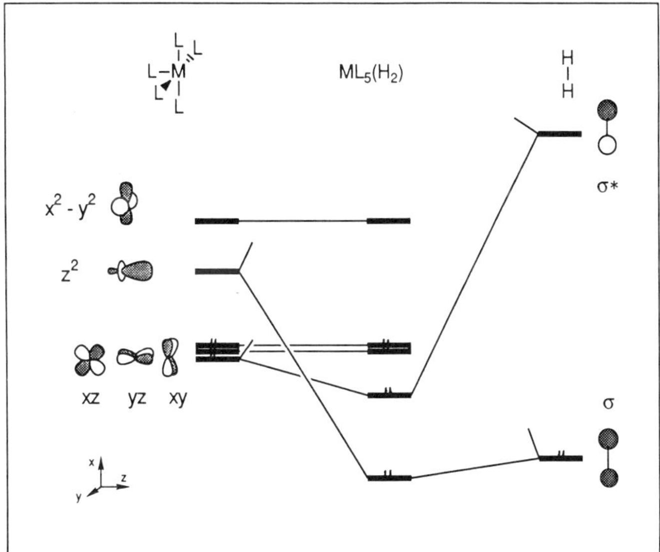

FIGURE 5.1
Qualitative molecular orbital diagram for the assembly of $ML_5(H_2)$ from the ML_5 fragment and H_2.

empty site of the pyramid. The next-higher orbital, $x^2 - y^2$ is antibonding with respect to the four ligands forming the basal plane and plays no role in the bonding of the metal fragment to the dihydrogen ligand because of its high energy and incorrect orientation. For a d^6 electron count, xy, yz, and xz are filled and σ is the lowest unoccupied molecular orbital (LUMO). The orbitals of H_2 are the occupied σ_{H_2} and empty $\sigma^*_{H_2}$.

Let us suppose first that H_2 lies in the xz plane. The calculations show a stabilization resulting from the interaction of occupied and empty orbitals of relevant symmetry on the ML_5 and H_2 fragments. The interaction of the metal σ with σ_{H_2} is responsible for the metal–(H_2) σ bond and for the forward electron transfer from the ligand toward the metal, as discussed in the previous section. This interaction cannot be at the origin of any orientational preference because this interaction is of cylindrical symmetry and will not be discussed further here. The interaction of the metal xz with the $\sigma^*_{H_2}$ orbital is responsible for the metal–H_2 π bond and for the back-donation from the metal to the ligand in a bonding mode analogous to that of Dewar, Chatt, and Duncanson. However, H_2 is a weak ligand when compared to a more conventional ligand (such as ethylene) on a metal, because the energy gap between the metal orbitals and the H_2 orbitals is large, that is, σ_{H_2} is a low-lying and $\sigma^*_{H_2}$ a high-lying orbital.

We wish to understand why H_2 is found to eclipse an L–M–L axis in two of these complexes but not the third. Similar eclipsing has been found in ethylene complexes for which the bonding scheme to the metal is similar to that of H_2 and the electronic reasoning for this preference has been well-developed [45]. In a complex in which the four ligands cis to the olefin are identical, the olefin π^* overlaps with either the xz or yz orbitals or the linear combination of them controlled by the geometry. This result leads to the prediction that there should be no rotation barrier because by symmetry the total overlap is independent of angle. However, the staggered geometry is disfavored by a four-electron destabilization due to the overlap of the occupied xy orbital with the occupied olefin π orbital (**16**). This will give rise to an electronic contribution to the rotational barrier even when all the ligands are equivalent. A similar effect should apply here too.

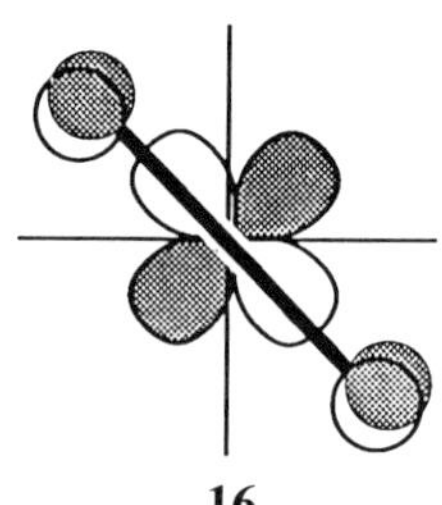

16

In the case of **13**, the xz and yz orbitals are not of equal energy because one (yz in our set of axes) is stabilized by three π^*_{CO} orbitals (**17**), whereas xz

is stabilized by only one π^*_{CO} orbital (**18**).

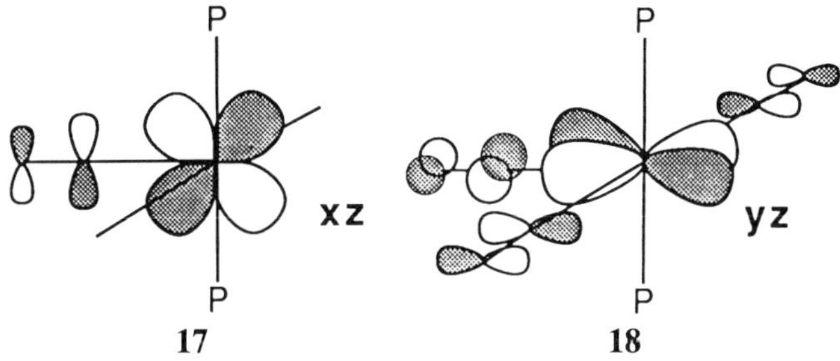

The best back-bonding is thus obtained when the highest-energy metal orbital, *xz*, overlaps with $\sigma^*_{H_2}$. For this reason H_2 aligns itself with the P–W–P axis, even though this orientation is the one that is least-favored on steric grounds.

In the case of **14**, the four cis ligands are of identical chemical nature but the environment is not isotropic. The pinching of two trans phosphine ligands (P_1 and P_3) toward the terminal H ligand causes a destabilization of *xz* because the phosphine lone pairs mix into *xz* in an antibonding way. In addition, *xz* becomes extended away from these two phosphine ligands as a result of *p–d* mixing [4a]. Both effects make such a distorted *xz* orbital (**19**), lying in the plane of the smaller transoid P—Fe—P, a better candidate for back-donation than the *yz* orbital.

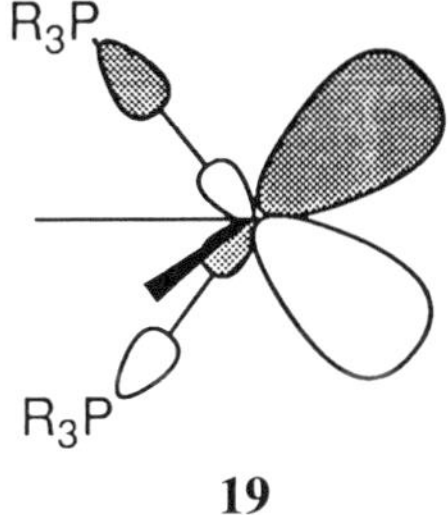

19

One question we need to answer is why, unlike olefins, H_2 sometimes can be staggered with respect to the cis ligands as in complex **15**. We mentioned previously that an olefin is energetically unfavorable when staggered with respect to the cis ligands. This originates from the overlap of the occupied π_{CC} with the occupied metal *xy* orbital as shown in **16**, giving rise to a four-electron destabilization. By rotating the olefin by 45°, the overlap between these two occupied orbitals goes to zero, which makes the eclipsed conformation more stable. However, the energy difference will be linked to the size of the overlap. What can be shown by calculation is that, in the case of H_2, the overlap between σ_{H_2} and *xy* is very small so that the staggered conformation is not so strongly disfavored in an electronic sense. In a complex rotation where the four ligands cis to H_2 are identical, no rotation barrier is calculated for the rotation of H_2 whereas the staggered conformation is calculated to be 10 kcal mol^{-1} higher than the eclipsed one in $Cr(CO)_5$ (C_2H_4) [45b]. This is an important difference between the bonding scheme of olefin and H_2.

However, the preceding arguments need to be expanded in order to provide a complete interpretation of the structure of **15**. Because the pinching of the two transoid phosphine ligands is considerable (149.7° in **15**, versus 172.1° in **14**) an eclipsed conformation should have been preferred. Another factor turns out to be important here. This is the interaction of $\sigma^*_{H_2}$ with an occupied orbital best described as representing the Fe—H σ bond (**20**).

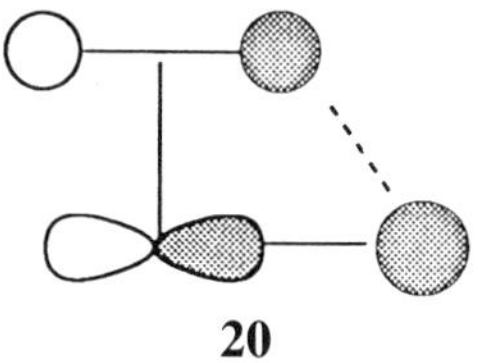

20

This attractive interaction is similar to the cis-effect that has been suggested to take place between an electron-rich ligand and a ligand bearing acceptor orbitals [46, 47] and, of course, is somewhat analogous to the anomeric effect in organic chemistry. This interaction reaches its maximal value when the H—H bond is parallel to the Fe—H axis. The actual orientation of H_2 is therefore the result of a compromise between the back-bonding interaction, which favors the alignment of H—H with the transoid P—Fe—P direction, and the cis interaction, which favors alignment with Fe—H.

Thus it appears that the preferred orientation of H_2 is very sensitive to subtle structure distortions at the metal center. This is evident in the fact that H_2 in complex **15** is staggered with respect to the phosphine ligands and the hydride in a specific manner. Turning the H_2 ligand by 90° to reach the other staggered orientation is energetically unfavorable. The nonequivalence of these two staggered orientations has been shown to be due to the structural distortion of the complex away from idealized O_h situation. This causes a spatial reorientation of the formally nonbonding "t_{2g}" set as well as raising the degeneracy between them. The factor mostly responsible for the nonequivalence of the two staggered structures is the difference in the cisoid P—Fe—P angles. The P_1—Fe—P_3 angle is larger (105.3°) than the P_1—Fe—P_2 angle (97.6°). Consequently, the lowest member of the t_{2g} set is tilted as shown in **21** to avoid interaction with the ligands.

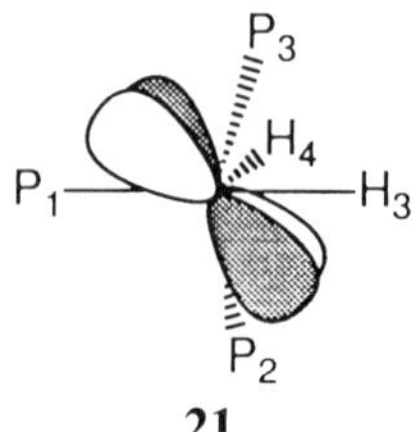

21

This orbital interacts best with $\sigma^*_{H_2}$ when the H—H bond is above the P_1—Fe—P_3 and P_2—Fe—H quarters. Thus, the experimental structure corresponds to a compromise between the optimal back-donation from the set of distorted metal orbitals and the cis-interaction. This attractive interaction found between an H_2 ligand and a cis Fe–H linkage has been shown to induce a vestigial bond between the closed nonbonded hydrogen centers (**20**) and may play some role in facilitating hydrogen exchange. As a matter of fact, many polyhydrides are reported to be highly fluxional on the NMR time scale.

Are there any other ligands that can force a similar cis interaction? From our present state of knowledge in this field it seems that the metal hydride is one of the best candidates, although calculations have indicated that it also occurs in cis carbene–olefine complexes [47], where it is responsible for the preference in some complexes for the conformation favoring the formation of a carbon–carbon bond as shown in **22**.

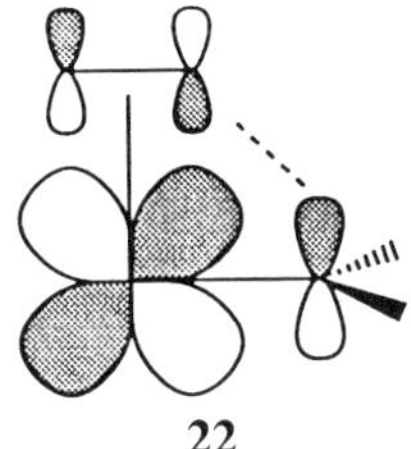

22

There are many reasons behind the fact that the metal hydride bond might play a privileged role. First, the σ M—H bond is strongly localized onto the hydride *s* orbital and is high in energy due to the accumulation of electronic density on the hydridic center. The spherical nature of the hydride *s* orbital also makes it possible to overlap with centers in its vicinity without cleaving the metal–hydride bond. For these reasons the interaction between a metal–hydride *s* orbital and a neighboring vacant orbital can play an important role. Bridging hydrides seem to behave similarly to terminal ones because analogous effects have been calculated in the case of an H_2 ligand coordinated to one Ru center in a dinuclear face sharing bioctahedron (P—N)Ru(η^2H_2)(μH)(μCl)$_2$Ru(PPh$_3$)$_2$H [48].

In electronic terms there are some restrictions on the type of ligand that can interact with the hydride in this way. Clearly, it needs to have an available empty orbital properly situated to overlap with the cis hydride center. H_2 is a good candidate because its proximity, to the metal–hydride bond makes the interaction possible despite the high energy of the $\sigma^*_{H_2}$ orbital. Another obvious candidate is an olefin. Its π^* orbital is ideally situated to make this type of interaction, as shown in **23**. Indeed it has been found that, in the *mer*-H(C_2H_4)$_2$Os(PMe$_2$Ph)$_3$$^+$ (**24**), one of the ethylenes orientates preferentially in the directions of the *cis*-Os—H linkage [49].

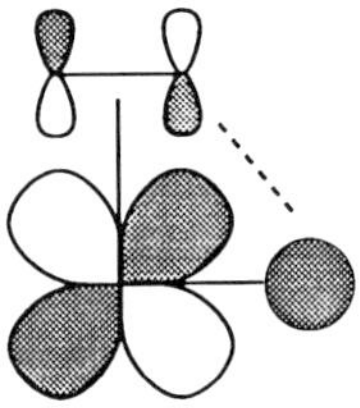

23

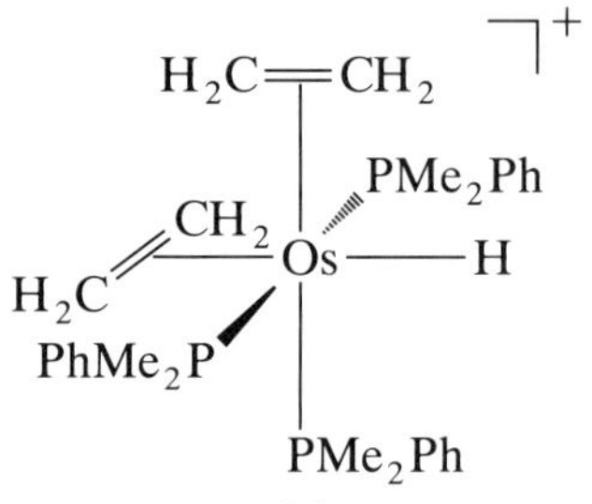

$H_2C{=}CH_2$ ⌉+
PMe_2Ph
CH_2
H_2C Os—H
$PhMe_2P$
PMe_2Ph

24

It is worth noting that this weak interaction between an occupied M—H σ bond and an empty neighboring π^*_{CC} orbital is nothing but an alternative representation of the agostic interaction [50]. In both cases a weak two-electron stabilization is at work between two centers. In the case of the cis effect, a rather active electron-donating center, the M—H bond, donates electrons to a weak acceptor orbital, the π^*_{CC} already involved in back-donation with the metal. In the agostic interaction, a very weak electron-donating bond, the C—H bond, donates electrons to a low-lying orbital centered on the metal and directed toward the missing ligand (**25**).

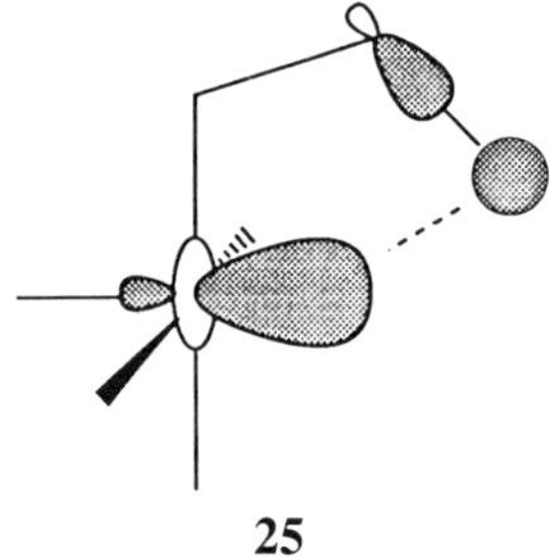

25

It is interesting to note that the same type of interaction stabilizes the reactants and the products of the M-ethyl–M—H-olefin exchange reaction. The interactions are sufficiently weak in both cases to be offset easily by other effects, but the interactions may well be very important in controlling the reactivity of these complexes.

Although all these calculations reproduce well the preferred conformation of H_2, a quantitative estimate of the rotation barrier has met with less success. This is not surprising considering that it is still difficult to reach such energetic accuracy especially in the case of transition metal complexes. Extended Hückel calculations on a model compound replacing all phosphine ligands by PH_3, assuming a rigid rotation of H_2 (situated at the experimental distance from the metal) about the metal mid-H_2 axis with no relaxation of the metal fragment, gives a rotation barrier of less than 2 kcal mol^{-1}, which is reasonably close to the experimental value [4a, 43]. Ab-initio calculations have had a variable success. In one set of calculations, Hay [44] has obtained W—H distances that are much too long when optimizing the position of H_2 attached to a frozen metallic fragment by ab-initio pseudopotential self-consistent field (SCF) methods. As a consequence, the rotation barrier of $W(CO)_3(PH_3)_2(H_2)$ was calculated to be too small (0.3 kcal mol^{-1}). Better values of the bond distances and therefore of the rotation barrier were obtained by improving the treatment of some internal orbitals of the metal [42b]. The influence of the nature of the phosphines also has been studied and it has been shown that there is no direct steric effect on the H_2 rotational barrier arising from the bulky ligands [42b]. Ab-initio calculations are underway in order to quantify the cis effect further [4b].

5.3.2. H_2 versus Dihydride

A crucial problem in this field is the question of the stability of the nonclassical dihydrogen complex versus that of classical dihydride. We have described in previous sections some of the factors that control this preference. By changing the metal and the spectator ligands, it is possible to shift the preference from one structure to the other. One especially interesting result is that it is possible to favor one structure over the other just by changing the site of coordination of H_2. Consider a trigonal d^8 pyramidal structure. Equatorial and apical ligands are symmetry-inequivalent and therefore there are a set of site preferences for ligands with given properties. It is known that the strongest π acceptor should occupy an equatorial site because the back-donation is significantly stronger here whereas a π donor prefers an apical site [51]. An olefin ligand, for instance, occupies the equatorial site where the C—C bond is significantly elongated. What would happen in the case of an H_2 ligand? To simplify the discussion, we will suppose that no other competing electron acceptor is present on the complex. If H_2 is located at the apical site, it will suffer a rather weak back-donation. If it is located at the equatorial site, back-donation is stronger, which although in principle it creates a more stable situation actually leads to a different result. Excessive electron transfer into $\sigma^*_{H_2}$ leads to elongation of the H—H, resulting in eventual cleavage of the dihydrogen ligand and transformation of the complex into a hexacoordinated d^6 dihydride complex. The same conclusions can be reached in a more formal way by looking at the interaction of C_{3v} and C_{2v} d^8 ML_4 fragments with an H_2 ligand. Ab-initio calculations have confirmed this qualitative picture [52a] and two such molecules have been characterized. A molecular H_2 ligand is coordinated at the apical site of the C_{3v} fragment d^8 $Rh(P(CH_2CH_2PPh_2)_3)^+$ in $Rh(P(CH_2CH_2PPh_2)_3)(H_2)^+$, whereas the dihydride isomer is known as C_{2v} $Rh(P(CH_2CH_2PPh_2)_3)H_2^+$ [52b].

5.4. POLYHYDROGEN STRUCTURES WITH ONE METAL ATOM

There are, as yet, no really well-established structures that contain coordinated H_n units with $n > 2$. There is the compound studied by Heinekey [53a], $CpIr(PMe_3)H_3$, and several related species, $CpRu(PCy_3)H_3$ [54a–54c] and Cp_2NbH_3 [54d], where open, allyl-like trihydrogen species were postulated on the basis of unusually large J_{HD} coupling constants in the NMR. However, a neutron-diffraction study of the iridium complex demonstrated that $CpIr(PMe_3)H_3$ has a trihydride structure in the solid state and debate has centered [53b–53e] around the interpretation of the very unusual NMR spectrum of these species. However, polyhydrogen species need to be considered as models for transition states [55] or intermediates in hydrogen exchange reactions and are thus worth exploring. Hydrogen exchange reactions have been observed by NMR in numerous mixed hydride–dihydrogen complexes such as $FeH_2(PEtPh_2)_3(H)_2$ [4a], $IrH(H_2)(bp)(PPh_3)_2^+$ [7], and $IrH_2(H_2)_2(PCy_3)_2^+$ [7b, 7c]. Isotopic exchange has been observed in experiments on $(CO)_3W(P\textit{i}\text{-}Pr_3)_2(H_2)$ compound [56] in both solution and the solid state at 20°C, giving a statistical mixture of H_2, HD, and D_2 species within a few days. This exchange does not involve dissociation of the phosphine ligands, which therefore precludes a mechanism in which both H_2 and D_2 are coordinated to two empty coordination sites. The cleanest evidence yet for exchange has been recorded in liquid xenon at −70°C during experiments with $Cr(CO)_{6-x}(H_2)_x$ species [5c]. Many processes possible in a more conventional solvent are excluded in these experiments. First, no exchange occurs with $Cr(CO)_5(H_2)$ because generation of this species photochemically in an H_2–D_2-containing solution leads only to $Cr(CO)_5(H_2)$, $Cr(CO)_5(D_2)$, and no $Cr(CO)_5(HD)$. However, it appears that the *bis*-dihydrogen complex is involved in hydrogen scrambling because after its generation in an H_2–D_2-containing environment, the molecule $Cr(CO)_5(HD)$ is generated thermally in solution. The simplest scheme to link these two results is the following:

$$Cr(CO)_6 \xrightarrow[H_2]{h\nu} Cr(CO)_5(H_2) \xrightarrow[D_2]{h\nu} Cr(CO)_4(H_2)(D_2)$$

$$\xrightarrow{\Delta} Cr(CO)_4(HD)_2 \xrightarrow{\Delta} Cr(CO)_5(HD)$$

It should be emphasized that the scrambled species $Cr(CO)_4(HD)_2$ is inferred and not positively identified. However, as we have described, no scrambling occurs with $Cr(CO)_5(H_2)$ with D_2 in solution. The implication that has been drawn is that scrambling only occurs when both H_2 and D_2 are coordinated to the same metal center. We now examine theoretically some of the structures that are geometrically feasible for the case of four hydrogen atoms, in order to cast light on the various possibilities for H–D exchange.

We begin by looking at some species [7] that could be intermediates for the scrambling reaction in $Cr(CO)_4(H_2)_2$, on the basis that the mechanism has to be consistent with the lack of observation of a similar reaction in $Cr(CO)_5(H_2)$. We exclude intermolecular processes such as that which could arise via attack of a hydride on coordinated (H_2), because this is unlikely in the liquid xenon medium used for the experimental studies. Some possible intermediates containing four hydrogen atoms are shown in **26–29**, three of which contain polyhydrogen units.

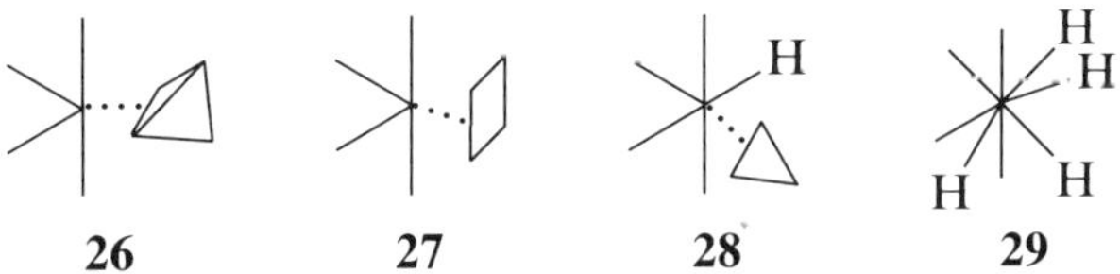

There are clearly orientational isomers for all of these species. Clusters of hydrogen atoms with these geometries are presently not known by chemists, although they are all known for main group atoms. The coordinated tetrahedron **26** is known for phosphorus [57, 58], and the other geometries are known for unsaturated hydrocarbon units. However, clusters of hydrogen atoms have been identified mass-spectrometrically [59] up to quite high molecular weights. Theoretical studies [60] have shown that the geometries expected for these clusters are very different indeed from those required here. The species H_5^+, for example, is predicted to have a distorted "spiropentahydrogen" structure. Simple one-electron calculations are insufficient to rank the species **26–29** accurately by energy; however, by making analogies with molecules containing electronically related ligands, we will be able to assess their structural stability. In principle, there are open (**30** and **31**) analogues of these closed coordinated H_3 and H_4 units (**26–28**).

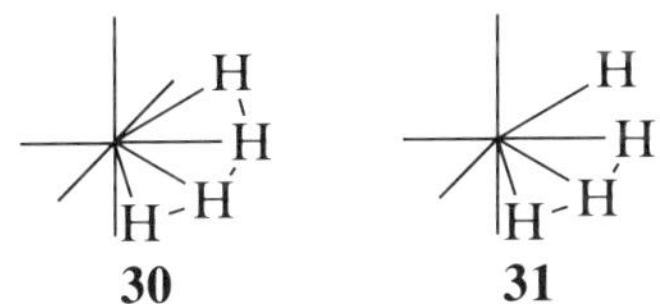

We report in Table 5.1 the results of ab-initio SCF calculations performed independently by Kober and Hay [30a] and Pacchioni [30b] on isomers of $Cr(CO)_4(H_2)_2$ along with results from EHT calculations [17]. Although the relative energy differences for these different calculations cannot be compared, it is clearly apparent that two of the isomers, the tetrahedral **26** and the square **27**, are unlikely to be viable intermediates. Rather large discrepancies in numerical values are found for the relative energies of the other isomers. Considering the well-known difficulties associated with these calculations, it is premature to draw definite conclusions from them. It should be noted that other isomers containing two hydrides and an

Table 5.1
Calculated Energy Differences between Some Tetrahydrogen Complexes

	Structure	Energy (kcal mol^{-1})		
		EHMO	Ab initio	
13	$Cr(CO)_4(H_2)_2$	0^a	0^b	0^c
	$Cr(CO)_4(H_2)H_2$	—	12	29
34	$Cr(CO)_4(H_4)$ (tetrahedral)	160	—	219
35	$Cr(CO)_4(H_4)$ (square)	86	54	130
36	$Cr(CO)_4(H_3)H$ (triangular)	35	76	133
39	$Cr(CO)_4(H_3)H$ (linear)	22	22	82
37	$Cr(CO)_4H_4$ (tetrahydride)	36	65	70
38	$Cr(CO)_4(H_4)$ (open)	17	23	51

[a]Reference 17.
[b]Reference 30a.
[c]Reference 30b.

H_2 ligand are calculated to have a lower energy than some of those mentioned in Table 5.1 and might play a role in the H exchange process. We will refer to them later on. For the moment we concentrate on the topological factors that make some structures viable as intermediates.

A first step is to recognize the level structure of the free ligands, shown in Scheme 5.9. The tetrahedron, triangle, and diatomic have only one bonding level (the next orbitals are higher in energy than that of an isolated

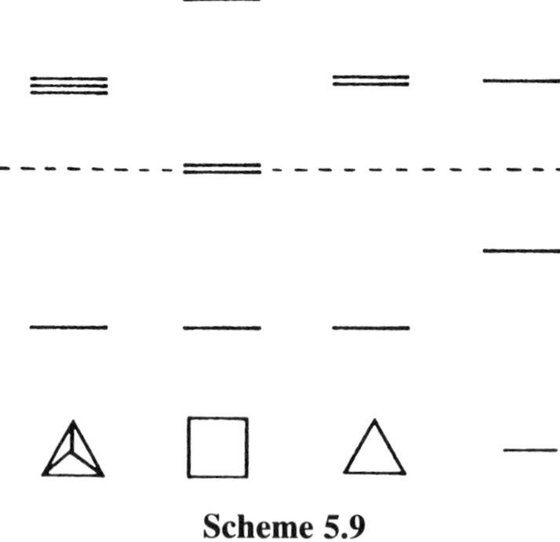

Scheme 5.9

H center in this simple analysis), and the square has one bonding and two nonbonding levels. Thus we can imagine that in $Cr(CO)_4(H_2)_2$ the four electrons of $(H_2)_2$ are located in deep-lying orbitals and mostly centered on σ_{H_2}, as they are too in $Cr(CO)_4H(H_3)$ (**28**). In **26** and **28** there is only one orbital to house four electrons, so two must lie in higher-energy ligand or metal-located orbitals. These are the intermediate structures calculated to be of highest energy. On our orbital model, a metal–hydride linkage results in a pair of electrons lying in a bonding orbital, and it is perhaps no surprise that **28** and **29** are close in energy. Thus our simple one-electron ideas would suggest that either of these two species, **28** or **29** is a viable intermediate. Better calculations than these are required to distinguish between them.

The calculated charge distribution for **26** is shown in **32**.

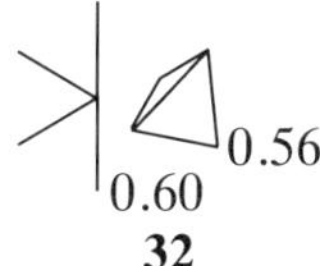

32

In accord with the idea that two electrons have to be placed in a high-energy orbital (in this case one that is metal-located) only two of the four hydrogen-fragment electrons are ligand-localized. An alternative structure could be via coordination of the tetrahedron via a face (rather than an edge) when the MH_4 skeleton has a trigonal bipyramidal shape with M located at the apical site of the trigonal bipyramid. A molecular orbital diagram for a trigonal bipyramidal H_5 unit in shown in Scheme 5.10. Although the real system will have a lower symmetry, this simplified model also shows that this structure is less stable than the one in which the four electrons of the H centers sit mostly in the two deep-lying σ_{H_2} orbitals as in $Cr(CO)_4(H_2)_2$. Basically, one pair of electrons has moved to a higher energy level in the tetrahydrogen complex **26**. The orbital diagram in Scheme 5.10 shows that a similar problem besets the stability via coordination of the H_4 unit by a face.

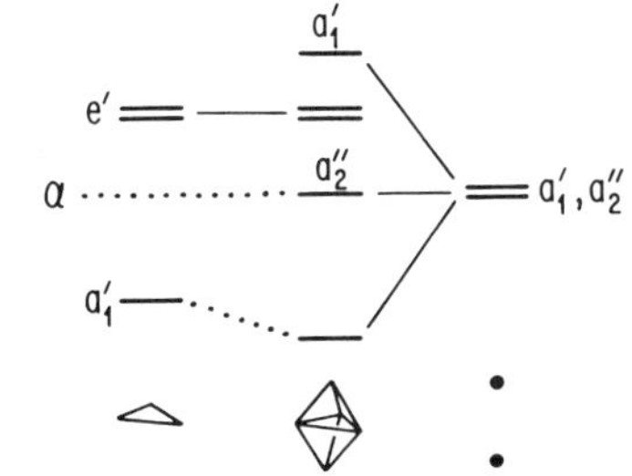

Scheme 5.10

Using the olefin analogy, square H_4 should correspond to cyclobutadiene. We can use this result to understand the electronic instability of $Cr(CO)_4(H_4)$ (**27**). The energetics of coordination of cyclic and acyclic conjugated hydrocarbons to transition metal fragments has been analyzed by Elian and Hoffmann [61]. They have explored the electronic details behind the observation that $Fe(CO)_3$ prefers to interact with a conjugated four-electron donor (such as cyclobutadiene) but that $Cr(CO)_4$ prefers two isolated (unconjugated) double bonds. The difference lies in the relative energies of the π levels of the C_nH_n units that control the energetics of interaction with the frontier orbitals on the metal. An interesting part of their study concerned the direction of charge transfer between metal and ligand as a function of the nature of the fragment and the organic ligand. Importantly, whereas the organic ligand is well-represented as $C_4H_4^{2-}$ in $Fe(CO)_3(C_4H_4)$, charge transfer occurs in the opposite direction in the $Cr(CO)_4$ analogue. The same result is obtained from a population analysis of $Cr(CO)_4(H_4)$ (**27**), which is shown in **33**,

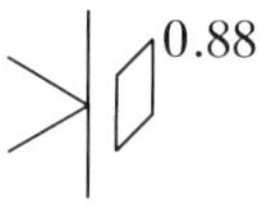

0.88

33

when these "ligand electrons" are transferred to the metal. In contrast, loss of one CO might lead to rather different structural preferences.

A coordinated H_3 unit was an obvious choice for an intermediate in the exchange involving hydride–dihydrogen ligands [4a, 7] such as $FeH_2(H_2)(PR_3)_3$, and a molecule that contains an H_3 unit and a hydride is a possible intermediate in the H–D scrambling in $Cr(CO)_4(H_2)_2$. $Cr(CO)_4(H_3)H$ (**28**) is predicted to be an energetically accessible structure on the basis of EHT calculations, although available ab-initio results place it somewhat higher in energy. The population analysis shown in **44**

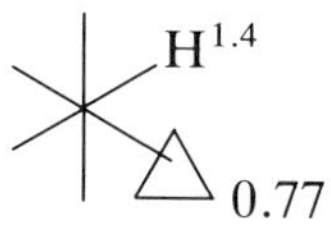

$H^{1.4}$

0.77

34

reinforces its description as one containing a coordinated hydride (H^-) and a triangular H_3^+ molecule. If H_3^+ is a simple two-electron donor (like cyclopropenium cation), then it is clear from a molecular orbital diagram (Scheme 5.11) how an 18-electron count results for the metal. It is also related in a simple manner to the $M(H_2)_2$ structure by formal transfer of a proton from one H_2 ligand to the other. This makes the electrons of the H_3^+ very stable without destabilizing to an unrealistic level the electrons centered on the single H center with respect to the ideal localization in the H—H bond.

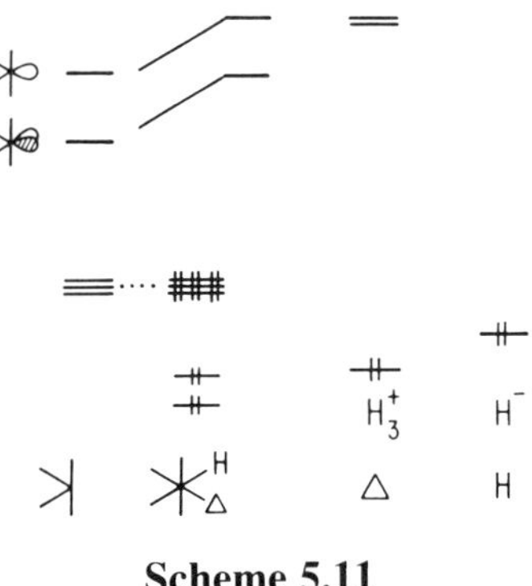

Scheme 5.11

There is another way to view the counting process that leads us to the hypothesis that some H_3 transition metal complexes might be viable intermediates. Just as the face-coordinated tetrahedron led to an MH_4 trigonal bipyramid, the MH_3 unit forms a tetrahedron, which requires two electrons for stability. The predicted structure for $V(CO)_5(H_3)$ (**35**)

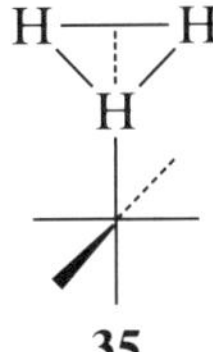

35

satisfies this criterion. In order to enhance the stability of such (H_3)-containing species, minimization of the extent of back-donation might be as important here as in the (H_2)-containing species. Thus, a molecule containing one or more nitrosyl groups in place of carbonyl might be a realistic synthetic goal.

Another molecule that may be understood electronically in the same way is $Co(CO)_3(H_3)$. Figure 5.2 shows a molecular orbital diagram for this species. At the relevant electron count there is a good HOMO–LUMO gap (2.8 eV). The picture is dominated by the large interaction between a filled a_1 orbital on H_3 and an empty orbital on $Co(CO)_3$, which justifies the view shown in **36**.

36

However, there is some stabilization of the degenerate metal HOMO (e) by

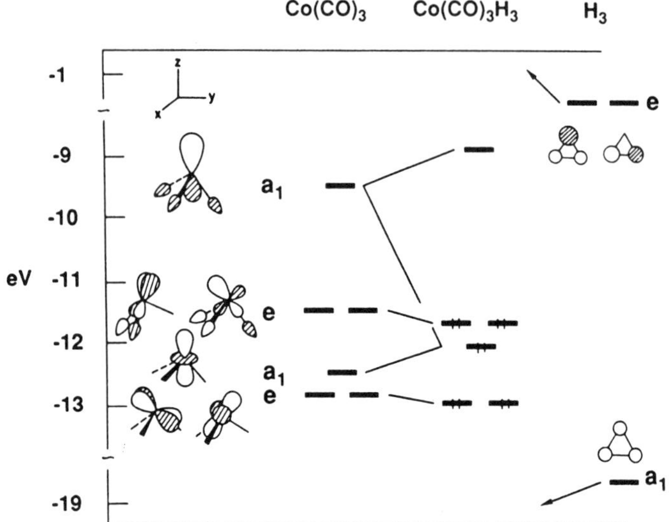

FIGURE 5.2
Assembly of the molecular orbital diagram for $Co(CO)_3(H_3)$ from those of fragments.

the empty e pair on H_3^+. If this stabilization is large, it will lead to instability of the coordinated H_3, through rupture of the H—H bonds in a way analogous to the interaction involving $\sigma^*_{H_2}$ in (H_2) complexes. There is also a small destabilization of the occupied $Co(CO)_3$ a_1 orbital.

Returning to the question of H–D exchange in $Cr(CO)_4(H_2)_2$, in Table 5.1 no species is calculated to lie particularly low in energy for facile exchange in liquid xenon. The two lowest-energy computed pathways are either via a $Cr(CO)_4(H_3)H$ species containing an open (H_3) unit or via the species $Cr(CO)_4(H_4)$ containing an open (H_4) unit. On its own the latter species cannot give rise to H–D exchange, but such a process is geometrically allowed if such an intermediate decays to $Cr(CO)_4(H_2)H_2$, where the (H_2) unit is derived from the central H_2 fragment of the open (H_4) unit. This species is computed to lie at a lower energy than the tetrahydrogen complex. We believe that both of these intermediates may be stabilized by the cis effect as described in Section 5.3.

Finally, one should note that exchange of H centers in a polyhydride compound does not necessarily require modification of the oxidation state of the metal center by the formation or cleavage of H—H bonds. It thus is possible to understand exchange reactions on d^0 metal centers that must participate in an exchange reaction without oxidatively cleaving the C—H bond [62]:

$$C'p_2M\text{—}CH_3 + {}^{13}CH_4 \rightarrow C'p_2M\text{—}CH_3 + CH_4 \qquad (M = Y, Lu)$$

5.5. INTERACTIONS BETWEEN NONBONDED HYDRIDE CENTERS

Here we ask if there is a significant amount of interaction between coordinated hydrogen centers even at distances that are significantly larger than those corresponding to full bonds. Our discussion illustrates some of the limits of the isolobal analogy. Let us compare the two molecules CH_4 and the known dihydride $Os(CO)_4H_2$, viewed as two H centers bonded to isolobal carbene or $Os(CO)_4$ fragments, respectively. A positive overlap population is calculated [63] between the two H centers of the Os complex but not in CH_4. By a careful examination of the interaction between symmetry-adapted orbitals, it appears that the main difference can be attributed to the topology of the acceptor orbitals of the CH_2 and $Os(CO)_4$ fragments. The H—H overlap population will be determined by the extent of the electronic occupation of in- and out-of-phase hydrogen 1*s* combinations. If, after combination with the correct metal orbitals, more electrons are located in the in-phase $\sigma^+_{H_2}$ than the out-of-phase $\sigma^-_{H_2}$ orbitals (Scheme 5.12), then the H—H overlap population will be positive, indicating an attraction between them. In the case of CH_4, the in-phase combination of H *s* orbitals interacts with an acceptor orbital which is made of *s* and *p* character in CH_2 (37a).

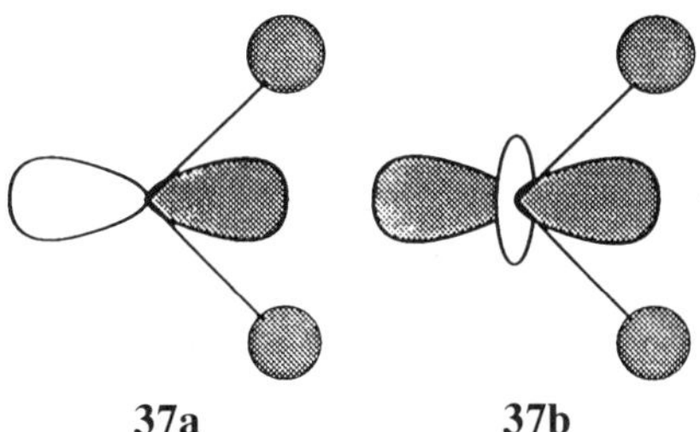

In contrast, the equivalent orbital for $Os(CO)_4$ is predominantly z^2 (with additional *s* and *p* contribution) **37b**. Because the H centers sit close to the nodal cone of the z^2 orbitals in **37b**, the overlap between the two fragment

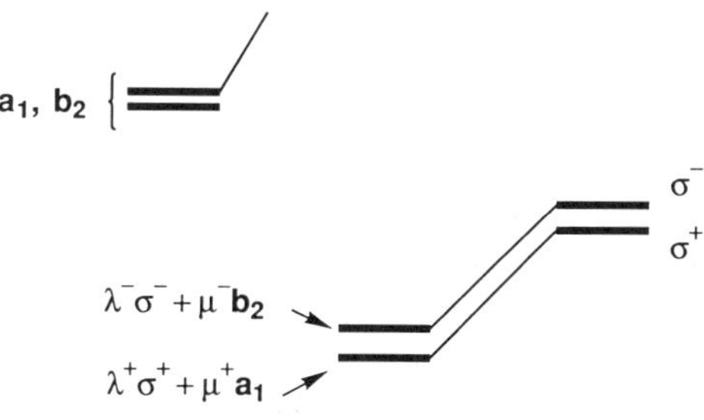

Scheme 5.12

orbitals is much less in the case of a metal than in the case of the main group element. As a consequence, more electrons remain in the in-phase $\sigma^+_{H_2}$ combination in $Os(CO)_4H_2$ (i.e., $\lambda^+_{Os} > \lambda^+_{CH_2}$) and a positive H–H interaction is calculated.

For transition metal hydrides this leads to two consequences: (1) There are usually less than two electrons per hydride center and (2) when two or more hydrides are located cis to each other on a mononuclear or polynuclear transition metal complex, a weak but significant positive overlap population is calculated between the adjacent hydride centers. This result contrasts with what is calculated in the main group chemistry, where as expected the overlap population between *cis*-H centers is calculated to be negative. It also correlates with the fact that transition metal hydrides are fluxional, whereas no equivalent behavior is reported for main group hydride compounds.

5.6. BINUCLEAR POLYHYDROGEN MOLECULES

5.6.1. Electronic Stability

Up to now we have studied the electronic requirements for the stability of a complex containing H_2, H_3, and H_4 complexes, not only by considering the stability of the coordinated H_n unit, but also by recognizing that the $ML_x(H_n)$ moiety itself forms a polyhedron. In this section we look at the electronic structure of $(ML_x)_yH_{n-y}$ complexes, where there is more than a single metal atom and together the metal and hydrogen atoms form a polynuclear cluster with n vertices. It is interesting to speculate just how large n can be. The stability of quite large systems is in principle already established; clusters of hydrogen atoms H_n^+, up to quite large n, have been known from mass-spectrometric studies for many years [59], and there are theoretical studies on the structures of the smaller clusters [60]. Our attention will lie with the case of $y = 2$, metal dimer molecules.

Recall that if $W(CO)_5(H_2)$ and $Co(CO)_2(NO)(H_2)$ are isoelectronic with H_3^+, then the molecules $V(CO)_5(H_3)$ (**35**) $Co(CO)_3(H_3)$ (**36**) are two species (hypothetical at this stage) that are isoelectronic with H_4^{2+}. The analogy relies on the presence of a single empty frontier orbital on the metal fragment, which may replace the single empty 1*s* orbital of H^+. Of course what we really are saying is that $Co(CO)_3^-$ and $V(CO)_5^-$ are isolobal with H^+. We need to be a little cautious in making such generalizations. Sometimes this approach, which only uses the a_1 orbital of the fragment, is too much of a simplification to describe these species accurately. Although it does allow a way to view the stability of these polyhedral species in different geometries, in reality a larger set of frontier fragment orbitals may be involved. Now we extend such ideas to the case of molecules containing two metal-atom fragments.

5.6.2. Trigonal Bipyramid

The energy-level diagram of a trigonal bipyramid was shown in Scheme 5.10. It differs from that of the triangle that it contains a nonbonding orbital. Electronically, we would propose that this geometry would be stable for both H_5^{3+} and H_5^+ (two or four electrons, respectively). As we have already mentioned, ab-initio molecular orbital calculations for H_5^+ show [60] that a distorted form of the spiro pentahydrogen structure is the most stable structure. We show elsewhere [64] how the ideas of Hückel theory simply demonstrate the stability of the spiro structure for H_5^{+3} versus the trigonal bibyramidal structure of H_5^+.

If we extend the orbital analogy between a $Co(CO)_2(NO)$ or $Co(CO)_3^-$ unit and H^+, the hypothetical species $Co_2(CO)_6H_3^-$, containing a sandwiched H_3 unit (with short H—H distances) (**38**) is isoelectronic with H_5^{3+}.

38

Figure 5.3 shows the construction of a molecular orbital diagram for this molecule from those of $Co_2(CO)_6$ and H_3. The M_2L_6 dimer and the $M_2L_6H_3$ and $M_2L_4H_4$ structures were studied several years ago by Hoffmann and co-workers [65, 66]. Their goal was the design of a molecule containing both M—H bonding and H · · · H nonbonding distances. The present molecules are different. Here we want the hydrogen atoms to come together as part of a polyhedral unit. Notice that the results of the calculations indicate structural stability for 34 electrons, to be contrasted with the 30 electrons needed in the (known) series of $M_2L_6H_3$ complexes containing three bridging hydrides, for example, $Os_2(PMe_2Ph)_6H_3^+$ [67]. The orbital picture for the $M_2L_6(H_3)$ complex in Figure 5.3 shows features similar to those for $Co(CO)_3H_3$ in Figure 5.2. A strong interaction is found between the a_1' level on H_3 and the in-phase combination of unoccupied a_1 $Co(CO)_3$ frontier orbitals of the same symmetry. There is a weaker stabilization of the symmetric e' pair [the in-phase partner of the degenerate e pair of $Co(CO)_3$ frontier orbitals] via overlap with the degenerate H_3^+ e' levels. The largest interaction, being with the a_1 orbitals leads to the description **39** of the structure, which highlights the analogy with H_5^{3+} and with **36**.

39

Notice the good HOMO–LUMO gap in Figure 5.3, computed [64] to be

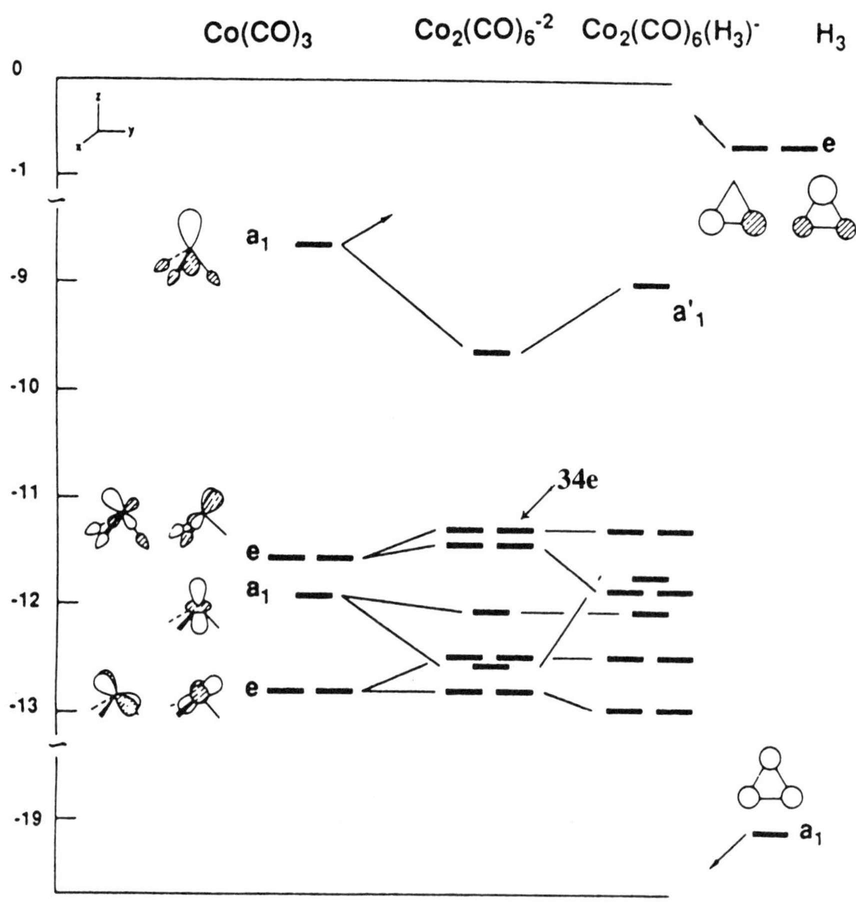

FIGURE 5.3
Assembly of the molecular orbital diagram for $Co_2(CO)_6(H_3)^-$ from those of the fragments.

2.3 eV at 34 electrons. The tightness of the H_3 triangle maintains a strong interaction with the metal a_1 orbital, but a smaller one with the metal e orbitals, an important consideration if the integrity of the triangle is to be maintained. We used similar arguments, minimizing the extent of σ^* interaction while strengthening that of σ interaction with the metal, in the case of the $ML_n(H_2)$ complexes. Inclusion of good π acceptors in the ML_n unit should help in this case as well; accordingly, the bond overlap populations for some species containing NO ligands, with a fixed electron count, are shown in Scheme 5.13. There is, however, another interesting result: The energy gap between the degenerate (e') HOMO and second-highest occupied molecular orbital (SHOMO) is reduced by the presence of the stronger π acceptor (a shift of 0.25 eV is calculated on going from $[Co(CO)_3]_2(H_3)^-$ to [Co

5.13

$Co(CO)_3H_3$	$Co_2(CO)_6H_3^-$	$Co_2(CO)_4(NO)_2H_3^+$	$Co_2(CO)_6H_4$
0.2961	0.2574	0.2570	0.2061

Scheme 5.13
Overlap population for H—H.

$(CO)_2(NO)]_2(H_3)^+$). Such a result encourages the formation of the 34-electron (polyhedral) compound by reduction of the 30 electron (bridging) compound.

As the H—H distance increases [see (5.1)], the polyhedral molecule eventually is converted into the conventionally bridged one.

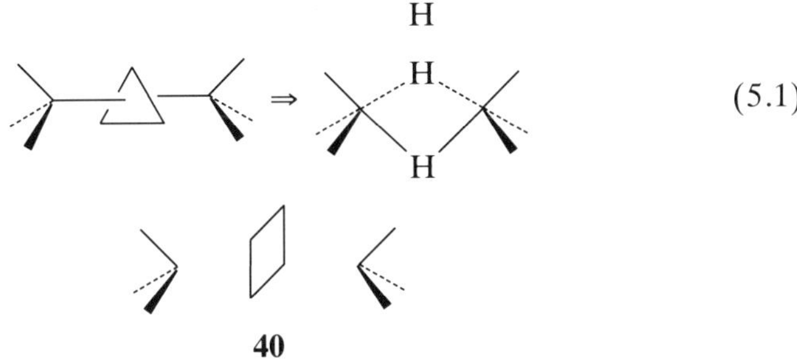

The orbital picture for this transformation is presented in Scheme 5.14. The *e″* orbitals, which are antisymmetric with respect to the two metal centers, go

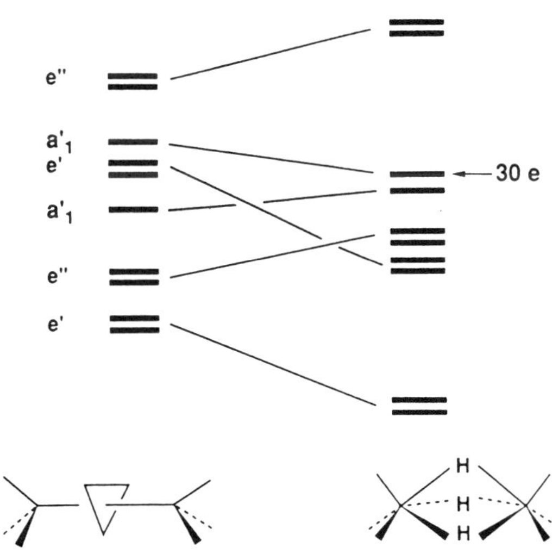

Scheme 5.14

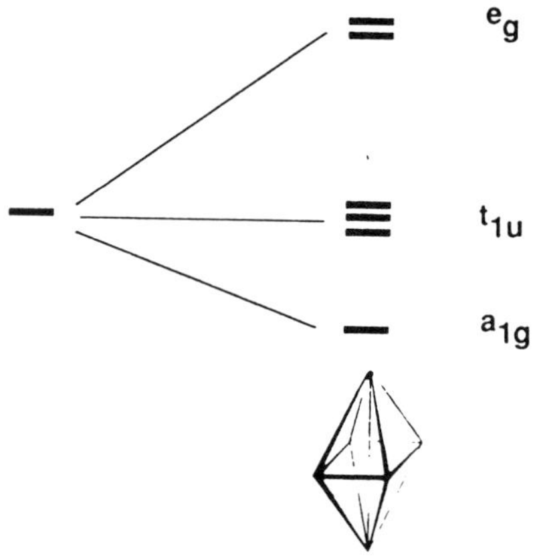

Scheme 5.15

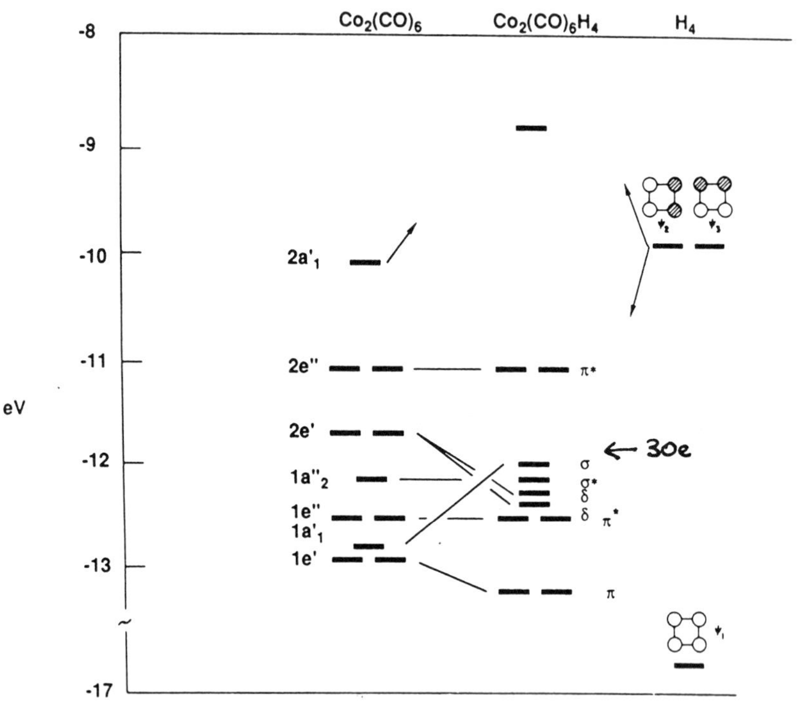

FIGURE 5.4
Assembly of the molecular orbital diagram for $Co_2(CO)_6(H_4)$ from those of fragments.

up in energy as the metal–metal bond distance is reduced. A reasonable energy gap in the bridging structure is found for 30 electrons (4 electrons fewer than the parent) and the hydrogen atoms are described as bridging hydrides. There are some similarities to the orbital picture for the oxidative addition of H_2 to a d^8 square-planar metal complex. Both transformations involve the replacement of H—H bonds by M—H bonds. There are several molecules with this geometry and electron count, for example, $H_3Ru_2(PMe_3)_6{}^+BF_4{}^-$ [68] and $H_3Fe_3[MeC(CH_2PPh_2)_3]_2{}^+PF_6{}^-$ [69].

5.6.3. Octahedron

There are three nonbonding orbitals and one bonding orbital in the level structure of the H_6 octahedron, shown in Scheme 5.15. By analogy with the trigonal bipyramid, we may envisage electronic stability for the two-electron $H_6{}^{4+}$ molecule. By analogy with the treatment of the trigonal bipyramid, the octahedral structure should be stable for the species $Co_2(CO)_6(H_4)$ (**40**). Electronically, we would describe this as in **41**,

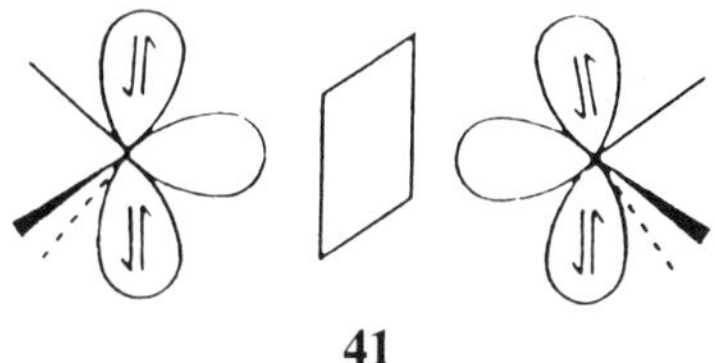

41

with the assembly of an orbital diagram for this molecule given in Figure 5.4. As yet no example of this type of system exists. Compressing the metal–metal bond

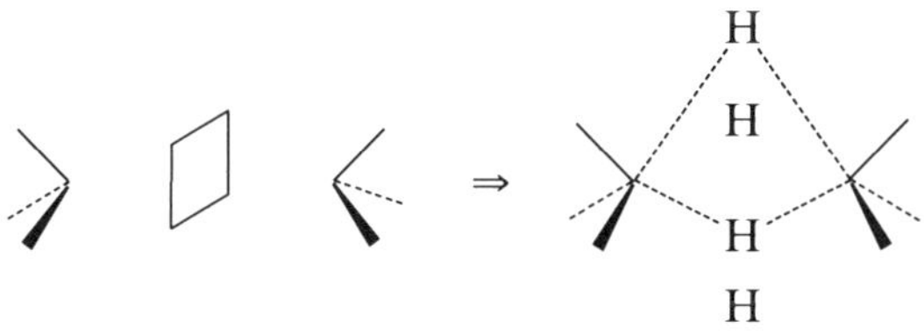

H H H H (5.2)

and thus increasing the H—H distance leads to a similar electronic situation to the one we showed, in the trigonal bipyramid case (Scheme 5.14) in that a gap opens at 30 electrons, 4 electrons fewer than the electron count needed for the polyhydrogen system. This is driven by destabilization of the degenerate HOMO of the 34-electron system. $H_4[Re(PEt_2Ph)_2H_2]_2$ is an example

[70] of a species with four bridging hydrogen atoms and this electron count, but this is a species that is not based on an M_2L_6 framework.

With an electron count of 32 electrons, the HOMO of molecule **40** is only half full and a Jahn–Teller distortion is predicted that would remove the degeneracy of the e' HOMO. The observed structure [71] of the 32-electron $Os_2(PMe_2Ph)_6H_4$,

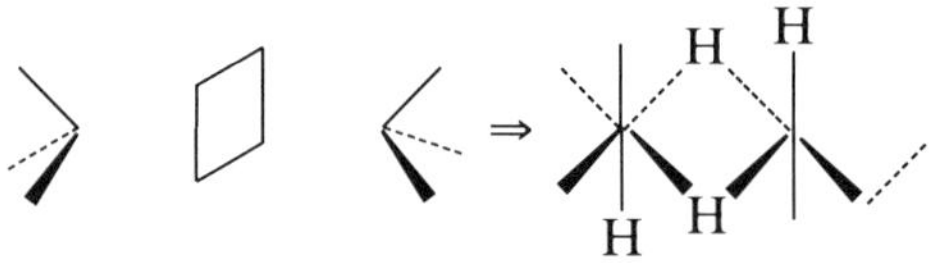

(5.3)

leads us to one possible distortion coordinate. A calculation [64] shows the opening of a significant gap at 32 electrons for the distorted molecule. Thus 30-, 32-, and 34-electron systems with different geometries are understandable, but only examples of the first two are known. It appears that, because the polyhedral molecule is stable when only the lowest, totally symmetric orbital of the skeleton is doubly occupied and thus is isoelectronic with $H_n^{(n-2)+}$, the bridging hydride is stable for 30 electrons and the polyhydrogen molecule stable for 34 electrons. There is an interesting aspect of this result. Often the opposite occurs—addition of electrons to a molecule leads to its breakup. In the series $P_4 \rightarrow B_4H_{10} \rightarrow P_4R_4 \rightarrow PCl_3$, for example (20 to 26 electrons), bonds are increasingly severed.

Of course, generation by calculation of a good HOMO–LUMO gap in an unknown molecule is no guarantee of stability. There are many molecules that should exist if this were the sole criterion. However, as in other areas of chemistry, such theoretical considerations may encourage synthetic efforts in this exciting field.

5.7. CONCLUSIONS

We have covered much ground in this review, giving a broad view of H_2 coordination. The competition between structures containing this unit and those processing simpler hydride or more complex polyhydrogen units has been an instructive area to pursue. We also have shown that even in simple hydrides there is often an attraction between formally nonbonded atoms, which may be important in determining the pathways for hydrogen exchange processes. Although most of our arguments have been concerned with qualitative symmetry and overlap ideas supported by numerical computations of the extended Hückel type, it is clear from Table 5.1 that quantitative ab-initio computations are not yet accurate enough to resolve the small details that determine the rather gross aspects of the chemistry of these systems.

ACKNOWLEDGMENTS

This research was supported by the National Science Foundation (grant DMR 84-14175) and the EEC (S. A. J., grant SC1000155). Some of this work was done during the stay of J. K. B. at Orsay. He thanks CNRS for making it possible. The Laboratoire de Chimie Théorique is associated with the CNRS (URA 506) and is a member of the ICMO and IPCM departments.

REFERENCES

1. Kubas, G. J.; Ryan, R. R.; Swanson, B. I.; Vergamini, P. J.; Wasserman, H. J. *J. Am. Chem. Soc.* **1984**, *100*, 451.
2. Kubas, G. J.; Ryan, R. R. *Polyhedron* **1986**, *5*, 473.
3. (a) Morris, R. H.; Sawyer, J. F.; Shiralian, M.; Zubkowski, J. D. *J. Am. Chem. Soc.* **1985**, *107*, 5581. (b) Ricci, J. S.; Koetzle, T. F.; Bautista, M. T.; Hofstede, T. M.; Morris, R. H.; Sawyer, J. F. *J. Am. Chem. Soc.* **1989**, *111*, 8823.
4. (a) Van Der Sluys, L. S.; Eckert, J.; Eisenstein, O.; Hall, J. H.; Huffman, J. C.; Jackson, S. A.; Koetzle, T. F.; Kubas, G. J.; Vergamini, P. J.; Caulton, K. G. *J. Am. Chem. Soc.* **1990**, *112*, 4831. (b) Riehl, J. F.; Eisenstein, O.; Pelissier, M. Private communication.
5. (a) Upmacis, R. K.; Gadd, G. E.; Poliakoff, M.; Simpson, M. B.; Turner, J. J.; Whyman, R.; Simpson, A. F. *J. Chem. Soc., Chem. Commun.* **1985**, 27. (b) Church, S. P.; Grevels, F.-W.; Hermann, H.; Shaffner, K. *J. Chem. Soc., Chem. Commun.* **1985**, 30. (c) Upmacis, R. K.; Poliakoff, M.; Turner, J. J. *J. Am. Chem. Soc.* **1986**, *108*, 3645.
6. Gadd, G. E.; Upmacis, R. K.; Poliakoff, M.; Turner, J. J. *J. Am. Chem. Soc.* **1986**, *108*, 2547.
7. (a) Crabtree, R. H., Lavin, M. *J. Chem. Soc., Chem. Commun.* **1985**, 794. (b) Crabtree, R. H.; Lavin, M. *J. Chem. Soc., Chem. Commun.* **1985**, 1661. (c) Crabtree, R., H.; Lavin, M.; Bonneviot, L. *J. Am. Chem. Soc.* **1986**, *108*, 4032.
8. Jia, G.; Meek, D. W. *J. Am. Chem. Soc.* **1989**, *111*, 757.
9. Chinn, M. S.; Heinekey, D. M. *J. Am. Chem. Soc.* **1987**, *109*, 5865.
10. (a) Jackson, S. A.; Hodges, P. M.; Poliakoff, M.; Turner, J. J.; Grevels, F.-W. *J. Am. Chem. Soc.* **1990**, *112*, 1221. (b) Hodges, P. M.; Jackson, S. A.; Jacke, J.; Poliakoff, M.; Turner, J. J.; Grevels, F.-W. *J. Am. Chem. Soc.* **1990**, *112*, 1234.
11. Mediati, M.; Tachibana, G. N.; Jensen, C. M. *Inorg. Chem.* **1990**, *29*, 3.
12. Hampton, C.; Cullen, W. R.; James, B. R.; Charland, J.-P. *J. Am. Chem. Soc.* **1988**, *110*, 6918.
13. Joshi, A. M.; James, B. R. *J. Chem. Soc., Chem. Commun.* **1989**, 1785.
14. (a) Collman, J. P.; Wagenknecht, P. S.; Hembre, R. T.; Lewis, N. S. *J. Am. Chem. Soc.* **1990**, *112*, 1294. (b) Nolan, S. P.; Marks, T. J. *J. Am. Chem. Soc.* **1989**, *111*, 8538.

15. (a) Kubas, G. J. *Acc. Chem. Res.* **1988**, *21*, 120. (b) Crabtree, R. H., Hamilton, D. G. *Adv. Organomet. Chem.* **1988**, *28*, 299. (c) Crabtree, R. H. *Acc. Chem. Res.* **1990**, *23*, 95.
16. (a) Crabtree, R. H.; Hamilton, D. G. *J. Am. Chem. Soc.* **1986**, *108*, 3124. Related results were found for $M(H_2)H(depe)_2^+$, M = Fe, Ru, Os. Bautista, M.; Earl, K. A.; Morris, R. H.; Sella, A. *J. Am. Chem. Soc.* **1987**, *109*, 3780. (b) Khalsa, G. R. K.; Kubas, G. J.; Unkefer, C. J.; Var Der Sluys, L. S.; Kubat-Martin, K. A. *J. Am. Chem. Soc.* **1990**, *112*, 3855. (c) Luo X.-L., Crabtree, R. H. *J. Chem. Soc., Chem. Commun.* **1990**, 189.
17. Burdett, J. K.; Phillips, J. R.; Pourian, M. R.; Poliakoff, M.; Turner, J. J.; Upmacis, R. K. *Inorg. Chem.* **1987**, *26*, 3054.
18. Dobson, G. R.; Brown, R. A. *Inorg. Chim. Acta.* **1972**, *6*, 65.
19. (a) Howdle, S. M.; Poliakoff, M. *J. Chem. Soc., Chem. Commun.* **1989**, 1099. (b) Howdle, S. M.; Healy, M. A.; Poliakoff, M. *J. Am. Chem. Soc.* **1990**, *112*, 4804.
20. Kubas, G. J.; Unkefer, C. J.; Swanson, B. I.; Fukushima, E. J. *J. Am. Chem. Soc.* **1986**, *108*, 7000.
21. Morris, R. H.; Earl, K. A.; Luck, R. L.; Lazarowych, N. J.; Sella, A. *Inorg. Chem.* **1987**, *26*, 2674.
22. Albright, T. A.; Burdett, J. K.; Whangbo, M.-H. *Orbital Interactions in Chemistry*, Wiley: New York, 1985.
23. Oka, T. *Phys. Rev. Lett.* **1980**, *45*, 531.
24. Burdett, J. K.; Lee, S. *J. Am. Chem. Soc.* **1985**, *107*, 3050, 3063.
25. Casida, M. E.; Chen, M. M. L.; MacGregor, R. D.; Schaefer, H. F. *Israel. J. Chem.* **1980**, *19*, 127.
26. Raghavachari, K.; Whiteside, R. A.; Pople, J. A.; Schleyer, P. v.R. *J. Am. Chem. Soc.* **1981**, *103*, 5649.
27. Gordon, M. S.; Gano, D. R.; Binkley, J. S.; Frisch, M. J. *J. Am. Chem. Soc.* **1986**, *108*, 2191.
28. Sosa, C.; Schlegel, H. B. *J. Am. Chem. Soc.* **1984**, *106*, 5847.
29. Bauschlicher, C. W.; Haber, K.; Schaefer, H. F.; Bender, C. F. *J. Am. Chem. Soc.* **1977**, *99*, 3610.
30. (a) Kober, E. M., Hay, P. J. *The Challenge of d and f Electrons, Theory and Computation*; ACS Symposium Series 394; American Chemical Society: Washington, DC, 1989; p. 92. (b) Pacchioni, G. *J. Am. Chem. Soc.* **1990**, *112*, 80.
31. Lammertsma, K.; Olah, G. A.; Barzaghi, M.; Simonnetta, M. *J. Am. Chem. Soc.* **1982**, *104*, 6851.
32. Saillard, J.-Y.; Hoffmann, R. *J. Am. Chem. Soc.* **1984**, *106*, 2006.
33. Sevin, A. *Nouv. J. Chim.* **1981**, *5*, 233.
34. Jean, Y.; Lledos, A. *Nouv. J. Chim.* **1986**, *10*, 635.
35. (a) Johnson, C. E.; Fisher, B. J.; Eisenberg, R. *J. Am. Chem. Soc.* **1983**, *105*, 7772. (b) Johnson, C. E.; Eisenberg, R. *J. Am. Chem. Soc.* **1985**, *107*, 3148.
36. van Beek, J. A. M.; van Koten, G.; Smeets, W. J. J.; Spek, A. L. *J. Am. Chem. Soc.* **1986**, *108*, 5010.
37. For example, see Salem, L. *Electrons in Chemical Reactions*; Wiley: New York, 1984.
38. Barton, T. J.; Grinter, R.; Thomson, A. J.; Davies, B.; Poliakoff, M. *J. Chem. Soc., Chem. Commun.* **1977**, 841.
39. Sweany, R. L. *J. Am. Chem. Soc.* **1981**, 103, 2410.
40. (a) Fredin, L.; Hauge, R. H.; Kafafi, Z. H.; Margrave, J. L. *J. Chem. Phys.* **1985**, *82*, 3542. (b) Sevin, A.; Chaquin, P.; Hamon, L.; Hiberty, P. C. *J. Am. Chem. Soc.* **1988**, *110*, 5681.
41. Poliakoff, M., Turner, J. J. *J. Chem. Soc., Dalton Trans.* **1974**, 2276.

42. (a) Gonzalez, A. A.; Zhang, K.; Nolan, S. P.; Lopez de la Vega, R.; Mukerjee, S. L., Hoff, C. D.; Kubas, G. J. *Organometallics* **1988**, *12*, 2429. (b) Eckert, J.; Kubas, G. J.; Hall, J. H.; Hay, P. J.; Boyle, C. M. *J. Am. Chem. Soc.* **1990**, *112*, 2324. (c) Eckert, J.; Blank, H.; Bautista, M. T.; Morris, R. H. *Inorg. Chem.* **1990**, *29*, 747.
43. Jean, Y.; Eisenstein, O.; Volatron, F.; Maouche, B.; Sefta, F. *J. Am. Chem. Soc.* **1986**, *108*, 6587.
44. (a) Hay, P. J. *Chem. Phys. Lett.* **1984**, *103*, 466. (b) Hay, P. J. *J. Am. Chem. Soc.* **1987**, *109*, 705.
45. (a) Bachmann, C.; Demuynck, J.; Veillard, A. *J. Am. Chem. Soc.* **1978**, *100*, 2366. (b) Albright, T. A.; Hoffmann, R.; Thibeault, J. C.; Thorn, D. L. *J. Am. Chem. Soc.* **1979**, *101*, 3801.
46. (a) Fenske, R. F.; DeKock, R. L. *Inorg. Chem.* **1970**, *9*, 1053. (b) Marsella, J. A.; Curtis, C. J.; Bercaw, J. E.; Caulton, K. G. *J. Am. Chem. Soc.* **1980**, *102*, 7244.
47. Volatron, F.; Eisenstein, O. *J. Am. Chem. Soc.* **1986**, *108*, 2173.
48. Jackson, S. A.; Eisenstein, O. *Inorg. Chem.*, in press.
49. Johnson, T. J.; Huffman, J. C.; Caulton, K. G.; Jackson, S. A.; Eisenstein, O. *Organometallics* **1989**, *8*, 2073.
50. (a) Brookhart, M.; Green, M. L. H. *J. Organomet. Chem.* **1983**, *250*, 395. (b) Eisenstein, O.; Jean, Y. *J. Am. Chem. Soc.* **1985**, *107*, 1177. Demolliens, A.; Jean, Y.; Eisenstein, O. *Organometallics* **1986**, *5*, 1457.
51. Rossi, A. R.; Hoffmann, R. *Inorg. Chem.* **1975**, *14*, 365.
52. (a) Maseras, F.; Duran, M.; Lledos, A.; Bertran, J. *Inorg. Chem.* **1989**, *28*, 2984. (b) Bianchini, C.; Mealli, C.; Peruzzini, M.; Zanobini, F. *J. Am. Chem. Soc.* **1987**, *109*, 5548.
53. (a) Heinekey, D. M.; Payne, N. G.; Schulte, G. K. *J. Am. Chem. Soc.* **1988**, *110*, 2303. (b) Zilm, K. W.; Heinekey, D. M.; Millar, J. M.; Payne, N. G.; Demou, P. *J. Am. Chem. Soc.* **1989**, *111*, 3088. (c) Jones, D. H.; Labinger, J. A.; Weitekamp, D. P. *J. Am. Chem. Soc.* **1989**, *111*, 3087. (d) Heinekey, D. M.; Millar, J. M.; Koetzle, T. F.; Payne, N. G.; Zilm, K. W. *J. Am. Chem. Soc.* **1990**, *112*, 909. (e) Zilm, K. W.; Heinekey, D. M.; Millar, J. M.; Payne, N. G.; Neshyba, S. P.; Duchamp. J. C.; Szczyrba, J. *J. Am. Chem. Soc.* **1990**, *112*, 920.
54. (a) Arligue, T.; Chaudret, B.; Devillers, J.; Poilblanc, R. *C. R. Acad. Sci. Paris, Série II* **1987**, *305*, 1523. (b) Chaudret, B.; Commenges, G.; Jalon, F.; Otero, A. *J. Chem. Soc., Chem. Commun.* **1989**, 210. (c) Arigue, T.; Border, C.; Chaudret, B.; Devillers, J.; Poilblanc, R. *Organometallics* **1989**, *3*, 1308. (d) Antinolo, A.; Chaudret, B.; Commenges, G.; Fajardo, M.; Jalon, F.; Morris, R. H.; Otero, A.; Schweltzer, C. T. *J. Chem. Soc., Chem. Commun.* **1988**, 1210.
55. Parkin, G.; Bercaw, J. E. *J. Chem. Soc., Chem. Commun.* **1989**, 255.
56. (a) Kubas, G. J.; Ryan, R. R.; Wrobleski, D. A. *J. Am. Chem. Soc.* **1986**, *108*, 1339. (b) Kubas, G. J.; Unkefer, C. J.; Swanson, B. I.; Fukushima, E. *J. Am. Chem. Soc.* **1986**, *108*, 7000.
57. Lindsell, W. E.; McCullough, K. J.; Welch, A. J. *J. Am. Chem. Soc.* **1983**, *105*, 4487.
58. These P_4 complexes have been analyzed in theoretical terms by Kang, S. K.; Albright, T. A.; Silvestre, J. *Croat. Chem. Acta.* **1984**, *57*, 1355.
59. Clampitt, R.; Gowland, L. *Nature* **1969**, *223*, 815.
60. Yamaguchi, Y.; Gaw, J. F.; Schaefer, H. F. *J. Chem. Phys.* **1983**, *78*, 4074.
61. Elian, M.; Hoffmann, R. *Inorg. Chem.* **1975**, *14*, 1058.
62. Rabaâ, H.; Saillard, J.-Y.; Hoffmann, R. *J. Am. Chem. Soc.* **1986**, *108*, 4327.
63. Jackson, S. A.; Eisenstein, O. *J. Am. Chem. Soc.*, in press.
64. Burdett, J. K.; Pourian, M. R. *Organometallics* **1987**, *6*, 1684.

65. Summerville, R. H.; Hoffmann, R. *J. Am. Chem. Soc.* **1979**, *101*, 3821.
66. Dedieu, A.; Albright, T. A.; Hoffmann, R. *J. Am. Chem. Soc.* **1979**, *101*, 3141.
67. Dahl, L. F. *Ann. N.Y. Acad. Sci.* (*USA*) **1983**, *415*, 43.
68. (a) Dapporto, P.; Midollini, S.; Sacconi, L. *Inorg. Chem.*, **1975** *14*, 1643.; (b) Dapporto, P.; Fallani, G.; Midollini, S.; Sacconi, L., *J. Am. Chem. Soc.* **1973**, *95*, 2021.
69. Jones, R. A.; Wilkinson, G.; Colquohoun, I. J.; McFarlane, W.; Galas, A. M. R.; Hursthouse, M. B. *J. Chem. Soc.*, *Dalton Trans.* **1980**, 2480.
70. (a) Bau, R.; Carroll, W. E.; Teller, R. G.; Koetzle, T. F. *J. Am. Chem. Soc.* **1977**, *99*, 3872. (b) Bau, R.; Carroll, W. E.; Hart, D. W.; Teller, R. G.; Koetzle, T. F. *Adv. Chem. Series* **1978**, *73*, 167.
71. Green, M. A.; Huffman, J. C.; Caulton, K. *J. Organomet. Chem.* **1978**, *243*, C78.

CHAPTER 6

Hydrogen Transfer From Transition Metal Hydrides: Theoretical Aspects

NOBUAKI KOGA and KEIJI MOROKUMA

Institute for Molecular Science, Myodaiji, Okazaki 444, Japan

6.1. INTRODUCTION

Intramolecular hydrogen transfer from a transition metal hydride to an unsaturated ligand [eq. (6.1)] is one of the most important processes in catalytic as well as stoichiometric reactions [1].

$$\begin{array}{l} \mathrm{H} \\ | \\ \mathrm{M{-}X} \end{array} \rightarrow \mathrm{M{-}X{-}H} \tag{6.1}$$

For instance, hydride migration to a coordinated olefin (X = olefin), so-called olefin insertion into an M—H bond, has been considered to be the key elementary step of olefin hydrogenation, olefin hydroformylation, olefin isomerization, and others. With this importance in mind, intramolecular hydride migrations to C_2H_4, CO, CO_2, CH_2, and CH_2O have been studied theoretically, using varieties of molecular orbital (MO) methods.

Intermolecular hydrogen transfer also has been known to take place, in which a hydrogen behaves like a hydride [2], a hydrogen atom [3], or a proton [4]. Although such processes have not been studied extensively by theoretical methods, the acidity of a hydride ligand in its transfer from transition metal hydrides has been discussed. Also, intramolecular hydrogen transfer in the dinuclear complex, which may be considered as a model of intermolecular hydrogen transfer, has been studied.

In this chapter, we review these theoretical studies. The theoretical methods used range from semi-empirical to ab-initio MO methods. As is shown later, these studies started at the semiempirical level, using the extended Hückel method (EH) [5] and the complete-neglect-of-differential-overlap method (CNDO) [6]. The EH method is useful when discussing the orbital interaction, because the total energy is the sum of the orbital energies, and the differential overlap (which is essential for description of the interaction between two fragments) is taken into account correctly. This method is not self-consistent and thus may underestimate the electrostatic effect and overestimate the polarization effect. The CNDO method is, on the other hand, self-consistent. However, the energetics of a reaction obtained with this method is often unreliable because of the lack of differential overlaps.

Recently, the more-reliable ab-initio MO method [7] has been used to study these reactions. Using the energy-gradient technique, which gives the force acting on atoms, one can determine with reasonable reliability the structures of the transition state (TS) as well as equilibrium molecules. The Hartree–Fock (HF) level of calculation, where the wave function is represented by a single Slater determinant, does not take into account the electron correlation effect, which is often required for the description of, for instance, partially formed or broken bonds at the TS and the interaction in a weak intermolecular complex. Therefore, to obtain more-reliable energetics of a reaction, the correlation effect must be taken into account properly. For this purpose, one adopts computational methods using the multideterminantal wave function, such as the configuration-interaction (CI) method [8], the Møller–Plesset (MP) perturbation method [9], and the multiconfiguration self-consistent-field (MCSCF) method [10]. The MCSCF method includes the complete active space self-consistent field (CASSCF) and the generalized valence bond (GVB) [11] methods.

The development of the effective core potential (ECP), sometimes called by different names such as the model potential and the pseudopotential, has made possible the ab-initio MO calculation of molecules containing heavy atoms such as transition metal complexes. The ECP approximation replaces the chemically inert core electrons by a potential and therefore reduces the calculation to a valence-electron problem with a substantial savings in computer time [12]. In addition, the relativistic effect, which affects the size and energy of atomic orbitals of a heavy atom, could be included in the ECP.

Some calculations also have been carried out with density functional (DF) theory, which includes the Hartree–Fock–Slater (HFS) method [13], also called the Xα method. The DF method can take into account the electron correlation effect at least partly. The energy-gradient method is utilized with the DF method as well [14].

This chapter starts with the review of theoretical studies of hydride migration to a coordinated olefin (Section 6.2). Intramolecular hydride migration to formaldehyde, carbon dioxide, carbon monoxide, and methylene is reviewed in Sections 6.3, 6.4, 6.5, and 6.6, respectively. Section 6.7 reviews theoretical studies on the acidity of hydrogen ligand. The final section contains concluding remarks.

6.2. HYDRIDE MIGRATION TO COORDINATED OLEFIN

Hydride migration to coordinated olefin, also called olefin insertion, is an important elementary step of catalytic reactions [1]. Although the equilibrium between an olefin hydride and the corresponding alkyl species has been observed only rarely, a few recent experiments have elucidated the energetics of the hydride migration to coordinated olefin. Roe [15] showed that the hydride migration [eq. (6.2)] of a four-coordinate d^8 Rh complex takes place with the activation enthalpy of 13.0 kcal mol^{-1}. Doherty and Bercaw [16] analyzed the electronic effect of substituents on the kinetics of hydride migration. They showed that the activation enthalpy of reaction (6.3) is 14.7 kcal mol^{-1} for ethylene.

$$cis\text{-HRh}(C_2H_4)[P(i-Pr)_3]_2 \rightarrow cis\text{-Rh}(C_2H_5)[P(i-Pr)_3]_2 \quad (6.2)$$

$$Cp^*{}_2NbH(CH_2{=}CHR) \rightarrow Cp^*{}_2Nb(CH_2CH_2R) \quad (6.3)$$

These reactions show that the hydride migration to coordinated olefin is intrinsically an easy reaction to take place.

Sakaki, Kato, Kanai, and Tarama [17] studied the ethylene insertion reaction into a Pt—H bond [eq. (6.4)] by the CNDO method.

$$trans\text{-HPtCl}(C_2H_4)(PH_3)_2 \rightarrow trans\text{-PtCl}(C_2H_5)(PH_3)_2 \quad (6.4)$$

The energetics they obtained is only qualitative, because they used the CNDO method and the structures were not optimized. Nevertheless, they found an important participation of the vacant d orbital. First, they considered the reaction system without $PtCl(PH_3)_2{}^+$:

$$H^- + C_2H_4 \rightarrow C_2H_5{}^- \quad (6.5)$$

The highest occupied molecular orbital (HOMO) of this uncatalyzed reacting system is **1**. The mixing of π^* into the antibonding combination of $1s_H$ and π (**2**) gives this HOMO as shown in eq. (6.6). This means that the four-electron repulsion between π and $1s_H$, represented by **2**, is relieved by this π^* mixing.

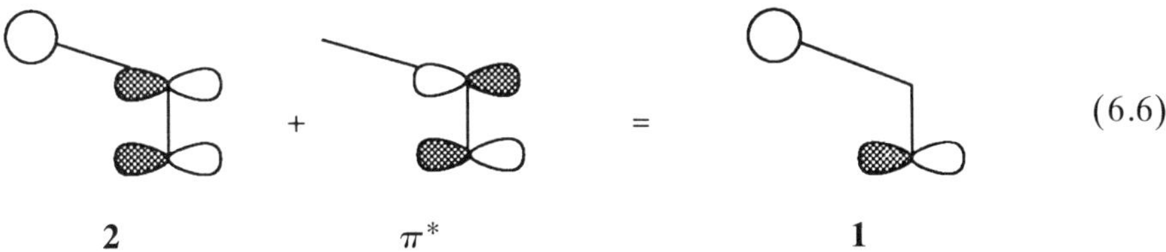

(6.6)

In the catalyzed system, **1** interacts with a vacant d orbital of Pt as shown in **3**, resulting in further stabilization. The orbital interaction **3** also suggests that the TS should be four-centered.

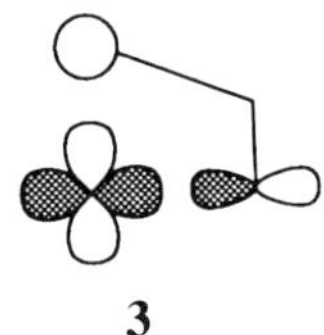

3

Fukui and Inagaki [18] analyzed the orbital interaction among three systems and applied their conclusions to olefin insertion. In olefin insertion, the interacting system is supposed to consist of a hydride anion, olefin, and a central transition metal. According to their analysis, the interaction among the HOMO of hydride ($1s_H$), the lowest unoccupied molecular orbital (LUMO) of unsaturated bond (π^*) and the LUMO of the central metal (d), shown in **4**, controls the reaction.

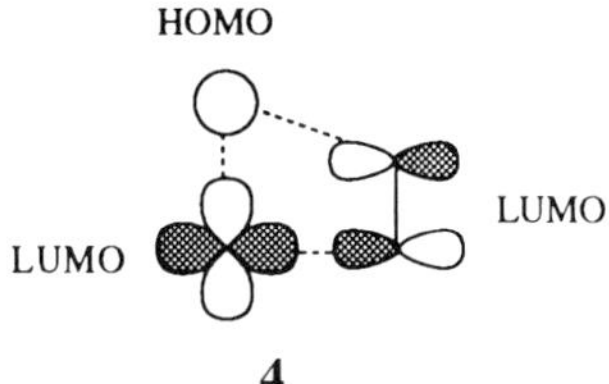

4

This interaction is similar to that obtained by Sakaki et al., shown in **3**. In **4**, the four-electron repulsion between π and $1s_H$ is neglected and only the bond formation is focused on. Tatsumi, Yamaguchi, and Fueno [19a] and Fujimoto, Yamasaki, Mizutani, and Koga [19b] also pointed out the importance of the vacant d orbital.

Thorn and Hoffman [20] extended these discussions. The interaction of the $H^- + C_2H_4$ system with a metal gives the bonding (**5**) and antibonding (**6**) combination between $1s_H$–π (**2**) and $d_{x^2-y^2}$. Mixing of π^* into **5** and **6**, in order to relieve the repulsion between $1s_H$ and π, gives **3** and **7** as shown in eqs. (6.7) and (6.8). In **7**, the interaction between M and C^α is strongly antibonding. Therefore, **7** should be vacant, which means that **6** and thus $d_{x^2-y^2}$ should be vacant.

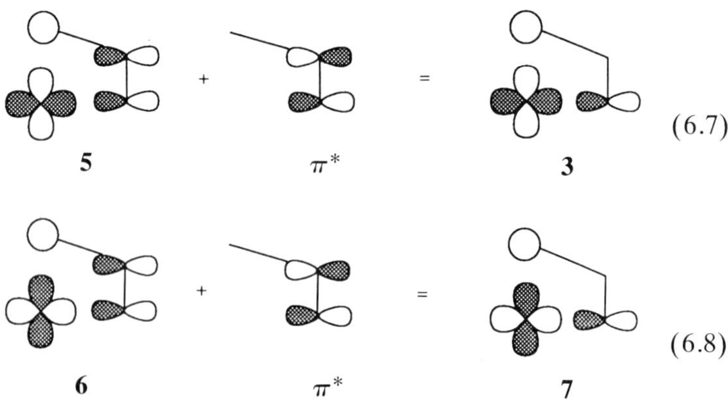

In the preceding discussion, the stabilizing effect of a transition metal on nucleophilic addition was emphasized. Accordingly, the newly formed C—H bond does not appear explicitly in the discussion. The interaction between $1s_H$ and π^* is responsible for this new bond.

When one regards the hydride migration as a $[2s + 2s]$ reaction, a different explanation is possible [21]. In general, in a multicentric reaction where two or more bonds are exchanged, both efficient electron donation and back-donation are required to make the barrier sufficiently low. In olefin insertion, the orbital interactions **8** and **9** have to take place.

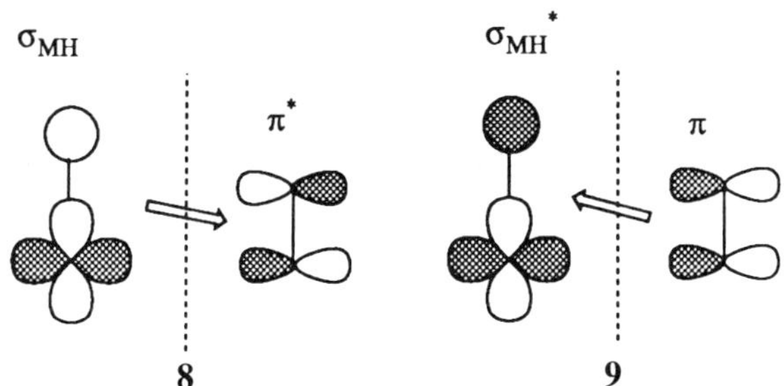

The electron donation from π to σ^*_{MH} and the back-donation from σ_{MH} to π^* facilitate the cleavage of the CC π and MH σ bonds and the formation of the MC σ and CH σ bonds. If $d_{x^2-y^2}$ is vacant, an appropriate occupied σ_{MH} and a vacant σ^*_{MH} orbital exist.

In the reaction of a d^8 four-coordinate complex, for instance, $HPt(C_2H_4)(PH_3)_2$, the $d_{x^2-y^2}$ is formally vacant; thus, olefin insertion is expected to be easy [20]. Koga and co-workers [22, 23] studied the olefin insertion and its reverse β-hydrogen elimination of d^8 group X complexes,

$$M(H)_2(C_2H_4)(PH_3) \rightarrow M(C_2H_5)(H)(PH_3) \qquad (6.9)$$

$$M = \begin{cases} \text{Ni} & \mathbf{10} & \mathbf{11} \\ \text{Pd} & \mathbf{12} & \mathbf{13} \\ \text{Pt} & \mathbf{14} & \mathbf{15} \end{cases}$$

They fully optimized the structures of the TSs as well as the reactants and the products, using the ab-initio energy gradient technique at the relativistic ECP-RHF level for M = Pt and the all-electron restricted Hartree–Fock (RHF) level for M = Pd and Ni. The optimized structures are shown in Figure 6.1 with relative energies.

All the TSs are four-centered and tight, as indicated by short M–C and M–H distances. There are, however, some differences among different metals. The Pd–H and the C–C distances at the TS for the Pd reaction (1.645 and 1.398 Å, respectively) are closer to those of **12** (1.593 and 1.336 Å, respectively) than **13** (2.127 and 1.530 Å); the TS is located near the reactant on the potential surface. The TS for the Pt reaction is located only slightly

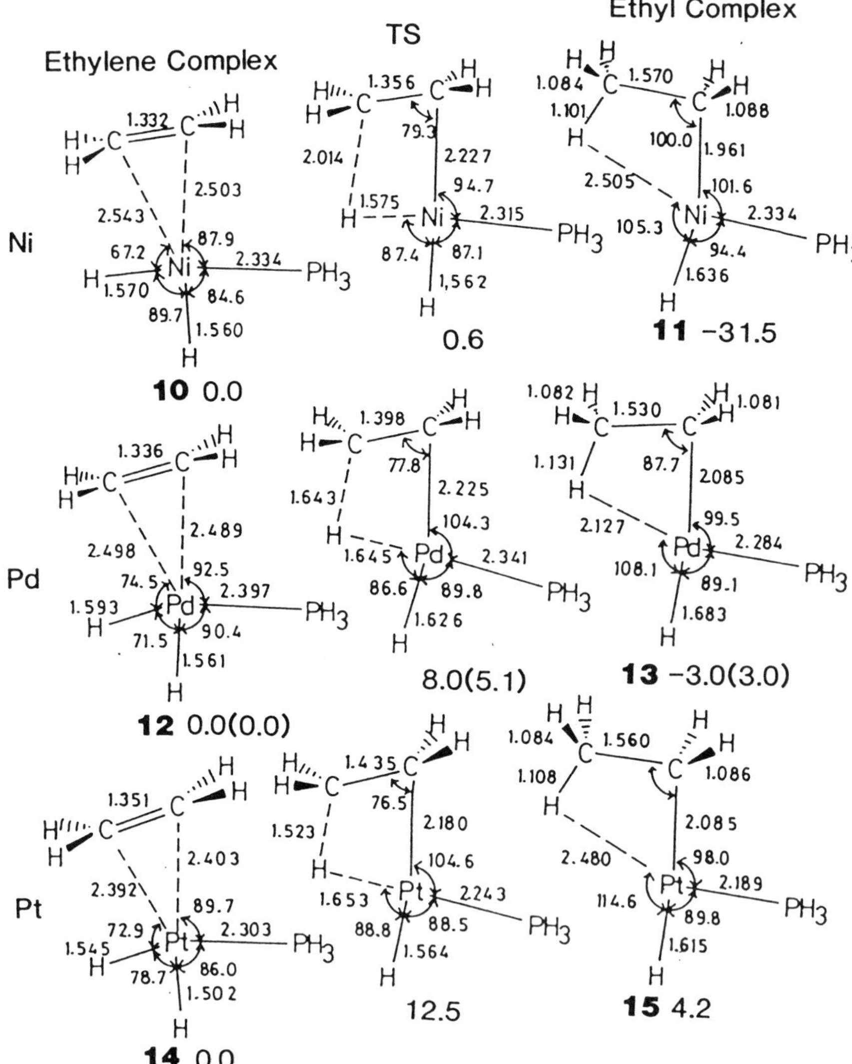

FIGURE 6.1
Structures (in angstroms and degrees), optimized at the RHF level, of reactants, products, and transition states for ethylene insertion,

$$M(H)_2(C_2H_4)(PH_3) \rightarrow M(C_2H_5)(H)(PH_3) \qquad (M = Ni, Pd, Pt)$$

Energies are RHF values (MP2 values in parentheses) in kilocalories per mole relative to ethylene complexes [22, 23].

closer to the corresponding ethyl complex (**15**) than that for the Pd reaction, as seen in the βCH and CC bond distances (1.523 and 1.435 Å for Pt and 1.643 and 1.398 Å for Pd). On the other hand, the TS of the Ni reaction is much closer to the reactant (**10**) in structure, compared with those of the Pd and the Pt reaction. Although the olefin insertion for the Pd and the Pt complexes is nearly thermoneutral, that for the Ni complex is very exothermic, making the corresponding TS reactant-like.

The activation barriers of the Pd and the Pt insertion are low. Because these reactions are almost thermoneutral, the low barriers are considered to reflect the intrinsic ease of olefin insertion of the four-coordinate d^8 olefin complex. The occupied orbitals in **12** and **14** related to the orbital interaction shown in **8** and **9** overlap nicely with the vacant partner orbitals to give rise to a low activation barrier.

As is well-known, Zeise's salt has an upright structure. The upright structures of $MCl_3(C_2H_4)^-$ have been calculated to be more stable than the in-plane structure by 7 and 15 kcal mol^{-1} for M = Pd and Pt, respectively [24]. Therefore, one might imagine that the reactants have upright structure and thus the activation barrier would be higher than that calculated. However, this is not true for the reactions (6.9); the upright structure is less stable by 2 and 5 kcal mol^{-1} at the RHF and MP2 levels for the Pd complex and 3 kcal mol^{-1} at the RHF level for the Pt complex [25]. In the present system, one of the ligands cis to ethylene is a small hydride and the in-plane ligands (PH_3 and H) make the back-donation to the in-plane ethylene stronger, slightly favoring the in-plane ethylene coordination.

Hydride migration creates a vacant coordination site into which a low-lying vacant d orbital extends. An electrical donation from the C—H σ orbital to a vacant metal d orbital can take place. In the optimized structures of these ethyl complexes, one can find structural features characteristic of the agostic interaction (intramolecular C—H····M interaction): the long C—H bond length, the small MCC angle and the short M····C—H distance [26]. The origin of this interaction is the electron donation from σ_{CH} to a metal vacant d [27–29]. The interaction between the migrating hydrogen and metal is maintained during the reaction and the product Pd and Pt ethyl complexes are rather similar in structure to the corresponding TSs because of this interaction. This interaction also activates the βC—H bond incipiently and thus lowers the activation energy of β-elimination reaction [22].

Bäckvall, Björkman, Pettersson, and Siegbahn [30] compared the reactivity of nucleophile migration to cis ethylene among several nucleophiles (Nu = H^-, CH_3^-, OH^-, and F^-) in the d^8 four-coordinate complex $Pd(Nu)_2(PH_3)(C_2H_4)$, based on the frontier orbital theory. They focused on the electron-donative interaction from σ_{PdNu} to π^* (**8**). Their calculations by the ab-initio RHF method with the double-zeta quality basis set and the ECP approximation have shown that the relatively small orbital energy difference between σ_{PdNu} and π^* for Nu = H^- and CH_3^- explains their easy migration to cis ethylene.

The d^6 $M(H)(C_2H_4)L_4$ is isolobal to d^8 $M(H)(C_2H_4)L_2$ and thus its hydride migration to coordinated ethylene is expected to take place easily. Dedieu [32] studied the hydride migration of $H_2RhCl(C_2H_4)(PH_3)_2$ **16** [eq. (6.10)] at the all-electron RHF level of calculation with the split-valence basis set. This reaction is a model of the olefin insertion step of the olefin hydrogenation catalytic cycle by the Wilkinson catalyst [33]. This step is believed to be the rate-determining step [34]. In Dedieu's study, only the essential geometrical parameters were changed in calculating the potential-energy surface for the reaction, whereas the structures of the reactant and the product were assumed.

$$\underset{\mathbf{16}}{H_2RhCl(C_2H_4)(PH_3)_2} \rightarrow \underset{\mathbf{17}}{HRhCl(C_2H_5)(PH_3)_2} \tag{6.10}$$

Three isomers of **17** were taken into account as shown in Scheme 6.1. His calculations indicated that **17b** is the most stable, whereas the direct product of hydride migration to the ethylene, **17a**, is the least stable. The reaction mechanism was shown to be ethylene insertion into the Rh—H bond up to the TS in which the formation of new C—H bond is almost completed (1.10 Å). After the TS is passed through, the ligand rearrangement takes place, to give the most stable **17b**. The total exothermicity was calculated to be 22 kcal mol^{-1} and the activation energy, 13 kcal mol^{-1}. Dedieu dis-

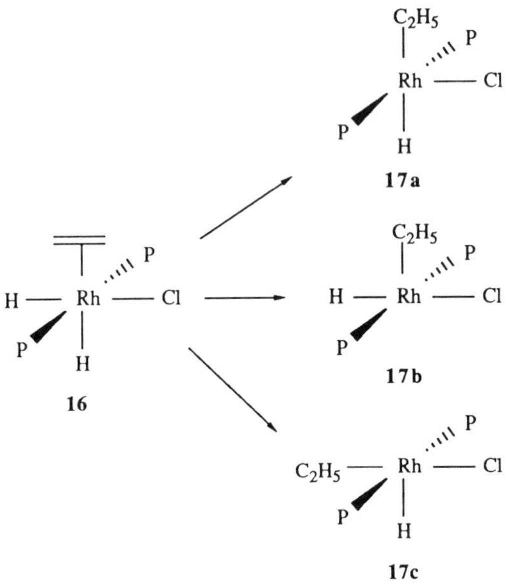

Scheme 6.1

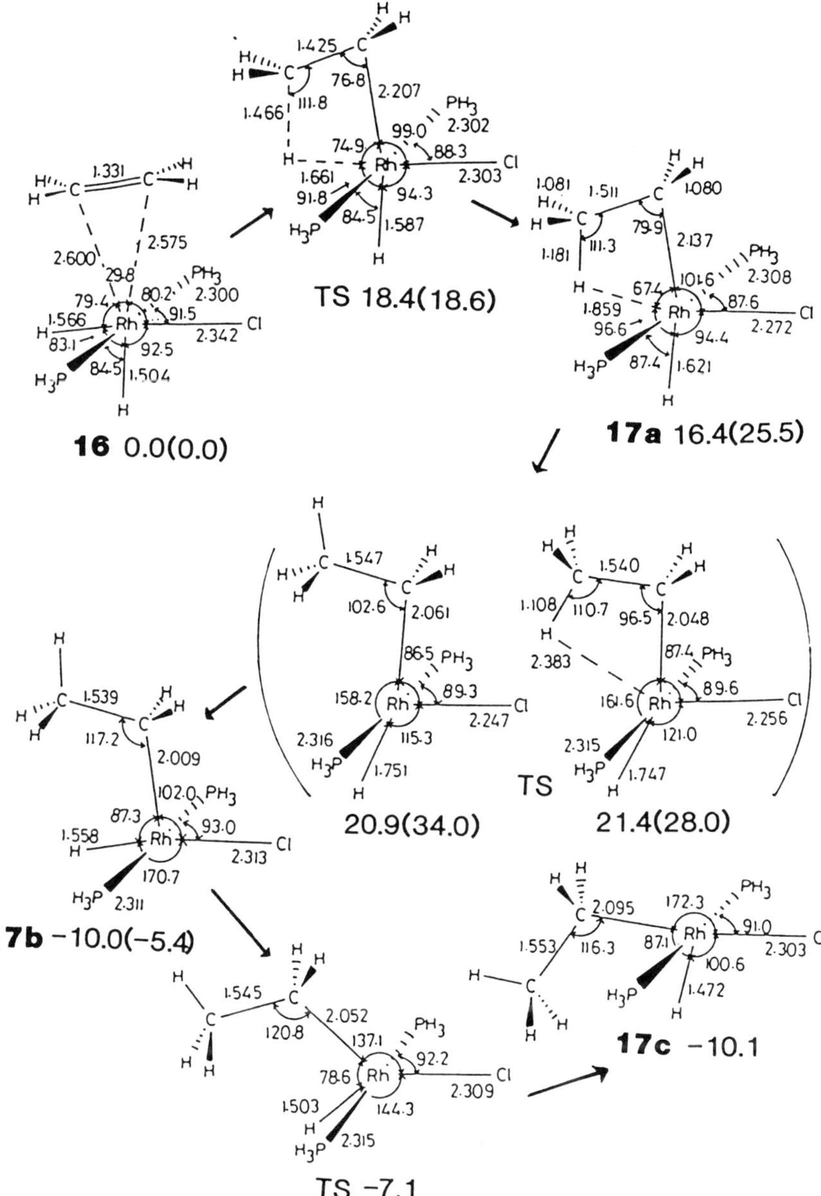

FIGURE 6.2
Structures (in angstroms and degrees), optimized at the RHF level, of reactants, products, and transition states for ethylene insertion of $H_2RhCl(C_2H_4)(PH_3)_2$ **16** and isomerization of $HRhCl(C_2H_5)(PH_3)_2$ **17**. Energies are RHF values (MP2 values in parentheses) in kilocalories per mole relative to ethylene complexes [31].

cussed the formation of new Rh—H and Rh—C bonds in a way similar to Sakaki et al. and Thorn and Hoffmann, described previously. In addition, he ascribed the origin of the activation barrier to the four-electron repulsions between the C_2H_4 π and the Rh—H σ orbitals and between the C—C σ orbital and the occupied *d* orbital. The results on the energetics and structure changes, although interesting, must be regarded as qualitative, because the structure optimization was not carried out.

More recently, Koga and co-workers [31, 35] carried out, with the ab-initio RHF energy-gradient method, the structure determination of (almost) all the intermediates and the TSs of the olefin hydrogenation catalytic cycle under the ECP approximation. The optimized structures of **16** and **17** and the TS between them are shown in Figure 6.2 with the relative energies. The energy profile and the structures should be compared with Dedieu's results.

The optimized Rh—C_2H_4 distance of the ethylene complex, **16**, is very long (with Rh—C distances of 2.58 and 2.60 Å), whereas the Rh—C distance assumed in Dedieu's study is 2.15 Å. This assumption would make the reaction more exothermic and the activation barrier lower. A detailed analysis showed that the long distance is ascribed to the cis effect of Cl and the trans effect of H, both of which make the Rh—C_2H_4 bond weaker.

One can see a very strongly agostic ethyl group in the product ethyl complex, **17a**: the small RhCC angle of about 80°, a short Rh····H distance of 1.86 Å, and the long βC—H bond of 1.18 Å. The electron-withdrawing Cl located trans to the vacant coordination site makes the donative interaction from the βC—H to the vacant *d* orbital stronger. In addition, the weak Rh—C_2H_5 bond due to the trans H and the cis Cl effects assists a large ethyl distortion. Although Dedieu did not note in his original paper, he later pointed out the agostic interaction in **17a** [36].

The TS obtained is four-centered, as expected from the preceding discussions, and is very different from the TS structure found by Dedieu with a partial optimization. It is not easy to determine the TS structure of a complicated system by partial optimization. The structure change and the approximate second-derivative matrix at this TS have indicated that the hydride migrates to ethylene, although up to the TS ethylene moves toward the hydride as if to catch it. This olefin insertion requires a substantial activation energy because of the large endothermicity, in contrast to the olefin insertion reactions of the Pd and the Pt complexes [eq. (6.9)].

Isomerization from the unstable trans ethyl hydride complex, **17a**, to more stable cis complexes is expected to take place easily. This is also necessary in the catalytic cycle, because the hydride and the ethyl ligands can eliminate reductively only from a cis complex. As shown in Figure 6.2, the cis ethyl hydride complexes, **17b** and **17c**, are more stable than the trans intermediate, **17a**, by 26.4 and 26.5 kcal mol^{-1}, respectively, at the RHF level. The lowest-energy path of isomerization from **17a** to **17b** was found to be the simultaneous hydride migration and ethyl CC internal rotation through the trigonal bipyramidal TS with a small activation energy of 5 kcal mol^{-1}

(RHF). The most favorable path to **17c** is chloride migration of **17b** with a small barrier of about 3 kcal mol^{-1} (RHF). The trans intermediate **17a** is in a shallow energy minimum and thus the insertion and isomerization may take place as one combined step with the effective overall barrier height of 21 kcal mol^{-1}, similar to Dedieu's results. An MP2 calculation at the RHF optimized geometries actually shows that **17a** is higher in energy than the TS connecting it and **16** and is a transient species.

Koga and co-workers also clarified the reason for the high endothermicity of reaction (6.10). Their comparison of the energy profile of the insertion between $H_2RhCl(C_2H_4)(PH_3)_2$ and $H_2RhH(C_2H_4)(PH_3)_2$ showed that Cl makes the ethylene insertion more endothermic; the olefin insertion of the latter complex is exothermic by 2 kcal mol^{-1}. Chlorine has a stronger cis effect and a weaker trans effect than H. At the positions cis to Cl, **16** has one Rh—H and one Rh—C_2H_4 bond, whereas **17a** has one Rh—H and one Rh—C_2H_5 bond, and therefore, the cis effect is more or less cancelled out between **16** and **17a**. The Rh—H bond trans to Cl in **16** is strong and requires a large energy to break, whereas in **17a** a vacant site is trans to Cl. Consequently, **16** is much more stable than **17a** and olefin insertion of **16** is endothermic. On the other hand, in $H_2RhH(PH_3)_2(C_2H_4)$ the Rh—H bond trans to H is weak, and therefore olefin insertion is not endothermic. The calculations showed that the Rh—H bond of **16** is 20 kcal mol^{-1} stronger than that of the $H_2RhH(PH_3)_2(C_2H_4)$. It has been clearly shown that the Cl effects play a key role in making the olefin insertion rate-determining.

Olefin insertion from a five-coordinate complex is assumed to take place in several catalytic cycles [37] and thus has been studied theoretically several times, as will be shown here. The olefin insertion reaction of the five-coordinate complex would be more complicated than that of the four- and six-coordinate complexes; one could imagine olefin insertion taking place from various isomers, because the intramolecular rearrangement takes place easily through a Berry pseudorotation (BPR) [38]. Thorn and Hoffman [20] discussed, with the EH method, the olefin insertion of the d^8 five-coordinate $HPtCl(PH_3)_2(C_2H_4)$ **18**. They showed that the most stable isomer is a trigonal bipyramidal (TBP) structure **18a** with axial H and equatorial (in-plane) ethylene, which is not suitable for olefin insertion.

H
P
Pt
P
Cl

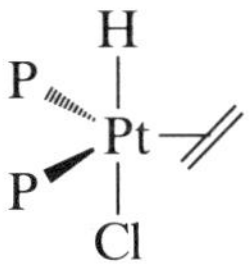

18a

According to their analysis, the structures suitable for olefin insertion are **18b** and **18c**. The vacant *d* orbitals of these structures are not very different from the d orbital appearing in **3–9**. The reactions via **18b** and **18c** require an activation energy from **18a** larger than 40 kcal mol^{-1}. These structures are suitable for the insertion but are unstable. On the other hand, Thorn and Hoffmann argued that the reaction from a square pyramidal (SP) isomer with the cis basal ethylene and hydride is virtually identical with that from a d^8 four-coordinate complex discussed previously.

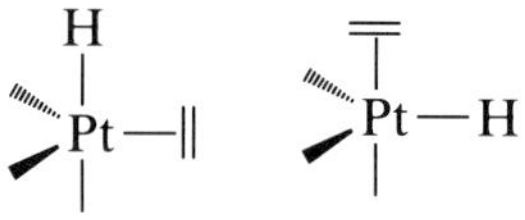

18b **18c**

Recently, Koga, Jin, and Morokuma [21] studied olefin insertion of five-coordinate d^8 Rh complex $HRh(CO)_2(C_2H_4)(PH_3)$ **19** [eq. (6.11)]. This is a model of an elementary step in the catalytic cycle of hydroformylation [39]. They optimized the structures of the isomers of **19** and **20** and those of the TSs for BPR of **19** and for olefin insertion at the ab-initio RHF level with the ECP approximation.

$$\underset{\mathbf{19}}{Rh(H)(CO)_2(PH_3)(C_2H_4)} \rightarrow \underset{\mathbf{20}}{Rh(CO)_2(PH_3)(C_2H_5)} \qquad (6.11)$$

The profile of the potential-energy surface of BPR is shown in Figure 6.3. Aside from the orientation of C_2H_4, formally, seven TBP structures as well as nine SP structures are possible. Five of the TBP structures (**19a–19e**) were found to be equilibrium structures and the other two (**19f** and **19g**) TSs for rearrangement. Relative energies show that H prefers the axial position and that C_2H_4 prefers equatorial over axial. Therefore, although the axial M—H bonds are shorter than the corresponding equatorial bonds, the axial M—C_2H_4 bond is longer than the equatorial bond. No TBP structure has vertical equatorial C_2H_4. All the four SP structures Koga et al. found are TSs for BPR. Paying attention to relatively thermoneutral rearrangements, **19d** ↔ **19c** and **19c** ↔ **19e**, one can find that the intrinsic barrier for BPR is very low. Of course, a more endothermic rearrangement, for instance, **19d** → **19f**, requires a much higher activation barrier, reflecting its endothermicity.

There may exist varieties of transition states for olefin insertion starting from varieties of conformers. After an extensive search, however, they found only three TSs, all being square pyramidal with basal H and basal C_2H_4, as shown in Figure 6.4. These transition states are on the reaction paths from the most stable isomers, **19a** and **19b**, both of which have axial H and equatorial ethylene. The reaction coordinate from the reactant up to the transition state is ethylene rotation, coupled with the skeletal change from TBP to SP. The ethylene rotation allows the C═C axis to become parallel to the Rh—H bond for a favorable four-center arrangement. The C—H bond

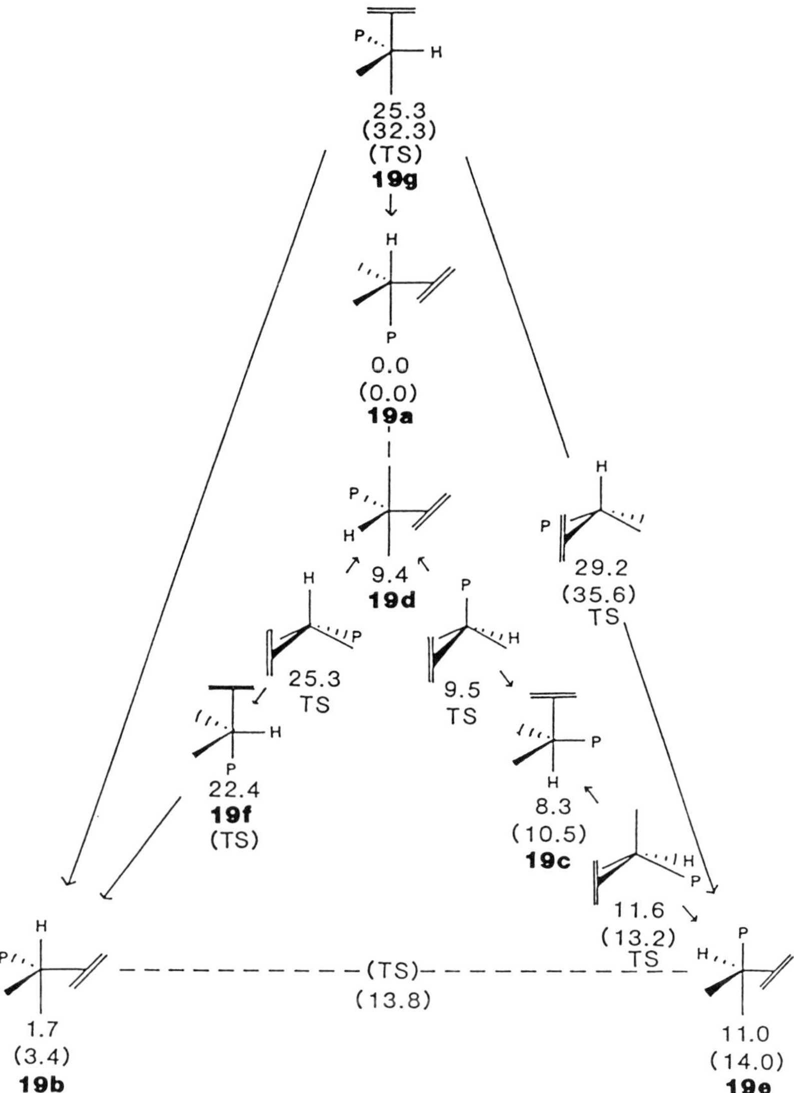

FIGURE 6.3
Potential-energy profile for the intramolecular rearrangement of $Rh(H)(CO)_2(PH_3)(C_2H_4)$ **19**, calculated in kilocalories per mole at the RHF level (at the MP2 level in parentheses) at the RHF optimized structures, relative to the most stable isomer **19a** [21].

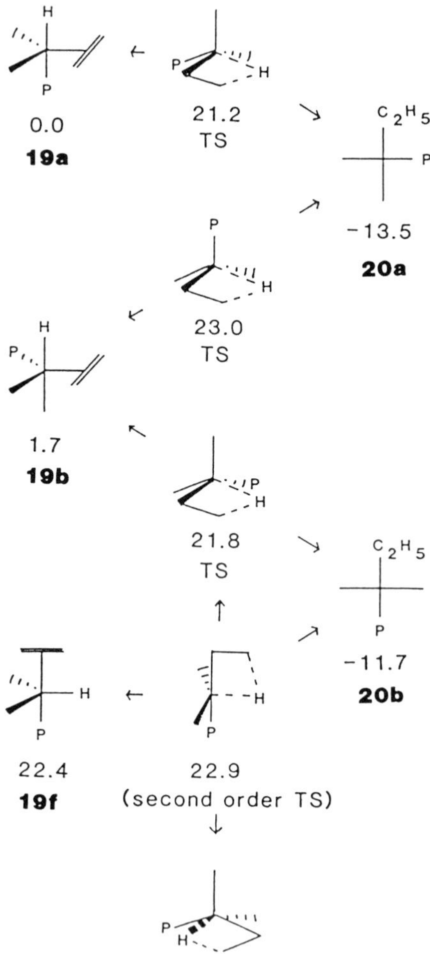

FIGURE 6.4
Potential-energy profile for ethylene insertion of $Rh(H)(CO)_2(PH_3)(C_2H_4)$ **19**, calculated in kilocalories per mole at the RHF level (at the MP2 level in parentheses) at the RHF optimized structures, relative to **19a** [21].

to be formed is still very long (1.57, 1.64, and 1.75 Å), showing that the bond exchange stage of reaction takes place after the TS is passed.

The corresponding barrier heights are calculated to be about 21 kcal mol^{-1} at the RHF level. The MP2 calculations at the RHF optimized structures gave similar activation barriers. These barriers indicate that the reaction takes place relatively easily. At the square pyramidal TS, $d_{x^2-y^2}$ is formally empty and can play the same role as $d_{x^2-y^2}$ at the TS of a d^8 four-coordinate complex. This is probably why olefin insertion of d^8 five-coordinate complex passes through a square pyramidal TS. In spite of this similarity, the activation energy of this olefin insertion reaction is higher than that of the same reaction of d^8 four-coordinate complexes discussed previously. The skeletal deformation required to reach the SP structure ready for olefin insertion has been found to be an important part of this higher activation energy.

The olefin insertion reaction of the hydroformylation catalytic cycle by Co catalysts has been studied with ab-initio RHF and HFS methods [40–42].

Antolovic and Davidson [41a] studied the reaction

$$\underset{\mathbf{21}}{HCo(CO)_3(C_2H_4)} \rightarrow \underset{\mathbf{22}}{Co(CO)_3(C_2H_5)} \qquad (6.12)$$

As shown in Figure 6.5, they determined at the RHF level the structures of four isomers of **21**, in which the hydride and ethylene are located cis to each other, facilitating the hydride migration, and of three isomers of **22** of the TBP type, in which one of the coordination sites is vacant. Antolovic and Davidson showed that in the RHF optimized structures of **21** the equatorial M—L bonds are generally shorter than the axial M—L bonds, different from the Rh complex, **19**. It should be noted here that the electron correlation effect, which is not taken into account at the RHF level, is important for the structure-determination of the first-row transition metal carbonyl complexes [43, 44]. Electron diffraction experiments show the C_{3v} structure of $HCo(CO)_4$ **23**, with the Co—C_{ax} bond being shorter than the Co—C_{eq} bond [45]. The RHF calculation, contrary to this, gives the Co—C_{ax} bond longer than the Co—C_{eq} bond [41b]. (An example of a second-row transition metal complex is $HRh(CO)_2(C_2H_4)(PH_3)$ **19**, in which Rh—C_{ax} is shorter than Rh—C_{eq} [21].) In addition, the C_{3v} structure **23** is 1.3 kcal mol^{-1} less stable than the C_{2v} structure **24** at the RHF level.

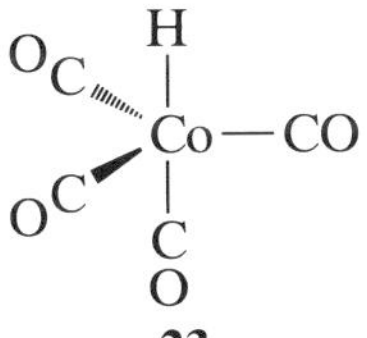

23 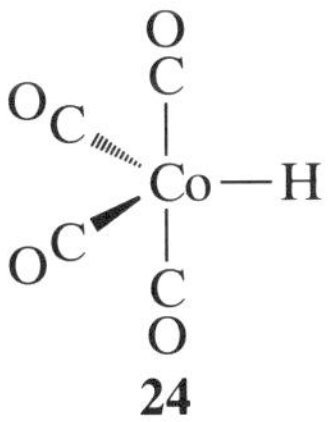 **24**

The electron correlation, however, improves the structure and relative stability. At the SDCI level of calculation, the C_{3v} structure is more stable, as in experiment, than the RHF-preferred C_{2v} structure [43].

The energetics of hydride migration (6.12) was calculated by using the linear synchronous transient (LST) method (also shown in Figure 6.5). The TS has not been optimized and the BPR could not take place during the insertion, because the C_s symmetry has been imposed. Due to these two

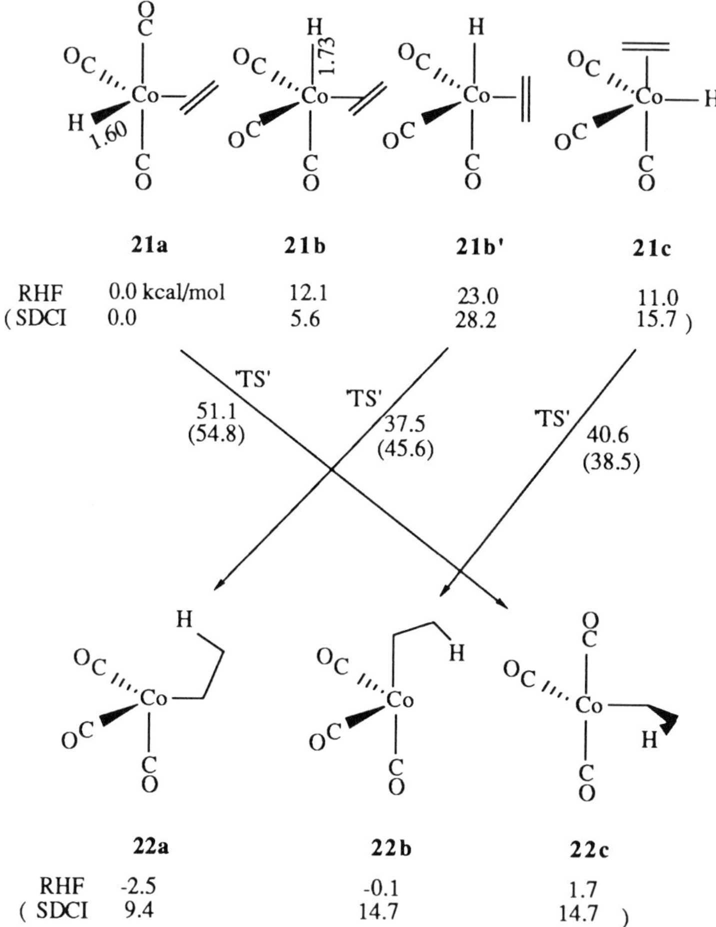

FIGURE 6.5
Potential-energy profile for ethylene insertion of $HCo(CO)_3(C_2H_4)$ **21**, calculated in kilocalories per mole at the RHF level (at the SDCI level in parentheses) at the RHF optimized structures, relative to **21a**. The transition states are assumed to be on the LST path. The Co—H bond distances in **21a** and **21b** are in angstroms [41].

restrictions, the activation energies Antolovic and Davidson obtained should be regarded as the upper limits. They concluded that the least stable isomer, **21b**′, is a good starting geometry for olefin insertion, because the lowest activation energy, 14 and 17 kcal mol^{-1} at the RHF and SDCI levels, respectively, was obtained for **21b**′ → **22a**.

Versluis [42] studied reaction (6.12) with the HFS method. At the HFS level, the hydride favors the axial position, consistent with the experiments; thus he considered the reaction from **21b** and **21b**′, with the axial hydride,

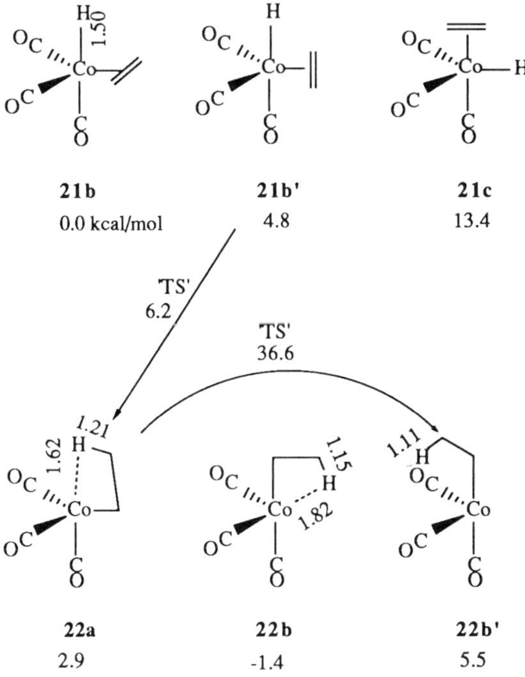

FIGURE 6.6
Potential-energy profile for ethylene insertion of $HCo(CO)_3(C_2H_4)$ **21**, calculated in kilocalories per mole at the HFS level relative to **21b**. The transition states are assumed to be on the LST path. The selected bond distances are shown in angstroms [42].

and **21c**. The reaction path from **21b′** was followed by the LST method. The HFS results shown in Figure 6.6 are very different from those shown in Figure 6.5: (i) **21b′** is not very unstable compared with **21b**, whereas at the RHF and SDCI levels the perpendicular ethylene destabilizes the system by 11 and 23 kcal mol^{-1}, respectively. (ii) The most stable product, **22b**, has an axial ethyl ligand, although the ab-initio RHF and SDCI calculations showed that the most stable isomer is **22a** with the equatorial ethyl ligand. (iii) The activation barrier from **21b′** is much smaller. (iv) The agostic interaction, which was not found by the RHF calculation, takes place in the HFS optimized structures, **22a** and **22b**.

These two studies of reaction (6.12) agree that **21b′** is a favorable structure for olefin insertion and that the activation energy from **21b′** is not very large. However, the quantitative results are very different from each other. More-sophisticated calculations would be required to solve the discrepancies, although the HFS results seem to be more reasonable. The olefin insertion of the five-coordinate complex would take place together with the BPR process to pass through a square pyramidal TS with the basal hydride and the basal ethylene as shown in reaction (6.11). At such a TS, there is no distinction between the axial and the equatorial positions. Therefore, the insertion would give the most stable isomer of **22**. Therefore, one has to relax the C_s symmetry constraint, imposed by the LST method, to take the BPR into account.

The regioselectivity of olefin insertion of $HCo(CO)_3(CH_2{=}CHCH_3)$, in which a linear or a branched alkyl complex is formed, has been discussed with the CNDO and the ab-initio RHF method by Grima, Choplin, and Kaufmann [40]. They showed that the atomic charge controls the regioselectivity.

Also, several theoretical calculations have been carried out on five-coordinate complexes by the CNDO method. Reactions (6.14) and (6.15) are models of heterogeneous processes.

$$HPtCl(C_2H_4)(PH_3)_2 \rightarrow PtCl(C_2H_5)(PH_3)_2 \qquad ([17,46]) \qquad (6.13)$$

$$CpCrH(OSi(OH)_3)(C_2H_4) \rightarrow CpCr(C_2H_5)(OSi(OH)_3) \qquad ([47]) \qquad (6.14)$$

$$HPd(Cl)_3(CH_2{=}CHPh)^{2-} \rightarrow Pd(Cl)_3(CH_2CH_2Ph)^{2-} \qquad ([48]) \qquad (6.15)$$

These CNDO calculations have shown that the olefin insertion is always downhill without activation barrier. However, the energetics of the CNDO calculation is not reliable. Ab-initio ECP calculations of reaction (6.16) have been carried out by Daudey, Jeung, Ruiz, and Novaro [49] to compare the ab-initio energetics of the reaction (6.16) with the CNDO energetics of reaction (6.15).

$$HPd(Cl)_3(C_2H_4)^{2-} \rightarrow Pd(Cl)_3(C_2H_5)^{2-} \qquad (6.16)$$

The reactant and the product were assumed to have square pyramidal and square planar structure, respectively. The activation barrier from the square pyramidal reactant was calculated to be 11 kcal mod^{-1}, different from the CNDO calculations of reaction (6.15). Differences between the CNDO and ab-initio calculations also were found in the theoretical studies of the Ziegler–Natta olefin insertion into $Ti{-}CH_3$ bond [50]. Looking at Daudey et al.'s potential energy curve, one can find a local minimum between the TS and the square pyramidal reactant with stabilization of about 18 kcal mol^{-1}. This local minimum would correspond to a TBP structure, from which the activation energy can be calculated to be 30 kcal mol^{-1}. This activation energy should be regarded as the upper limit, because the C_s structure has been assumed. As shown previously, the olefin insertion from the five-coordinate d^8 would pass through an SP transition state with the basal olefin and hydride.

6.3. HYDRIDE MIGRATION TO COORDINATED FORMALDEHYDE

Hydride migration to the coordinated formaldehyde is considered to take place in the homogeneous catalyzed hydrogenation of carbon monoxide, which transforms synthesis gas into monoalcohols and polyalcohols [51]. This hydride migration could give two intermediates, hydroxymethyl and methoxy complexes, from which ethylene glycol and methanol are obtained as the final products [eq. (6.17)].

$$\begin{array}{ccccc} & \nearrow & HOCH_2ML_n & \longrightarrow & (CH_2OH)_2 \\ HM(H_2CO)L_n & & & \searrow & \\ & \searrow & CH_3OML_n & \longrightarrow & CH_3OH \end{array} \qquad (6.17)$$

The hydride migration to the coordinated formaldehyde is similar to that to the coordinated olefin, because the formaldehyde has π and π^* orbital as the ethylene does. Consequently, it is expected that the formaldehyde insertion could take place easily and would pass through the four-centered TS. The ab-initio RHF and HFS calculations have been carried out to elucidate the difference of the two modes [eq. (6.17)] of insertion reaction [52, 42].

Nakamura and Morokuma [52] determined the structures of the stationary points of reaction (6.18) by the ab-initio RHF energy-gradient method with the ECP approximation.

$$\begin{array}{ccc} & \nearrow & \underset{\mathbf{24}}{HRu(CH_2OH)(CO)_3} \\ \underset{\mathbf{23}}{H_2Ru(CH_2O)(CO)_3} & & \\ & \searrow & \underset{\mathbf{25}}{HRu(OCH_3)(CO)_3} \end{array} \qquad (6.18)$$

The results they obtained are shown in Figure 6.7. The methoxy formation, **23a** → **25b**, is much more exothermic than the hydroxymethyl formation, **23b** → **24b**. This difference was attributed to the M—O bond, which is stronger than the M—C bond. The TSs for both reactions are four-centered, as expected. In spite of this structural similarity, the activation energy for hydroxymethyl formation is much higher, because of the unfavorable overlap

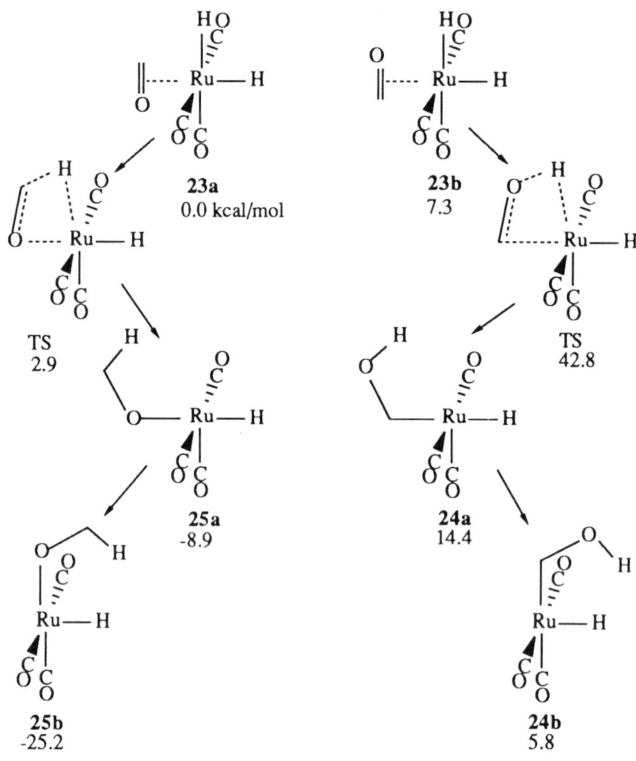

FIGURE 6.7
Potential energy profile for formaldehyde insertion of $H_2Ru(CH_2O)(CO)_3$ **23**, calculated in kilocalories per mole at the RHF level, relative to **23a** [52].

between σ_{MH} and the polarized π^* orbital, as shown in **26**.

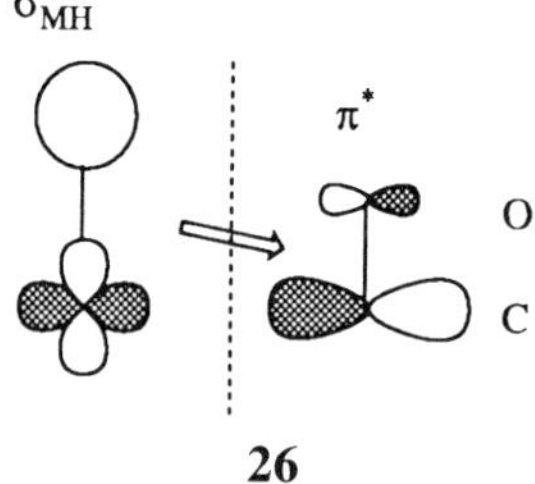

26

In addition to the large exothermicity and the smaller activation energy, the fact that the reactant of the methoxy formation, **23a**, is more stable than that of the hydroxymethyl formation, **23b**, makes the methoxy formation much more favorable, in agreement with experimental observation [53].

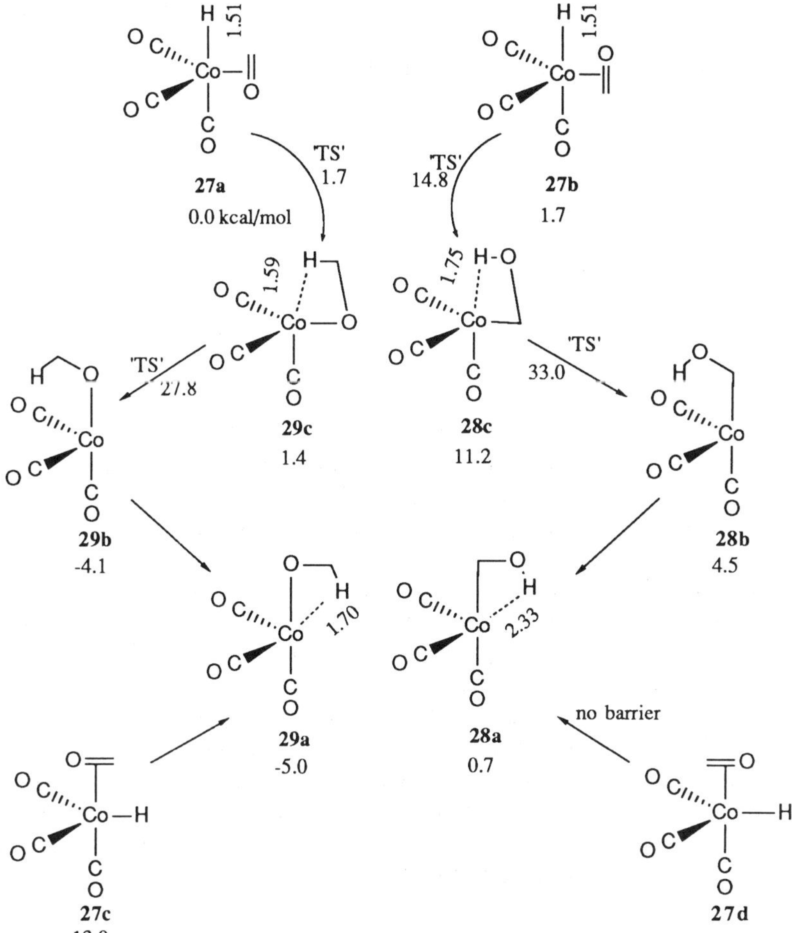

FIGURE 6.8
Potential-energy profile for formaldehyde insertion of $HCo(CO)_3(CH_2O)$ **27**, calculated in kilocalories per mole at the HFS level, relative to **27a**. The transition states are assumed to be on the LST path. The selected bond distances are shown in angstroms [42].

Versluis [42] compared methoxy and hydroxymethyl formation in reaction (6.19) with the HFS method, as shown in Figure 6.8.

$$\begin{array}{ccc} & \nearrow & \underset{\mathbf{28}}{Co(CO)_3(CH_2OH)} \\ \underset{\mathbf{27}}{HCo(CO)_3(CH_2O)} & & \\ & \searrow & \underset{\mathbf{29}}{Co(CO)_3(OCH_3)} \end{array} \tag{6.19}$$

In the reactants (the **27**s), the C=O bond was assumed to be parallel to the Co—H bond, because this structure is favorable for the reaction. The reaction was followed by the LST method. Similar to the RHF results of reaction (6.18), the methoxy formation is easier than the hydroxymethyl formation; the former is more exothermic and the corresponding activation energy is lower. The more exothermic methoxy formation was ascribed to the stronger Co—O bond [54]. In the products, when a vacant coordination site cis to the methoxy or the hydroxymethyl ligand is available, the agostic interaction was found to take place.

27 is a d^8 five-coordinate complex and thus the hydride migration would pass through a square pyramidal TS, which has been precluded by the use of the LST method. In such a case, as discussed in the preceding section, the rearrangement to the most stable isomers could take place during the insertion, although Verluis also studied the isomerization of the products. Therefore, further investigations should be needed.

6.4. HYDRIDE MIGRATION TO CARBON DIOXIDE

The introduction of CO_2 into the organic substrate has drawn substantial attention, and the CO_2 insertion into the M—X bond (X = H^-, CH_3^-, OR^-) has been studied experimentally [55]. CO_2 has π and π^* orbitals available for the bond exchange and thus can participate in a four-centered transition state.

Sakaki and Ohkubo [56] studied the CO_2 insertion into a Cu—H bond [eqs. (6.20) and (6.21)] at the ab-initio RHF and RMP2 levels.

$$CO_2 + \underset{\mathbf{30}}{CuH(PH_3)_2} \rightarrow \underset{\mathbf{31}}{Cu(PH_3)_2(HCO_2)} \tag{6.20}$$

$$CO_2 + \underset{\mathbf{32}}{CuH(PH_3)_3} \rightarrow \underset{\mathbf{33}}{Cu(PH_3)_3(HCO_2)} \tag{6.21}$$

They determined at the ECP-RHF level the structures of the reactant and the product for reaction (6.20), using the energy-gradient method, and those for reaction (6.21), by parabolic fitting of the potential-energy curve along some critical geometrical parameters. The reaction schemes studied are shown in Figure 6.9. The Cu–H distance was taken as the reaction coordinate and the structures at several Cu–H distances were determined, to calculate the potential-energy curves for these reactions. The energy calculations also were carried out at the ECP-MP2 and the all-electron RHF levels. The TSs for both reactions were located at the Cu–H distance of 1.6 Å at the ECP-RHF level as shown in Figure 6.9, whereas a change of the basis set and the electron correlation effect slightly shift the position of the TSs. The TSs for both reactions at the all-electron RHF level have longer C—H bonds [1.8 and 2.0 Å for reactions (6.20) and (6.21), respectively]. The electron correlation makes the C—H bond of the TS for reaction (6.21) shorter (1.3 Å). The activation barriers calculated at each position are shown in Figure 6.9.

Although Sakaki and Ohkubo also considered η^1-COOH coordinated structures of **31**, such structures are less stable than the η^1-OCOH coordinated structures, analogous to the difference between the hydroxymethyl and the methoxy complex shown in the preceding section. Although the η^1-COOH structures with a positive charge on both Cu and C are electrostatically less favorable, the η^1-OCOH structures are favorable with the positive Cu and the negative O interacting. In addition, HCO_2^- is more stable than $HOCO^-$. A CNDO calculation for CS_2 and CO_2 insertion [eq. (6.22)] also has shown the preference of η^1-XCXH (X = S, O) coordination [46].

$$\textit{trans}\text{-HPtCl(PH}_3)_2 + \text{CX}_2 \rightarrow \textit{trans}\text{-PtCl(CX}_2\text{H)(PH}_3)_2 \qquad (\text{X} = \text{O, S}) \tag{6.22}$$

Sakaki and Ohkubo found that $Cu(\eta^2\text{-}O_2CH)(PH_3)_2$ **31a** is the most stable isomer of **31** and thus the thermodynamic product. There are two conformations for the $Cu(\eta^1\text{-OCOH})(PH_3)_2$, H being syn (**31c**) or anti (**31b**) with respect to the central Cu atom. The positive hydrogen prefers to be far away from the positive Cu; **31c** is not an equilibrium structure and collapses into **31b**. The activation barrier for the rearrangement from **31b** to **31a** of the final product was calculated to be only a few kilocalories per mole and thus this step is not rate-determining.

Opposite to what was found for reaction (6.20), the η^2-O_2CH product, **33a**, of reaction (6.21) is slightly less stable than the η^1-OCOH isomer, **33b**. In the Cu(I) complex, only *s* and *p* orbitals are available for bonding and thus the η^2 structure having five Cu—L bonds is not favorable. Thus **33b** is the final product of reaction (6.21).

Reaction (6.20) is exothermic with a low activation barrier, as shown in Figure 6.9. Reaction (6.21) is also exothermic and requires a small activation barrier. Both reactions pass through a four-centered transition state with a

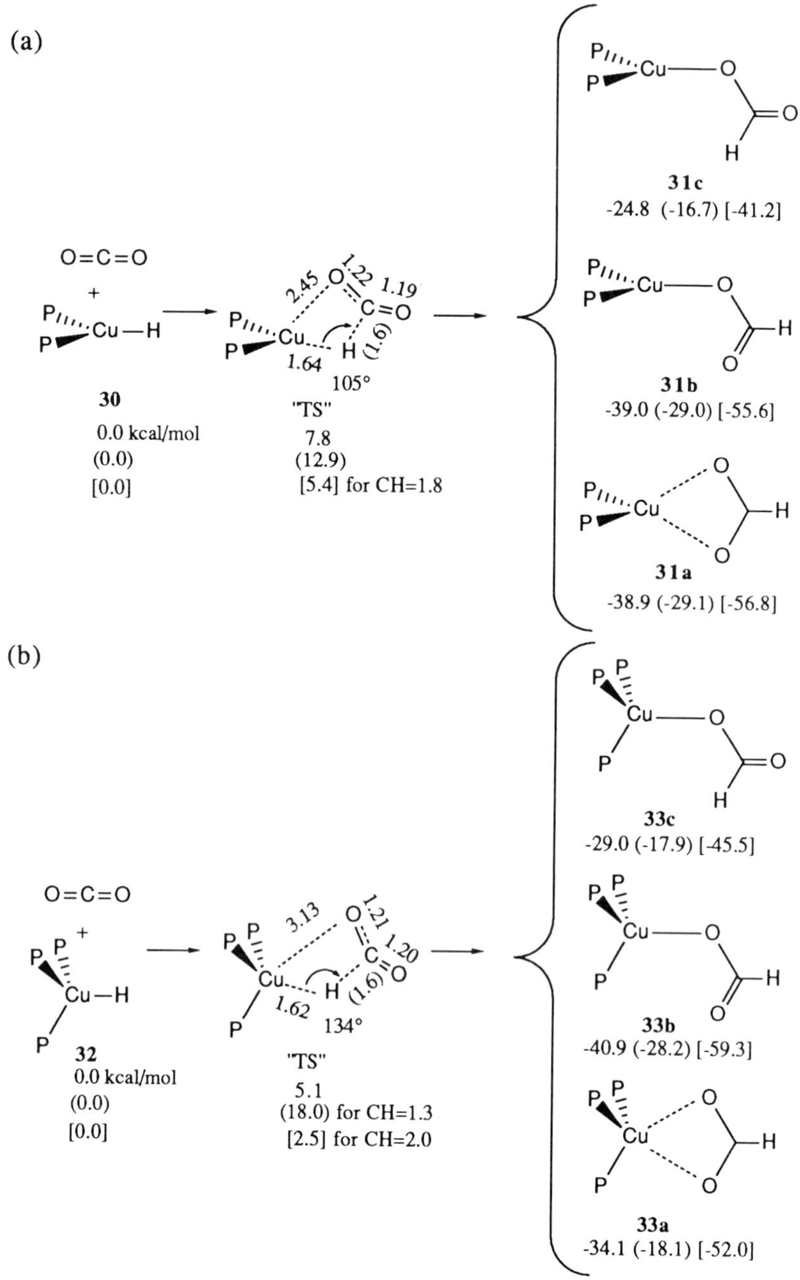

(a)

O=C=O + $CuH(PH_3)$

30
0.0 kcal/mol
(0.0)
[0.0]

"TS"
7.8
(12.9)
[5.4] for CH=1.8

31c
-24.8 (-16.7) [-41.2]

31b
-39.0 (-29.0) [-55.6]

31a
-38.9 (-29.1) [-56.8]

(b)

O=C=O + $CuH(PH_3)_3$

32
0.0 kcal/mol
(0.0)
[0.0]

"TS"
5.1
(18.0) for CH=1.3
[2.5] for CH=2.0

33c
-29.0 (-17.9) [-45.5]

33b
-40.9 (-28.2) [-59.3]

33a
-34.1 (-18.1) [-52.0]

FIGURE 6.9
Potential-energy profiles for CO_2 insertion to (a) $CuH(PH_3)$ **30** and (b) $CuH(PH_3)_3$ **32**, calculated in kilocalories per mole at the ECP-RHF level, the ECP-RMP2 level (in parentheses) and at the all-electron RHF level (in square brackets), relative to the reactants [56]. The "transition-state" structures (in angstroms and degrees) are those on the ECP-RHF potential-energy curve and are optimized with the C—H distance fixed to be 1.6 Å. At the levels of calculation where the C—H distance at the corresponding TS is not 1.6 Å, the C—H distances are shown together with the activation energies.

long $Cu–O^1$ distance. Sakaki and Ohkubo found that, at the early stage of reaction, the reaction (6.21) is easier than the reaction (6.20). At the later stage of reaction including the TS, the RHF and the MP2 calculations have given a relative energy for reaction (6.21) lower and higher, respectively, than that for reaction (6.20). Therefore, the difference between the two reactions at the later state has not been discussed.

They concluded that the origin of the activation barrier is the deformation of CO_2 and the exchange repulsion between $CuH(PH_3)_n$ and CO_2 and that the charge transfer from $CuH(PH_3)_n$ to CO_2 is important for stabilizing the reaction system. Also, the polarization within $CuH(PH_3)_n$ and CO_2 moieties and the electrostatic interaction between O^1 and Cu cooperate in the formation of new bonds. Such interactions are shown in **34**.

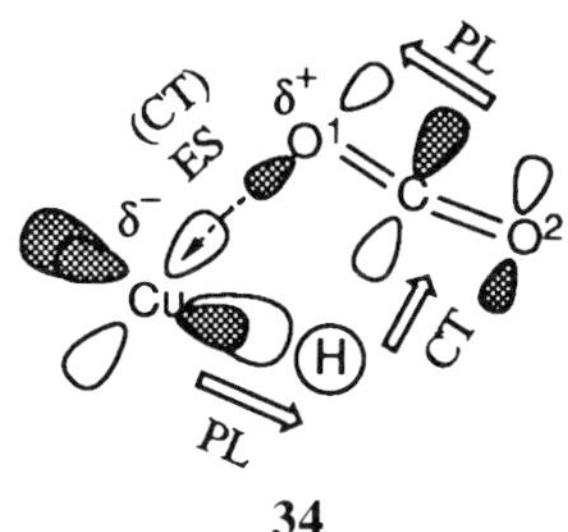

34

From the viewpoint of the orbital interaction, the O^1–Cu interaction is not favored because of the out-of-phase combination. Remember that Cu is d^{10} and thus no vacant d orbital is available. Therefore, the $Cu–O^1$ distance is much longer than the Cu–C distance, as shown in Figure 6.9. The additional phosphine in reaction (6.21) favors the charge transfer from the CuH moiety to CO_2, resulting in a lower energy for reaction (6.21) in the early stage of reaction.

Bo and Dedieu [57] studied the interaction of CO_2 with $HCr(CO)_5^-$ and carried out a preliminary calculation for the insertion reaction (6.23) at the RHF level with a split-valence basis set.

$$\underset{\mathbf{35}}{CO_2 + HCr(CO)_5^-} \rightarrow \underset{\mathbf{36}}{Cr(COOH)(CO)_5^-} \tag{6.23}$$

The interaction of CO_2 with $HCr(CO)_5^-$ that results in the formation of an adduct, **37**, has been their main interest. The geometrical parameters of HCO_2 moiety in **37** have been optimized with the frozen ideal structure of $Cr(CO)_5$ and are shown in Figure 6.10. The short C–H distance of 1.15 Å and the long Cr–H distance of 1.88 Å indicate that the HCO_2 moiety has the character of the formate ligand. The charge distribution also agrees with this character. The RHF interaction energy corrected for the basis-set superposition error has been calculated to be 8 kcal mol^{-1}. The orbital interaction favoring the adduct formation is shown in **38**. The electron donation from

134.5° 1.15 170° 1.88 -CO +CO

37 0.0 kcal/mol | 39 3.0 | 36a -26.4

40 -18.1 | 36b -30.4

FIGURE 6.10
RHF optimized structure of $HCr(CO_2)(CO)_5^-$ **37** (in angstroms and degrees) and potential-energy profile for CO_2 insertion of $HCr(CO_2)(CO)_5^-$ **37**, calculated in kilocalories per mole at the RHF level [56].

$d_{z^2} + 1s_H$ to the π^* of CO_2 takes place.

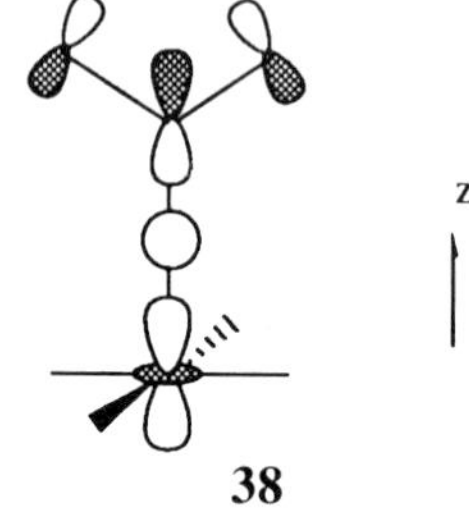

The reaction that Bo and Dedieu studied is shown in Figure 6.10, in which CO dissociation and association are involved. Only a few key geometrical parameters of the formate complexes were optimized. The assumed η^2-HOCO structure, **39**, which is considered not to be very far from the TS, is not very unstable. **39** has a vacant *d* orbital, facilitating the bond exchange as discussed in the section of hydride migration to olefin. The insertion reaction from **37** coupled with the CO dissociation may take place easily, to lead to an η^2-O_2CH complex, **40**. The additional CO coordination to **40** would give **36**.

6.5. HYDRIDE MIGRATION TO CARBONYL LIGAND

The hydride migration to the carbonyl ligand [eq. (6.24)], which is also called carbonyl insertion into an M—H bond, is believed to take place in homogeneous catalytic reactions of the reductive hydrogenation of carbon monoxide [58], although this hydride migratory insertion has been known only for a few cases in transition metal complexes [58, 59].

$$M(H)(CO) \rightarrow M-\overset{O}{\overset{\|}{C}}-H \qquad (6.24)$$

One of the known examples is the formyl formation from the rhodium hydride complex [eq. (6.25)] [60]. The formyl group is, however, formed through a radical chain process [61]. In the actinide complex the hydride migration can take place easily, as in the example of eq. (6.26), which has an activation free energy of 9 kcal mol^{-1} and an exothermicity of 2 kcal mol^{-1} [62]. This exothermicity has been ascribed to the oxophilicity of Th, which favors the η^2-formyl complex.

$$Rh(OEP)(H) + CO \rightarrow Rh(OEP)(CHO) \qquad (6.25)$$

$$Cp^*ThH(OR) + CO \rightarrow Cp^*Th(\eta^2\text{-}O{=}C{-}H)(OR) \qquad (6.26)$$

On the other hand, the alkyl group migratory insertion is observed in many organometallic reactions and catalytic processes [1]. The difficulty of hydride migratory insertion is believed to be due to the stronger M—H bond, which makes the hydride migratory insertion endothermic [63]. Although it is difficult to study the hydride migratory insertion experimentally, it is possible to study theoretically and compare the hydride migratory insertion with the alkyl migratory insertion.

There are two possible reaction pathways for the reaction: (i) hydride insertion into the M—CO bond or (ii) hydride migration to CO. As shown later, theoretical calculations have shown that the hydride migrates to CO in late or middle transition metal complexes.

$$H-M-C{\equiv}O \;\xrightarrow{(i)}\; H(O{=})C-M \qquad H-M-CO \;\xrightarrow{(ii)}\; M-C(H){=}O$$

Table 6.1
Energy of Reaction (ΔE) and Activation Energy ($\Delta E^{\ddagger}$) (kcal mol^{-1}) of Carbonyl Insertion of $HMn(CO)_5$.[a]

Method	$\Delta E^{\ddagger}$	$\Delta E(\eta^1)$	$\Delta E(\eta^2)$	Ref.
EHT	16 (20)			[65]
RHF[b]	14	11		[66a]
SDCI[b]	39	38		[66a]
HFS	40 (21)	38 (18)	22 (−1)	[71]
PRDDO	20 (18)	14 (1)	−5 (−20)	[75]
RHF[c]		18 (3)		[75]
RHF[d]		26 (10)	15 (−1)	[75]

[a]The numbers in parentheses are for CH_3 migration.

[b]The basis functions used are $[5s3p3d]/(13s8p6d)$ for Mn, $[3s2p]/(9s5p)$ for C and O, and $[3s]/(6s)$ for H.

[c]The basis set used is double-zeta for $3s$, $3p$, $4s$, and $4p$ and triple-zeta for $3d$ for Mn and the 4-31G for C, O, and H.

[d]The d polarization functions on C and O are added to the last basis set.

The reaction mechanism, the substituent effects, and the energetics of the alkyl group migration to a carbonyl ligand of an Mn complex [eq. (6.27)] have been extensively studied experimentally [64]. These experiments have motivated theoreticians to study the methyl and the hydride migration in Mn complexes [eqs. (6.27) and (6.28)], as summarized in Table 6.1.

$$\underset{\mathbf{41}}{CH_3Mn(CO)_5} \rightarrow \underset{\mathbf{42}}{CH_3COMn(CO)_4} \tag{6.27}$$

$$\underset{\mathbf{43}}{HMn(CO)_5} \rightarrow \underset{\mathbf{44}}{HCOMn(CO)_4} \tag{6.28}$$

Berke and Hoffmann [65] focused their attention on the methyl migration [eq. (6.27)] and searched for a reaction path with the EH method (Scheme (6.2). The calculated activation barrier is 20 kcal mol^{-1}. They discussed this low activation barrier in terms of the orbital interaction between CH_3CO^- and $Mn(CO)_4{}^+$ as shown in **45**. In the uncatalyzed $CH_3{}^- \cdots CO$ system, the σ_{CO}–$\sigma_{CH_3{}^-}$ exchange repulsion raises the HOMO (σ_{CO}–$\sigma_{CH_3{}^-}$) high in energy, destabilizing the system. The interaction of this orbital with

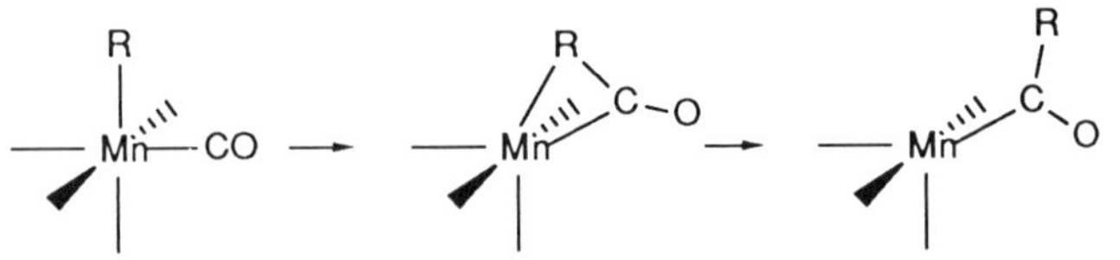

R
Mn—CO → Mn (R, C–O) → Mn (C(R), O)

Scheme 6.2

the vacant d orbital of $Mn(CO)_4^+$ results in the stabilization around the TS region.

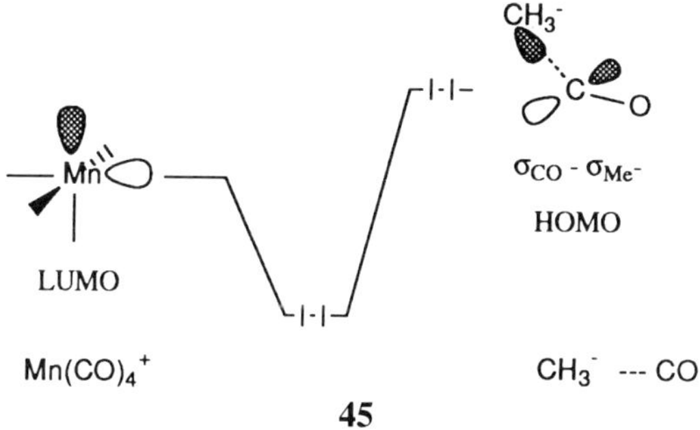

45

In order to compare the effect of a migrating group, Berke and Hoffman calculated the hydride migration and found an activation energy of 16 kcal mol^{-1}, 4 kcal mol^{-1} smaller than that of the methyl migration. This lower barrier was ascribed to the stabilization of the orbital that develops into the new C—H bond. According to their calculation, the η^2-acetyl complex does not exist. Several other aspects of migratory insertion were discussed as well in this EH study. Although the conclusions obtained can explain qualitatively what is happening and are consistent with those of the following, more accurate ab-initio studies, this is not the case for the energetics. As shown in Table 6.1, the hydride migration (6.28) is very endothermic and requires a large activation energy at the ab-initio level, as discussed later.

Nakamura and Dedieu [66] carried out ab-initio SCF and CI calculations with the split-valence basis functions for reaction (6.28). They determined the reaction path at the SCF level by partial geometry optimization without using the energy-gradient technique, to find that the mechanism is a hydride migration. They did not consider the possibility of an η^2-formyl complex. At the SCF level the barrier is 14 kcal mol^{-1} and the endothermicity is 11 kcal mol^{-1}. At the SDCI level the barrier increases to 39 kcal mol^{-1} and the endothermicity to 38 kcal mol^{-1}. This more-reliable CI calculation suggests that hydride migration is quite difficult and, even if the hydride migrates, the reverse reaction would take place with a very small barrier. The reason the hydride migration rather than the CO insertion takes place is explained by looking at the HOMO shown in **46a** [66b].

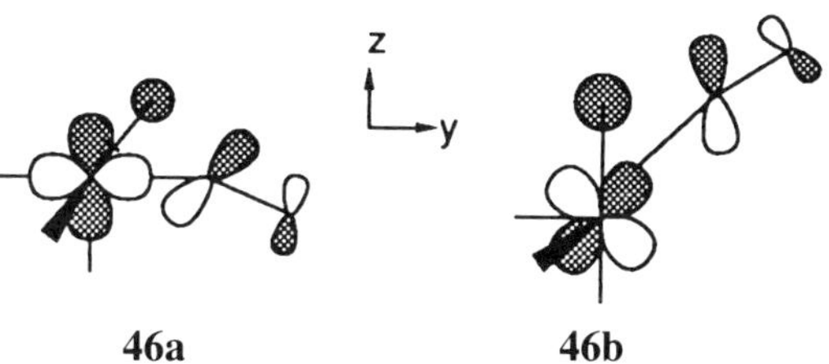

46a **46b**

Table 6.2
Energy of Reaction (kcal mol^{-1}) of Carbonyl Insertion at Several Levels of Calculation

Reactant	SCF	CASSCF4	CASSCF10	MP2	SDCI	Ref.
HMn(CO)	2.3[a]	15.1	35.5	35.5	23.3	[68]
	11.5[b]				27.4	[70]
	13.9[c]					[70]
	7.4–15.2[d]					[69]
	19.5–25.4[e]					[69]
$CH_3Mn(CO)$	−1.8[a]			36.4		[68]
$HPd(CO)^+$	−10.2[a]	−21.4	−33.6	−15.8		[68]

[a]The basis functions used are $[5s3p3d]/(13s8p6d)$ for Mn, $[6s4p5d]/(15s9p8d)$ for Pd, $[3s2p]/(9s5p)$ for C and O, and $[3s]/(6s)$ for H (basis set I).

[b]One set of p and d polarization functions on H, C, and O are added to the basis set I.

[c]Two sets of p and d polarization functions on H, C, and O are added to the basis set I.

[d]The basis sets used do not include the polarization function.

[e]The basis sets used include the polarization function.

In this HOMO, the bonding combination of the hydride s orbital and the carbonyl π^*, somewhat destabilized by $5\sigma_{CO}$, interacts mainly with the empty $d_{y^2-z^2}$ metal orbital into which metal s and p orbitals mix (spd hybrid). When CO instead of the hydride migrates, this bonding combination interacts with the occupied d_{yz} orbital as shown in **46b**, resulting in destabilization. The same result also has been obtained in methyl migration [eq. (6.31)] by Koga and Morokuma [67].

The large energetics difference between SCF and CI potential-energy surfaces was ascribed to the unbalanced description of π back-donation between the reactant Mn—CO bond and the product Mn—CHO bond at the SCF level. Dedieu, Sakaki, Strich, and Siegbahn [68] carried out an ab-initio CASSCF and MP2 calculation for reaction (6.29) with an assumed structure in order to investigate this difference in more detail (Table 6.2). The electron correlation effect in reaction (6.28) was reproduced qualitatively in reaction (6.29).

$$\mathrm{RMn(CO)} \rightarrow \mathrm{Mn(RCO)} \qquad (\mathrm{R} = \mathrm{H}, \mathrm{CH_3}) \qquad (6.29)$$

Analysis of the CASSCF wave function demonstrated that the correct description of π back-donation in the Mn complexes requires a multi-determinantal wave function such as in CI, CASSCF, and Møller–Plesset perturbation theory. The Hartree–Fock method does not describe the π back-donation in the Mn complexes properly. Although both in-plane and out-of-plane π back-donations take place in the reactant, only out-of-plane π back-donation can take place in the product, because the in-plane CO π^* orbital changes to the C—H σ bond orbital during the reaction. In addition, the formyl ligand in the product is negatively charged and thus back-donation is weak. This crucial difference in π back-donation between the reactant and the product cannot be described by the HF method. The HF level of calculations artificially favors the product.

Dedieu et al. [68] also carried out an ab-initio calculation for a Pd reaction [eq. (6.30)], as shown in Table 6.2, using an assumed structure.

$$HPd(CO)^+ \rightarrow Pd(CHO)^+ \quad (6.30)$$

The reason they studied this reaction is that Koga and Morokuma [67] found a correlation effect on the CH_3 migratory insertion reaction (6.31), as shown in Table 6.3, which is different from the preceding Mn reaction (6.29).

$$M(CH_3)(H)(CO)(PH_3) \rightarrow M(COCH_3)(H)(PH_3) \qquad (M = Pd, Pt) \quad (6.31)$$

The electron correlation effect on reaction (6.31) makes the reaction more exothermic, as in the case for Table 6.2. Dedieu et al. found that the electron correlation effect on Pd—H and Pd—CHO σ bonds is essential and that the π back-donation is not as important in the Pd complex as in the Mn complex, because the *d* orbital of Pd(II) is more stable than that of Mn(I) due to the higher oxidation number. They concluded that the correlation effect may vary according to the nature and oxidation state of the transition metal atom.

In the preceding ab-initio calculations of reactions (6.28)–(6.30), the basis functions do not include polarization functions that are *d* functions on C and O and *p* functions on H. Axe and Marynick [69] carried out the RHF calculations for the hydride migration [eq. (6.29) with R = H] with several larger polarized basis sets. The exothermicity calculated with unpolarized basis sets is between 9 and 15 kcal mol^{-1}, whereas that calculated with polarized basis sets is between 20 and 25 kcal mol^{-1}. The latter value is closer to the SDCI value calculated with an unpolarized basis set [68]. Polarization functions improve the π back-donation as well. Therefore, the electron correlation effect found by Dedieu et al. [68] might be overestimated.

Dedieu [70] later calculated the energy of hydride migration [eq. (6.29)] by the SDCI method with a polarized basis set as shown in Table 6.2. The electron correlation effect calculated with an unpolarized basis set has been found actually to be overestimated. However, the electron correlation effect still amounts to 16 kcal mol^{-1} with the polarized basis set. Both polarization functions and the correlation effect are indispensable for a quantitative

Table 6.3
Energy of Reaction (ΔE) and Activation Energy ($\Delta E^{\ddagger}$) (kcal mol^{-1}) of Carbonyl Insertion[a] for
(i) $Pd(CH_3)(H)(CO)(PH_3) \rightarrow Pd(COCH_3)(H)(PH_3)$,[b]
(ii) $Pd(CH_3)(H)(CO)(PH_3) \rightarrow Pd(CH_3)(CHO)(PH_3)$, and
(iii) $Pd(H)_2(CO)(PH_3) \rightarrow Pd(CHO)(H)(PH_3)$[c]

				$\Delta E(\eta^2)$	
	Method	$\Delta E^{\ddagger}$	$\Delta E(\eta^1)$	**a**	**b**
(i)	RHF/I	25.7	19.1		
	RHF/II	18.8	5.0		
	MP2/II	13.5	8.8		
	RHF/III	22.0	10.6		
	MP2/III	10.2	4.3		
(ii)	RHF/I	26.7	23.3	17.6	20.4
	RHF/II	16.3	8.6		
	MP2/II	10.4	8.1		
	RHF/III	18.5	15.2	13.2	25.1
	MP2/III	7.4	6.7	7.1	22.8
(iii)	RHF/I	26.9	25.4		
	RHF/III	19.2	17.7		
	MP2/III	8.2	9.4		

[a]The basis sets used consist of the following: (I) $[2s2p2d]/(3s3p4d)$ and effective core potential for Pd, 3-21G for CH_3, CO, and hydride, and STO-2G for PH_3; (II) $[2s2p2d]/(3s3p4d)$ and effective core potential for Pd, $[3s2p]/(9s5p)$ for C and O, $[2s]/(4s)$ for all the hydrogens, and $[6d4p]/(11s6p)$ for P; and (III) $[2s2p3d]/(3s3p4d)$ and effective core potential for Pd, $[3s2p1d]/(9s5p1d)$ for C and O, $[6d4s]/(11s6p)$ for P, $[3s1p]/(5s1p)$ for the hydride, and $[2s]/(4s)$ for the other hydrogens. The structures were optimized at the RHF/I level.
[b]Reference 67.
[c]Reference 76.

calculation of the Mn reaction through a correct description of π-back donation.

Ziegler, Versluis, and Tschinke [71] followed Berke and Hoffmann's reaction path for the hydride migration as well as the methyl migration of $RMn(CO)_5$ [eqs. (6.27) and (6.28)] by the HFS method. The structure of the formyl ligand as well as the acetyl ligand was optimized, although the structure of the $Mn(CO)_4$ fragment was taken from the experiments. Ziegler et al. obtained the endothermicity of 38 kcal mol^{-1} for the formation of the η^1 product (η^1-**44**) with the activation barrier of 40 kcal mol^{-1}, as shown in Table 6.1. It is interesting to see that this endothermicity coincides with that calculated at the SDCI level by Nakamura and Dedieu [66a], noting that the HFS method in part takes the electron correlation into account. The hydride migration is much more endothermic than the methyl migration, which has a calculated endothermicity of 18 kcal mol^{-1}. The difference in endothermicity between the two reactions was ascribed to the difference in both strength between the Mn—CH_3 and Mn—H bonds: The latter has been calculated to be 12 kcal mol^{-1} stronger than the former with the HFS method. This is in good agreement with the experimental difference of 14 kcal mol^{-1} [63b].

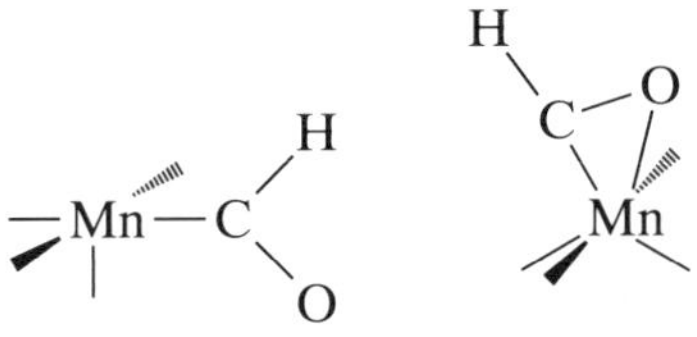

η^1-**44** η^2-**44**

Contrary to Berke and Hoffmann, Ziegler et al. found that η^2-formyl complex (η^2-**44**) is more stable than the η^1-**44** by 16 kcal mol^{-1}. The Mn—O distance (2.30 Å) was calculated to be substantially longer than the Mn—C bond (1.90 Å), suggesting relatively weak η^2-interaction. This is different from early transition metal complexes, which have been studied theoretically [72, 73] and experimentally [74]. In the early transition metal complex, the M–O interaction would be stronger because of its electron deficiency, as will be shown in the reaction of a Sc complex.

A similar η^2 complex has been found by Axe and Marynick [75], using the partial retention of diatomic differential overlap (PRDDO) and the ab-initio RHF methods. They studied the substituent effects on reaction (6.27), where they determined the structures of **43**, η^1- and η^2-**44**, and the TS connecting **43** and η^1-**44** for reaction (6.28) as well as those of the corresponding stationary points for the methyl migration [eq. (6.27)] by the PRDDO method, an approximate ab-initio method. The results are also shown in Table 6.1. They also carried out ab-initio RHF calculations for better energetics. The results show that the hydride migration is more endothermic, which is ascribed to the stronger Mn—H bond than the Mn—CH_3 bond. The former is calculated to be stronger by 20 and 19 kcal mol^{-1} with an unpolarized and a polarized basis set, respectively. They also found the η^2 product is more stable than the η^1 product.

One could conclude based on the preceding theoretical studies that the hydride migration of $HMn(CO)_5$ is very endothermic and that, in addition to the strong Mn—H bond, the change in π back-donation during the reaction, that is, from out-of-plane and in-plane π back-donations in the reactant to only in-plane π back-donation in the product, is essentially responsible to this large endothermicity. Berke and Hoffman [65] also suggested that, in the complex of heavier metals with more-diffuse and less-stable d orbitals, the π back-donation would make a large contribution to the M—CO bonding and thus the loss of in-plane π back-donation would affect the energetics.

Koga and Morokuma [67] adopted the ab-initio RHF energy-gradient method for reaction (6.31) to determine the structures of the reactant, the product, and the transition state, in which the CH_3 and H are cis to the CO. The difference in energetics between reaction (6.29) with R = H and reaction (6.30) is quite large; the Pd reaction is much more exothermic, suggesting that the hydride migration of $Pd(CH_3)(H)(CO)(PH_3)$ **47** might not be difficult. Therefore, for comparison, Koga and Morokuma [76] studied the reactions

$$\underset{\mathbf{47}}{Pd(CH_3)(H)(CO)(PH_3)} \rightarrow \underset{\mathbf{48}}{Pd(CH_3)(CHO)(PH_3)} \qquad (6.32)$$

$$\underset{\mathbf{49}}{Pd(H)_2(CO)(PH_3)} \rightarrow \underset{\mathbf{50}}{Pd(CHO)(H)(PH_3)} \qquad (6.33)$$

The fully optimized structures at the RHF level are shown in Figure 6.11. The energetics is shown in Table 6.3. The TS structure indicates that the hydride migration takes place. Although they did not take into account an η^2-acetyl complex in the previous calculation on methyl migration (6.31), they considered an η^2-formyl complex for reaction (6.32). At the highest level of calculation (RMP2/III), the η^2 product is less stable than the η^1 product.

The methyl migration (6.31) and the hydride migration (6.33) were compared, because the ligand trans to the migrating group is the same in both migrations. The hydride migration (6.33) is more endothermic at the MP2 level. This larger endothermicity of reaction (6.33) has been ascribed to the H—Pd bond, which is stronger by 5 kcal mol^{-1} than the CH_3—Pd bond, although this difference is much smaller than the difference in the strength between the CH_3—Mn and the H—Mn bond shown previously.

At the MP2 level, the TS for the hydride migration (6.33) disappears. Around the "transition-state" region, the hydride migration is more favorable than the methyl migration. The three-centered interaction at the TS can be expressed by the orbital interaction, which is similar to **46a**. The relatively more stable transition state for hydride migration may be attributed to the nondirectionality of the hydride 1*s* orbital; the spherical 1*s* orbital can strongly interact simultaneously with the π^* orbital of CO and the *spd* hybrid orbital of a transition metal.

The fact that the hydride migration is uphill may be the reason the formyl complex is not detected. However, the endothermicity of hydride migration is low enough for the formyl complex to be an important transient species in catalytic processes.

Axe and Marynick [75] compared, for reaction (6.27), the orbital interaction between the migrating-group orbital and metal *spd* hybrid orbital in the reactant with that between the migrating-group orbital and the CO π^* orbital in the product. The *spd* hybrid is more stable than CO π^*. If the π^* orbital is more stable, for instance, like that of CS, or the *spd* hybrid is less stable, for instance, as in Co, the interaction of the product is strengthened, resulting in the lower energy required. Theoretical studies of the hydride

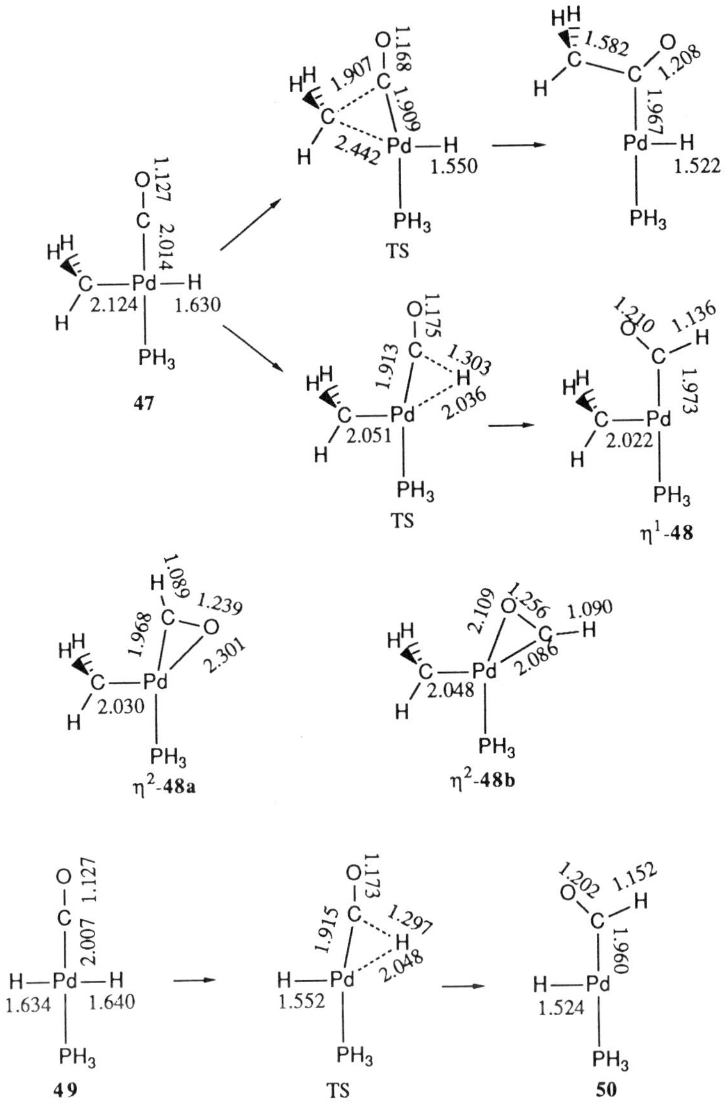

FIGURE 6.11
Structures (in angstroms and degrees), optimized at the RHF level, of reactants, products, and transition states for methyl and hydride migration to CO of $Pd(CH_3)(H)(CO)(PH_3)$ **47** and those for hydride migration to CO of $Pd(H)_2(CO)(PH_3)$ **49** [67, 76].

migration to CS and to CO in a Co complex have been carried out, as shown later.

Ziegler et al. [71] also reached the same consideration of the π^* orbital energy in their studies of reactions (6.27) and (6.28) and they have compared the hydride migration to CS [eq. (6.34)] with that to CO [eq. (6.28)] by the HFS method. Experimentally, it has been found that the migration to a CS ligand is preferred to the migration to a CO ligand [77]. Even the hydride has been observed to migrate to CS in an Os complex [78].

$$\underset{\mathbf{51}}{HMn(CO)_4(CS)} \rightarrow \underset{\mathbf{52}}{HCSMn(CO)_4} \quad (6.34)$$

Reaction (6.34), giving the η^1-thioformyl complex (η^1-**52**) was calculated to be 17 kcal mol^{-1} endothermic, compared with the endothermicity of 38 kcal mol^{-1} for reaction (6.28). Because the Mn—H bond is broken in both reactions (6.28) and (6.34), the difference is expected to originate from the difference in the H—CX bond strength (X = O or S). Ziegler et al.'s calculations showed that the H—$CSMn(CO)_4$ bond is 20 kcal mol^{-1} stronger than the H—$COMn(CO)_4$ bond, consistent with the stronger interaction of H 1s with CS π^* discussed previously. Calculations have also shown that the rearrangement from η^1-**52** to η^2-**52** is 31 kcal mol^{-1} exothermic and thus the formation of η^2-**52** from **51** is 14 kcal mol^{-1} exothermic overall, supporting the experimental suggestion of a stable η^2-thioformyl complex [78c].

Reaction (6.35) is considered to be a step of the hydroformylation catalytic cycle (R = alkyl) [79].

$$RCo(CO)_4 \rightarrow RCOCo(CO)_3 \quad (6.35)$$

Antolovic and Davidson [41] studied the insertion reaction of R = H with the ab-initio RHF method, and Ziegler et al. [80] carried out the HFS calculations for R = CH_3 as well as R = H and compared the reactivity between methyl and hydride.

Anotlovic and Davidson determined the structures of $HCo(CO)_4$ **53** and $HCOCo(CO)_3$ **54** by the RHF energy-gradient method [41], although there are some problems in such a procedure for the Co carbonyl complex as shown in the Section 6.2. They further followed the reaction path according to the LST procedure. The structures of **54** with relative energies are shown in Figure 6.12. The **54**s have a deformed TBP structure with one vacant coordination site. The energetics of the three reactions in which the C_s symmetry is enforced and thus the BPR process is neglected is also shown in Figure 6.12. The activation barrier is higher than the binding energy of CO. One of the reasons for this from the computational aspect is the lack of electron correlation and the other is that the structure is not optimized and the C_s LST path was followed. In ethylene insertion of a five-coordinate $RhH(C_2H_4)(CO)_2(PH_3)$ [eq. (6.11)], BPR and insertion couple strongly [21].

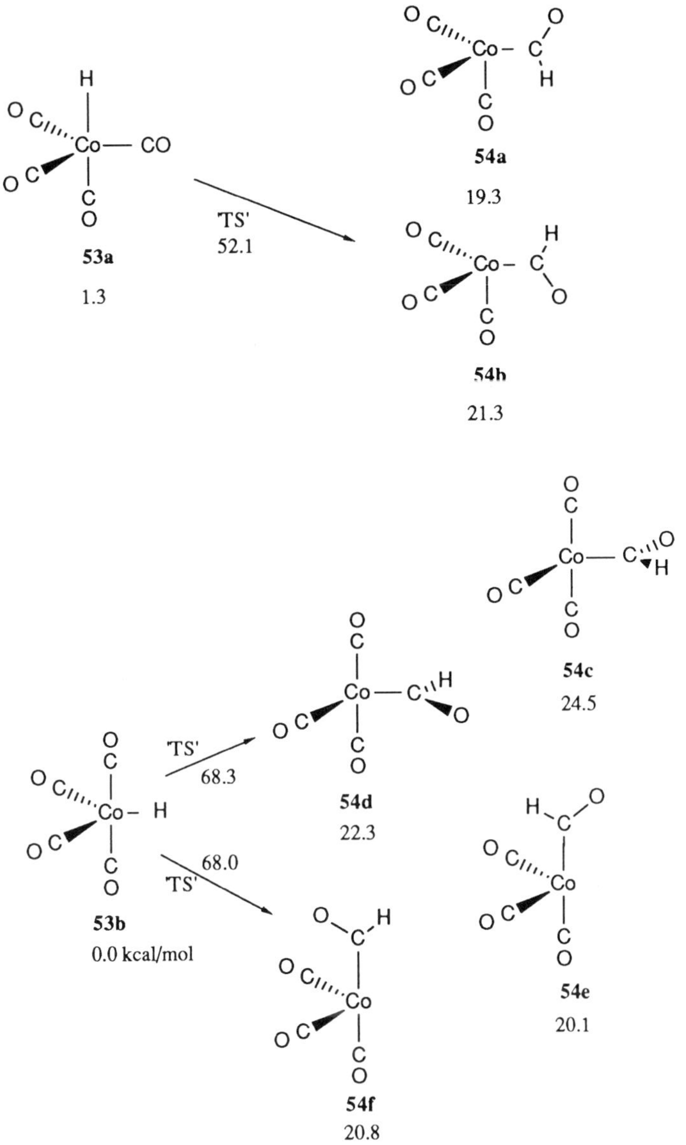

53a 1.3

'TS' 52.1

54a 19.3

54b 21.3

54c 24.5

'TS' 68.3

54d 22.3

53b 0.0 kcal/mol

68.0 'TS'

54e 20.1

54f 20.8

FIGURE 6.12
Potential-energy profile for CO insertion of $HCo(CO)_4$ **53**, calculated in kilocalories per mole at the RHF level, relative to **53b**. The transition states are on the LST path [41].

The BPR, which was neglected in Antolovic and Davidson's study, has to be taken into account; the reaction may pass through a non-least-motion path without any symmetry. Koga, Ding, and Morokuma [81] in fact found such a reaction path for ethyl migratory insertion to CO of $RhH(C_2H_5)(CO)_2(PH_3)$. The difficulty of H migration has been ascribed to the strong M—H bond. The path **53a** → **54b** is the most favorable because of the weakness of the axial Co—H bond. However, the RHF calculations may give an incorrect trend of relative bond strength, suspected from the bond distance argument shown in Section 6.2. Therefore, these results are not conclusive.

In the HFS calculation of Ziegler et al. [80], the structures of **53** and **54** were optimized and the reaction path was followed by the LST method as well. The electron correlation effect is taken effectively into account in their calculation through the HFS correction term. Therefore, their results are different from those of Antolovic and Davidson. The C_{3v} structure of $HCo(CO)_4$, **53a** (**23** in Section 6.2), is more stable than the C_{2v} structure, **53b** (**24** in Section 6.2), by 15 kcal mol^{-1} and the Co—C_{eq} bond is slightly longer than the Co—C_{ax}, in agreement with the experiment but in contrast with the RHF calculation.

The reaction steps they studied are shown in Figure 6.13. It was found that neither **54b** nor **54f** (kinetic product) exists, in contrast to the Antolovic

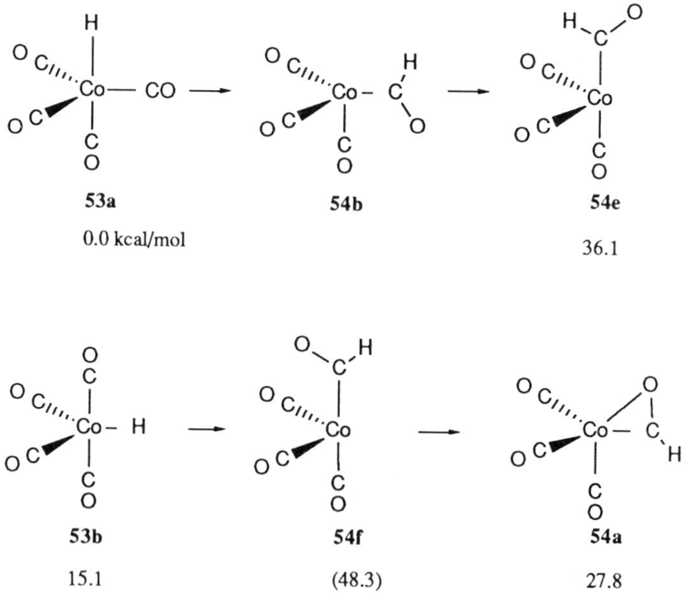

FIGURE 6.13
Potential-energy profile for formaldehyde insertion of $HCo(CO)_4$ **53**, calculated in kilocalories per mole at the HFS level relative to **53a**. The transition states are assumed to be on the LST path [80].

and Davidson RHF calculation. The HFS geometry optimization with the C_s symmetry constraint led to a hydride carbonyl complex, indicating that these formyl complexes are kinetically unstable. The equilibrium structures of the **54**s determined by Ziegler et al. have the vacant coordination site trans to the (migrating) hydrogen of the formyl group and, thus, are not kinetic products. The kinetic products, **54b** and **54f**, could transform into more-stable isomers through formyl migration **54b** → **54e** or **54f** → **54a** or Co—CHO rotation **54b** → **54a** or **54f** → **54e** (the authors did not discuss the latter process). **54a** is more stable than **54e**, because in **54a** the η^2 interaction takes place. Again, the HFS structure of **54a** is different from the RHF structure, in which such an interaction has not been found. The **53a** → **54e** and **53b** → **54a** are 36 and 13 kcal mol^{-1} endothermic, respectively, which are more endothermic than the corresponding methyl migration. The higher endothermicity was ascribed to the Co—H bond, which is stronger than the Co—CH_3 bond. The HFS results seem to be more reasonable than the ab-initio RHF results.

Rappé [72] studied the hydride migratory insertion to CO in an early transition metal complex [eq. (6.36)] and found that the η^1-formyl complex is not an equilibrium structure. This model reaction was chosen because of its simplicity.

$$\underset{\textbf{55}}{Cl_2ScH} + CO \rightarrow \underset{\textbf{56}}{Cl_2Sch(CO)} \rightarrow \underset{\textbf{57}}{Cl_2ScCHO} \qquad (6.36)$$

Preceding the CO insertion, the CO coordination takes place, to give **56**. The structures of **55**, **56**, **57**, and the TS between **56** and **57** were determined by the RHF energy-gradient method. The energy calculations were carried out at the GVB and CI level of calculations with the valence double-zeta plus polarization basis functions. The energetics is shown in Figure 6.14. For **57**, only the η^2 structure with a shorter Sc—O bond than the Sc—C bond exists. This is clearly in contrast to the late transition metal formyl complex. Scandium is electron-deficient and thus the bonding interaction between the lone pair of formyl oxygen and the empty d orbital of Sc was found to favor

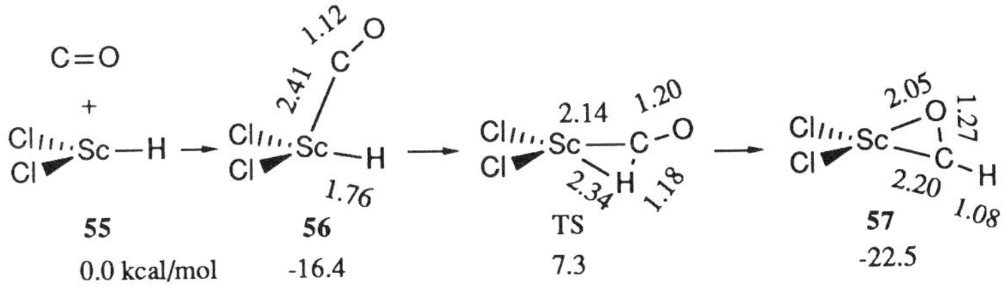

FIGURE 6.14
Potential-energy profile for CO insertion, Cl_2ScH (**55**) + CO, calculated in kilocalories per mole at the GVB-CI level, relative to **55**. The selected bond distances are shown in angstroms [72].

the η^2 coordination shown in **58**. This is in agreement with the previous EH studies of an η^2-acyl complex [72, 83].

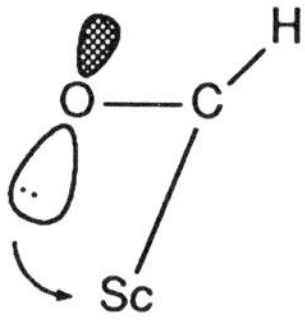

H
O—C
Sc
58

The reaction from the CO complex was calculated to require an activation barrier of 24 kcal mol^{-1} with an exothermicity of 6 kcal mol^{-1}. The TS is structurally similar to what would be expected for an η^1-formyl complex. An analysis of the GVB wave function showed that the CO lone pair and the Sc—H bond are converted smoothly into the Sc—C and C—H bonds. The C—H bond length of 1.18 Å at the TS shows that the C—H bond is nearly formed. The activation barrier is relatively high, because of the instability of an η^1-formyl complex.

Also, semiempirical methods have been adopted for study of the hydride migration to a carbonyl ligand [83]. In these studies, the difficulty of hydride migration was discussed in terms of the charge on the migrating group instead of the M—R bond strength. Saddei, Freund, and Hohlneicher [83b] studied reactions (6.27) and (6.28) by the CNDO method, to find the difference in nucleophilicity between H and CH_3 in the charge distribution. Blyholder, Zhao, and Lawless studied the reaction of $HFe(CO)_4^-$ by the modified intermediate neglect of differential overlap (MINDO) method to ascribe the endothermicity to the weak formyl C—H bond [83a]. They found that the HOMO of the product is strongly C—H antibonding and that the negative charge accumulates on the formyl hydrogen. Our [84] ab-initio RHF calculation on this complex has not shown such a negative charge and the antibonding character. Probably, it would be necessary to confirm the results with an ab-initio calculation.

The preceding theoretical studies focussed on the hydride migration from transition metals to carbonyls in the transition metal complexes. Also of interest are reactions on the surface in catalytic processes. Several studies [85–87] have appeared in which the surface is modelled by a single transition metal.

McKee, Dai, and Worley [86] studied the reaction (6.37). Worley and co-workers [88] experimentally focussed on the intermediates in the reactions of hydrogens with CO over supported rhodium films. A key intermediate on the Rh surface is the carbonyl hydride [88c].

$$\underset{\mathbf{59}}{HRh(CO)} \rightarrow \underset{\mathbf{60}}{Rh(CHO)} \tag{6.37}$$

Table 6.4
Energy of Reaction (ΔE) and Activation Energy ($\Delta E^{\ddagger}$) (kcal mol^{-1}) of Hydride Migration to Carbonyl, $HMCO^{n+} \rightarrow MCHO^{n+}$

M	n	Method	$\Delta E^{\ddagger}$	ΔE	Ref.
Rh	0	RHF	22.2	13.5	[86]
		MP3[a]	20.1	26.3	
Rh	0	HF	20.6	5.9	[87]
		SDCI	16.3	3.6	
Rh	1	HF	20.9	12.2	[87]
		SDCI	18.1	5.6	
Pd	0	HF	16.1	−2.0	[87]
		SDCI	22.9	6.3	
Pd	1	HF	15.6	3.5	[87]
		SDCI	11.6	−1.8	

[a]Zero-point vibration correction is included.

They determined linear and nonlinear equilibrium structures of **59**, the η^1 structure of **60**, and the TS between them, using the energy-gradient technique at the RHF level under the ECP approximation. Energy calculations were carried out at the MP3 level with a larger basis set as shown in Table 6.4. The linear **59** is less stable than the bent form.

$$\begin{array}{c} H \\ | \\ Rh-C-O \end{array} \qquad H-Rh-C-O$$

bent **59** linear **59**

The hydride migratory insertion from this bent **59** was calculated to be 29 kcal mol^{-1} endothermic and the conversion from the formyl complex to carbonyl hydride requires no activation energy, similar to the theoretical results for the hydride migration of $Pd(H)_2(CO)(PH_3)$ **49**, mentioned previously, and in agreement with the experimental fact that the formyl species has not been observed for supported Rh catalysts [89]. Nevertheless, they noted that the possibility of transient formyl species produced with an activation energy of 29 kcal mol^{-1} is not ruled out.

Pacchioni, Fantucci, Koutecký, and Ponec [87] studied the reaction

$$HM(CO)^n \rightarrow M(CHO)^n \qquad (M = Rh, Pd, n = 0, +1) \qquad (6.38)$$

They carried out SCF and CI energy calculations with partial geometry optimization by using the ECP approximation. The ECP used [90] is different from those used in the studies mentioned previously. The results are summarized also in Table 6.4. The energetics for HRh(CO) is quite different from that of McKee at al. In addition, Pacchioni et al. showed that the linear

H—M—CO is more stable than the bent structure. This would come from the difference in the ECP used. The M—CO binding energy in monocarbonyl complexes was calculated to be 4 and 7 kcal mol^{-1} for Rh and Pd, respectively, at the MRDCI level with their ECP [91]. These binding energies are very small, compared with the other calculations; the Rh—CO binding energy is 20–40 kcal mol^{-1} (MP3, ECP) [86], the Pd—CO binding energy is 37 kcal mol^{-1} (MP2, ECP) [92], and 34 and 22 kcal mol^{-1} (all-electron modified coupled pair functional calculation with and without the first-order relativistic correction, respectively) [93] for the Pd—CO bond. Also, the results of M = Pd and $n = +1$ should be compared with the energies shown in Table 6.2. Judging from the fact that their Pd—CO binding energy is quite different from the results of all-electron calculations, there may be some problems in the ECP used. Thus, their results should be considered qualitative. Nevertheless, it is interesting that the hydride migration in a Pd cationic system is easier than in the neutral system [85, 87], because this is in agreement with the experimental fact that the ion concentration is correlated with the catalytic activity [94].

6.6. HYDRIDE MIGRATION TO COORDINATED CARBENE

The hydride or alkyl migration to carbene ligand has been considered to be a key step of catalytic reactions [95]. Theoretical calculations on reactions (6.39) and (6.40) have been carried out [71, 96].

$$\mathrm{RMn(CO)_4(CH_2)} \rightarrow \mathrm{RCH_2Mn(CO)_4} \qquad (\mathrm{R} = \mathrm{H}, \mathrm{CH_3}) \tag{6.39}$$

$$\mathrm{ClRu(H)(CH_2)} \rightarrow \mathrm{ClRu(CH_3)} \tag{6.40}$$

Ziegler, Versluis, and Tschinke [71] compared the CH_2 insertion into an Mn—H bond with that into an Mn—CH_3 bond and with the CO insertion shown in Section 6.5, using the HFS method. They calculated exothermicities of 27 and 17 kcal mol^{-1} for H and CH_3 migrations, respectively; these migrations are much more exothermic than the migrations to CO and CS. This was ascribed to the stronger C—R bond in $RCH_2Mn(CO)_4$ than in $RCXMn(CO)_4$ (X = O or S). For instance, at the HFS level, a C—H bond energy in $CH_3Mn(CO)_4$ was calculated to be 82 kcal mol^{-1}, whereas that in $HCOMn(CO)_4$ and $HCSMn(CO)_4$ is 19 and 38 kcal mol^{-1}, respectively. Although the p_π orbital of CH_2 plays the same role as the π^* of CO, it is localized on C more than π^* of CO and CS and is lower in energy, resulting in the stronger C—H bonding interaction in the product. In addition, the occupied π orbital of CO and CS makes the new bond weaker, because of the four-electron repulsion between the C—H bond and the π orbital.

Carter and Goddard [96] have studied reaction (6.40) at the HF, GVB, GVB-CI, and CASSCF levels of calculation. At the highest level, six-electron and six-orbital CASSCF, they found an activation barrier of 12 kcal mol^{-1} with an exothermicity of 7 kcal mol^{-1}. The TS is three-centered, similar to that for the hydride migration to CO. They concluded that the hydride migration to CH_2 is feasible for the late transition metal complex. The GVB wave function obtained at the structure near the transition state shows that the Ru—H bond smoothly converts into the C—H bond with the aid of the CH_2 p_π orbital.

6.7. ACIDITY OF HYDROGEN IN TRANSFER

As mentioned in the introduction, the hydrogen behaves as a hydride, a hydrogen atom, or a proton during the transfer from a transition metal atom. How the hydrogen behaves depends on nature of the transition metal, its oxidation state, ligands, and so forth. Here, two theoretical studies will be reviewed.

Using the Fenske–Hall method, Bursten and Gatter [97] analyzed the factor that may control the character of the hydrogen in $CpML_nH$ systems. They discussed the acidity in terms of the orbital energies of deprotonated $CpML_n^-$. The frontier orbital energy levels for $CpCr(CO)_3^-$ **61** and $CpCr(NO)_2^-$ **62** are shown in Scheme 6.3. Both the deprotonation energy and population analysis have shown that the hydrogen in $CpCr(NO)_2H$ **63** is more acidic than in $CpCr(CO)_3H$ **64**. However, the unstable HOMO for **62** suggests the instability of **62** with respect to the removal of two electrons, which gives $CpCr(NO)_2^+$. Therefore, the hydrogen of **63** should be hydridic. They proposed that the energy difference between HOMO and the next HOMO is an indicator of the stability of the deprotonated anion: If the difference is larger, the loss of electrons from the anion HOMO that gives the corresponding cation would take place to stabilize the system and thus the hydrogen is hydridic; if the difference is small, the anion is stable and thus the hydrogen is acidic.

Branchadell and Dedieu [98] studied the hydrogen transfer from one metal to the other metal in a dinuclear Rh complex (**65**) with the ab-initio CASSCF method, using an assumed structure. **65** has a formal electron count of d^6 and d^8 for Rh^1 and Rh^2, respectively. Three reaction pathways they discussed are shown in Figure 6.15. In reaction path A, the hydrogen transfers from Rh^1 to Rh^2 as a radical. The product, **66**, has a formal electron count of d^7–d^7 and is thus a biradical. The potential surface for this reaction path is uphill and the endothermicity was calculated to be 39 kcal mol^{-1}. This endothermicity was attributed to the repulsion between the doubly occupied d_{z^2} of Rh^2 and the migrating hydrogen $1s$ orbital. The hydrogen atom transfer is not favorable. In reaction path B, before migration the ligand rearrangement around Rh^2 takes place, giving a trigonal pyramid

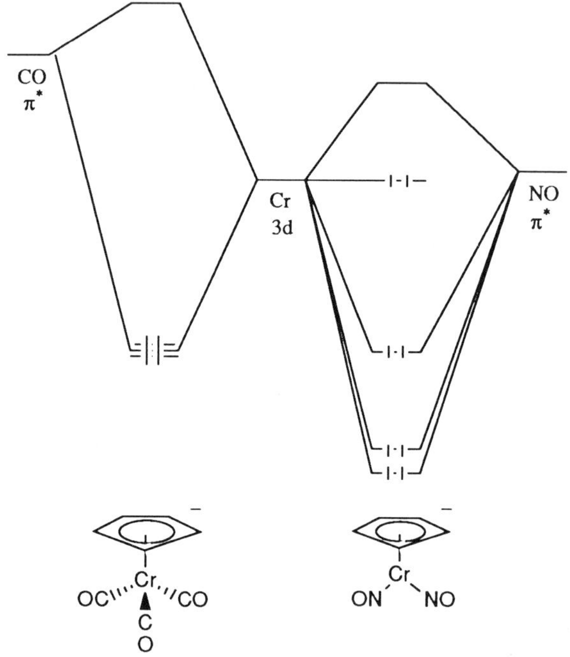

Scheme 6.3

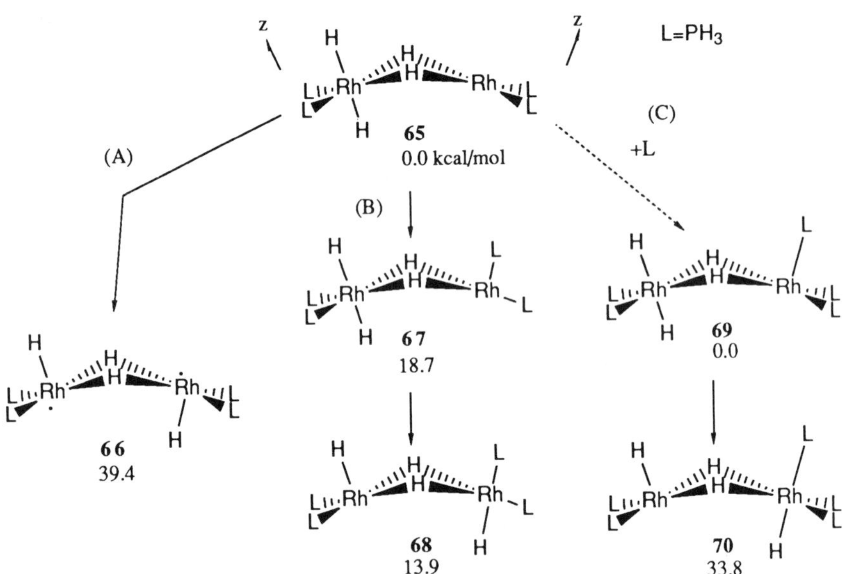

FIGURE 6.15
Potential energy profile for hydrogen migration in Rh dinuclear complex. The energies for the reactions A and B calculated at the CASSCF level are relative to **65**. Those for the reaction C are relative to **69** [98].

structure, **67**. As a result of this rearrangement, d_{z^2} of Rh^2 becomes vacant and thus accepts electrons, favoring the hydride transfer. The calculated results shown in Figure 6.15 demonstrated that this migration is feasible with the activation barrier and endothermicity of approximately 20 and 14 kcal mol^{-1}, respectively. In reaction path C, the additional ligand coordinates to Rh^2, leading to **69**, prior to the transfer. Although reaction path C was calculated to be endothermic by 34 kcal mol^{-1} with $L = PH_3$, Branchadell and Dedieu suggested that if the ligand of Rh^2 is a strong σ donor, the proton transfer could take place.

6.8. CONCLUSION

In this chapter we reviewed theoretical studies of hydrogen migration from transition metal complexes. They mainly cover the intramolecular hydride migration to coordinated unsaturated ligands such as C_2H_4, CO, CO_2, H_2CO, and CH_2. They could be divided into two classes: the migrations via a four-centered transition state, which is favored by the electron donation and back-donation as shown in **8** and **9**, and those via a three-centered transition state, which is caused by the interaction among $1s_H$, π^*, and d orbitals as shown in **46a**. The former holds true for hydride migrations to C_2H_4, CO_2, and H_2CO; the hydride migration to CO and CH_2 comes under the latter classification. At these multicenter transition states, the spherical $1s$ orbital can interact strongly with the transition metal d orbital and π^* orbital of coordinated unsaturated ligands simultaneously. Accordingly, the activation barrier was calculated to be low, when the reactionis nearly thermoneutral, for instance, for the hydride migration to C_2H_4 in four-coordinate d^8 Pd and Pt complexes [reaction (6.9)] and the hydride migration to CO in the four-coordinate Pd complex [reaction (6.33)]. Even in ethyl complexes, the products of reactions (6.9), the hydrogens remain to interact with the central transition metal, which results in the agostic interaction. In several cases of the hydride migration to CO, corresponding transition states disappeared; the potential energy surfaces from the hydride carbonyl complexes to the formyl complex are uphill and thus the endothermicity would control the migration. This endothermicity varies with the change of the central transition metal and its oxidation number. Although some formyl complexes are kinetically unstable, the endothermicity of the formyl complex formation is not too large.

In the case of the migration between two transition metal atoms (Figure 6.15), the hydride migration could take place easily if the vacant d orbital of the transition metal atom, which accepts the hydrogen, is available. The vacant d orbital interacts with the $1s_H$ orbital like the π^* of C_2H_4 and CO, preferring the hydride transfer to the proton and the hydrogen atom transfer.

Although some important electronic aspects of reactions can be elucidated by using the semiempirical method, recent ab-initio calculations have given more reliable and quantitative results. Such studies have led to a deeper understanding of the hydrogen transfer reactions and thus are expected in the near future to give more information on how to control the reactions.

REFERENCES

1. For instance, see the following. (a) Collman, J. P.; Hegedus, L. S.; Norton, J. R.; Finke, R. G. In *Principles and Applications of Organotransition Metal Chemistry*; University Science Books: Mill Valley, CA, 1987; pp. 355. (b) Cotton, F. A.; Wilkinson, G. In *Advanced Inorganic Chemistry*; Wiley: New York, 1988; p. 1199. (c) Yamamoto, A. In *Organotransition Metal Chemistry*; Wiley: New York, 1986; p. 361.
2. (a) Venanzi, L. M. *Coord. Chem. Rev.* **1982**, *43*, 251. (b) Kao, S. C.; Spillett, C. T.; Ash, C.; Lusk, R.; Park, Y. K.; Darensbourg, M. Y. *Organometallics* **1985**, *4*, 83, and references cited therein. (c) Martin, B. D.; Warner, K. E.; Norton, J. R. *J. Am. Chem. Soc.* **1986**, *108*, 33.
3. (a) Byers, B. H.; Brown, T. L. *J. Am. Chem. Soc.* **1977**, *99*, 2527. (b) Wegman, R. W.; Brown, T. L. *J. Am. Chem. Soc.* **1980**, *102*, 2494. (c) Nappa, M. J.; Santi, R.; Halpern, J. *Organometallics* **1985**, *4*, 34. (d) Jacobsen, E. N.; Bergman, R. G. *J. Am. Chem. Soc.* **1985**, *107*, 2023. (e) Edidin, R. T.; Norton, J. R. *J. Am. Chem. Soc.* **1986**, *108*, 948.
4. (a) Sartain, W. J.; Selegue, J. P. *J. Am. Chem. Soc.* **1985**, *107*, 5818. (b) Longato, B.; Martin, B. D.; Norton, J. R.; Anderson, O. P. *Inorg. Chem.* **1985**, *24*, 1389.
5. Hoffmann, R. *J. Chem. Phys.* **1963**, *39*, 1397.
6. Pople, J. A.; Segal, G. A. *J. Chem. Phys.* **1966**, *44*, 3289.
7. Hehre, W. J.; Radom, L.; Schleyer, P. v. R.; Pople, J. A. *Ab Initio Molecular Orbital Theory*; Wiley: New York, 1986.
8. See for instance, Shavitt, I. In *Methods of Electronic Structure Theory*; Shaefer, H. F., Ed.; Plenum: New York, 1977; p. 189.
9. (a) Møller, C.; Plesset, M. S. *Phys. Rev.* **1934**, *46*, 618. (b) Pople, J. A.; Binkley, J. S.; Seeger, R. *Int. J. Quantum Chem.* **1976**, *10*, 1.
10. See, for instance, Wahl, A. C.; Das, G. In *Methods of Electronic Structure Theory*; Schaefer, H. F., Ed.; Plenum: New York, 1977, p. 51.
11. (a) Roos, B. O. *Adv. Chem. Phys.* **1987**, *69*, 399. (b) Bobrowicz, F. W.; Goddard, W. A. In *Methods of Electronic Structure Theory*; Schaefer, H. F.; Ed.; Plenum: New York, 1977; p. 79.
12. (a) Hay, P. J.; Wadt, W. R. *J. Chem. Phys.* **1985**, *82*, 270. Wadt, W. R.; Hay, P. J. *J. Chem. Phys.* **1985**, *82*, 284. Hay, P. J.; Wadt, W. R. *J. Chem. Phys.* **1985**, *82*, 299. (b) Sakai, Y.; Miyoshi, E.; Klobukowski, M.; Huzinaga, S. *J. Comp. Chem.* **1987**, *8*, 226, 256.
13. Slater, J. C. *Adv. Quantum Chem.* **1972**, *6*, 1.
14. (a) Satoko, C. *Chem. Phys. Lett.* **1981**, *83*, 111. (b) Versluis, L.; Ziegler, T. *J. Chem. Phys.* **1988**, *88*, 322, and references cited therein.
15. Roe, D. C. *J. Am. Chem. Soc.* **1983**, *105*, 7770.
16. Doherty, N. M.; Bercaw, J. E. *J. Am. Chem. Soc.* **1985**, *107*, 2670.
17. Sakaki, S.; Kato, H.; Kanai, H.; Tarama, K. *Bull. Chem. Soc. Jpn.* **1975**, *48*, 813.
18. Fukui, K.; Inagaki, S. *J. Am. Chem. Soc.* **1975**, *97*, 4445.
19. (a) Tatsumi, K.; Yamaguchi, K.; Fueno, T. *J. Mol. Catal.* **1977**, *2*, 437. (b) Fujimoto, H.; Yamasaki, T.; Mizutani, H.; Koga, N. *J. Am. Chem. Soc.* **1985**, *107*, 6157.

20. Thorn, D. L.; Hoffmann, R. *J. Am. Chem. Soc.* **1978**, *100*, 2079.
21. Koga, N.; Jin, S.-Q.; Morokuma, K. *J. Am. Chem. Soc.* **1988**, *110*, 3417.
22. Koga, N.; Kitaura, K.; Obara, S.; Morokuma, K. *J. Am. Chem. Soc.* **1985**, *107*, 7109.
23. Koga, N.; Morokuma, K. In *Quantum Chemistry: The Challenge of Transition Metals and Coordination Chemistry*; Veillard, A., Ed.; NATO ASI Series; Reidel: Dordrecht, 1986, Vol. 176, p. 351.
24. Hay, P. J. *J. Am. Chem. Soc.* **1981**, *103*, 1390.
25. Koga, N.; Morokuma, K. Unpublished results.
26. Brookhart, M.; Green, M. L. H. *J. Organomet. Chem.* **1983**, *250*, 395.
27. (a) Koga, N.; Obara, S.; Morokuma, K. *J. Am. Chem. Soc.* **1984**, *106*, 4625. (b) Obara, S.; Koga, N.; Morokuma, K. *J. Organomet. Chem.* **1984**, *270*, C33.
28. Koga, N.; Morokuma, K. *J. Am. Chem. Soc.* **1988**, *110*, 108.
29. Cotton, F. A.; LaCour, T.; Stanislowski, G. *J. Am. Chem. Soc.* **1974**, *96*, 754.
30. Bäckvall, J.-E.; Björkman, E. E.; Pettersson, L.; Siegbahn, P. *J. Am. Chem. Soc.* **1984**, *106*, 4369.
31. (a) Koga, N.; Daniel, C.; Han, J.; Fu, X. Y.; Morokuma, K. *J. Am. Chem. Soc.* **1987**, *109*, 3455. (b) Daniel, C.; Koga, N.; Han, J.; Fu, X. Y.; Morokuma, K. *J. Am. Chem. Soc.* **1988**, *110*, 3773.
32. Dedieu, A. *Inorg. Chem.* **1981**, *20*, 2803.
33. (a) Osborn, J. A.; Jardine, F. H.; Young, J. F.; Wilkinson, G. *J. Chem. Soc. A* **1966**, 1711. (b) Jardine, F. H.; Osborn, J. A.; Wilkinson, G. *J. Chem. Soc. A* **1967**, 1574. (c) Montelatici, S.; van der Ent, A.; Osborn, J. A.; Wilkinson, G. *J. Chem. Soc. A* **1968**, 1054.
34. (a) Halpern, J.; Wong, C. S. *J. Chem. Soc., Chem. Commun.* **1973**, 629. (b) Halpern, J. In *Organotransition Metal Chemistry*; Ishii, Y., Tsutsui, M., Eds.; Plenum: New York, 1975; p. 109. (c) Halpern, J.; Okamoto, T.; Zakhariev, A. *J. Mol. Catal.* **1976**, *2*, 65.
35. Koga, N.; Morokuma, K. In *The Challenge of d and f Electrons: Theory and Computation*; Salahub, D. R.; Zerner, M. C., Eds.; ACS Symposium Series 394; American Chemical Society: Washington, DC, 1989; p. 77.
36. Dedieu, A. *Top. Phys. Organomet. Chem.* **1985**, *1*, 1.
37. For instance, see reference 1.
38. (a) Ugi, I.; Marquarding, D.; Klusacek, H.; Gillespie, P. *Acc. Chem. Res.* **1971**, *4*, 288. (b) Holmes, R. R. *Acc. Chem. Res.* **1972**, *5*, 296. (c) Berry, R. S. *J. Chem. Phys.* **1960**, *32*, 933. (d) Muetterties, E. L. *J. Am. Chem. Soc.* **1969**, *91*, 1636, 4115. (e) Ugi, I.; Marquarding, D.; Klusacek, H.; Gokel, G.; Gillespie, P. *Angew. Chem. Int. Ed. Engl.* **1970**, *9*, 703.
39. Evans, D.; Osborn, J. A.; Wilkinson, G. *J. Chem. Soc. A* **1968**, 3133.
40. Grima, J. Ph.; Choplin, F.; Kaufmann, G. *J. Organomet. Chem.* **1977**, *129*, 221.
41. (a) Antolovic, D.; Davidson, E. R. *J. Am. Chem. Soc.* **1987**, *109*, 5828. (b) Antolovic, D.; Davidson, E. R. *J. Am. Chem. Soc.* **1987**, *109*, 977.
42. Versluis, L. Ph.D. Thesis, The University of Calgary, 1989.
43. Antolovic, D.; Davidson, E. R. *J. Chem. Phys.* **1988**, *88*, 4967.
44. Lüthi, H. P.; Siegbahn, P. E. M.; Almlöf, J. *J. Phys. Chem.* **1985**, *89*, 2156.
45. McNeill, E. A.; Scholer, F. R. *J. Am. Chem. Soc.* **1977**, *99*, 6243.
46. Armstrong, D. R.; Fortune, R.; Perkins, P. G. *J. Catal.* **1976**, *41*, 51.
47. Armstrong, D. R.; Fortune, R.; Perkins, P. G. *J. Catal.* **1976**, *42*, 435.
48. Armstrong, D. R.; Novaro, O.; Ruiz-Vizcaya, M. E.; Linarte, R. *J. Catal.* **1977**, *48*, 8.
49. Daudey, J. P.; Jeung, G.; Ruiz, M. E.; Novaro, O. *Mol. Phys.* **1982**, *46*, 67.
50. Novaro, O.; Blaisten-Barojas, E.; Clementi, E.; Giunchi, G.; Ruiz-Vizcaya, M. E. *J. Chem. Phys.* **1978**, *68*, 2337.
51. Dombek, B. D. *Adv. Catal.* **1983**, *32*, 325.

52. Nakamura, S.; Morokuma, K. *Abstracts*, 33d Symposium on Ortganometallic Chemistry, Japan; Tokyo, October 1986; paper A109.
53. Dombek, B. D. *J. Am. Chem. Soc.* **1980**, *102*, 6855.
54. Ziegler, T.; Tschinke, V.; Versluis, L.; Baerends, E. J.; Ravenek, W. *Polyhedron* **1988**, *7*, 1625.
55. (a) Darenbourg, D. J.; Kudaroski, R. A. *Adv. Organomet. Chem.* **1983**, *22*, 129. (b) Palmer, D. A.; van Eldik, R. *Chem. Rev.* **1983**, *83*, 651. (c) Walther, D. *Coord. Chem. Rev.* **1987**, *79*, 135. (d) Braunstein, P.; Matt, D.; Nobel, D. *Chem. Rev.* **1988**, *88*, 747.
56. Sakaki, S.; Ohkubo, K. *Inorg. Chem.* **1988**, *27*, 2020; **1989**, *28*, 2583.
57. Bo. C.; Dedieu, A. *Inorg. Chem.* **1989**, *28*, 304.
58. (a) Herrmann, W. A. *Angew. Chem. Int. Ed. Engl.* **1982**, *21*, 117. (b) Blackborow, J. R.; Daroda, R. J.; Wilkinson, G. *Coord. Chem. Rev.* **1982**, *43*, 17.
59. (a) Grimmett, D. L.; Labinger, J. A.; Bonfiglio, J. N.; Masuo, S. T.; Shearin, E.; Miller, J. S. *Organometallics* **1983**, *2*, 1325. (b) Masters, C. *Adv. Organomet. Chem.* **1979**, *17*, 61. (c) Fahey, D. R. *J. Am. Chem. Soc.* **1981**, *103*, 136.
60. (a) Wayland, B. B.; Woods, B. A. *J. Chem. Soc., Chem. Commun.* **1981**, 700. (b) Farnos, M.D.; Woods, B. A.; Wayland, B. B. *J. Am. Chem. Soc.* **1986**, *108*, 3659.
61. Paonessa, R. S.; Thomas, N. C.; Halpern, J. *J. Am. Chem. Soc.* **1985**, *107*, 4333.
62. Moloy, K. G.; Marks, T. J. *J. Am. Chem. Soc.* **1984**, *106*, 7051.
63. (a) See reference 1(a), p. 369. (b) Conner, J. A.; Zafarani-Moattar, M. T.; Bickerton, J.; El Saied, N. I.; Suradi, S.; Carson, R.; Al Takhin, G.; Skinner, H. A. *Organometallics* **1982**, *1*, 1166. (c) Lane, K. R.; Sallans, L.; Squires, R. R. *Organometallics* **1984**, *4*, 408.
64. (a) Coffield, T. H.; Kozikowski, J.; Closson, R. D. *J. Org. Chem.* **1957**, *22*, 598. (b) Calderazzo, F.; Cotton, F. A. *Inorg. Chem.* **1962**, *1*, 30. (c) Keblys, K. A.; Filbey, A. H. *J. Am. Chem. Soc.* **1960**, *82*, 4204. (d) Mawby, R. J.; Basolo, F.; Pearson, G. *J. Am. Chem. Soc.* **1964**, *86*, 3994. (e) Calderazzo, F.; Noack, K. *Coord. Chem. Rev.* **1966**, *1*, 118. (f) Noack, K.; Calderazzo, F. *J. Organomet. Chem.* **1967**, *10*, 101. (g) Cawse, J. N.; Fiato, R. A.; Pruett, R. L. *J. Organomet. Chem.* **1979**, *172*, 405. (h) McHugh, T. M.; Rest, A. J. *J. Chem. Soc., Dalton Trans.* **1980**, 2323. (i) Flood, T. C.; Jensen, J. E.; Statler, J. A. *J. Am. Chem. Soc.* **1981**, *103*, 4410.
65. Berke, H.; Hoffmann, R. *J. Am. Chem. Soc.* **1978**, *100*, 7224.
66. (a) Nakamura, S.; Dedieu, A. *Chem. Phys. Lett.* **1984**, *111*, 243. (b) Dedieu, A.; Nakamura, S. In *Quantum Chemistry: The Challenge of Transition Metals and Coordination Chemistry*; Veillard, A., Ed.; NATO ASI Series; Reidel: Dordrecht, 1986; Vol. 176, p. 277.
67. Koga, N.; Morokuma, K. *J. Am. Chem. Soc.* **1986**, *108*, 6136.
68. Dedieu, A.; Sakaki, S.; Strich, A.; Siegbahn, P. E. M. *Chem. Phys. Lett.* **1987**, *133*, 317.
69. Axe, F. U.; Marynick, D. S. *Chem. Phys. Lett.* **1987**, *141*, 455.
70. Dedieu, A. In *The Challenge of d and f Electrons Theory and Computation*; Salahub, D. R., Zerner, M. C., Eds.; ACS Symposium Series 394; American Chemical Society: Washington, DC, 1989; p. 58.
71. Ziegler, T.; Versluis, L.; Tschinke, V. *J. Am. Chem. Soc.* **1986**, *108*, 612.
72. Rappé, A. K. *J. Am. Chem. Soc.* **1987**, *109*, 5605.
73. Tatsumi, K.; Nakamura, A.; Hofmann, P.; Stauffert, P.; Hoffmann, R. *J. Am. Chem. Soc.* **1985**, *107*, 4440.
74. (a) Fachinetti, G.; Fochi, G.; Floriani, C. *J. Chem. Soc., Dalton Trans.* **1977**, 1946. (b) See reference 62 and references cited therein.
75. Axe, F. U.; Marynick, D. S. *J. Am. Chem. Soc.* **1988**, *110*, 3728.
76. Koga, N.; Morokuma, K. *New J. Chem.* in press.

77. Clark, G. B.; Collins, T. J.; Marsden, K.; Roper, W. R. *J. Organomet. Chem.* **1983**, *259*, 215.
78. (a) Collins, T. J.; Roper, W. R. *J. Chem. Soc., Chem. Common.* **1976**, 1044. (b) Roper, W. R.; Town, K. G *J. Chem. Soc., Chem. Commun.* **1977**, 781. (c) Collins, T. J.; Roper, W. R. *J. Organomet. Chem.* **1978**, *159*, 73.
79. Heck, R. F.; Breslow, D. S. *J. Am. Chem. Soc.* **1961**, *83*, 4023.
80. Versluis, L.; Ziegler, T.; Baerends, E. J.; Ravenek, W. *J. Am. Chem. Soc.* **1989**, *111*, 2018.
81. Koga, N.; Ding, Y. B.; Morokuma, K.
82. Curtis, M. D.; Shiu, K.-B.; Butler, W. M. *J. Am. Chem. Soc.* **1986**, *108*, 1550.
83. (a) Blyholder, G.; Zhao, K. M.; Lawless, M. *Organometallics*, **1985**, *4*, 1371, 2170. (b) Saddei, D.; Freund, H. J.; Hohlneicher, G. *J. Organomet. Chem.* **1980**, *186*, 63.
84. Koga, N.; Morokuma, K. Unpublished data.
85. Anikin, N. A.; Bagatur'yants, A. A.; Zhidomirov, G. M.; Kazanskii, V. B. *Zh. Fiz. Khim.* **1983**, *57*, 653.
86. McKee, M. L.; Dai, C. H.; Worley, S. D. *J. Phys. Chem.* **1988**, *92*, 1056.
87. Pacchioni, G.; Fantucci, P.; Koutecký, J.; Ponec, V. *J. Catal.* **1988**, *112*, 34.
88. (a) Worley, S. D.; Mattson, G. A.; Caudill, R. *J. Phys. Chem.* **1983**, *87*, 1671. (b) Dai, C. H.; Worley, S. D. *J. Phys. Chem.* **1986**, *90*, 4219. (c) Henderson, M. A.; Worley, S. D. *J. Phys. Chem.* **1985**, *89*, 392, 1417.
89. Yates, J. T.; Cavanagh, R. R. *J. Catal.* **1982**, *74*, 97.
90. Durand, Ph.; Barthelat, J. C. *Theoret. Chim. Acta* **1975**, *38*, 283.
91. Koutecký, J.; Pacchioni, G.; Fantucci, P. *Chem. Phys.* **1985**, *99*, 87.
92. Rhohlfing, C. M.; Hay, P. J. *J. Chem. Phys.* **1985**, *83*, 4641.
93. Blomberg, M. R. A.; Lebrilla, C. B.; Siegbahn, P. E. M. *Chem. Phys. Lett.* **1988**, *150*, 522.
94. (a) van der Lee, G.; Schuller, B.; Post, H.; Favre, T. L. F.; Ponec, V. *J. Catal.* **1986**, *98*, 522. (b) van der Lee, G.; Ponec, V. *J. Catal.* **1986**, *99*, 511.
95. Canestrari, M.; Green, M. L. H. *J. Chem. Soc., Dalton Trans.* **1982**, 1789.
96. Carter, E. A.; Goddard, W. A. *J. Am. Chem. Soc.* **1987**, *109*, 579.
97. (a) Bursten, B. E.; Gatter, M. G. *J. Am. Chem. Soc.* **1984**, *106*, 2554. (b) See also Bursten, B. E.; Gatter, M. G. *Organometallics* **1984**, *3*, 895, 941.
98. Branchadell, V.; Dedieu, A. *New J. Chem.* **1988**, *12*, 443.

CHAPTER 7

Theoretical Studies of The Photochemistry of Transition Metal Hydrides

C. DANIEL and A. VEILLARD

Laboratoire de Chimie Quantique, E.R. 139 du CNRS, Institut Le Bel, F-67000 Strasbourg, France

7.1. INTRODUCTION

Although studies of the photochemistry of transition metal hydrides were not very common before 1975, several qualitative aspects of their photoreactivity began to emerge in the late 1970s [1–4]. Monohydride complexes were shown to undergo a variety of photochemical reactions [4], with experimental evidence obtained for both the homolysis of the metal–hydrogen bond [5–7] and the photoinduced ligand dissociation [8–13], the latter being the dominant process. The ligand field excited states were supposed to be responsible for the ligand dissociation, the nature of the excited states giving rise to the metal–hydrogen bond homolysis being unresolved. The photoelimination of alkanes from monohydride complexes of tungsten had been postulated as the primary process in the generation of metallocenes [14–16]. Also, some evidence was reported that elimination of H_2 is likely upon photolysis of mononuclear dihydride and polyhydride complexes, leading to highly reactive intermediates [17]. The nature of the excited states responsible for this reaction was not clearly defined, but careful mechanistic studies pointed to a concerted elimination process [18, 19]. The few polynuclear polyhydride complexes that had been investigated had shown ligand loss or metal–metal bond homolysis rather than molecular hydrogen elimination [20, 21].

The field of transition metal hydride photochemistry has expanded tremendously during the last decade, mostly in connection with the fact that their irradiation generates unsaturated species with a role in many catalytic processes and chemical reactions [22–38]. For instance, the photoreductive elimination of molecular hydrogen from dihydride complexes has been used to generate intermediates that are able to activate the carbon–hydrogen bonds in saturated organic compounds [22–28], and this has provided an impetus to new investigations of the reaction mechanism. Recent flash-photolysis studies have led to the conclusion that there are at least two mechanisms operative for the photoelimination of H_2 from dihydride complexes: One being the concerted loss of H_2 from the initial complex, as previously proposed [18, 19], the other being the prior dissociation of another ligand to give an unsaturated intermediate that undergoes rapid elimination of molecular hydrogen [39, 40]. A few examples of ligand loss upon photolysis of transition metal polyhydrides also have been reported [41, 42]. Although most of the polynuclear hydrides undergo ligand loss or metal–metal bond homolysis upon irradiation [43–46], examples of H_2 photoelimination have been described for dinuclear platinum hydride complexes [47, 48]. This recent surge of activity concerning the monohydride complexes [49–62] as well as the dihydride and polyhydride systems [54, 63–68] is partly a consequence of the development of new identification techniques that are able to detect and to characterize very reactive intermediates [69]. Recent experiments in dihydrogen-containing matrices [70] have allowed the characterization of complexes with coordinated molecular hydrogen and concurrently have pointed to the nature of the main primary process upon irradiation of these complexes, namely, the photoelimination of molecular hydrogen [71]. In spite of this experimental activity, many unknowns still remain regarding the sequence of events occurring between the initial excitation of the reactant and the formation of the primary products in their ground states. Most of the time, the electronic spectra of the reactants are poorly resolved and detailed assignment is usually lacking. The general situation is well-described by the following statement from a recent paper [72]. "There is an extensive photochemistry of transition metal hydrido complexes, but it has generally been based only on product and quantum yields studies. There is little insight in this area into excited-state dynamics, detailed photochemical mechanisms, or the nature of the excited state or states responsible for the photochemistry."

Even if a quantitative theoretical study of the dynamics of these photochemical reactions is still beyond reach, modern quantum chemical methods are able to establish a useful unifying framework in our understanding of the photoreactivity of organometallic molecules. The goal of a qualitative theoretical study is to elucidate the mechanism of the primary process and to identify the photoactive excited states responsible for the photoreactivity of the molecule. Understanding the occurrence of different reactions (ligand loss vs. metal–hydrogen bond homolysis or ligand loss vs. elimination of H_2) represents a fundamental goal in the study of the photoreactivity of transition metal hydrides, the ultimate purpose being the calculation of the quantum yields for the different pathways. The current theoretical interpretations of the photochemical reactions were based on a simple analysis in terms of bonding and antibonding orbitals [3]. More recently, the use of state correlation diagrams, which point to the possible pathways interconnecting the reactants to the products, was able to throw some light on the mechanism of the primary photochemical process [73–76]. This approach appears still useful for rationalizing the photoreactivity of large systems. A particularly powerful means of understanding the photochemical processes relies on the potential-energy surfaces (PES) or the more-readily visualized potential-energy curves (PEC) that connect the reactants to the primary products [77]. Because this method requires important computational resources, it can be used only for selected systems among the simplest [78–83].

The first part of this chapter is devoted to a survey of the different theoretical methods. The next two sections summarize the results obtained for the excitation energies of $HCo(CO)_4$, $HMn(CO)_5$, and $H_2Fe(CO)_4$ and for the potential-energy curves associated with the photochemical reactions of these systems:

$$HCo(CO)_4 \xrightarrow{h\nu} H + Co(CO)_4 \tag{7.1}$$

$$HCo(CO)_4 \xrightarrow{h\nu} CO + HCo(CO)_3 \tag{7.2}$$

$$HMn(CO)_5 \xrightarrow{h\nu} H + Mn(CO)_5 \tag{7.3}$$

$$HMn(CO)_5 \xrightarrow{h\nu} CO + HMn(CO)_4 \tag{7.4}$$

$$H_2Fe(CO)_4 \xrightarrow{h\nu} H_2 + Fe(CO)_4 \tag{7.5}$$

The final section deals with the use of state correlation diagrams applied to a variety of photochemical reactions.

7.2. THEORETICAL METHODS

Until recently, current understanding of the photochemical reactions of organometallics [3] was based on molecular orbital diagrams coupled with an analysis in terms of the bonding and antibonding character. It usually is assumed that photodissociation is the consequence of exciting an electron from a bonding orbital to the corresponding antibonding orbital. For instance, homolysis of the metal–metal bond in $Mn_2(CO)_{10}$ takes place upon excitation to the $^1B_2$ state, which corresponds to the promotion of an electron from the σ to the σ^* orbital (σ and σ^* denote the molecular orbitals that are, respectively, bonding and antibonding with respect to the metal–metal bond). Labilization of the Re—CH_3 bond in $Re(CO)_5CH_3$ has been traced [3] to the excitation of an electron from a σ_b level (mainly localized on the methyl, and bonding with respect to the metal–carbon bond) to a $d_{z^2}(\sigma^*)$ orbital (mainly metal in character, and antibonding with respect to the metal–carbon bond). This analysis, although conceptually appealing and certainly useful, has a number of drawbacks and may turn partly erroneous in some cases:

1. It cannot explain easily the observation of concurrent photochemical reactions at a given wavelength. It has been pointed out [3] that, in $Re(CO)_5CH_3$, the $d_{z^2}(\sigma^*)$ orbital is also antibonding with respect to the Re—CO σ bond and that carbonyl dissociation might compete with the homolysis of the metal–methyl bond. The same explanation appears to be valid for the concurrent reactions of $HCo(CO)_4$ [79, 81] and $Mn_2(CO)_{10}$ [84], although the coefficients of the carbonyl ligands turn out to be rather small in the σ^* orbital.
2. A related difficulty concerns, for instance, the photosubstitution reaction of the complexes $Fe(CO)_3$(R—DAB), where R—DAB = 1,4 diaza-1,3-butadiene, with the loss of a carbonyl ligand observed upon an internal transition of the Fe—(R—DAB) metallacycle [[85]].
3. There are a number of reports in the literature of a given photochemical reaction occurring upon excitation at different wavelengths, with the quantum yield relatively insensitive to the wavelength [86]–[88]. This seems hard to reconcile with the idea that the photocleavage of a bond results from exciting an electron from the orbital that is bonding (with respect to this bond) to the orbital that is antibonding, because this should correspond to a transition into a well-defined, unique excited state.
4. Photolysis of the metal–metal bond in $Mn_2(CO)_{10}$ is observed upon excitation into the $^1B_2$ state (corresponding to the $\sigma \rightarrow \sigma^*$ excitation). The potential-energy curve (with respect to the metal–metal distance) of this $^1B_2$ state is not dissociative! In fact, the potential-energy curve for this state should be similar to the curve for the state $^1\Sigma_u^+$ of H_2, which is bonding and dissociates to ionic products, not to radicals [89]!
5. Photolysis of the metal–hydrogen bond in $HMn(CO)_5$ occurs upon irradiation at 51,800 cm^{-1} [90]. Calculations indicate that the singlet state corresponding to the $\sigma \rightarrow \sigma^*$ excitation lie at much higher energy, above 60,000 cm^{-1} [83b, 91].

A step forward in the understanding of the mechanism of photochemical reactions was the use of state correlation diagrams [73–76]. In this approach, one identifies first the ground states and low-lying excited states of the reactants and of the primary products, on the basis of either a qualitative energy level scheme or ab-initio calculations. One assumes that some symmetry element is retained during the course of the reaction. Then one sets up a state correlation diagram, based on the spin and symmetry properties, between the reactants and the primary products (rules for the construction of a state correlation diagram may be found in reference 77). A photochemical reaction is expected to occur either if the excited molecule goes directly on a single surface (without any appreciable barrier) to the products or if internal conversion may be achieved easily between the excited state and the ground state surfaces, usually as a consequence of an avoided crossing. State correlation diagrams have helped to identify the symmetry of the excited states involved in a number of photochemical reactions (for instance, the $^3E'$ state in the photodissociation of $Fe(CO)_5$ [73], the $^3A_1$ state in the photolysis of the Co—H bond in $HCo(CO)_4$ [73], and the $^3B_2$ state in the photodissociation of $Mn_2(CO)_{10}$ [74]. They are subject to the following limitations:

1. State correlation diagrams are not very useful for identifying the spin multiplicity of the excited states involved, because efficient intersystem crossing is relatively common for organometallics [85, 92, 93].
2. They do not say anything about the relative stability of the reactant and products, and they ignore the energy barriers that do not result from avoided crossings.
3. In the absence of a wave function, one has to make some assumption on the formal oxidation states of the atoms. For instance, we assumed in our earlier work [73] that the Co atom in $HCo(CO)_4$ is d^8, in accordance with the common belief [94, 95] that this is a hydride compound. This resulted in an avoided crossing between the two states $^1A_1$. Subsequent work [79, 81] has shown that the Co — H bond in this compound should be described as covalent (in agreement with more-recent descriptions of the bonding [95, 96] and that the potential energy curve of the $^1A_1$ ground state does not show any avoided crossing.

State correlation diagrams are just a mere approximation to the potential-energy surfaces that connect the reactants to the primary photochemical products. The calculation of these PESs represents a more realistic approach, and in principle this could be achieved through a variety of quantum chemical approaches. In reality, the situation is less favorable and, for instance, none of the semiempirical approaches seems very useful. Extended Hückel theory (EHT), which has been extremely powerful for mapping PESs of organometallics in the ground state [97], is not very appropriate for excited states because it ignores the spin. The INDO/S CI formalism yields excitation energies that are not always accurate [98]. Other semiempirical methods, such as MINDO and AM1, appear to be oriented toward ground-state properties [99] and have not been parametrized for transition metals so far. Density-functional methods encounter some problems in the description of bond making and bond breaking and in the treatment of excited states [100]. In the hierarchy of ab-initio methods, self-consistent field (SCF) theory does not describe properly the dissociation of covalent bonds and yields only qualitative results for the excited states of organometallics [101] whereas configuration interaction (CI) based on SCF reference wave functions shows a slow convergence of the results as a function of the length of the CI expansion, resulting sometimes in relatively inaccurate excitation energies [102]. So far, the best approach (which has been tested extensively [81–83, 91, 98, 103]) seems to perform contracted CI calculations [104] based on complete active space SCF (CASSCF) reference wave functions [105]. It has been found that this type of calculation, referred to as MRCISD (multireference CI with all single and double excitations within a given space) with a CASSCF reference wave function, represents a good approximation to a full CI calculation [106].

The potential-energy curves calculated are cross sections of the many-dimensional potential-energy surface, obtained with a number of restrictive assumptions:

- The bond lengths are kept constant (except for the bond that dissociates) and usually are fixed to the experimental values [81, 83].
- The highest possible symmetry is maintained along the reaction path. For instance, for the photodissociation of $HCo(CO)_4$, it has been assumed that C_{3v} symmetry is retained along the reaction paths corresponding to the dissociation of the Co—H and Co—$(CO)_{axial}$ bonds [81].
- Relativistic effects are not included. (For this reason, PESs have been computed so far only for systems with a metal of the first transition series, where these effects are comparatively less important [107, 108].)

Technical details regarding the choice of the basis set, the reference wave function, and the extent of configuration interaction may be found in the original papers [81–83].

None of the theoretical studies of the excited states presented in Section 7.3 has included a calculation of the oscillator strengths. This is certainly unfortunate, because in some cases their estimate could help in assigning the bands observed in the experimental absorption spectrum. A typical example corresponds to $HCo(CO)_4$, with the experimental absorption spectrum beginning at about 36,000 cm^{-1} and with only one resolved, intense band at 44,000 cm^{-1} [50]. The lowest singlet state is calculated to be a ^1E state corresponding to a $d_\delta \rightarrow \sigma^*$ excitation [81], which is neither a pure $d \rightarrow d$ band nor a pure (metal to hydrogen) charge-transfer (CT) band. It is presently unclear whether the band at 44,000 cm^{-1} corresponds to this ^1E state or to a (metal to carbonyl) charge-transfer state (in this case, the band corresponding to the $^1A_1 \rightarrow {}^1E$ excitation must be relatively weak and hidden below the intense band at 44,000 cm^{-1}) [81, 103].

7.3. EXCITED STATES OF THE REACTANTS

The rigorous way to calculate the excitation energies consists of carrying out for each electronic state a CASSCF calculation followed by a contracted CI calculation [109]. However, this procedure is too expensive for organometallic molecules, due to the size of the system and to the large number of low-lying states. A more economical approach that we have used consists of performing a CASSCF calculation for a particular state and using this CASSCF reference wave function for the contracted configuration interaction (CCI) calculations of all the states of interest. Because we focus on the lowest excited states that are of the $d \rightarrow d$, $d \rightarrow \sigma^*$, and $\sigma \rightarrow \sigma^*$ type, this CASSCF calculation was carried out for a selected high-spin state (usually hypothetical) in which the occupation number of the orbitals d and σ^* vacant in the ground state is 1 (σ and σ^* denote the molecular orbitals that are, respectively, bonding and antibonding with respect to the metal–hydrogen bond). The CASSCF active space was limited to the $3d$ orbitals, the $4d$ orbitals that correlate them, and σ and σ^* orbitals. For each electronic state, two CCI calculations were performed. The first has only one reference configuration corresponding to the required state; the second is a multireference calculation including all the configurations that appear with a coefficient larger than 0.08 in the monoreference CI wave function. All the $3d$ electrons and those of the metal–hydrogen bonds are correlated in the CCI calculations. Single and double excitations to all virtual orbitals, except the counterparts of the carbonyl $1s$ and of the metal $1s$, $2s$, and $2p$ orbitals are included in the CCI calculations.

7.3.1. $HCo(CO)_4$ Excited States

The calculated excitation energies to the lowest excited states of $HCo(CO)_4$ are reported in Table 7.1 [81]. The lowest part of the electronic spectrum is characterized by two triplet states corresponding to a $^3E\ d_\delta \rightarrow \sigma^*$ excitation calculated at 25,900 cm^{-1} and to a $^3A_1\ \sigma \rightarrow \sigma^*$ excitation calculated at 34,600 cm^{-1}. The lowest excited singlet state corresponding to a $^1E\ d_\delta \rightarrow \sigma^*$ excitation is calculated at 36,000 cm^{-1}. The high value calculated for the excitation energy to the state $^1A_1\ \sigma \rightarrow \sigma^*$ (> 60,000 cm^{-1}) is a consequence of the ionic character of this state.

Table 7.1
Calculated CCI Excitation Energies (in cm^{-1}) for $HCo(CO)_4$[a]

	One-Electron Excitation in the Main Configuration	
$^1A_1 \rightarrow {}^1E$	$d_\delta \rightarrow \sigma^*$	36,000
$^1A_1 \rightarrow {}^3A_1$	$\sigma \rightarrow \sigma^*$	34,600
$^1A_1 \rightarrow {}^3E$	$d_\delta \rightarrow \sigma^*$	25,900 (24,250)[b]

[a] For each electronic state, a multireference CCI calculation was performed on a CASSCF wave function optimized for the state $^3A_1$ as reference wave function [81].

[b] Value obtained from CCI calculations performed on the CASSCF wave functions optimized separately for the states $^1A_1$ and ^3E from reference 103.

The comparison with the experimental spectrum of $HCo(CO)_4$ [50], characterized by an absorption beginning around 36,000 cm^{-1} and only one resolved band centered at 44,000 cm^{-1}, leads to two possible interpretations of this spectrum. The first hypothesis considers that the band at 44,000 cm^{-1} corresponds to the $d \rightarrow \pi^*_{CO}$ excitations and covers the $d_\delta \rightarrow \sigma^*$ absorption located somewhere between 36,000 and 44,000 cm^{-1}. The second interpretation, based on the hybrid nature of the $d_\delta \rightarrow \sigma^*$ band, which is neither a pure ligand-field band nor a pure charge-transfer band, considers that this absorption either corresponds to the band at 44,000 cm^{-1} or is at higher energy [110]. In this last case, the calculated excitation energy is in error by 1 eV or more. However, similar calculations on the related compounds $Fe(CO)_5$ and $HMn(CO)_5$ [91, 98] have yielded excitation energies with an accuracy better than half an eV.

7.3.2. $HMn(CO)_5$ Excited States

The calculated excitation energies to the lowest excited states of $HMn(CO)_5$ are reported in Table 7.2 [82, 83b]. The electronic spectrum of $HMn(CO)_5$ is characterized by a high density of states between 25,000 and 45,000 cm^{-1}, these states corresponding to either $d \rightarrow \sigma^*$ or $d \rightarrow d$ excitations. This is a consequence of the $d^6\sigma^2$ ground-state electronic configuration of $HMn(CO)_5$ with two low-lying vacant orbitals, one $3d$ orbital and one σ^* orbital. The experimental absorption spectrum of $HMn(CO)_5$ [112] in the gas phase exhibits one broad band centered around 46,700 cm^{-1} with two shoulders at 34,500 and 51,300 cm^{-1}. The intense band at 46,700 cm^{-1} and the shoulder at 51,300 cm^{-1} correspond probably to several allowed CT transitions

Table 7.2
Calculated CCI Excitation Energies (in cm^{-1}) for $HMn(CO)_5$ [a]

	One-Electron Excitation in the Main Configuration	
$^1A_1 \rightarrow {}^3A_1$	$\sigma \rightarrow \sigma^*$	61,670
$^1A_1 \rightarrow c\,{}^1E$	$d_\pi \rightarrow \pi^*$	53,900
$^1A_1 \rightarrow b\,{}^1E$	$d_\pi \rightarrow \sigma^*$	43,150
$^1A_1 \rightarrow {}^1A_2$	$d_{xy} \rightarrow d_{x^2-y^2}$	38,500 (36,700)[b]
$^1A_1 \rightarrow a\,{}^1E$	$d_\pi \rightarrow d_{x^2-y^2}$	33,580
$^1A_1 \rightarrow {}^3A_2$	$d_{xy} \rightarrow d_{x^2-y^2}$	33,020
$^1A_1 \rightarrow b\,{}^3E$	$d_\pi \rightarrow \sigma^*$	32,940
$^1A_1 \rightarrow a\,{}^3E$	$d_\pi \rightarrow d_{x^2-y^2}$	23,940

[a] For each electronic state, a multireference CCI calculation was performed on a CASSCF wave function optimized for a state $^5A_2$ as reference wave function [82, 83b].

[b] Value obtained from CCI calculations performed on the CASSCF wave functions optimized separately for the states $^1A_1$ and $^1A_2$ from reference 91.

Table 7.3
Calculated CCI Excitation Energies (in cm^{-1}) for $H_2Fe(CO)_4$ [a]

	One-Electron Excitation in the Main Configuration	
$a\,^1A_1 \rightarrow b\,^1B_2$	$\sigma_u \rightarrow \sigma_g^*$	53,560
$a\,^1A_1 \rightarrow b\,^3B_2$	$\sigma_u \rightarrow \sigma_g^*$	45,560
$a\,^1A_1 \rightarrow b\,^1A_1$	$3d \rightarrow \sigma_g^*$	39,760
$a\,^1A_1 \rightarrow a\,^1B_2$	$3d \rightarrow \sigma_u^*$	39,000
$a\,^1A_1 \rightarrow a\,^3B_2$	$3d \rightarrow \sigma_u^*$	33,240
$a\,^1A_1 \rightarrow b\,^3A_1$	$3d \rightarrow \sigma_g^*$	26,290

[a] For each electronic state, a multireference CCI calculation was performed on a CASSCF wave function optimized for a state $^5A_1$ as reference wave function [82, 83a].

($^1A_1 \rightarrow {}^1A_1$ or $^1A_1 \rightarrow {}^1E$) resulting from $d_\pi \rightarrow \sigma^*$ and $d_\pi \rightarrow \pi^*$ excitations. The shoulder of low intensity at 34,500 cm^{-1} has been assigned [91] to the allowed $^1A_1 \rightarrow a\,^1E$ ligand-field (LF) transition, corresponding to a $d_\pi \rightarrow d_{x^2-y^2}$ excitation calculated at 33,580 cm^{-1}.

Because our interest centers mostly on the LF, $d \rightarrow \sigma^*$ and $\sigma \rightarrow \sigma^*$ excitations that are responsible for the photoreactivity of $HMn(CO)_5$ [80, 82, 83b] only one of the many $d \rightarrow \pi^*_{CO}$ excitations has been calculated.

The excitation energies to the pair of low-lying ^3E states that play a key role in the photochemistry of $HMn(CO)_5$ [82, 83b] and correspond, respectively, to $d \rightarrow d$ and $d \rightarrow \sigma^*$ excitations are calculated at 24,000 and 33,000 cm^{-1}.

7.3.3. $H_2Fe(CO)_4$ Lowest Excited States

The calculated values for the excitation energies to the lowest excited states corresponding to some of the symmetry-allowed transitions for $H_2Fe(CO)_4$ are reported in Table 7.3 [82, 83a]. (The notation σ_g and σ_g^* denotes the bonding combinations of the two hydrogen s orbitals that are, respectively, bonding and antibonding with respect to the M—H bonds, σ_u and σ_u^* being the antibonding combinations of the two hydrogen s orbitals that are, respectively, bonding and antibonding with respect to the M—H bonds). The electronic spectrum of this dihydride is characterized by two sets of low-lying states, denoted $b\,^{1,3}A_1$ and $a\,^{1,3}B_2$, between 26,000 and 40,000 cm^{-1} and resulting from $d \rightarrow \sigma_g^*$ and $d \rightarrow \sigma_u^*$ excitations. The next states of symmetry B_2 (denoted $b\,^{1,3}B_2$) are calculated at 45,560 cm^{-1} (triplet component) and 53,560 cm^{-1} (singlet component) and correspond to a $\sigma_u \rightarrow \sigma_g^*$ excitation. The excitation energies calculated for the states corresponding to the $\sigma_g \rightarrow \sigma_u^*$ and $\sigma_g \rightarrow \sigma_g^*$ excitations are high ($> 60,000$ cm^{-1}) because the promotion of an electron from the totally bonding σ_g orbital corresponding to the three-center interaction in $H_2Fe(CO)_4$ should be energetically expensive.

The experimental excitation energy (254 nm, or 39,400 cm^{-1}) necessary to induce the elimination of molecular hydrogen from $H_2Fe(CO)_4$ is high enough to populate the lowest $a\,^1B_2$ and $b\,^1A_1$ singlet states calculated at 39,000 and 39,760 cm^{-1}, respectively. Moreover, the calculated excitation energies to these states agree rather well with the experimental absorption spectrum of $H_2Fe(CO)_4$ [111], which shows one shoulder around 37,000 cm^{-1}.

7.4. POTENTIAL-ENERGY SURFACES AND REACTION MECHANISMS

A detailed description of the method used to obtain the potential-energy curves that connect the ground and excited states of the reactant to the ground and excited states of the photoproducts may be found in the original papers [80–83].

7.4.1. Potential-Energy Surfaces for $HCo(CO)_4$

Upon irradiation at 254 nm in a low-temperature matrix [49, 50] $HCo(CO)_4$ undergoes two different photochemical reactions: the homolysis of the metal–hydrogen bond

$$HCo(CO)_4 \xrightarrow{h\nu} H + Co(CO)_4 \tag{7.6}$$

and the heterolytic loss of carbonyl ligand

$$HCo(CO)_4 \xrightarrow{h\nu} CO + HCo(CO)_3 \tag{7.7}$$

Figure 7.1 shows a set of potential-energy surfaces for the ground and excited states of $HCo(CO)_4$ obtained from the potential-energy curves calculated under C_{3v} constraint for the loss of either the hydrogen or the axial carbonyl ligand [81]. These potential-energy surfaces turn out to be relatively simple (without any avoided crossing or any energy barrier), but show one important feature, namely, the intersection of the $^3A_1$ and the ^3E surfaces. The dissociative character of the $^3A_1$ $\sigma \rightarrow \sigma^*$ and of the ^3E $d_\delta \rightarrow \sigma^*$ states will govern the photoreactivity of $HCo(CO)_4$. The surfaces of Figure 7.1 form the basis for a qualitative understanding of the mechanism of the photochemistry of $HCo(CO)_4$.

Irradiation at the experimental wavelength of 254 nm, or 39,400 cm^{-1}, will bring the molecule into the ^{1}E state corresponding to the $d_\delta \rightarrow \sigma^*$ excitation (calculated at 36,000 cm^{-1}, see Table 7.1). After intersystem crossing to the $^3A_1$ $\sigma \rightarrow \sigma^*$ state, the molecule may evolve along three different channels corresponding to the following:

1. The dissociation of the Co—H bond along the $^3A_1$ curve, with formation of the products H and $Co(CO)_4$ in their ground states.

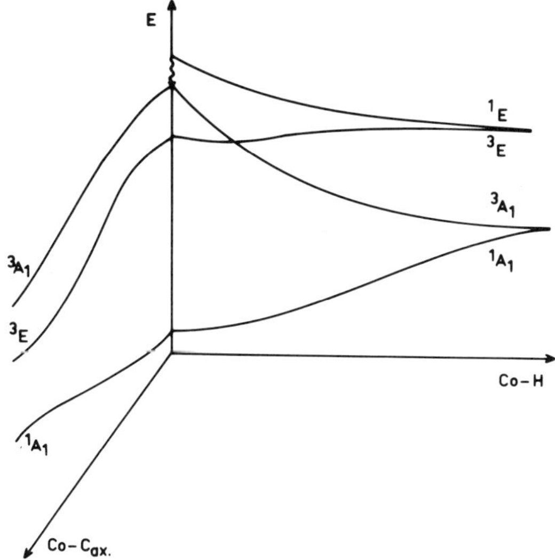

FIGURE 7.1
Potential-energy surfaces corresponding to the loss of hydrogen and axial carbonyl ligand in $HCo(CO)_4$.

2. The loss of a carbonyl ligand along the $^3A_1$ curve, giving rise to the products CO and $HCo(CO)_3$, the latter in the excited state $^3A_1$.
3. The formation of the same products CO and $HCo(CO)_3$, but with the latter now in the excited state ^3E, after crossing from the $^3A_1$ to the ^3E surface.

The potential-energy surfaces of Figure 7.1 account for the observation of two concurrent reactive channels upon irradiation of $HCo(CO)_4$ at a unique wavelength with the state $^1E\ d_\delta \rightarrow \sigma^*$ being the photoactive state responsible for both photochemical processes.

7.4.2. Potential-Energy Surfaces for $HMn(CO)_5$

Irradiation of $HMn(CO)_5$ in a low-temperature matrix leads to two different photochemical reactions [8, 90]: the heterolytic loss of a carbonyl ligand upon irradiation at 229 nm,

$$HMn(CO)_5 \xrightarrow{h\nu} CO + HMn(CO)_4 \tag{7.8}$$

and the homolysis of the metal–hydrogen bond upon irradiation at 193 nm,

$$HMn(CO)_5 \xrightarrow{h\nu} H + Mn(CO)_5 \tag{7.9}$$

A set of potential-energy surfaces for the ground and excited states of $HMn(CO)_5$ deduced from the potential-energy curves calculated under C_{4v}

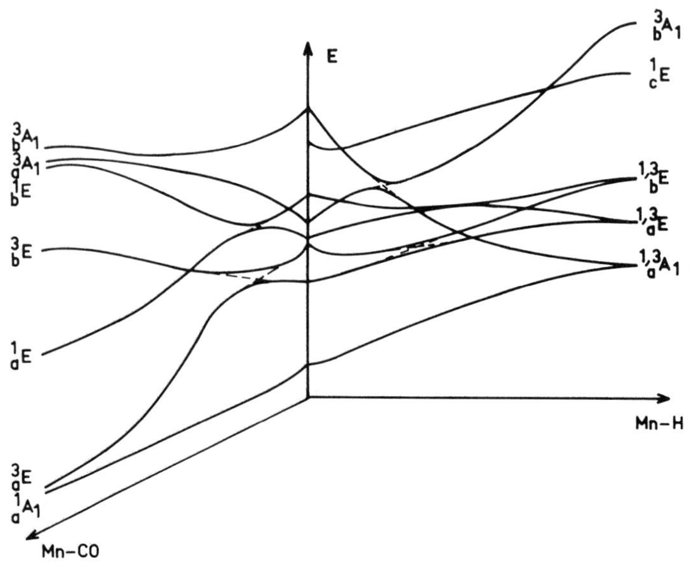

FIGURE 7.2
Potential-energy surfaces corresponding to the loss of hydrogen and axial carbonyl ligand in $HMn(CO)_5$.

constraint for the loss of either the hydrogen or the axial carbonyl ligand [82, 83b] is reported in Figure 7.2. If the potential-energy surfaces appear relatively simple for $HCo(CO)_4$ (Figure 7.1), they look much more complicated (with the presence of avoided crossings and energy barriers) in the case of $HMn(CO)_5$. This is a consequence of the high density of states in the low-energy region of the electronic spectrum of $HMn(CO)_5$ (Table 7.2). The ${}^3A_1$ $\sigma \rightarrow \sigma^*$ state is dissociative with respect to the metal–hydrogen bond homolysis but not with respect to the loss of the carbonyl ligand. On the contrary, the lowest LF excited states $a\,{}^{1,3}E$ are dissociative only with respect to the carbonyl loss. The potential-energy surfaces of Figure 7.2 account for the occurrence of different reactive channels upon irradiation of $HMn(CO)_5$ at different wavelengths:

(i) Irradiation at the experimental wavelength of 229 nm, or 43,700 cm^{-1}, will bring the molecule into the state $b\,{}^1E$ corresponding to the excitation $d_\pi \rightarrow \sigma^*$ calculated at 43,150 cm^{-1}, (see Table 7.2). From there, the molecule goes down along the $b\,{}^1E$ potential-energy curve corresponding to a Mn—CO elongation until it reaches a potential well corresponding to an avoided crossing. At this point, the system evolves to the $a\,{}^1E$ state through internal conversion and dissociation will then occur along the $a\,{}^1E$ curve with the formation of the products CO and $HMn(CO)_4$ in this excited state, the latter as a square pyramid with H apical. A subsequent Berry pseudorotation probably yields $HMn(CO)_4$ as a square pyramid with H basal (this structure corresponds to the isomer identified experimentally [90]);

(ii) Irradiation at the experimental wavelength of 193 nm, or 51,800 cm^{-1}, will bring the molecule into the highest $c\,^1E$ state corresponding to a $d_\pi \rightarrow \pi^*$ excitation (calculated at 53,900 cm^{-1}, see Table 7.2). Intersystem crossing at a metal–hydrogen distance of about 1.70 Å will bring the system first into the state $b\,^3A_1$ and next through internal conversion into the $a\,^3A_1$ $\sigma \rightarrow \sigma^*$ state. From there the molecule will dissociate along this $a\,^3A_1$ potential-energy curve to the products H and $Mn(CO)_5$ in their ground state $^{1,3}A_1$.

From a look at the potential-energy surfaces of Figure 7.2, one can understand why a high excitation energy is necessary to cleave the metal–hydrogen bond in $HMn(CO)_5$. The only potential-energy curves that are dissociative with respect to the metal–hydrogen bond in the vicinity of the equilibrium distance correspond to the $b\,^3A_1$ or the $c\,^1E$ states, both calculated above 50,000 cm^{-1}.

7.4.3. Potential-Energy Surfaces for $H_2Fe(CO)_4$

When $H_2Fe(CO)_4$ is irradiated in a low-temperature matrix, the primary process is the loss of dihydrogen [113]:

$$H_2Fe(CO)_4 \xrightarrow{h\nu} H_2 + Fe(CO)_4 \qquad (7.10)$$

The potential-energy curves corresponding to the elimination of molecular hydrogen under C_{2v} constraint are represented in Figure 7.3 [82, 83a]. The curves originating from the states $a\,^{1,3}B_2$ $(3d \rightarrow \sigma_u^*)$ and $b\,^1A_1$ $(3d \rightarrow \sigma_g^*)$ are dissociative with respect to the elimination of molecular hydrogen. The presence of avoided crossings between the two sets of curves $a\,^{1,3}B_2$ and $b\,^{1,3}B_2$ is responsible for the existence of two energy barriers around 1.30 Å. These potential-energy curves form the basis for a qualitative understanding of the photochemistry of $H_2Fe(CO)_4$.

Irradiation at the experimental wavelength of 254 nm or 39,400 cm^{-1} will bring the molecule into the nearly degenerate states $a\,^1B_2$ and $b\,^1A_1$ calculated at 39,000 and 39,760 cm^{-1}, respectively (Table 7.3). From there the system has the choice between two reactive channels:

1. The dissociation along the $b\,^1A_1$ potential-energy curve, with formation of the products H_2 and $Fe(CO)_4$, the latter in the excited state $b\,^1A_1$.
2. The population of the $a\,^3B_2$ state after intersystem crossing from the $a\,^1B_2$ state followed by the elimination of molecular hydrogen, occurring along the $a\,^3B_2$ potential-energy curve with a small energy barrier (of the order of 10 kcal mol^{-1}) yielding the products in their ground state.

One important feature of Figure 7.3 is the nature of the $a\,^3B_2$ curve in its dissociative part that corresponds to a $\sigma_u \rightarrow \sigma_g^*$ excitation. The promotion of

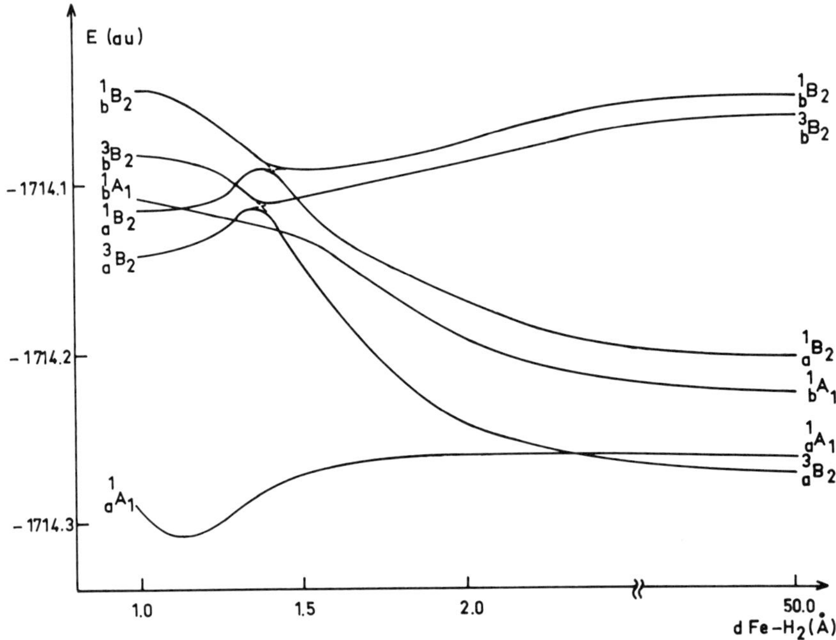

FIGURE 7.3
Potential-energy curves corresponding to the elimination of molecular hydrogen in $H_2Fe(CO)_4$.

an electron from a molecular orbital (MO) that is H—H antibonding and M—H bonding to a MO that is H—H bonding and M—H antibonding leads to the photoelimination of H_2; this seems to be a general feature of the photochemistry of the transition metal dihydrides and polyhydrides [114].

7.5. STATE CORRELATION DIAGRAMS

We recall that these state correlation diagrams are a tentative substitute for the potential-energy curves. They are based exclusively on spin and symmetry properties, usually without any hypothesis about the relative stability of the reactants and products. In a number of cases, we shall complement them with qualitative potential-energy curves that are probably closer to reality. The diagrams presented will be restricted to systems with a metal atom from the first or second transition series. In the case of the third transition series, spin–orbit coupling removes the spin and symmetry restrictions [75, 115] (This is already true, in some cases, even for the systems with metals of the first and second series [115].)

7.5.1. The Elimination of H_2 from $MoCp_2H_2$

This reaction, probably close to concerted, is well-documented [18, 116]:

$$MoCp_2H_2 \xrightarrow{h\nu} MoCp_2 + H_2 \tag{7.11}$$

$MoCp_2H_2$ has a structure of C_{2v} symmetry and it is assumed that this symmetry is retained during the course of the reaction. According to both EHT calculations [117] and ab-initio SCF calculations [73], there is one occupied orbital of symmetry b_2 that is metal–hydrogen bonding and hydrogen–hydrogen antibonding, and one empty orbital a_1 that is metal–hydrogen antibonding and hydrogen–hydrogen bonding. One-electron excitation from b_2 to a_1 leads to excited states ${}^{1,3}B_2$ and the corresponding potential-energy curves must be dissociative with respect to the formation of $MoCp_2 + H_2$. The ground state of the primary product $MoCp_2$ is a ${}^3E_{2g}$ state [115] or ${}^3B_2 + {}^3A_1$ in the C_{2v} symmetry. A state correlation diagram is shown in Figure 7.4a, with a qualitative set of potential-energy curves drawn in Figure 7.4b. The excited state ${}^3B_2$ of $MoCp_2H_2$ correlates directly with the ground state of the products; this triplet state could be reached through intersystem crossing after excitation into the corresponding ${}^1B_2$ state.

Geoffroy and Bradley [18] pointed out that, in the absorption spectrum of $MoCp_2H_2$, the breadth of the band at 310 nm is consistent with an excited

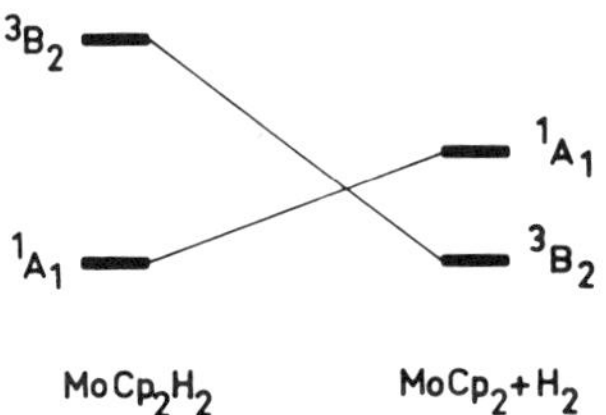

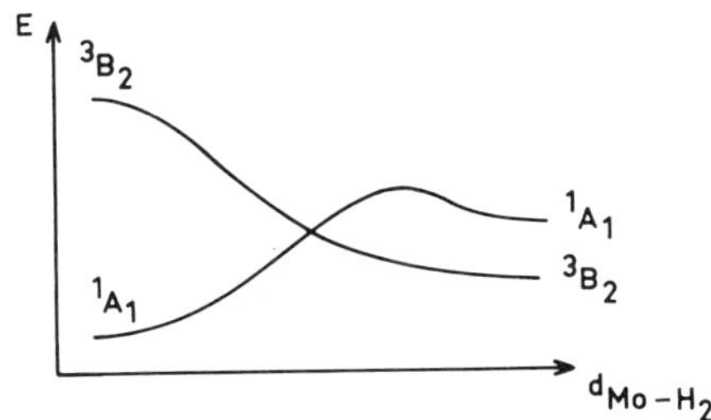

FIGURE 7.4
(a) State correlation diagram for the dissociation of H_2 from $MoCp_2H_2$. (b) Qualitative potential-energy curves for the dissociation of H_2 from $MoCp_2H_2$.

state that is dissociative in nature (with the irradiation being conducted at 366 nm). Geoffroy and Bradley assumed that the elimination of molecular hydrogen results from the population of the orbital that they denote $2b_1$ (d_{xy}—H_2^*) (b_2 in this work and in reference 117), on the basis that this orbital is strongly antibonding between the metal and the two hydrogens [18]. However, this orbital is hydrogen–hydrogen antibonding and it seems dubious that excitation into this orbital will lead to H_2 elimination.

Grebenik, Grinter, and Perutz also discussed the electronic factors that favor this reaction [118]. They state correctly that "population of the a_1^* level favours concerted M—H bond cleavage and H—H bond formation." They mention that "the low-lying $^3A_1$ and $^3B_2$ excited states of Cp_2MH_2 correlate with the ground state of Cp_2M" but improperly refer to a figure that is not a state correlation diagram but a molecular orbital diagram. In the same figure, they assign the transitions $a_1 \rightarrow b_2$ and $a_1 \rightarrow a_1$ as those that labilize the ligands, both involving the excitation of an electron from the nonbonding $3d$ orbital a_1. We note that the transition $a_1 \rightarrow b_2$ corresponds again to the excitation of an electron into an orbital that is hydrogen–hydrogen antibonding and as such should not lead to H_2 elimination.

7.5.2. Elimination of H_2 from Dihydride Complexes of d^6 Ru and Rh

Ultraviolet irradiation of $RuH_2(CO)(PPh_3)_3$ leads to elimination of molecular hydrogen [13],

$$RuH_2(CO)(PPh_3)_3 \xrightarrow{h\nu} H_2 + Ru(CO)(PPh_3)_3 \tag{7.12}$$

A similar photoreaction has been reported for the rhodium dihydride $RhH_2Cl(PPh_3)_3$ [40],

$$RhH_2Cl(PPh_3)_3 \xrightarrow{h\nu} H_2 + RhCl(PPh_3)_3 \tag{7.13}$$

We shall derive a state correlation diagram for the hypothetical reaction

$$MH_2L_4 \xrightarrow{h\nu} ML_4 + H_2 \tag{7.14}$$

MH_2L_4 being a cis dihydride of a metal d^6, with the assumption that C_{2v} symmetry is retained along the reaction path. An energy level scheme is easily obtained for this type of system [19, 73, 119] and is shown in Figure 7.5. Both the reactant and the products (ML_4 being supposed planar [120]) have a ground state $^1A_1$. In analogy with $H_2Fe(CO)_4$, the reactant MH_2L_4 has a set of low-lying excited states $^{1,3}B_2$ corresponding to the $\sigma_u \rightarrow \sigma_g^*$ excitation and these $^{1,3}B_2$ potential-energy curves must be dissociative, because σ_u (shown in **1**) is metal–hydrogen bonding and hydrogen–hydrogen antibond-

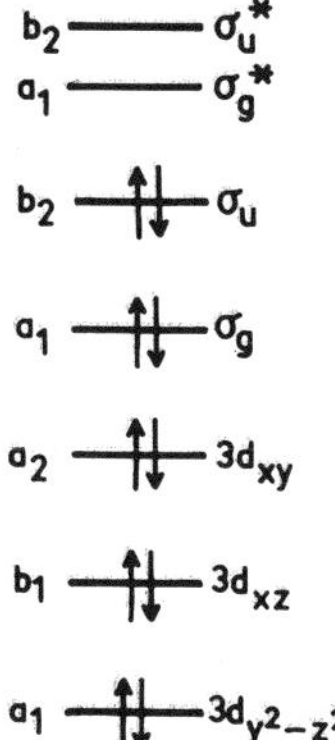

FIGURE 7.5
An energy level scheme for the system MH_2L_4 of C_{2v} symmetry (the choice of axis is similar to the one used for $H_2Fe(CO)_4$ [75]).

ing, whereas σ_g^* (shown in **2**) is metal–hydrogen antibonding and hydrogen–hydrogen bonding.

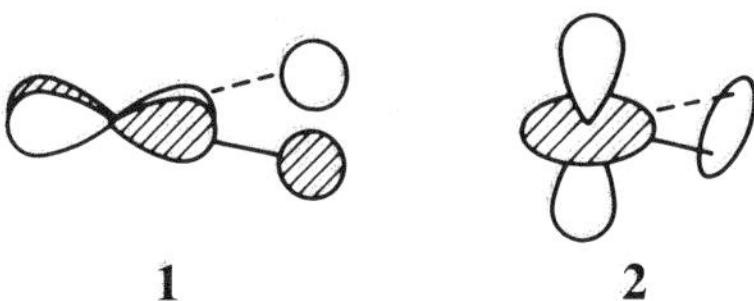

Thus a one-electron excitation from σ_u to σ_g^* will favor the elimination of H_2. The corresponding state correlation diagram is shown in Figure 7.6. We

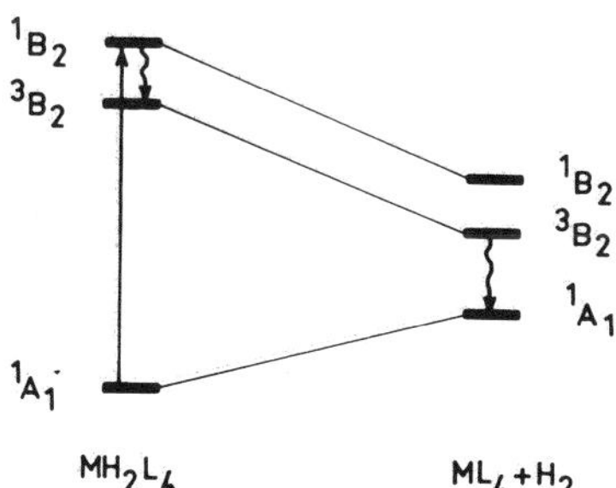

FIGURE 7.6
State correlation diagram for the dissociation of H_2 from ML_4H_2.

can deduce some qualitative features from this diagram:

1. The direct thermal reaction is allowed, probably thermodynamically unfavorable [121].
2. Photochemical elimination of dihydrogen probably will occur through excitation into the $^1B_2$ state followed by conversion to the $^3B_2$ state through intersystem crossing.
3. The products should be formed initially into the $^3B_2$ excited state (chemiluminescence should be observed for the photochemical elimination of H_2 in the gas phase).

More-realistic potential-energy curves probably will show the following, in analogy with the potential-energy curves for $H_2Fe(CO)_4$ [82, 83a]:

1. A barrier for the thermal elimination of dihydrogen (which may be responsible for the absence of thermal recombination) [121].
2. Two avoided crossings between the potential-energy curves of the states $^{1,3}B_2$ (corresponding to the one-electron excitation from $3d_{y^2-z^2}$ to σ_u^*) and of the states $^{1,3}B_2$ (corresponding to the one-electron excitation from σ_u to σ_g^*).

One will notice that our analysis, which traces the elimination of H_2 to the dissociative character of the $^{1,3}B_2$ potential-energy curves (corresponding to the one-electron excitation from σ_u to σ_g^*) is different from the analysis given in reference 19, where the elimination of H_2 was traced to the excitation $\sigma_{x^2-y^2} \rightarrow \sigma_{x^2-y^2}^*$ ($\sigma_u \rightarrow \sigma_u^*$ in our notation). This excitation populates an orbital that is H—H antibonding and as such should not lead to H_2 elimination. Furthermore, for $H_2Fe(CO)_4$, the corresponding states $^{1,3}A_1$ have been found at higher energy than the states $^{1,3}B_2$ [83a].

7.5.3. Elimination of H_2 from $[MoH_4(diphos)_2]$ and $[MoH_4(PPh_2Me)_4]$

Irradiation of $[MoH_4(diphos)_2]$ and $[MoH_4(PPh_2Me)_4]$ leads to the elimination of H_2 and probably to the formation of the unstable species $Mo(diphos)_2$ and $Mo(PPh_2Me)_4$ [123],

$$[MoH_4(diphos)_2] \xrightarrow{h\nu} 2H_2 + [Mo(diphos)_2] \tag{7.15}$$

$$[MoH_4(PPh_2Me)_4] \xrightarrow{h\nu} 2H_2 + [Mo(PPh_2Me)_4] \tag{7.16}$$

On the basis of the following discussion, we believe that there is no need to postulate that this reaction proceeds in a stepwise fashion via initial generation of $[MoH_2(diphos)_2]$ or $[MoH_2(PPh_2Me)_4]$ [123].

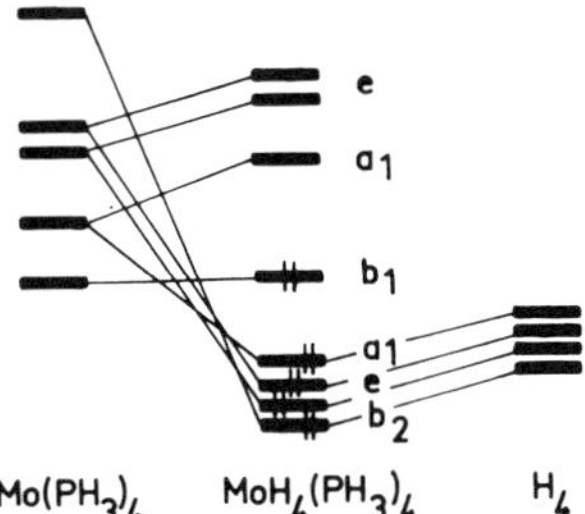

FIGURE 7.7
An energy-level scheme for the system $MoH_4(PH_3)_4$ (with D_{2d} symmetry).

In what follows, we shall mimic these systems through the model compound $[MoH_4(PH_3)_4]$, assuming a structure of D_{2d} symmetry [124], and we shall derive a state correlation diagram for the reaction

$$[MoH_4(PH_3)_4] \xrightarrow{h\nu} 2H_2 + [Mo(PH_3)_4] \tag{7.17}$$

assuming that D_{2d} symmetry is retained along the reaction path. An energy level scheme is easily obtained for $[MoH_4(PH_3)_4]$ [125] and is shown in Figure 7.7, with a ground state $^1A_1$. There is a relatively low-lying set of excited states $^{1,3}E$ corresponding to the one-electron excitation $e \rightarrow a_1$, where e denotes the occupied MO that is metal–hydrogen bonding and hydrogen–hydrogen antibonding and a_1 denotes an empty orbital that is metal–hydrogen antibonding and hydrogen–hydrogen bonding, with the corresponding potential-energy curves being dissociative with respect to the elimination of two hydrogen molecules. Regarding the product $[Mo(PH_3)_4]$, it seems that both the square-planar and tetrahedral coordinations are possible for a transition metal d^6 with a coordination number of 4 [120]. Let us assume first a square-planar structure for $[Mo(PH_3)_4]$, with two low-lying

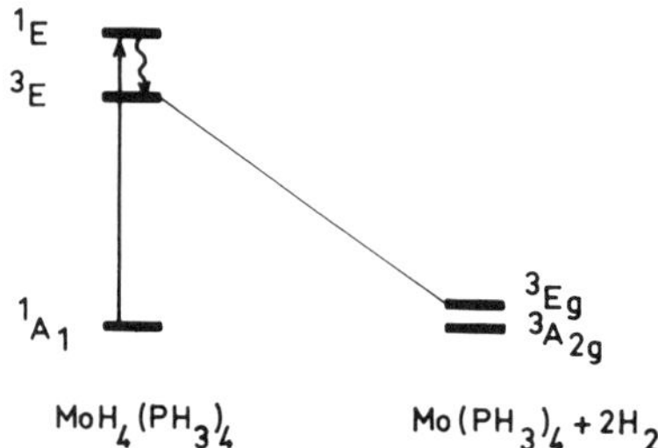

FIGURE 7.8
State correlation diagram for the photoelimination of H_2 from $MoH_4(PH_3)_4$ [assuming a square-planar structure for $Mo(PH_3)_4$].

states $^3A_{2g}$ and ^3E_g close in energy [126] (we make no assumption on their relative position). The corresponding state correlation diagram is shown in Figure 7.8. A similar diagram would be obtained with the assumption of a tetrahedral structure for $Mo(PH_3)_4$, because the corresponding ground state $^3T_1$ correlates with the states $^3A_{2g} + ^3E_g$ of the square-planar structure.

7.5.4. Elimination of H_2 from $(C_5Me_5)Rh(PMe_3)H_2$

Ultraviolet irradiation of $(C_5Me_5)Rh(PMe_3)H_2$ results in the elimination of H_2 [127],

$$(C_5Me_5)Rh(PMe_3)H_2 \xrightarrow{h\nu} H_2 + (C_5Me_5)Rh(PMe_3) \qquad (7.18)$$

According to extended Hückel calculations [128], the reactant has one occupied orbital that is metal–hydrogen bonding and hydrogen–hydrogen antibonding and one of a pair of near-degenerate, empty orbitals (corresponding to the e_g orbitals in O_h symmetry) that is metal–hydrogen antibonding but hydrogen–hydrogen bonding. Excitation from the former to the latter will lead to excited states with the corresponding potential-energy curves being dissociative with respect to the elimination of H_2. In fact, the potential-energy curves for the elimination of H_2 from this system should be qualitatively similar to the potential-energy curves obtained for $H_2Fe(CO)_4$ [82, 83a], because in both systems the metal atom may be viewed formally as d^6 and hexacoordinate. One difference may concern the ground state of the product, which is a triplet for $Fe(CO)_4$ but might be a singlet for $(C_5Me_5)Rh(PMe_3)$ (the possibility of a singlet rather than triplet ground state has been discussed for the d^8 CpML fragment [75, 129]).

7.5.5. Photoelimination of H_2 from dihydrogen complexes: $Cr(CO)_5(H_2)$ and $Ni(CO)_3(H_2)$

$Cr(Co)_5(H_2)$, a system with coordinated molecular dihydrogen, has a broad near-ultraviolet absorption at 370 nm and is destroyed by UV light of this wavelength [71a],

$$Cr(CO)_5(H_2) \xrightarrow{h\nu} Cr(CO)_5 + H_2 \qquad (7.19)$$

A molecular orbital scheme has been reported for $Cr(CO)_5(H_2)$ [130], the molecule having the C_{2v} symmetry with a ground state $^1A_1$ and low-lying excited states of $^{1,3}A_1$ symmetry corresponding to an excitation from the MO that is metal–(H_2) bonding into the MO that is metal–(H_2) antibonding. A state correlation diagram is shown in Figure 7.9. The potential-energy curves for the states $^{1,3}A_1$ must be dissociative with respect to the elimination of H_2. The products will be formed in an excited state, most probably the $^3A_1$ state. The state correlation diagram for the photoelimination of H_2 from

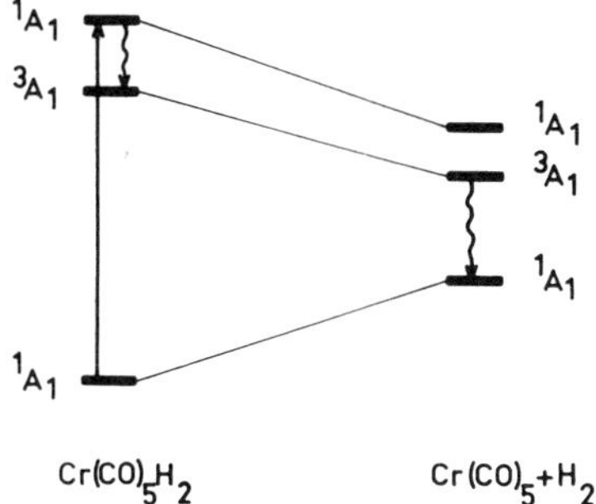

FIGURE 7.9
State correlation diagram for the dissociation of H_2 from $Cr(CO)_5(H_2)$.

$Ni(CO)_3(H_2)$ [71b],

$$Ni(CO)_3(H_2) \xrightarrow{h\nu} Ni(CO)_3 + H_2 \qquad (7.20)$$

should be similar (an energy level scheme for $Ni(CO)_3(H_2)$ may be found in reference 131).

7.5.6. Cu + CH_4 Reaction

A qualitative state correlation diagram has been proposed [132] for the sequence of reactions

$$Cu + CH_4 \longrightarrow CH_3CuH \longrightarrow CH_3 + CuH \qquad (7.21)$$

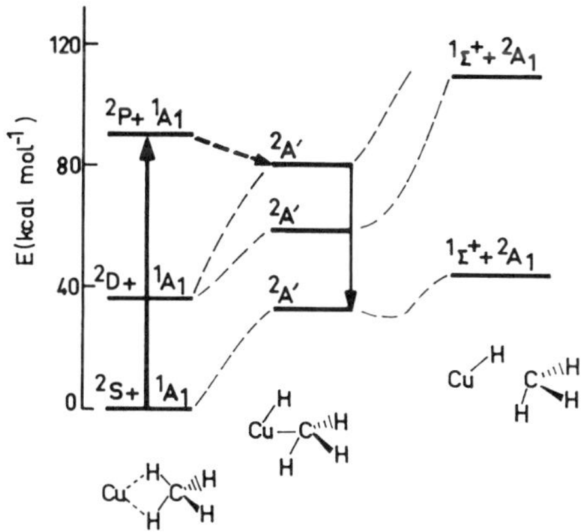

FIGURE 7.10
State correlation diagram for the reaction Cu + CH_4 → CH_3CuH → CH_3 + CuH. (Reproduced from Reference 132.)

and is reproduced in Figure 7.10. Potential-energy curves have been reported for the latest reaction [133].

7.5.7. Photochemistry of MoH_2

Potential-energy curves have been reported for the low-lying electronic states of MoH_2 [134]. Excitation of MoH_2 from the ground state $^5B_2$ into the excited state $^7A_1$ will result in the elimination of molecular hydrogen.

7.6. CONCLUSION

Geoffroy has noticed that "elimination of H_2 is a very general photoreaction for dihydrides and polyhydrides and has now been demonstrated for compounds of V, Mo, W, Re, Fe, Ru, Co and Ir that contain a diverse array of ligands. Indeed one can say with some confidence that elimination of H_2 is likely to obtain upon photolysis of any monomeric di- or polyhydride complex [4]. Similarly, we believe that the homolysis of the M—H bond in monohydrides is a rather general reaction, although it may be obscured in some systems by the competition with the photoelimination of other ligands. The general occurrence of these two photoreactions is certainly not fortuitous, but results from the dissociative character of the potential-energy curve associated with a certain type of excited state. It is clear, from the preceding examples, that the homolysis of the metal–hydrogen bond results from the dissociative character of the potential-energy curve for the triplet corresponding to the $\sigma_{M-H} \rightarrow \sigma^*_{M-H}$ excitation. Also, the elimination of H_2 is a consequence of the dissociative character of the potential-energy curve for the state resulting from the $\sigma_{MH_2} \rightarrow \sigma^*_{MH_2}$ excitation, where σ_{MH_2} is metal–hydrogen bonding and hydrogen–hydrogen antibonding, whereas $\sigma^*_{MH_2}$ is metal–hydrogen antibonding and hydrogen–hydrogen bonding. This does not imply that the reactive excited state is necessarily the state corresponding to the $\sigma_{M-H} \rightarrow \sigma^*_{M-H}$ or $\sigma_{MH_2} \rightarrow \sigma^*_{MH_2}$ excitation, as one can see it from the homolysis of the Co—H bond in $HCo(CO)_4$, with the $^3A_1$ $\sigma \rightarrow \sigma^*$ curve being reached through intersystem crossing after excitation into the ^1E state. In fact, the calculation of the potential-energy curves corresponding to the lowest excited states is a requisite to the elucidation of the detailed reaction mechanism and there is clearly a need for further work along this line.

The mechanisms that we have proposed for the homolysis of a metal–hydrogen bond and for the elimination of H_2 from a dihydride are certainly not restricted to transition metal–hydride systems. For instance, the reductive elimination of ethane through the photolysis of $(C_5H_5)CoMe_2PPh_3$ [135],

$$CpCoMe_2PPh_3 \xrightarrow{h\nu} CpCoPPh_3 + C_2H_6 \tag{7.22}$$

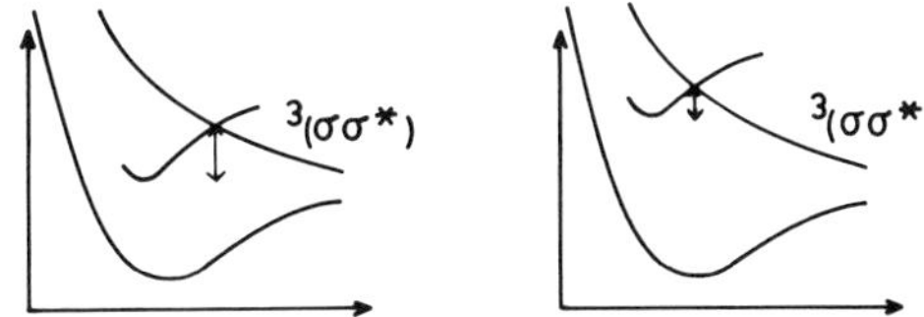

FIGURE 7.11
The influence of the position of an excited state on the barrier to homolysis of a M—H or M—R bond.

certainly bears some analogy to the elimination of H_2 from $(C_5Me_5)Rh(PMe_3)H_2$. The dissociation of the metal–hydrogen bond along the potential-energy curve of the $^3(\sigma\sigma^*)$ state represents a general mechanism that also must be operative for the photolysis of metal–alkyl [136], metal–silyl [137], and metal–metal bonds [74, 138]. The $^3(\sigma\sigma^*)$ potential energy curve plays the same role in the photochemistry of hydride and alkyl compounds of the main group metals [139].

Ultimately, changes in the photochemical behavior of related systems may be explained through their potential-energy curves. Meyer [72] noticed that the complexes $[(bpy)_2M(CO)H]^+$, where M = Ru, Os, are photochemically inert with respect to the homolysis of the metal–hydrogen bond whereas, in the related complexes $[(bpy)Re(CO)_3H]$, a relatively efficient Re—H photochemistry does exist, with the MLCT excited state being higher in energy in the latter system. As shown in Figure 7.11, raising an excited state may lower the energy barrier preventing internal conversion or intersystem crossing to the $^3(\sigma\sigma^*)$ state.

More work will be needed before the mechanism of these photochemical reactions is completely understood. The key role that we have assigned to the $^3(\sigma\sigma^*)$ state in the homolysis of the metal–hydrogen bond is in agreement with the conclusion that, in the photolysis of methylaquacobaloxime, the fission of the Co—CH_3 bond proceeds through the excited triplet state [140]. Photochemical studies in the gas phase certainly would be useful, as would be experiments on the wavelength dependence of the photolytic processes. Theoretical evaluations of the probability of nonradiative processes together with estimates of the lifetime of the excited states should provide some information on the relative efficiency of the possible photoprocesses.

REFERENCES

1. Geoffroy, G. L.; Lehman, J. R. *Adv. Inorg. Chem. Radiochem.* **1977**, *20*, 189.
2. Geoffroy, G. L.; Bradley, M. G.; Pierantozzi, R. Advances in Chemistry Series 167, American Chemical Society: Washington, DC, 1978; p. 181.
3. Geoffroy, G. L.; Wrighton, M. S. *Organometallic Photochemistry*; Academic: New York, 1979.

4. Geoffroy, G. L. *Prog. Inorg. Chem.* **1980**, *27*, 123.
5. Hoffman, N. W.; Brown, T. L. *Inorg. Chem.* **1978**, *17*, 613.
6. Endicott, J. F.; Wong, C. L.; Inoue, T.; Natarajan, P. *Inorg. Chem.* **1979**, *18*, 450.
7. Grebenik, P.; Downs, A. J.; Greens, M. L. H.; Perutz, R. N. *J. Chem. Soc., Chem. Commun.* **1979**, 742.
8. Rest, A. J.; Turner, J. J. *J. Chem. Soc., Chem. Commun.* **1969**, 375.
9. Mills, W. J.; Clark, R. J. *Inorg. Chem.* **1968**, *7*, 1801.
10. Beyers, B. H.; Brown, T. L. *J. Am. Chem. Soc.* **1977**, *99*, 2527.
11. Werner, P.; Ault, B. S.; Orchin, M. *J. Organomet. Chem.* **1978**, *162*, 189.
12. Corder, H. L.; Courtney, A. R.; DeMarco, D. *J. Am. Chem. Soc.* **1978**, *101*, 1606.
13. Geoffroy, G. L.; Bradley, M. G. *Inorg. Chem.* **1977**, *16*, 744.
14. Berry, M.; Cooper, N. J.; Green, M. L. H.; Simpson, S. J. *J. Chem. Soc., Dalton Trans.* **1980**, 29.
15. Berry, M.; Elmitt, K.; Green, M. L. H. *J. Chem. Soc., Dalton Trans.* **1979**, 1950.
16. Cooper, N. J.; Green, M. L. H.; Mahtab, R. *J. Chem. Soc., Dalton Trans.* **1979**, 1557.
17. For a complete review see reference 4 and references therein.
18. Geoffroy, G. L.; Bradley, M. G. *Inorg. Chem.* **1978**, *17*, 2410.
19. Geoffroy, G. L.; Pierantozzi, R. *J. Am. Chem. Soc.* **1976**, *98*, 8054.
20. Epstein, R. A.; Gaffney, T. R.; Geoffroy, G. L.; Gladfelter, W. L.; Henderson, R. *J. Am. Chem. Soc.* **1979**, *101*, 3847.
21. Johnson, B. F. G.; Lewis, J.; Twigg, M. V. *J. Organomet. Chem.* **1974**, *67*, C75.
22. Janowiz, A. H.; Bergman, R. G. *J. Am. Chem. Soc.* **1982**, *104*, 352; **1983**, *105*, 3929.
23. Wenzel, T. J.; Bergman, R. G. *J. Am. Chem. Soc.* **1986**, *108*, 4856.
24. Fisher, B. J.; Eisenberg, R. *Organometallics* **1983**, *2*, 764.
25. Wu, J.; Bergman, R. G. *J. Am. Chem. Soc.* **1989**, *111*, 7628.
26. Peerlans, R. A.; Bergman, R. G. *Organometallics* **1984**, *3*, 508.
27. Wax, M. J.; Strykler, J. M.; Buchanan, J. M.; Kovac, C. A.; Bergman, R. G. *J. Am. Chem. Soc.* **1984**, *106*, 1121.
28. Jones, W. D.; Feher, F. J. *J. Am. Chem. Soc.* **1982**, *104*, 4240.
29. Matsui, Y.; Orchin, M. *J. Organomet. Chem.* **1982**, *236*, 381.
30. Orchin, M. *Acc. Chem. Res.* **1981**, *14*, 259.
31. Halpern, J. *Inorg. Chim. Acta.* **1981**, *50*, 11.
32. Mutterties, E. L. *Inorg. Chim. Acta.* **1981**, *50*, 1.
33. Alt, H. G.; Eichner, M. E. *Angew. Chem. Int. Ed. Engl.* **1982**, *21*, 78.
34. Kazlanskas, R. J.; Wrighton, M. S. *J. Am. Chem. Soc.* **1980**, *102*, 1727.
35. Kazlanskas, R. J.; Wrighton, M. S. *J. Am. Chem. Soc.* **1982,** *104*, 6005.
36. Peterson, J. R.; Bennett, D. W.; Spicer, L. D. *J. Catal.* **1981**, *71*, 223.
37. Kunin, A. J.; Eisenberg, R. *Organometallics* **1988**, *7*, 2124.
38. Darsillo, M. S.; Gafney, H. D.; Paquette, M. S. *Inorg. Chem.* **1988**, *27*, 2815.
39. Wink, D. A.; Ford, P. C. *J. Am. Chem. Soc.* **1985**, *107*, 5566.
40. Wink, D. A.; Ford, P. C. *J. Am. Chem. Soc.* **1986**, *108*, 4838.
41. Green, M. A.; Huffman, J. L.; Caulton, K. G. *J. Am. Chem. Soc.* **1981**, *103*, 695.
42. Roberts, D. A.; Geoffroy, G. L. *J. Organomet. Chem.* **1988**, *214*, 221.
43. Foley, H. C.; Geoffroy, G. L. *J. Am. Chem. Soc.* **1981**, *103*, 7176.
44. Graff, J. L.; Wrighton, M. S. *J. Am. Chem. Soc.* **1980**, *102*, 2123.
45. Bentsen, J. G.; Wrighton, M. S. *Inorg. Chem.* **1984**, *23*, 512.
46. Bentsen, J. G.; Wrighton, M. S. *J. Am. Chem. Soc.* **1984**, *106*, 4041.
47. Foley, H. C.; Morris, R. H.; Targos, T. S.; Geoffroy, G. L. *J. Am. Chem. Soc.* **1981**, *103*, 7337.
48. Hill, R. H.; de Mayo, P.; Puddephatt, R. J. *Inorg. Chem.* **1982**, *21*, 3642.
49. Sweany, R. L. *Inorg. Chem.* **1980**, *19*, 3512.

50. Sweany, R. L. *Inorg. Chem.* **1982**, *21*, 752.
51. Church, S. P.; Poliakoff, M.; Timney, J. A.; Turner, J. J. *J. Am. Chem. Soc.* **1981**, *103*, 7515.
52. Mahmoud, K. A.; Rest, A. J.; Alt, H. G. *J. Organomet. Chem.* **1983**, *246*, C37.
53. Mahmoud, K. A.; Rest, A. J.; *J. Chem. Soc., Dalton Trans.* **1984**, *187*.
54. Baynham, R. F. G.; Chetwynd-Talbot, J.; Grebenik, P.; Perutz, R. N.; Powell, M. H. A. *J. Organomet. Chem.* **1985**, *284*, 229.
55. Hoyano, J. K.; Graham, W. A. G. *J. Am. Chem. Soc.* **1982**, *104*, 3722.
56. Bradley, M. G.; Roberts, D. A.; Geoffroy, G. L. *J. Am. Chem. Soc.* **1981**, *103*, 379.
57. Chamberlain, L. R.; Rothwell, A. P.; Rothwell, I. P. *J. Am. Chem. Soc.* **1984**, *106*, 1847.
58. (a) Onishi, M.; Hiraki, K.; Matsuda, M.; Fukunaga T. *Chem. Soc. Jpn., Chem. Lett.*, **1983**, 261. (b) Onishi, M.; Hiraki, K.; Ishida, Y.; Dakeshita, K. *Chem. Soc. Jpn., Chem. Lett.* **1986**, 333. (c) Onishi, M.; Oishi, S.; Sakaguchi, M.; Takaki, I.; Hiraki, K. *Bull. Chem. Soc. Jpn.* **1984**, 607.
59. Nugel, P. D.; Brown, T. L. *J. Am. Chem. Soc.* **1984**, *106*, 3474.
60. Amadelli, R.; Carassiti, V.; Maldotti, A.; Aime, S.; Osella, D.; Milone, M. *Inorg. Chim. Acta*. **1984**, *81*, 111.
61. Sweany, R. L. *J. Am. Chem. Soc.* **1986**, *108*, 6986.
62. Sweany, R. L.; Russel, F. N. *Organometallics* **1988**, *7*, 719.
63. Bradley, M. G.; Roberts, D. A.; Geoffroy, G. L. *J. Am. Chem. Soc.* **1981**, *103*, 379.
64. Pomeroy, R. K. *J. Organomet. Chem.* **1981**, *221*, 323.
65. Green, M. A.; Huffman, J. L.; Caulton, K. G. *J. Organomet. Chem.* **1983**, *243*, C78.
66. Cloke, F. G. N.; Green, J. G.; Green, M. L. H.; Morley, C. P. *J. Chem. Soc., Chem. Commun.* **1985**, 945.
67. Brewer, K. J.; Murphy, W. R., Jr.; Moore, K. J.; Eberle, E. C.; Petersen, J. D. *Inorg. Chem.* **1986**, *25*, 2470.
68. Bloyce, P. E.; Rest, A. J.; Whitwell, I.; Graham, W. A. G.; Holmes-Smith, R. *J. Chem. Soc., Chem. Commun.* **1988**, 846.
69. Poliakoff, M. Weitz, E. *Adv. Organomet. Chem.* **1986**, 277.
70. Sweany, R. L. *J. Am. Chem. Soc.* **1985**, *107*, 2374. Gadd, G. E.; Upmacis, R. K.; Poliakoff, M.; Turner, J. J. *J. Am. Chem. Soc.* **1986**, *108*, 2547.
71. (a) Upmacis, R. K.; Gadd, G. E.; Poliakoff, M.; Simpson, M. B.; Turner, J. J.; Whyman, R.; Simpson, A. F. *J. Chem. Soc., Chem. Commun.* **1985**, 27. (b) Sweany, R. L.; Polito, M. A.; Moroz, A. *Organometallics* **1989**, *8*, 2305.
72. Megehee, E. G.; Meyer, T. J. *Inorg. Chem.* **1989**, *28*, 4084.
73. Veillard, A. *Nouv. J. Chim.* **1981**, *5*, 599.
74. Veillard, A.; Dedieu, A. *Nouv. J. Chim.* **1983**, *7*, 683.
75. Veillard, A.; Dedieu, A. *Theor. Chim. Acta*. **1983**, *63*, 339.
76. Daniel, C.; Veillard, A. *Nouv. J. Chim.* **1986**, *10*, 83.
77. Turro, N. J. *Modern Molecular Photochemistry*; Benjamin/Cummings: Menlo Mark, CA., 1978
78. Daniel, C.; Bénard, M.; Dedieu, A.; Wiest, R.; Veillard, A. *J. Phys. Chem.* **1984**, *21*, 4805.
79. Daniel, C.; Hyla-Kryspin, I.; Demuynck, J.; Veillard, A. *Nouv. J. Chim.* **1985**, *9*, 581.
80. Veilard, A.; Daniel, C.; Strich, A. *Pure Appl. Chem.* **1988**, *60*(2), 215.
81. Veillard, A.; Strich, A. *J. Am. Chem. Soc.* **1988**, *110*, 3793.
82. Daniel, C. *Coord. Chem. Rev.* **1990**, *97*, 141.
83. (a) Daniel, C. *J. Phys. Chem.* **1991**, *95*, 2394. (b) Daniel, C. to be published.
84. Veillard, A. Unpublished results.

85. van Dijk, H. K.; Stufkens, D. J.; Oskam, A. *J. Am. Chem. Soc.* **1989**, *111*, 541.
86. Tero-Kubota, S.; Hoshino, N.; Kato, M.; Goedken, V. L.; Ito, T. *J. Chem. Soc., Chem. Commun.* **1985**, 959.
87. Kobayashi, T.; Ohtani, H.; Noda, H.; Teratani, S.; Yamazaki, H.; Yasufuku, **X**. *Organometallics* **1986**, *5*, 110.
88. Prinslow, D. A.; Vaida, V. *J. Am. Chem. Soc.* **1987**, *109*, 5097.
89. Salem, L.; Rowland, C. *Angew. Chem. Int. Ed., Engl.* **1972**, *11*, 92.
90. Church, S. P.; Poliakoff, M.; Timney, J. A.; Turner, J. J. *Inorg. Chem.* **1983**, *22*, 3259.
91. Veillard, A.; Strich, A.; Daniel, C.; Siegbahn, P. E. M. *Chem. Phys., Lett.* **1987**, *141*, 329.
92. Meyer, T. J. *Pure Appl. Chem.* **1986**, *58*, 1193.
93. Manuta, D. M.; Lees, A. J. *Inorg. Chem.* **1986**, *25*, 1354.
94. James, B. R. *Homogeneous Hydrogenation*; Wiley: New York, 1973; p. 7.
95. Pearson, R. G. *Chem. Rev.* **1985**, *85*, 41.
96. Crabtree, R. H. *The Organometallic Chemistry of the Transition Metals*; Wiley: New York, 1988; p. 55.
97. Albright, T. A.; Burdett, J. K.; Whangbo, M. H. *Orbital Interactions in Chemistry*; Wiley: New York, 1988, and references therein.
98. Marquez, A.; Daniel, C. Fernandez Sanz, J. submitted to *J. Phys. Chem.*
99. Dewar, M. J. S.; Zoebisch, E. G.; Healy, E. F.; Stewart, J. J. P. *J. Am. Chem. Soc.* **1985**, *107*, 3902.
100. Salahub, D. R.; Zerner, M. C. In *The Challenge of d and f Electrons*; Salahub, D. R., Zerner, M. C., Eds.; ACS Symposium Series 394; American Chemical Society: Washington, DC, 1989; p. 1.
101. Veillard, A.; Demuynck, J. In *Modern Theoretical Chemistry, 4: Applications of Electronic Structure Theory*; Schaefer, H. F., Ed.; Plenum: New York, 1977; p. 187.
102. Daniel, C.; Veillard A. In *Quantum Chemistry: The Challenge of Transition Metals and Coordination Chemistry*, Veillard, A., Ed.; NATO ASI Series; Reidel: Dordrecht, 1986; p. 363.
103. Veillard, A.; Daniel, C.; Rohmer, M. M. *J. Phys. Chem.*, 1990, *94*, 5556.
104. Siegbahn, P. E. M. *Int. J. Quantum Chem.* **1983**, *23*, 1869.
105. Siegbahn, P. E. M.; Almlof, J.; Heiberg, A.; Roos, B. O. *J. Chem. Phys.* **1981**, *74*, 2384.
106. Bauschlicher, C. W.; Langhoff, S. R. *J. Chem. Phys.* **1987**, *86*, 5595.
107. Martin, R. L.; Hay, P. J. *J. Chem. Phys.* **1981**, *75*, 4539.
108. Pitzer, K. S. *Acc. Chem. Res.* **1979**, *12*, 271.
109. See, for instance, Matos, J. M. O.; Roos, B.; Malmquist, P. A. *J. Chem. Phys.* **1987**, *86*, 1458.
110. R. L. Sweany argues [50, 111] that weak singlet–singlet absorption should have been observable between 36,000 and 44,000 cm^{-1}.
111. Sweany, R. L. Chapter 2 of this volume.
112. Blackney, G. B.; Allen, W. F. *Inorg. Chem.* **1971**, *10*, 2763.
113. Sweany, R. L. *J. Am. Chem. Soc.* **1981**, *103*, 2410.
114. Veillard, A. *Chem. Phys. Lett.*, **1990**, *170*, 441.
115. Cox, P. A.; Grebenik, P.; Perutz, R. N.; Robinson, M. D.; Grinter, R.; Stern, D. R. *Inorg. Chem.* **1983**, *22*, 3614.
116. Chetwynd-Talbot, J.; Grebenik, P.; Perutz, R. N. *Inorg. Chem.* **1982**, *21*, 3647.
117. Lauher, J. W.; Hoffmann, R. *J. Am. Chem. Soc.* **1976**, *98*, 1729.
118. Grebenik, P.; Grinter, R.; Perutz, R. N. *Chem. Soc. Rev.* **1988**, *17*, 453.
119. Dedieu, A.; Strich, A. *Inorg. Chem.* **1979**, *18*, 2940.
120. Cotton, F. A.; Wilkinson, G. *Advanced Inorganic Chemistry*; Wiley: New York, 1988; p. 710ff.

121. Tolman, C. A.; Meakin, P. Z.; Lindner, P. L.; Jesson, J. P. *J. Am. Chem. Soc.* **1974**, *96*, 2762.
122. Daniel, C. Unpublished results.
123. Pierantozzi, R.; Geoffroy, G. L. *Inorg. Chem.* **1980**, *19*, 1821.
124. Meakin, P.; Guggenberger, L. J.; Peet, W. G.; Muetterties, E. L.; Jesson, J. P. *J. Am. Chem. Soc.* **1973**, *95*, 1467.
125. Burdett, J. K.; Hoffmann, R.; Fay, R. C. *Inorg. Chem.* **1978**, *17*, 2553.
126. Rohmer, M. M. In *Quantum Chemistry: The Challenge of Transition Metals and Coordination Chemistry*; Veillard, A., Ed.; NATO ASI Series; Reidel: Dordrecht, 1986; p. 363.
127. Jones, W. D.; Feher, F. J. *J. Am. Chem. Soc.* **1984**, *106*, 1650.
128. Hoffmann, R. *J. Chem. Phys.* **1963**, *39*, 1397.
129. Hofmann, P.; Padmanabhan, M. *Organometallics* **1983**, *2*, 1273.
130. Saillard, J. Y.; Hoffmann, R. *J. Am. Chem. Soc.* **1984**, *106*, 2006.
131. Burdett, J. K.; Phillips, J. R.; Pourian, M. R.; Poliakoff, M.; Turner, J. J.; Upmacis, R. *Inorg. Chem.* **1987**, *26*, 3054.
132. Ozin, G. A.; McCaffrey, J. G.; Parnis, J. M. *Angew. Chem. Int. Ed. Engl.* **1986**, *25*, 1072.
133. Castillo, S.; Poulain, E.; Novaro, O. *Int. J. Quantum. Chem., Quantum Chem. Symp.* **1989**, *23*, 509.
134. Li, J.; Balasubramanian, K. *J. Phys. Chem.* **1990**, *94*, 545.
135. Becalska, A.; Hill, R. H. *J. Am. Chem. Soc.* **1989**, *111*, 4346.
136. Alt, H. G. *Angew. Chem. Int. Ed. Engl.* **1984**, *23*, 766.
137. Reichel, C. L.; Wrighton, M. S. *Inorg. Chem.* **1980**, *19*, 3858.
138. Meyer, T. J.; Caspar, J. V. *Chem. Rev.* **1985**, *85*, 187.
139. Rohmer, M. M. *Chem. Phys. Lett.* **1989**, *157*, 207.
140. Sakaguchi, Y.; Hayashi, H.; I'Haya, Y. J. *J. Phys. Chem.* **1990**, *94*, 291.

CHAPTER 8

Isotope Effects in Reactions of Transition Metal Hydrides

R. MORRIS BULLOCK

Department of Chemistry, Brookhaven National Laboratory, Upton, New York 11973

8.1. INTRODUCTION

Both kinetic and equilibrium (or thermodynamic) isotope effects can provide critical information about reaction mechanisms. When used in conjunction with other experimental data (activation parameters, solvent effects, etc.), a careful study of isotope effects can provide mechanistic information that may be unavailable from other methods. Many fundamentally important details of the reaction mechanisms of transition metal hydrides have been delineated by experiments involving the use of isotopic labels.

Studies of kinetic and equilibrium isotope effects in reactions of transition metal hydrides are reviewed in this chapter. A review of kinetic isotope effects in metal clusters was published recently [1]. Most of the reactions discussed in this chapter involve mononuclear complexes containing terminal M—H bonds, but several examples where the metal–hydrogen interaction is in the form of dihydrogen complexes $M(\eta^2\text{-}H_2)$ and agostic (C—H····M) interactions [2] also are included. The site of deuteration usually is the M—H bond, but is in some cases a C—H that becomes an M—H in the reaction. When the rate of reaction of an unlabelled compound is faster than for the corresponding deuterated complex ($k_H/k_D > 1$), this is referred to as a "normal" isotope effect; note that the word normal only indicates the *direction* of the isotope effect and does not indicate that the *magnitude* of the effect is necessarily normal. Inverse isotope effects are those in which $k_H/k_D < 1$.

Bigeleisen's [3] theoretical treatment of kinetic isotope effects was developed over 40 years ago. Wolfsberg, Stern, Weston, and others have reported theoretical calculations for model systems [4]. A review by Westheimer [5] and the book by Melander and Saunders [6] provide thorough treatments of the important concepts related to the theory of isotope effects. A theoretical treatment of kinetic isotope effects based on quantum mechanics also has been presented [7], but is not discussed here.

The complete theoretical treatment provides an equation [3, 6] for prediction of isotope effects that contains a product of terms due to mass and moment of inertia (MMI), vibrational excitation (EXC), zero-point energy (ZPE), and quantum mechanical tunnelling. Substitution of D for H that is bonded to a heavy atom (such as a metal) generally has little effect on the mass or moment of inertia, so deviations of the MMI term from unity are often neglected. The spacing between vibrational levels of the molecules under consideration here normally are much greater than $k_B T$ (k_B = Boltzmann's constant). The fact that $h\nu \gg k_B T$ means that there will be little population of vibrationally excited states, so the EXC term also is approximated by unity in most instances. Therefore the zero-point energy terms are almost invariably the dominant influence on isotope effects. Neglecting terms other than the zero-point energy term, the isotope effect predicted by semiclassical theory [3–6] is given by eq. (8.1), where N is the number of atoms:

$$\frac{k_H}{k_D} = \exp\left\{\frac{-hc}{2k_B T}\left[\sum_i^{3N^{\ddagger}-7}\left(\bar{\nu}^{\ddagger}_{i(H)} - \bar{\nu}^{\ddagger}_{i(D)}\right) - \sum_i^{3N-6}\left(\bar{\nu}_{i(H)} - \bar{\nu}_{i(D)}\right)\right]\right\} \quad (8.1)$$

Note that the summation is over one fewer vibrational modes in the transition state ($3N^{\ddagger} - 7$) compared to the ground state (3N − 6), because one of the vibrational degrees of freedom becomes a translation in the transition state.

8.2. PROTON TRANSFER REACTIONS

Many of the general principles of experimental versus theoretical isotope effects are illustrated by the proton transfer reaction shown in eq. (8.2).

$$[CpM(CO)_3]^- + Cp\mathbf{M}(CO)_3H \rightleftharpoons CpM(CO)_3H + [Cp\mathbf{M}(CO)_3]^- \quad (8.2)$$

Norton and co-workers [8] reported isotope effects for this self-exchange reaction, in which a degenerate proton transfer occurs between a metal hydride and its conjugate base, the metal anion. The theoretical treatment [3] predicts the largest isotope effect when there is no thermodynamic driving force for the reaction ($\Delta G^\circ = 0$) and when the transition state is symmetric. For the reaction considered here this would mean that, at the transition state, the proton should be located at an equal distance from the two metals. Lower isotope effects normally are observed when the transition state is more reactant-like (an early transition state) or more product-like (a late transition state) compared to a symmetric transition state. Furthermore, a linear arrangement of the hydrogen and the metals (or carbons) between which the transfer is taking place is another requirement for observation of isotope effects as large as those predicted by theory. Isotope effects are generally lower when geometrical or steric constraints prevent a linear transition state [9]. All of these requirements appear to be met by this proton transfer self-exchange reaction. In the calculation of the expected isotope effect for eq. (8.2), the initial M—H stretching vibration is taken to be replaced by a translation in the decomposition of the (symmetric and linear) M····H····M transition state. Further simplifications to eq. (8.1) are often made, because in few if any organometallic examples are *all* of the frequencies of *all* vibrational modes known. In most cases, calculations of the expected isotope effect consider only the stretching vibrations of the bonds involving the isotopically labelled atom. Other vibrational modes would normally experience minimal change in going from the ground state to the transition state, and all bending modes are also normally neglected. (If bending modes in the ground state are allowed to relax in the transition state, the effect will be to raise the calculated isotope effect.) After making these approximations, application of eq. (8.1) to the proton transfer reaction [eq. (8.2)] allows the expected kinetic isotope effect to be calculated using eq. (8.3).

$$\frac{k_{\mathrm{H}}}{k_{\mathrm{D}}} = \exp\left\{\frac{hc}{2k_{\mathrm{B}}T}(\bar{\nu}_{\mathrm{MH}} - \bar{\nu}_{\mathrm{MD}})\right\} \tag{8.3}$$

The values of the M—H and M—D stretching frequencies were determined by IR spectroscopy, and substitution of appropriate values into eq. (8.3) gives the calculated values (Table 8.1).

The reasonably good agreement between the experimental values and those calculated by semiclassical theory does not rule out the presence of some contribution from quantum mechanical tunnelling in these proton transfers [10]. As mentioned previously, isotope effects as large as the calculated values are only expected for symmetric, linear transition states, where $\Delta G^\circ = 0$; if these conditions are met, a *lower* isotope effect usually is observed. Tunnelling [11, 12], in contrast, can *raise* the observed isotope effect. Because the relative importance of tunnelling increases with decreas-

Table 8.1
Calculated and Experimentally Determined Kinetic Isotope Effects for Proton Transfer Self-Exchange of Metal Hydrides [eq. (8.2)]

Metal Hydride (Deuteride)	Calculated k_H/k_D	Experimental k_H/k_D
$HCr(CO)_3Cp$	3.39	3.6 ± 0.2
$HMo(CO)_3Cp$	3.38	3.7 ± 0.2
$HW(CO)_3Cp$	3.53	3.7 ± 0.2

ing temperatures, tunnelling can in principle be detected by concave-upward curvature in Arrhenius plots of $\ln k$ versus $1/T$ [13]. Although this may appear straightforward, in practice it is difficult to obtain unambiguous evidence for tunnelling using this criterion. A formidable challenge is the problem of obtaining sufficiently precise kinetic data over a large enough range of temperatures (ideally a 100° range or larger) and at sufficiently low temperatures for the curvature to become apparent. Furthermore, in order to be convinced that tunnelling is in fact the cause of any observed curvature, it is necessary to rule out other possible sources of curvature, such as experimental error or a change in reaction mechanism.

Because tunnelling is far more pronounced for H than for D, the temperature dependence of k_H/k_D is potentially a sensitive test of tunnelling. In the absence of tunnelling, the temperature dependence of the kinetic isotope effect is expected to follow eq. (8.4),

$$\frac{k_H}{k_D} = \frac{A_H}{A_D} \exp\left(\frac{E_a^D - E_a^H}{RT}\right) \tag{8.4}$$

Bell [14] predicted that, in the absence of tunnelling, the ratio of the Arrhenius preexponential factors would be $A_H/A_D > 0.5$ with A_H/A_D normally being close to unity. Model calculations [15] have suggested an effective lower limit of about $A_H/A_D \approx 0.7$. Thus, observed ratios of the A_H/A_D lower than about 0.5 generally are taken as evidence for the operation of tunnelling. However, Stern and Weston [16] showed that experimental values of A_H/A_D *within* the range of $0.5–\sqrt{2}$ do not exclude the possibility of tunnelling. The differences in activation energies ($\Delta E_a = E_a^D - E_a^H$) also can be useful in a consideration of whether tunnelling occurs. In the absence of tunnelling, the maximum expected difference in observed activation energy is equal to the difference in zero-point energies of the M—H (vs. M—D) bond being broken. Using approximate values of $\nu_{MD} = 1400\ \text{cm}^{-1}$ and $\nu_{MH} = 1900\ \text{cm}^{-1}$ in eq. (8.5) gives ~ 0.7 kcal mol^{-1} for the difference in zero-point energy of M—H versus M—D,

$$\Delta E_0 = \frac{Nhc}{2}(\bar{\nu}_{MH} - \bar{\nu}_{MD}) \tag{8.5}$$

where N is Avogadro's number. For reactions involving cleavage of an M—H(D) bond, experimentally determined values of $E_a^D - E_a^H$ significantly higher than 0.7 kcal mol^{-1} indicate that tunnelling may be involved. For comparison, the higher C—H stretching frequencies compared to M—H stretching frequencies result in a difference in zero-point energies for C—H versus C—D of about 1.1 kcal mol^{-1}. Accordingly, predicted isotope effects normally will be smaller for reactions involving M—H bond cleavage compared to reactions involving C—H bond cleavage.

Ingold and co-workers [17] reported detailed studies of intramolecular hydrogen atom transfers in a series of sterically hindered aryl radicals. Strong evidence for tunnelling was obtained, including curved Arrhenius plots, low A_H/A_D values, and large values of $E_a^D - E_a^H$. A spectacular $k_H/k_D = 13{,}000$ (at $-150°C$) was observed in one of these examples. In contrast to the temperature-dependent isotope effects observed for this and other organic reactions [18], there are presently very few examples in transition metal hydride chemistry where the temperature-dependence of isotope effects has been thoroughly evaluated.

Roecker and Meyer [19] reported the kinetics of oxidation of alcohols by $[(bipy)_2(py)Ru{=}O]^{2+}$, and they found $k_H/k_D = 50$ at 25°C for oxidation of $PhCH_2OH$ versus $PhCD_2OH$. A significant amount of tunnelling occurs in the step of this reaction that involves breaking the C—H bond and formation of an O—H bond. There is compelling evidence for tunnelling in many organic and inorganic reactions involving H transfer. Because tunnelling is far more important for reactions involving transfer of H than for heavier elements, it is quite possible that tunnelling could be of considerable importance in some reactions of metal hydrides. Future experimental and theoretical work on isotope effects in metal hydride reactivity may help answer these questions.

8.3. SPECTROSCOPIC STUDIES OF EQUILIBRIUM ISOTOPE EFFECTS

Most isotope effects are obtained by comparing the kinetics of reactions of deuterium-labelled complexes with the kinetics of unlabelled complexes and are therefore *observed* as *kinetic* isotope effects (k_H/k_D). In many cases to be discussed, however, observed kinetic isotope effects are not due to a difference in rate of reaction for deuterium-labelled and unlabelled compounds, but instead have their origin in a thermodynamic isotope effect (K_H/K_D) that occurs prior to the rate-determining step and manifests itself in the observed kinetics. Observed kinetic isotope effects that are interpreted as being due to equilibrium isotope effects are particularly common in the case of inverse isotope effects, which have been encountered in numerous organometallic reactions involving rupture of an M—H bond and formation of a C—H bond. In some cases, observable changes in equilibrium constants upon substitution of H for D in organometallic complexes can be observed directly by ^1H NMR. The classic example of this in organometallic chemistry is a study of an Os_3 cluster reported by Shapley and co-workers [20]. The solid-state structure of $Os_3(CO)_{10}(\mu\text{-}H)_2(\mu\text{-}CH_2)$ was determined by neutron diffraction to have two bridging hydrides and a bridging methylene group. ^1H NMR spin saturation transfer experiments indicated that in solution, the methylene dihydride was in equilibrium with another isomer that had a bridging methyl group and only one bridging hydride. The ^1H NMR spectra of a partially deuterated sample of this compound gave convincing evidence for asymmetry in the bridging methyl group. Normally, geminal substitution of D for H causes only a small shift (about 0.01 ppm) in the ^{1}H NMR spectrum [21]. In this case, however, large separations between the resonances of the CH_2D and CHD_2 groups were observed, with the separation increasing at lower temperatures.

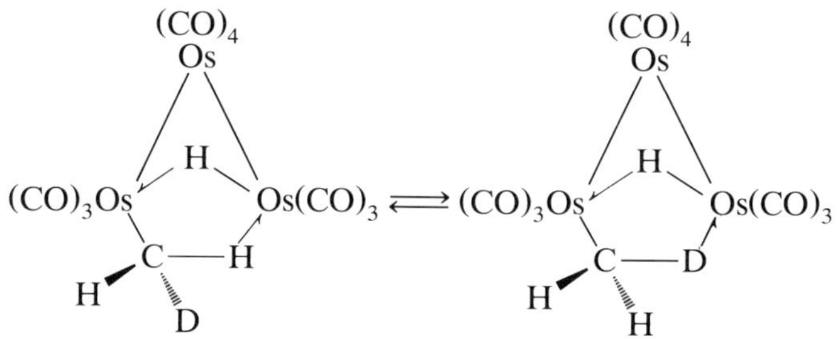

$$(\Delta H° = 0.13 \text{ kcal mol}^{-1}) \quad (8.6)$$

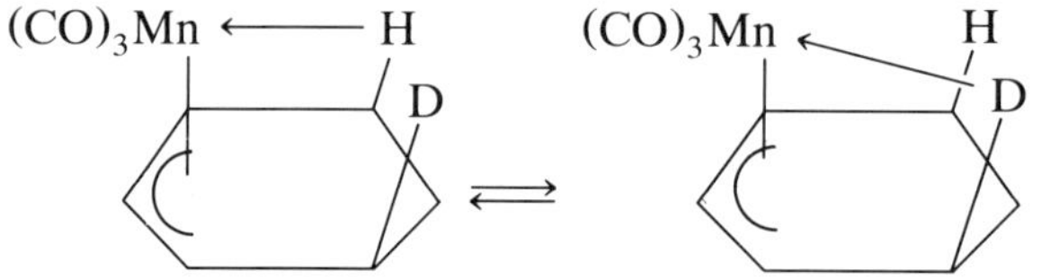

$$(\Delta H° = 0.22 \text{ kcal mol}^{-1}) \quad (8.7)$$

For example, at 35°C, the peak due to the CHD_2 appeared at 0.39 ppm higher field than the peak due to the CH_2D; this separation increased to 0.68 ppm at −76°C. This unusual effect can be explained in terms of the asymmetry in the bridging partially deuterated methyl group. As shown in eq. (8.6), distinct species result, depending on whether the D occupies a terminal position in the methyl group or a bridging position. The potential-energy well is comparatively more shallow for the bridging C—H(D)····M position, compared to the terminal C—H(D) position. Consequently, the difference in zero-point energy (the amount of stabilization of C—D compared to C—H) is smaller for the bridging position compared to the terminal position. Thus, the lowest-energy arrangement results when C—D is in the terminal position and C—H is in the bridging position. Furthermore, the H in the bridging position resonates at a higher field in the ^{1}H NMR spectrum (it is more like a metal hydride and less like an alkyl group compared to the terminal C—H), accounting for the upfield shift of CH_2D compared to CH_3. The preference for occupation of the bridging site by H instead of D becomes even more pronounced at lower temperatures, accounting for the increased differences in chemical shifts at low temperature. Evaluation of the temperature dependence of the equilibrium constants showed that the H-bridged form was favored over the D-bridged form by 0.13 kcal mol^{-1}. A somewhat greater equilibrium isotope effect ($\Delta H° = 0.22$ kcal mol^{-1}) was found by Brookhart and co-workers [22] for the preference of H compared to D bridging in the agostic manganese complex shown in eq. (8.7). Several examples of other complexes containing agostic bridging groups have been shown to exhibit a preference of bridging C—H and terminal C—D, as determined by similar NMR experiments [23].

8.4. HYDROGEN ATOM TRANSFER FROM METAL HYDRIDES

Homolytic cleavage of an M—H bond resulting in hydrogen atom transfer is an important reactivity pattern exhibited by many metal hydrides. Studies of this reaction illustrate some of the general principles that are applicable to an interpretation of isotope effects (particularly inverse isotope effects) discussed later for other systems. A well-studied class of reactions involving hydrogen atom transfer is the hydrogenation of C=C bonds of substituted styrenes, anthracenes, allenes, and dienes by metal hydrides. The classic study by Sweany and Halpern [24] of the hydrogenation of α-methylstyrene by $HMn(CO)_5$ [eq. (8.8)] demonstrated that this stoichiometric hydrogenation involved sequential hydrogen atom transfers to the organic substrate.

$$H_2C{=}C(Ph)(CH_3) + 2HMn(CO)_5 \longrightarrow HC(Ph)(CH_3)(CH_3) + Mn_2(CO)_{10} \quad (8.8)$$

Initial hydrogen atom transfer from the metal hydride to the alkene generates a carbon-centered radical, which then forms the hydrogenated product by abstracting a hydrogen atom from a second equivalent of the metal hydride.

Table 8.2 shows the inverse isotope effects found for eq. (8.8) and other hydrogenation reactions that proceed by sequential hydrogen atom transfers from metal hydrides to unsaturated substrates. These hydrogen atom transfer reactions will be discussed in terms of the generalized mechanism of Scheme 8.1, illustrated for α-methylstyrene as a typical substrate. For the mechanism shown in Scheme 8.1 (using [C=C] to represent the concentration of the unsaturated substrate to be hydrogenated by MH), the rate law is $-d[C{=}C]/dt = k_{obs}[C{=}C]$, with k_{obs} given by

$$k_{obs} = \frac{k_1 k_2[MH]^2}{k_{-1}[M\cdot] + k_2[MH]} \quad (8.9)$$

When the two terms in the denominator, $k_2[MH]$ and $k_{-1}[M\cdot]$, are of similar

Table 8.2
Kinetic Isotope Effects in Hydrogenation of Unsaturated Substrates by Hydrogen Atom Transfer

Substrate + Metal Hydride ⟶ Major Product(s)	k_{MH}/k_{MD}	T (°C)	Ref.
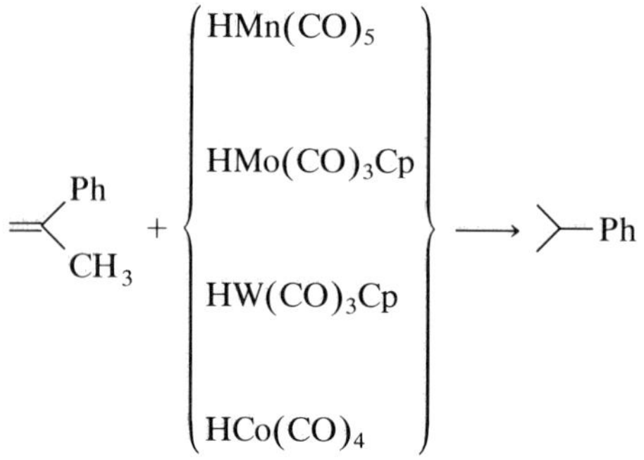 α-methylstyrene (Ph, CH_3) + $HMn(CO)_5$ ⟶ cumene (Ph)	0.4	65	[24]
α-methylstyrene (Ph, CH_3) + $HMo(CO)_3Cp$ ⟶ cumene (Ph)	0.48	65	[25]
α-methylstyrene (Ph, CH_3) + $HW(CO)_3Cp$ ⟶ cumene (Ph)	0.65	100	[25]
α-methylstyrene (Ph, CH_3) + $HCo(CO)_4$ ⟶ cumene (Ph)	0.68	0	[26]
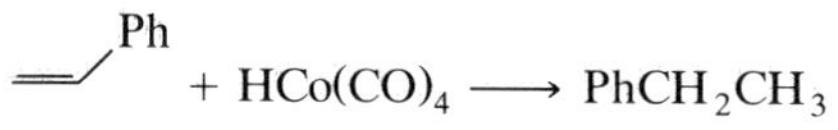 styrene (Ph) + $HCo(CO)_4$ ⟶ $PhCH_2CH_3$	0.45	0	[27]
 α-cyclopropylstyrene (Ph) + $HMn(CO)_5$ ⟶ Ph-substituted isopropylcyclopropane	0.4	25	[28]
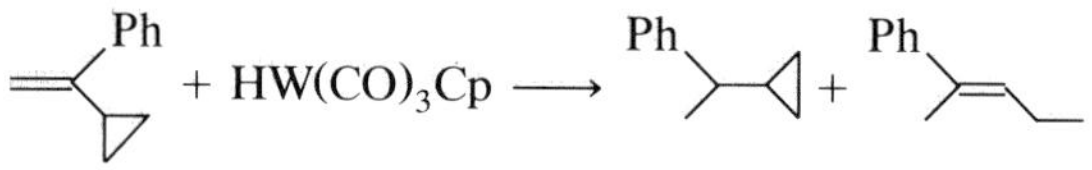 α-cyclopropylstyrene (Ph) + $HW(CO)_3Cp$ ⟶ Ph-substituted isopropylcyclopropane + Ph-substituted ring-opened alkene	0.55	100	[29]
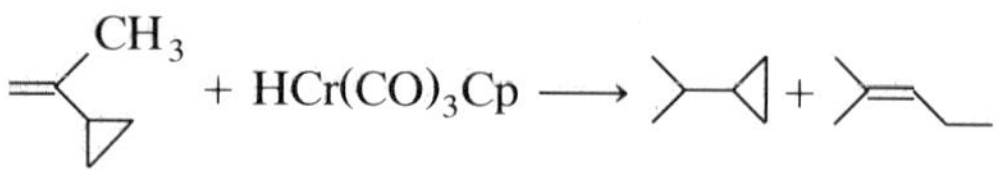 2-cyclopropylpropene (CH_3) + $HCr(CO)_3Cp$ ⟶ isopropylcyclopropane + ring-opened alkene	0.45	68	[29]
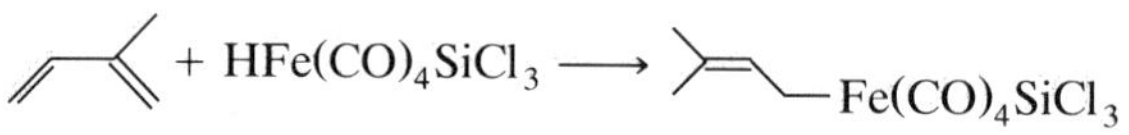 isoprene + $HFe(CO)_4SiCl_3$ ⟶ allyl–$Fe(CO)_4SiCl_3$	0.71	20	[30]
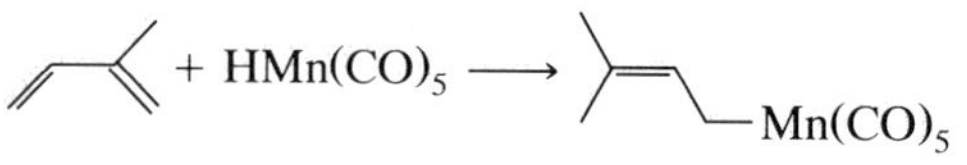 isoprene + $HMn(CO)_5$ ⟶ allyl–$Mn(CO)_5$	0.57	55	[31]
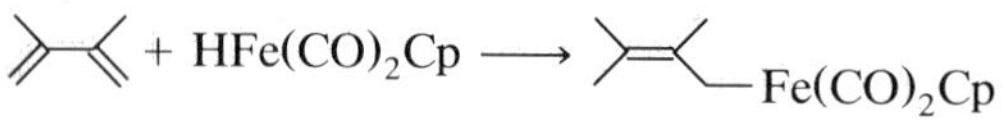 2,3-dimethylbutadiene + $HFe(CO)_2Cp$ ⟶ allyl–$Fe(CO)_2Cp$	0.86	21	[32]
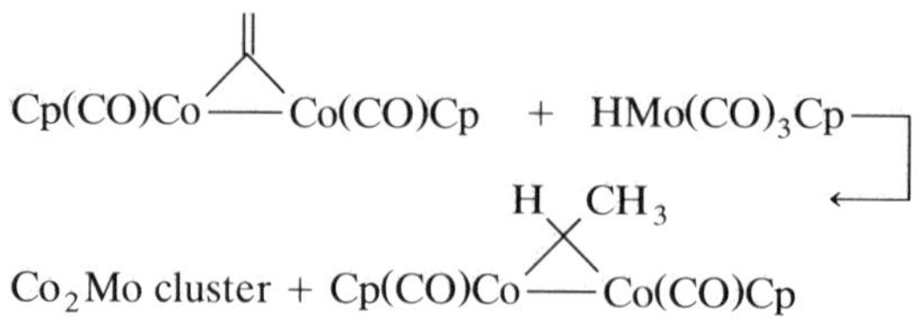 $Cp(CO)Co$–(μ-$C{=}CH_2$)–$Co(CO)Cp$ + $HMo(CO)_3Cp$ ⟶ Co_2Mo cluster + $Cp(CO)Co$–(μ-C(H)(CH_3))–$Co(CO)Cp$	0.63	60	[33]

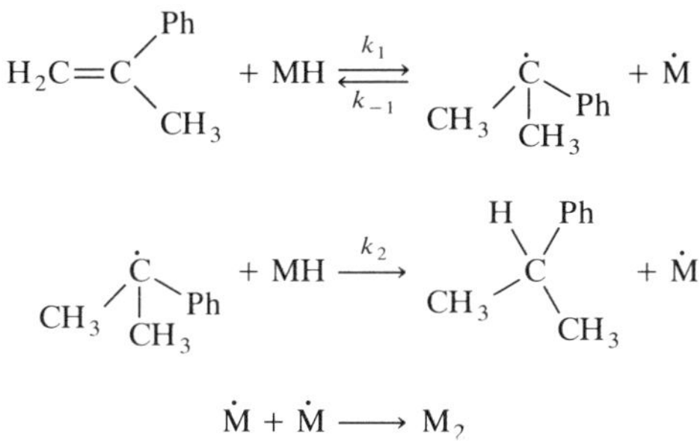

$$H_2C{=}C(Ph)(CH_3) + MH \underset{k_{-1}}{\overset{k_1}{\rightleftharpoons}} (CH_3)_2\dot{C}Ph + \dot{M}$$

$$(CH_3)_2\dot{C}Ph + MH \xrightarrow{k_2} (CH_3)_2CHPh + \dot{M}$$

$$\dot{M} + \dot{M} \longrightarrow M_2$$

Scheme 8.1

magnitude, then the rigorous expression for k_{obs} must be used. However, it is usually the case that $k_2[MH] \gg k_{-1}[M\cdot]$ and, when this is true, eq. (8.9) simplifies to $k_{obs} = k_1[MH]$. The rate law then becomes $-d[C{=}C]/dt = k_1[C{=}C][MH]$. This situation is applicable to all of the free-radical reactions listed in Table 8.2, and second-order kinetics were observed in each of these cases.

The possibility of inverse isotope effects had been predicted [3, 34, 35] prior to their observation in reactions of transition metal hydrides. Melander [35] discussed the influence of transition-state symmetry on isotope effects and said that "there seems to be nothing preventing the equilibrium isotope effect from being in either direction. Thus, indeed, a very product-like transition state could give rise to an inverse kinetic isotope effect provided that the equilibrium effect is sufficiently strong and in the proper direction. In general, this would require that a strongly endothermic reaction leads to a product in which the frequencies concerned with the atom transferred are higher than those of the reactant." The transfer of a hydrogen atom from a metal hydride to an alkene apparently fulfills these specific conditions, because it is a reaction in which an M—H bond ($\nu_{MH} \sim 2000\ cm^{-1}$) is broken and a C—H bond ($\nu_{CH} \sim 3000\ cm^{-1}$) is formed. Although the C—H bond being formed is significantly stronger than the M—H bond being ruptured, this first hydrogen atom transfer is nevertheless endothermic, because a C=C double bond is converted into a C—C single bond. The second hydrogen atom transfer (from the metal hydride to the carbon-centered radical) is exothermic, thus providing (along with the formation of the M—M bond in the organometallic product) the driving force that makes the overall hydrogenation reaction favorable.

The first hydrogen atom transfer (k_1) shown in Scheme 8.1 is not an elementary step but actually involves reversible formation of a caged radical pair as shown in Scheme 8.2 (where k_{1A}, k_{-1A}, and k_{1B} in Scheme 8.2 are the elementary steps comprising the overall k_1 process of Scheme 8.1). The

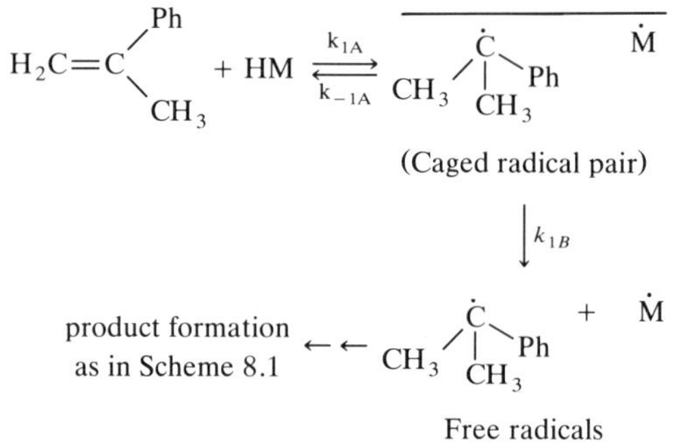

$$H_2C{=}C(Ph)(CH_3) + HM \underset{k_{-1A}}{\overset{k_{1A}}{\rightleftharpoons}} \overline{CH_3\dot{C}(CH_3)Ph \quad \dot{M}}$$

(Caged radical pair)

$$\downarrow k_{1B}$$

product formation as in Scheme 8.1 $\leftarrow\leftarrow$ $CH_3\dot{C}(CH_3)Ph + \dot{M}$

Free radicals

Scheme 8.2

reversibility of the caged radical pair formation raises questions regarding whether these observed inverse isotope effects are unequivocally indicative of inverse kinetic isotope effects for elementary, single-step reactions. In several cases discussed later in this chapter, inverse isotope effects have been interpreted in terms of an inverse thermodynamic isotope effect on a pre-equilibrium that occurs prior to the rate-determining step of the reaction. The observation of an inverse equilibrium isotope effect ($K_H/K_D < 1$) does *not* require that the isotope effect for either of the individual elementary steps (k_f and k_r for the forward and reverse elementary steps, $K_{eq} = k_f/k_r$) be inverse. An inverse equilibrium isotope effect can result when the isotope effect on each elementary step is normal, as long as the magnitude of the isotope effect for the reverse step (k_r) is greater than that for the forward step (k_f). Note that when $k_r^H/k_r^D > k_f^H/k_f^D > 1$, then $K_H/K_D < 1$. In a few cases discussed later [eqs. (8.35) and (8.38)], separate determinations of normal kinetic isotope effects for the forward and reverse elementary steps have been reported for equilibria exhibiting inverse equilibrium isotope effects.

Scheme 8.3 shows a qualitative energy level diagram for the hydrogen atom transfer reaction under consideration here. The difference in zero-point energy between M—H and M—D is less than that between C—H and C—D ($\Delta_{MH/MD} < \Delta_{CH/CD}$ in Scheme 8.3). An inverse *equilibrium* isotope effect therefore can result as a consequence of the fact that $\Delta G^\circ_H > \Delta G^\circ_D$. If $k_{-1A} \gg k_{1B}$, then the equilibrium that forms the caged radical pair would be established. In this case the isotope effect would be properly interpreted as being reflective of an equilibrium isotope effect in which the equilibrium constant for generation of the radical pair is more favorable for D than for H. Although this interpretation may indeed be appropriate for many of the reactions shown in Table 8.2, a complication arises from the difficulty of knowing with certainty whether the k_{1A}/k_{-1A} equilibrium is actually established or not.

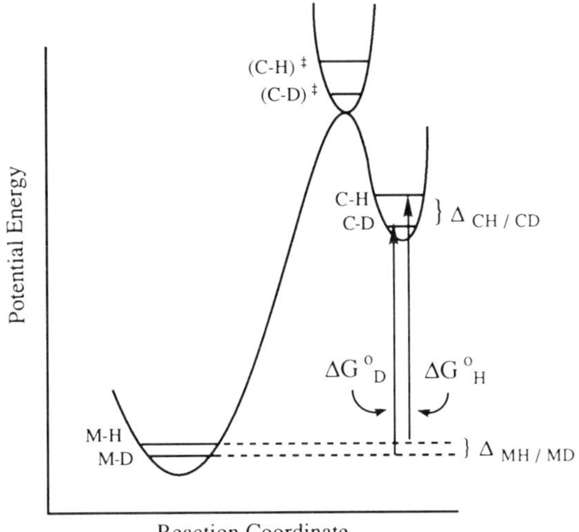

M-H + >C=C< ⇌ M• + >C•–C(H)<

Scheme 8.3

For several of the examples listed in Table 8.2, the reversibility of the initial hydrogen atom transfer was confirmed by incorporation of deuterium into recovered alkene (and concomitant conversion of metal deuterides to metal hydrides) in experiments using metal deuterides. The observation of this isotopic exchange unambiguously indicates that the reaction is reversible but does not always guarantee that the equilibrium is established. Jacobsen and Bergman [33] found that H–D exchange between $DMo(CO)_3(C_5H_5)$ and a Co_2-vinylidene complex (see final entry of Table 8.2) is extremely rapid, occurring over 100 times faster than the hydrogenation reaction. This very rapid H–D exchange provided evidence that the preequilibrium formation of caged radicals was established prior to the rate-determining step. The inverse isotope effect was interpreted as being controlled by an inverse equilibrium isotope effect in this case. Furthermore, they suggested that in the reaction of $DMo(CO)_3(C_5H_5)$ and the Co_2-vinylidene complex, the hydrogen atom back-transfer (analogous to k_{-1A} in Scheme 8.2) could occur with rates greater than 10^{13} s^{-1}. Although it is clear that methyl group rotation occurs faster than H–D exchange in this particular example, this might not always be the case. The possibility of an extremely large rate constant for this hydrogen atom back-transfer means that, in other cases shown in Table 8.2, the approximate time scale for this type of hydrogen atom transfer (from carbon to metal radical) may be roughly similar to the time scale for rotation of a methyl group. This is significant because it raises the possibility that the reversible hydrogen atom transfer could conceivably occur so fast in some cases that it might not be accompanied by detectable H–D exchange. In other words, if transfer of D· from MD to an unsaturated substrate to give a CH_2D group could be reversed faster than the CH_2D group could rotate, then the k_{1A}/k_{-1A} equilibrium might, in fact, be established, but it would not be evident, due to a lack of extensive H–D exchange in ^1H NMR experiments. These considerations accentuate the difficulty in reaching a definitive conclusion regarding whether an observed inverse isotope effect is properly interpreted as being due to a direct kinetic effect or to an equilibrium effect. As depicted in Scheme 8.3, an inverse isotope effect on a preequilibrium results from $\Delta G^\circ_D < \Delta G^\circ_H$. However, if the transition state for the first hydrogen atom transfer is product-like, the late transition state will have a zero-point-energy stabilization [$(C—D)^\ddagger$ vs. $(C—H)^\ddagger$ as shown in Scheme 8.3] that is similar to that of the product. In this case the inverse isotope effect will be caused by $\Delta G^\ddagger_D < \Delta G^\ddagger_H$. Although the details of interpretation differ somewhat for these two cases, the important point is that an inverse isotope effect will be observed. Furthermore, the fact that an inverse isotope effect can (and almost invariably does) result from hydrogen atom transfer reactions from metal hydrides to alkenes and related substrates, makes their observation quite valuable in mechanistic studies.

Orchin and co-workers [36] reported isotope effects for hydrogenation of substituted styrenes and substituted fluorenes (Table 8.3). They found that the rate of reaction of $HCo(CO)_4$ with a given substrate was at least two orders of magnitude higher than the analogous reaction with $HMn(CO)_5$. Despite the considerable differences in rates, the isotope effects for hydrogenation of a given substrate by the Co and Mn hydrides are similar. This illustrates another aspect of the utility of isotope effect studies: Because the activation energy for a reaction does not appear in the equation for prediction of isotope effects [eq. (8.1)], a comparison of isotope effects for mechanistically related reactions with different activation energies can be informative (although temperature dependence of the isotope effect could affect comparisons of this type). An interesting aspect of the data in Table 8.3 is a comparison of the isotope effects observed for the reactions of $HCo(CO)_4$ with ethylidenefluorene, benzylidenefluorene, and bifluorenylidene. The interpretation given by Orchin is that the progression from inverse to normal isotope effects was due to a shift in the position of the transition state along the reaction coordinate. The structure of bifluorenylidene has a considerable amount of twist, indicating some radical character in the ground state. The normal isotope effect observed in the reaction of this compound was interpreted in terms of a transition state that resembled the starting materials more than in the case of the other substrates. As noted earlier, the conversion of the C=C double bond of alkenes to the C—C single bond in

Table 8.3
Kinetic Isotope Effects in Hydrogen Atom Transfer Reactions as a Function of Change in Substrate Structure[a]

Substrate	+ Metal Hydride ⟶	Product	k_H/k_D	T (°C)
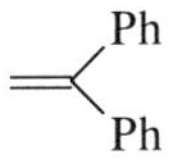 Ph, Ph	+ $HMn(CO)_5$ $HCo(CO)_4$ ⟶	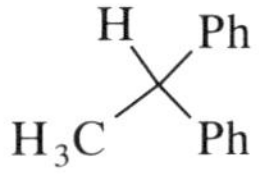 H, Ph, H_3C, Ph	0.58 0.58	31.5 0
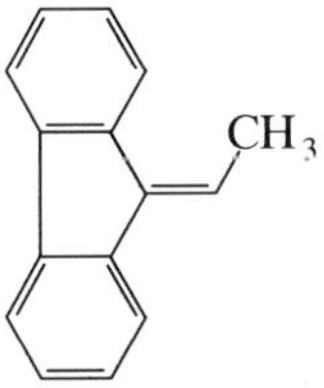 CH_3	+ $IIMn(CO)_5$ $HCo(CO)_4$ ⟶	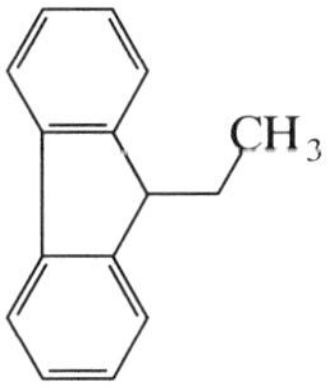 CH_3	0.60 0.43	31.5 −10
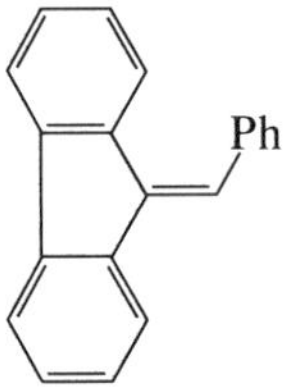 Ph	+ $HCo(CO)_4$ ⟶	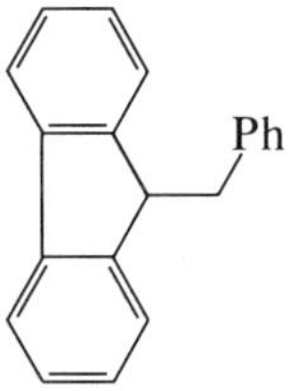 Ph	1.22	0
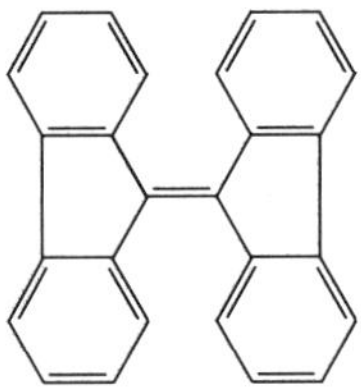	+ $HMn(CO)_5$ $HCo(CO)_4$ ⟶	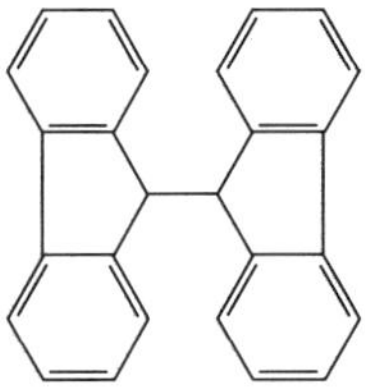	2.39 2.01	31.5 0

[a] Reference 36.

the carbon-centered radical intermediate is largely responsible for making the first hydrogen atom transfer endothermic in hydrogenations of this type. Consequently, the hydrogen atom transfer to bifluorenylidene would be less endothermic than hydrogen atom transfer to ethylidenefluorene. The shift to a more reactant-like transition state was suggested to account for the normal isotope effect.

The reaction of α-cyclopropylstyrene with metal hydrides gives a mixture of two hydrogenation products, due to the reversible ring-opening of the intermediate substituted cyclopropylmethyl radical. By studying the kinetics and product distributions of the reaction of α-cyclopropylstyrene with a series of metal hydrides, Bullock and Samsel [29] determined the relative rates of hydrogen atom transfer from various metal hydrides to α-cyclopropylstyrene and to the unrearranged and rearranged carbon-centered radicals. In the case of the reaction of $HCr(CO)_3(C_5H_5)$ with α-cyclopropylstyrene an expression similar to eq. (8.9) had to be used, because the concentration

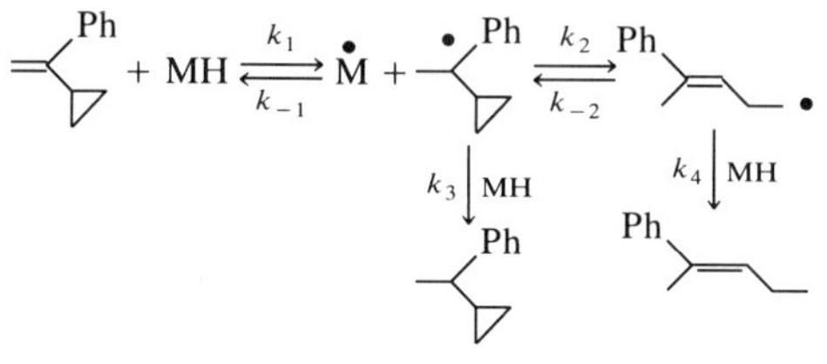

Scheme 8.4

of the organometallic radical $(C_5H_5)(CO)_3Cr\cdot$ was high enough to affect the kinetics. In this case [i.e., Scheme 8.4 for $MH = (C_5H_5)(CO)_3CrH$] an equilibrium constant of $K_{eq} = k_1/k_{-1} \approx 10^{-12}$ was determined. For other metal hydrides [e.g., $HMo(CO)_3(C_5H_5)$, $HW(CO)_3(C_5H_5)$, $HMo(CO)_2(PMe_3)(C_5H_5)$, $Cp^*Fe(CO)_2H$] second-order kinetics were obtained. In these cases the second-order rate constants roughly correlated with published bond-dissociation energies of the M—H bond of the metal hydrides, with the hydrides having the weakest M—H bonds exhibiting the largest rate constants. Comparison of the kinetics and the product ratios for the reaction of α-cyclopropylstyrene with $HW(CO)_3(C_5H_5)$ [vs. $DW(CO)_3(C_5H_5)$] at 100°C gave the isotope effects for k_1, k_3, and k_4 (Scheme 8.4). As for the related examples listed in Table 8.2, the overall reaction rate exhibited an inverse isotope effect ($k_H/k_D = 0.55 \pm 0.1$). The product-forming steps, hydrogen atom transfer from the tungsten hydride (or deuteride) to the carbon-centered radicals, exhibited normal isotope effects ($k_H/k_D = 2.2 \pm 0.2$ for k_3 and $k_H/k_D = 1.8 \pm 0.3$ for k_4), suggesting an early transition state for this exothermic reaction.

Norton and co-workers [37] carried out a detailed study of the kinetics of hydrogen atom transfer from metal hydrides to a stable carbon-centered radical [eq. (8.10)]. The second-order rate constant determined at 25°C for the reaction of this radical with $HMn(CO)_5$ was $k = 6.4 \times 10^2\ M^{-1}\ s^{-1}$.

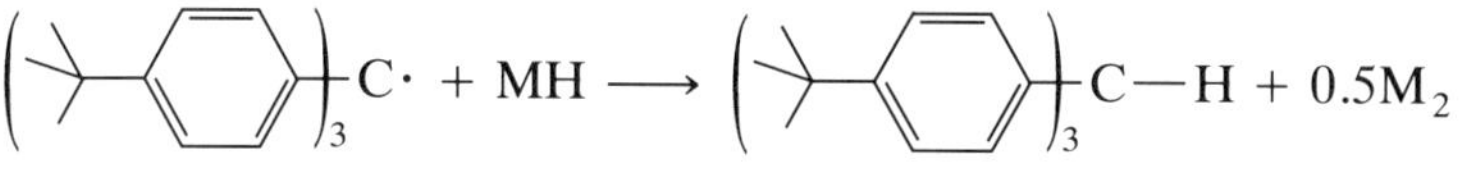

$$(\text{Ar})_3C\cdot + MH \longrightarrow (\text{Ar})_3C\text{—}H + 0.5M_2 \quad (8.10)$$

This relatively small rate constant reflects both the severe steric constraints for the hydrogen atom transfer and the relatively weak C—H bond being formed in eq. (8.10). The expected isotope effect on this reaction can be calculated. If the transition state were very late and resembled the product, then the C—H(D) bond stretching frequency of the product could be used as an estimate of the stretching frequency of the transition state.

$$\frac{k_H}{k_D} = \exp\left\{\frac{-hc}{2k_BT}\left[(\bar{\nu}_{CH} - \bar{\nu}_{CD}) - (\bar{\nu}_{MH} - \bar{\nu}_{MD})\right]\right\} \quad (8.11)$$

Substitution of the known Mn—H(D) and C—H(D) frequencies into eq.

(8.11) leads to a predicted k_H/k_D of 0.55. The kinetic isotope effect expected for a very early transition state was calculated using the same equation, but neglecting the terms involving C—H(D) stretches, because an early transition state would not be influenced by the developing C—H(D) bond but only by the ground state M—H(D) terms. This equation then becomes identical with eq. (8.3), and this calculation predicts $k_H/k_D = 3.3$. The experimentally determined value of $k_{HMn}/k_{DMn} = 2.8$ for eq. (8.10) is lower than the theoretical maximum, but still consistent with a relatively early transition state for the hydrogen atom transfer.

The temperature dependence of the kinetic isotope effect has been examined for eq. (8.10) with $HMn(CO)_5$. The isotope effect was found to decrease with increasing temperature, with the resulting difference in Arrhenius activation energy being $[\Delta E_a]^{H,D} = 1.62 \pm 0.09$ kcal mol^{-1} [37a]. This may indicate some tunnelling in this hydrogen atom transfer reaction, because the observed difference in activation energy is significantly higher than that expected based on the differences in zero-point energies of H—Mn versus D—Mn.

8.5. FORMATION OF METAL HYDRIDES BY HYDROGEN ATOM TRANSFER REACTIONS

Although the reactions discussed previously involve hydrogen atom transfer *from* metal hydrides, hydrogen atom transfer reactions can also be utilized to form transition metal hydrides. Gray and co-workers [38] carried out extensive photochemical studies of the "platinum pop" complex, tetrakis(pyrophosphito)diplatinate(II), $[Pt_2(P_2O_5H_2)_4]^{4-}$. The $d\sigma^*p\sigma$ triplet excited state of this complex abstracts hydrogen atoms from Et_3SiH, Ph_3GeH, n-Bu_3SnH, and other hydrogen-atom donors to give a Pt_2H_2 complex. The isotope effect for hydrogen atom transfer from n-Bu_3SnH (vs. n-Bu_3SnD) was $k_H/k_D = 1.7$ at 25°C. A larger isotope effect ($k_H/k_D = 4.7$) was found for hydrogen atom transfer from the α-(C—H) bond of $PhCH(OH)CH_3$.

8.6. INSERTION OF ALKENES INTO M—H BONDS

The insertion of an alkene into an M—H bond is a critical step in many catalytic reactions, including homogeneous hydrogenation and hydroformylation of alkenes. Direct study of this reaction is difficult because the actual insertion step is, in many cases, reversible and fast compared to other steps in the overall reaction. Doherty and Bercaw [39] examined the insertion of alkenes into the Nb—H bond of some permethylniobocene derivatives. The reaction of $Cp^*{}_2NbD_3$ with ethylene-d_4 resulted in some incorporation of D into the Cp* groups (and corresponding depletion of deuteration in the ethylene ligand). Despite this complication, it was determined that the thermodynamic isotope effect for the partially labelled system favored D in the ethylene and H in the Nb—H site [$\Delta E = 0.35 \pm 0.1$ kcal mol^{-1} for eq. (8.12)], as expected in view of the relative zero-point energies.

$$Cp^*_2Nb(H)(CX_2{=}CD_2) \rightleftharpoons Cp^*_2Nb(D)(CX_2{=}CHD) \quad (X = H \text{ or } D) \qquad (8.12)$$

The $Cp^*_2NbCH_2CH_3$ intermediate resulting from insertion of ethylene into the Nb—H bond could be trapped by reaction with CO or MeNC. A direct determination of the forward rate of this insertion was made by magnetization transfer techniques. A kinetic isotope effect of $k_H/k_D = 1.1 \pm 0.4$ was determined for the conversion of $Cp^*_2Nb(H)(CH_2{=}CH_2)$ to $Cp^*_2NbCH_2CH_3$ [vs. $Cp^*_2Np(D)(CD_2{=}CD_2)$]. It was noted that this isotope effect gives little detailed information about the transition state because it is small and represents the combined influence of the primary isotope effect and four secondary isotope effects.

Halpern, Okamoto, and Zakhariev [40] pointed out the difficulty of obtaining definitive mechanistic conclusions from kinetic studies of the overall reaction rate of complicated multistep catalytic reactions such as the hydrogenation of alkenes by Wilkinson's catalyst, $RhCl(PPh_3)_3$. However, the kinetics of individual steps (when they are amenable to study) can be informative. A kinetic isotope effect of $k_H/k_D = 1.15$ at 25°C was obtained in a study of the hydrogenation of cyclohexene by $RhH_2Cl(PPh_3)_3$ (vs. RhD_2). The rate-determining step was proposed to be the insertion of the alkene into the Rh—H bond [eq. (8.13)].

$$RhH_2Cl(PPh_3)_2(C_6H_{10}) \longrightarrow RhH(C_6H_{11})Cl(PPh_3)_2 \qquad (8.13)$$

8.7. β-HYDRIDE ELIMINATION FROM METAL ALKYL COMPLEXES

Many metal alkyl complexes with β-hydrogens decompose by β-hydride elimination to give metal hydrides and olefins. This process is the microscopic reverse of the insertion of an olefin into a metal–hydrogen bond, and both of these processes are fundamentally important reactions in organometallic chemistry. Metal hydrides are not always the ultimate organometallic product observed or isolated from the reaction, but studies of these reactions have provided insights regarding the forming and breaking of the metal–hydrogen bond.

Schwartz et al. [41] reported an isotope effect of $k_H/k_D = 2.28 \pm 0.20$ for β-hydride elimination from $(PPh_3)_2(CO)Ir(n\text{-octyl})$. This was done by determining the amount of 1-octene and 1-octene-d_1 formed in the decomposition of $(PPh_3)_2(CO)IrCH_2CHD(n\text{-}C_6H_{13})$. No scrambling of the deuterium labels was observed. A similar kinetic isotope effect of $k_H/k_D = 2.30 \pm 0.05$ at 15°C was reported by Ikariya and Yamamoto [42] for the thermal decomposition of $(PMe_2Ph)_2(acac)Co(CH_2CH_3)_2$ [vs. $Co(CH_2CD_3)_2$]. In both of these cases, β-hydride elimination was proposed as the rate-determining step.

Bercaw et al. [43] carried out a thorough study of the insertion of ethylene (and alkynes) into the Sc—C bond of $(C_5Me_5)_2Sc{-}R$ complexes. An isotope effect of $k_H/k_D = 2.0 \pm 0.3$ was found at room temperature for β-hydride elimination from $(C_5Me_5)_2ScCH_2CHDPh$. These experiments were carried out using a large excess of 2-butyne to insert into the Sc—H bond of $(C_5Me_5)_2ScH$, which was the initial organometallic product of the β-hydride elimination. Thus the reinsertion of styrene into the Sc—H bond was precluded, so that the observed rate constant was that of β-hydride elimination, with the k_H/k_D being determined from the ratio of styrene to styrene-d_1.

Whitesides et al. [44] carried out detailed studies on decomposition mechanisms of platinum alkyl complexes. Comparison of the rates of decomposition of *cis*-$(PEt_3)_2Pt(CH_2CH_3)_2$ versus $(PEt_3)_2Pt(CD_2CD_3)_2$ gave a negligible isotope effect of $k_H/k_D = 1.0 \pm 0.3$ at 118°C. This indicated that the C—H bond was not being broken in the rate-determining step and suggested phosphine dissociation (k_1 in Scheme 8.5) as the rate-determining step. Experiments carried out in the presence of 0.3 M added PEt_3 not only resulted in a much slower rate, but a significant isotope was observed as well ($k_H/k_D = 3.3 \pm 0.5$ at 131°C). Experiments on $(PEt_3)_2Pt(CH_2CD_3)_2$ showed extensive scrambling of isotopic labels in the recovered starting material [eq. (8.14)], indicating that β-hydride elimination was reversible and was not the

$$Et_3P{-}Pt(CH_2CH_3)_2(PEt_3) \underset{k_{-1}}{\overset{k_1}{\rightleftharpoons}} PEt_3 + \left[Pt(CH_2CH_3)_2(PEt_3)\right] \underset{k_{-2}}{\overset{k_2}{\rightleftharpoons}} Et_3P{-}Pt(H)(CH_2CH_3)(H_2C{=}CH_2)$$

$$Et_3P{-}Pt(H)(CH_2CH_3)(H_2C{=}CH_2) \underset{k_{-3}}{\overset{k_3}{\rightleftharpoons}} \left[Et_3P{-}Pt{-}\|(CH_2{=}CH_2)\right] + CH_3CH_3$$

$$\left[Et_3P{-}Pt{-}\|(CH_2{=}CH_2)\right] \xrightarrow{PEt_3} (PEt_3)_2Pt{-}\|(CH_2{=}CH_2)$$

Scheme 8.5

rate-determining step.

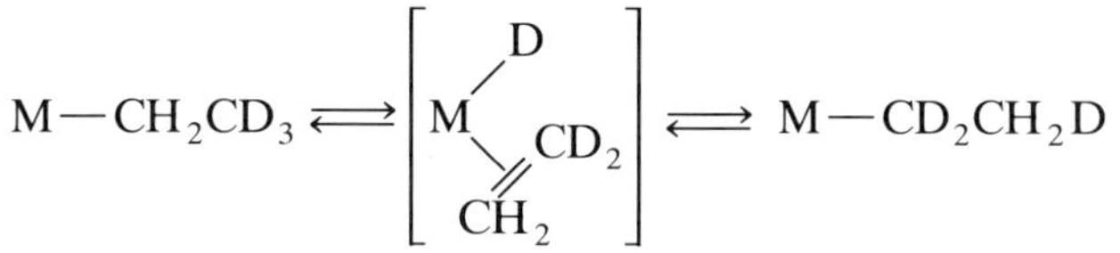

$$\mathrm{M{-}CH_2CD_3} \rightleftharpoons \left[\begin{array}{c} \mathrm{D} \\ \mathrm{M} \quad \mathrm{CD_2} \\ \mathrm{CH_2} \end{array}\right] \rightleftharpoons \mathrm{M{-}CD_2CH_2D} \qquad (8.14)$$

Phosphine dissociation was reversible in the presence of 0.3 M added PEt_3, and the rate-determining step under these conditions was proposed to be reductive elimination of ethane from the *monophosphine* intermediate, $(PEt_3)Pt(H)(Et)(CH_2{=}CH_2)$ (k_3 in Scheme 8.5).

A kinetic isotope effect of 1.7 ± 0.3 was observed for experiments carried out at 157°C in the presence of 1.64 M added PEt_3. This indicated that a third mechanism was operative, and it was proposed that the transition state involved two phosphines on Pt, because the kinetic isotope effect did not change appreciably when the added PEt_3 concentration was varied from 1.64 to 6.9 M. The data obtained at the higher phosphine concentrations did not distinguish unambiguously between two possible transition states for the *diphosphine* intermediate. One possibility is that reductive elimination of ethane from the intermediate, $(PEt_3)_2Pt(H)(Et)(CH_2{=}CH_2)$, was rate-determining (upper pathway, Scheme 8.6). The difference in the kinetic isotope effect observed for the rate-determining loss of ethane from the diphosphine intermediate (Scheme 8.6) compared to the monophosphine intermediate (Scheme 8.5) was rationalized on the basis of an early transition state for the former; that is, an earlier transition state for M—H bond cleavage exhibiting a lower isotope effect in comparison to a later transition state where the M—H bond cleavage was more complete. The other possibility is that the rate-determining step was loss of ethylene, shown as the lower pathway in Scheme 8.6. [This would give the organometallic intermediate $(PEt_3)_2Pt(H)Et$, which would then undergo rapid elimination of ethane to produce $[(PEt_3)_2Pt]$, which would form the observed product by reassociation of ethylene.] Because rate-determining loss of ethylene would involve no M—H or C—H

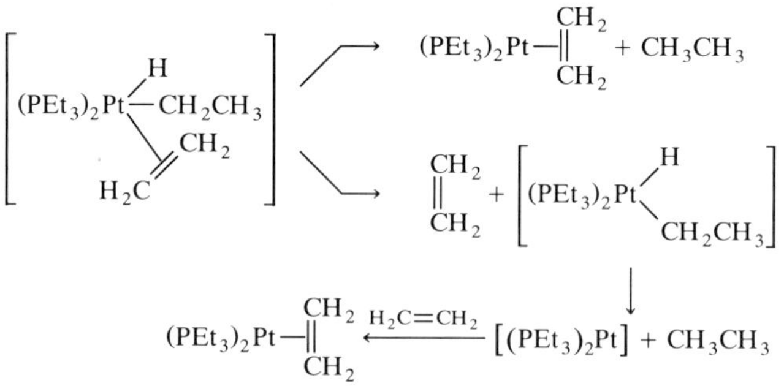

$$\left[(PEt_3)_2Pt(H)(CH_2CH_3)(H_2C{=}CH_2)\right] \longrightarrow (PEt_3)_2Pt(CH_2{=}CH_2) + CH_3CH_3$$

$$\left[(PEt_3)_2Pt(H)(CH_2CH_3)(H_2C{=}CH_2)\right] \longrightarrow CH_2{=}CH_2 + \left[(PEt_3)_2Pt(H)(CH_2CH_3)\right]$$

$$\left[(PEt_3)_2Pt(H)(CH_2CH_3)\right] \longrightarrow [(PEt_3)_2Pt] + CH_3CH_3$$

$$(PEt_3)_2Pt(CH_2{=}CH_2) \xleftarrow{H_2C=CH_2} [(PEt_3)_2Pt] + CH_3CH_3$$

Scheme 8.6

bonds being made or broken, this interpretation would be rationalized on the basis of a combination of secondary isotope effects on the rate-determining step, coupled with preequilibrium isotope effects prior to the rate-determining step.

Yamamoto and co-workers [45] compared the rates of decomposition of *trans*-$(PMePh_2)_2Pd(CH_2CH_3)_2$ with that of the β-deuterium substituted analog $L_2Pd(CH_2CD_3)_2$. A kinetic isotope effect of 1.4 ± 0.1 at 27°C was observed for this Pd system. There are several significant differences between the decompositions of the Pt and Pd alkyl complexes. In contrast to the reaction of $(PEt_3)_2Pt(CH_2CD_3)_2$, no scrambling of the isotopic labels was observed in $(PMePh_2)_2Pd(CH_2CD_3)_2$ recovered after partial reaction, indicating that β-hydride elimination was irreversible in the Pd alkyls. In the case of the Pd complexes, only a minor retardation was observed in the presence of added phosphine, and it was concluded that most of the reaction occurred from a diphosphine intermediate, even in the absence of added phosphine. The activation energy for the β-elimination is much higher for the Pt reaction compared to Pd reaction. It was proposed that the transition state involved a five-coordinate diphosphine $L_2Pd(H)(CH_2CH_3)(CH_2{=}CH_2)$ intermediate. This proposed intermediate is analogous to the $(PEt_3)_2Pt(H)(Et)(CH_2{=}CH_2)$ intermediate suggested by Whitesides and co-workers for decomposition of their *cis*-L_2PtR_2 alkyl complexes in the presence of high concentrations of added phosphine, and the observed isotope effects for the Pt and Pd reactions are similar. The differences in stereochemistry of the starting materials (trans for Pd but cis for Pt) do not preclude similar geometries in the proposed five-coordinate intermediates.

In contrast to the normal isotope effects observed for β-hydride eliminations in the Pt and Pd complexes discussed previously, Yamamoto et al. [46] found no appreciable isotope effect ($k_H/k_D = 1.01$ at 100°C) in the thermal decomposition of $(PPh_3)_2Pt(CH_2CH_3)(n\text{-Pr})$ versus $(PPh_3)_2Pt(CH_2CD_3)(n\text{-Pr})$. It was suggested that olefin loss was the rate-determining step in the mechanism. The reason for the substantial difference in isotope effect observed for $(PPh_3)_2Pt(CH_2CH_3)(n\text{-Pr})$ compared to the related decomposition of $(PEt_3)_2PtEt_2$ is not clear.

Kinetic and equilibrium isotope effects have been reported [47] for the decomposition of $(PEt_3)_2PtClCH_2CH_3$ [vs. $(PEt_3)_2PtCl(CD_2CD_3)$] [eq. (8.15)]. The kinetic isotope effect ($k_H/k_D = 2.5 \pm 0.2$ at 158°C) was similar to the equilibrium isotope effect ($K_H/K_D = 1.9 \pm 0.4$ at 158°C). In this reaction, the stretching vibrations of the product ethylene ($\nu_{CH} = 3080, 2972$ cm^{-1}) have slightly higher frequencies than those of the initial Pt—C_2H_5 complex ($\nu_{CH} = 2951$–2857 cm^{-1}).

$$\mathrm{Cl{-}\overset{\displaystyle PEt_3}{\underset{\displaystyle PEt_3}{Pt}}{-}CH_2CH_3 \rightleftharpoons Cl{-}\overset{\displaystyle PEt_3}{\underset{\displaystyle PEt_3}{Pt}}{-}H + H_2C{=}CH_2} \tag{8.15}$$

However, the observed normal isotope effect indicates that this effect is overwhelmed by the decrease in frequency from a C—H bond of $(PEt_3)_2PtClCH_2CH_3$ to the significantly lower stretching frequency of the Pt—H ($\nu_{PtH} = 2198$ cm^{-1}) that is formed in the reaction. Brainard and Whitesides [47] called attention to the similarity of their kinetic and equilibrium isotope effects in this reaction and suggested that ethylene loss, rather than β-hydride elimination is responsible for the isotope effect in this reaction (and possibly others as well).

Photolysis of the tungsten alkyl complex $(C_5H_5)(CO)_3WCH_2CH_3$ gives loss of CO and generates $(C_5H_5)(CO)_2WCH_2CH_3$ in which there is an agostic interaction between the β-(C—H) and the W [48]. The lack of an observable isotope effect for the deuterium-labelled $(C_5H_5)(CO)_3WCH_2CD_3$ indicates the C—H(D) bond is not being broken in the rate-determining step. It was proposed that the rate-determining step was cis to trans isomerization of $(C_5H_5)(CO)_2W(CH_2{=}CH_2)H$ [which forms by β-hydride elimination from $(C_5H_5)(CO)_2WCH_2CH_3$]. Cis–trans isomerizations of other metal hydrides exhibited no significant isotope effects; Faller and Anderson [49] found $k_H/k_D = 1.00 \pm 0.05$ for cis–trans isomerization of the molybdenum hydride (and deuteride) $(C_5H_5)(CO)_2(PPh_3)MoH$.

β-Hydride elimination cannot occur from metal alkyls (such as $M{-}CH_2CMe_3$) containing no hydrogens, but other pathways involving metal hydride intermediates have been observed. Whitesides and co-workers [50] observed $k_H/k_D \approx 3$ in kinetic studies of formation of four-, five-, and six-membered platinacycles by thermolysis of various $(PEt_3)_2PtR_2$ species.

Although β-hydride elimination from metal alkyls is common observed, few examples of β-hydride eliminations from metal vinyl complexes have been reported. McDade and Bercaw [51] found that the thermolysis of a bis(propenyl) zirconium complex gives an isolable zirconacyclopentene (Scheme 8.7). Substitution of D on the β-position of the propenyl group resulted in $k_H/k_D = 4.9 \pm 0.5$ at 24°C, whereas substitution at the α-posi-

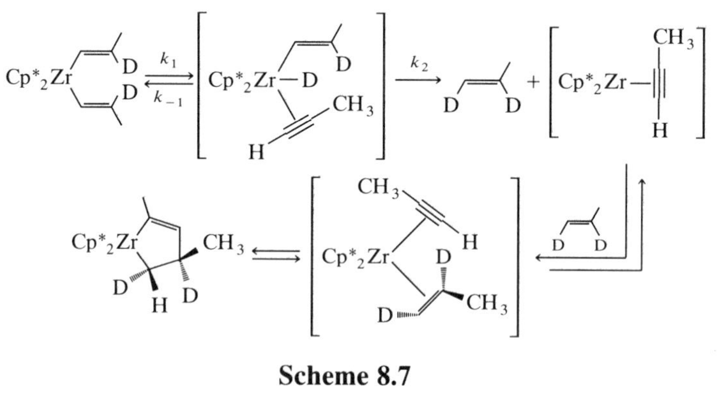

Scheme 8.7

tion gave a negligible effect ($k_H/k_D = 1.03 \pm 0.1$). The isotope effect was interpreted in terms of the combined effects of an equilibrium isotope effect (k_1/k_{-1}) on the β-hydride elimination and the kinetic isotope effect on the slow reductive elimination of propene (k_2 in Scheme 8.7).

8.8. REACTIONS OF METAL ALKYL HYDRIDE COMPLEXES

Reductive elimination of alkane from unobserved alkyl hydride intermediates was proposed as the rate-determining step in some of the mechanistic studies of β-hydride elimination from metal alkyl complexes discussed previously. Alkane eliminations from directly observable (and in many cases isolable) alkyl hydride complexes also have been examined, with isotope effects again playing a key role. As shown in Table 8.4, kinetic isotope effects for the elimination of alkane from three *cis*-$L_2Pt(R)H$ complexes [vs. *cis*-$L_2Pt(R)D$] are normal ($k_H/k_D > 1$).

Norton and co-workers [55] made extensive use of metal hydride isotope effects in a study of the reductive elimination of methane from $Cp_2W(CH_3)H$. At low concentrations (< 1 mM), methane was formed by an intramolecular reductive elimination, as shown by the observation of CH_3D and CD_3H from thermolysis (82.5°C) or photolysis of a solution containing $Cp_2W(CH_3)D$ and $Cp_2W(CD_3)H$ [eq. (8.16)].

$$Cp_2W(H)(CD_3) + Cp_2W(D)(CH_3) \xrightarrow[\text{concentration}]{\text{low}} CD_3H + CH_3D$$

$$\xrightarrow{\text{higher concentration}} CH_4 + CH_3D + CH_2D_2 + CHD_3 + CD_4 \quad (8.16)$$

However, thermolysis of $Cp_2W(CH_3)D$ and $Cp_2W(CD_3)H$ at higher concen-

Table 8.4
Isotope Effects for Reductive Elimination of RH from $L_2Pt(R)H$

$L_2Pt(R)H$	k_H/k_D	T (°C)	Ref.
$(PPh_3)_2Pt(H)CH_3$	3.3	−25	[52]
$(PPh_3)_2Pt(H)CH_2CF_3$	2.2	40	[53]
$(Cy_2PCH_2CH_2PCy_2)Pt(H)CH_2CMe_3$	1.5	69	[54]

trations gave significant amounts of CH_4, CH_2D_2, and CD_4 in addition to CH_3D and CD_3H, indicating the operation of intermolecular processes and an *apparent* intermolecular methane elimination. A combination of experiments, including double-labelling experiments using D and ^{13}C, led to the conclusion that *intra*molecular site exchange of methyl and hydride ligands, *inter*molecular exchange of hydride ligands, and *intra*molecular elimination of methane all were occurring in the $Cp_2W(CH_3)H$ system. These processes are not limited to the $Cp_2W(CH_3)H$ system, and each of these reactions will be discussed.

Separate experiments on $Cp_2W(CD_3)H$ and $Cp_2W(CH_3)D$ indicated that label exchange between the methyl and hydride ligands occurred *prior* to elimination of methane. The equilibrium constant determined for eq. (8.17), $K_{eq} = 1.4 \pm 0.2$, is *less* than the equilibrium constant of 3 (which would be predicted solely on the basis of statistics) and is thus reflective of the operation of an equilibrium isotope effect.

$$Cp_2W\begin{matrix} CD_3 \\ H \end{matrix} \underset{}{\overset{K_{eq}}{\rightleftharpoons}} Cp_2W\begin{matrix} CD_2H \\ D \end{matrix} \qquad (K^{calc} = 1.84,\ K^{expt} = 1.4) \quad (8.17)$$

$$Cp_2W\begin{matrix} CH_3 \\ D \end{matrix} \underset{}{\overset{K_{eq}}{\rightleftharpoons}} Cp_2W\begin{matrix} CH_2D \\ H \end{matrix} \qquad (K^{calc} = 4.88,\ K^{expt} = 4.9) \quad (8.18)$$

The equilibrium is shifted to the left in eq. (8.17), which is the direction allowing D to experience the higher force constant. The expected value of the equilibrium isotope effect can be predicted from eq. (8.19), with the factor of 3 at the start of the equation being the equilibrium constant predicted on the basis of statistics, because there are three ways to make CD_2H. The contributions from bending force constants (which are unknown but presumably small compared to the stretching vibrations) are neglected, and the calculation uses only those terms that arise from zero-point energy differences.

$$K_{eq} = 3\exp\left\{\frac{-hc}{2k_BT}\left[(\bar{\nu}_{WD} + \bar{\nu}_{CH}) - (\bar{\nu}_{WH} + \bar{\nu}_{CD})\right]\right\} \quad (8.19)$$

Using the values of W—H, W—D, C—H, and C—D stretching frequencies obtained from the IR spectra, the predicted equilibrium constant is 1.84, in reasonable agreement with the experimental value of 1.4. A similar treatment for eq. (8.18) accounts for the fact that the experimental value of $K = 4.9$ is *greater* than the statistical value of 3. The experimental value of 4.9 agrees well with the predicted value of 4.88.

Bergman and co-workers [56, 57] observed a similar label scrambling between alkyl and hydride ligands in the iridium and rhodium complexes $(C_5Me_5)(PMe_3)Ir(H)(cyclohexyl)$, $(C_5Me_5)(PMe_3)Ir(H)(CH_3)$,

$(C_5Me_5)(PMe_3)Rh(H)(CH_2CH_3)$, and $(C_5Me_5)(PMe_3)Rh(H)$(2,2-dimethylcyclopropyl). They suggested the reversible formation of a σ-complex, in which the C—H bond of the alkane acts as a ligand to the metal, similar to the three-center–two-electron bonding in agostic M····H—C systems. Migration of the metal from one C—H (or C—D) bond to another in the alkane complex accomplishes the scrambling of labels.

$$[Rh](^{13}CH_2CH_3)(D) \rightleftharpoons [Rh]\text{—}D\text{—}{}^{13}C(H)(H)(CH_3) \rightleftharpoons [Rh]\text{—}H\text{—}{}^{13}C(D)(H)(CH_3) \rightleftharpoons [Rh](^{13}CHDCH_3)(H) \quad (8.20)$$

$$[Rh] = (C_5Me_5)Rh(PMe_3)$$

Similar incorporation of deuterium into the methyl ligand of $Cp^*_2W(CH_3)D$ was observed by Parkin and Bercaw [58]. Jones and Feher [59] suggested an analogous label scrambling to account for unsuccessful attempts to prepare isotopically pure $Cp^*(PMe_3)Rh(D)CH_3$. In the case of the $[Cp_2Re(CD_3)H]^+$, Gould and Heinekey [60] determined an equilibrium constant of K = 1.0 ± 0.1 for eq. (8.21) at −30°C.

$$\left[Cp_2Re\begin{matrix}CD_3\\H\end{matrix}\right]^+ \rightleftharpoons \left[Cp_2Re\begin{matrix}CD_2H\\D\end{matrix}\right]^+ \qquad (K_{eq} = 1.0) \quad (8.21)$$

Because site exchange was much slower than methane elimination in this system, it was possible to measure the kinetics of approach to equilibrium for eq. (8.21), and activation parameters of $\Delta H^\ddagger = 22.3 \pm 1.9$ kcal mol^{-1} and $\Delta S^\ddagger = 24.9 \pm 9$ cal K^{-1} mol^{-1} were reported.

The importance of these σ-complexes lies not only in the way in which their reactivity accomplishes the intramolecular site exchange process, but also in the way by which their intermediacy explains the isotope effects observed in these systems for reductive elimination of alkane. In contrast to the *normal* isotope effects shown in Table 8.4 for the *cis*-L_2PtRH complexes, kinetic isotope effects for alkane elimination from other alkyl hydride complexes are *inverse* ($k_H/k_D < 1$). Table 8.5 lists the inverse isotope effects for alkane elimination from several alkyl hydride complexes. In all of these examples, the isotope effect was determined by comparing the rate of decomposition of the M(R)H complex with the rate of decomposition of the deuterated analog in which D labels were present in the alkyl ligand (R) as well as in place of the hydride. This meant that isotopic scrambling between the alkyl and hydride positions did not deplete the isotopic purity of the

Table 8.5
Isotope Effects for Reductive Elimination of RH from M(R)H

M(R)H	k_H/k_D	T (°C)	Ref.
$(C_5Me_5)(PMe_3)Rh(CH_2CH_3)H$	0.5	−30	[57]
$(C_5Me_5)(PMe_3)Ir(cyclohexyl)H$	0.7	130	[56]
$(C_5H_5)_2W(CH_3)H$	0.75	72.6	[55]
$(C_5Me_5)_2W(CH_3)H$	0.70	100	[58]
$[(C_5H_5)_2Re(CH_3)H]^+$	0.8	9	[60]

deuterated complex. Although these isotope effects are *measured* as kinetic effects on the rates of alkane elimination, they are often *interpreted* as being equilibrium isotope effects resulting from a two-step mechanism for alkane elimination. The formation of the alkane complex occurs in a preequilibrium step that is subject to an inverse isotope effect. That is, the equilibrium constant for formation of the $M(\eta^2\text{-}CD_4)$ intermediate from the $M(CD_3)D$ complex is higher than the equilibrium constant for production of the $M(\eta^2\text{-}CH_4)$ intermediate from $M(CH_3)H$. The interpretation for the higher equilibrium constant for the formation of the deuterated species is analogous to that discussed earlier in connection with inverse isotope effects in hydrogen atom transfer reactions, in which an M—H(D) bond of relatively low stretching frequency is ruptured and a C—H(D) bond of comparatively high stretching frequency is formed. The isotope effect on loss of alkane from these η^2-alkane intermediates [k_2 in eq. (8.22)] is expected to be small, because the C—H bond is already formed in th σ-complex.

Parkin and Bercaw [58] called attention to the apparent dichotomy involving the observed isotope effects in the reductive elimination of alkanes from alkyl hydride complexes (normal isotope effects in Table 8.4 compared to inverse isotope effects in Table 8.5). The normal isotope effects are found in *cis*-$L_2Pt(R)H$ systems where the alkyl hydride complexes are generally thermodynamically unstable (or, in other words, the cases where oxidative addition of C—H bonds is not thermodynamically favorable). In contrast, the inverse isotope effects listed in Table 8.5 are found in cases where the metal–hydrogen and metal–carbon bonds are stronger than for Pt, because in these systems C—H bond activation is thermodynamically favorable. It was proposed [58] that σ-complexes are intermediates in *all* of these reactions, with the critical feature noted by Parkin and Bercaw being the "energy of the transition state for dissociation of alkane relative to that of the transition state for σ-complex formation from the alkyl hydride."

$$M\begin{matrix} \diagup H \\ \diagdown R \end{matrix} \underset{k_{-1}}{\overset{k_1}{\rightleftharpoons}} \left[M - \begin{matrix} R \\ | \\ H \end{matrix} \right] \xrightarrow{k_2} [M] + \begin{matrix} R \\ | \\ H \end{matrix} \qquad (8.22)$$

In terms of eq. (8.22), an inverse isotope effect will be observed in cases where $k_{-1} \gg k_2$, whereas a normal isotope effect on reductive elimination of alkane is expected in cases where $k_{-1} \ll k_2$.

The intermediacy of σ-complexes in the reductive elimination of alkanes from alkyl hydride complexes accounts for the label scrambling between hydride and alkyl ligands as well as the inverse isotope effects observed in the reductive elimination reaction. These intramolecular processes satisfactorily account for the results of the isotopic labelling experiments in all cases except one. An additional complication was present in the $Cp_2W(CH_3)H$ system [55], because the formation of all possible methane isotopomers (CH_nD_{4-n}, $n = 0$–4) in the thermolysis [eq. (8.16)] of high concentrations of $Cp_2W(CH_3)H$ and labelled derivatives meant that some *intermolecular* reaction was taking place in addition to the *intramolecular* processes already discussed. The fact that the methane elimination *appeared* to be intermolecular at the higher concentrations was strongly suggestive of a bimolecular reaction, and exchange between the hydride ligands of different molecules was found to account for the intermolecularity of the reaction. For example, reaction of $(C_5H_5)_2W(H)CH_3$ and $(C_5H_5)_2WD_2$ resulted in deuterium incorporation into the hydride position of the alkyl hydride complex and incorporation of H into the dideuteride complex. Rate constants also were determined for H–D exchange between $(C_5H_5)_2WD_2$ and $(C_5H_5)_2WPhH$, $(CO)_4OsH_2$, and $(C_5H_5)(CO)_3WH$. The combination of all these results now collectively accounts for the fact that alkane elimination from $(C_5H_5)_2W(H)CH_3$ was intramolecular at low concentrations but appeared to be intermolecular at high concentrations. At lower concentrations, elimination of methane from $(C_5H_5)_2W(H)CH_3$ occurred before intermolecular hydride exchange, whereas at higher concentrations hydride exchange took place faster than methane elimination.

8.9. ISOTOPE EFFECTS IN C—H BOND ACTIVATION

The formation of alkyl hydride complexes from the reaction of organometallic complexes with alkanes is the microscopic reverse of the reductive elimination reaction from alkyl hydride complexes that was discussed in the previous section. Several recent studies on activation of C—H bonds by organometallic complexes have made use of isotope effects in experiments that have contributed greatly to an understanding of intermediates in C—H activation. Two key experiments by Jones and Feher [59] involved a determination of the isotope effects involved in the activation of labelled benzene with the intermediate Rh complex generated by photolysis of $(C_5Me_5)(PMe_3)RhH_2$. Competition of this highly reactive intermediate for

C_6D_6 versus C_6H_6 gave a small intermolecular isotope effect of only 1.05 ± 0.06 at 10°C [eq. (8.23)].

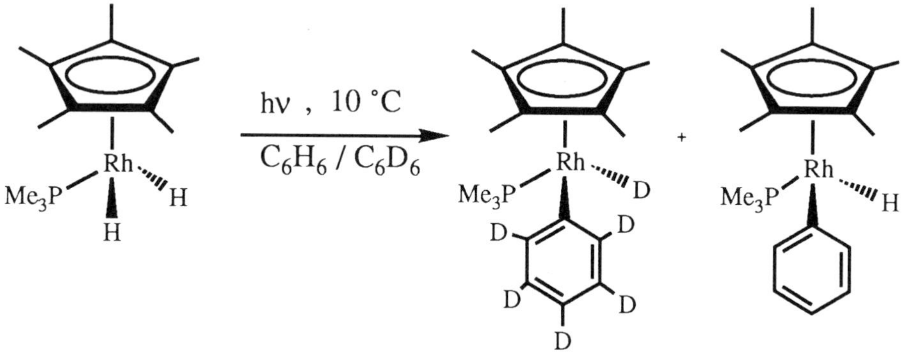

$$k_H/k_D = 1.05 \tag{8.23}$$

A larger intramolecular isotope effect of $k_H/k_D = 1.4 \pm 0.1$ was obtained in the reaction shown in eq. (8.24), where the selectivity of the Rh intermediate for a C—H or a C—D bond in the same arene was determined. The fact that different isotope effects were obtained from the intermolecular competition experiment [eq. (8.23)] and the intramolecular competition experiment [eq. (8.24)] requires the presence of an intermediate in the formation of $(C_5Me_5)(PMe_3)Rh(Ph)H$.

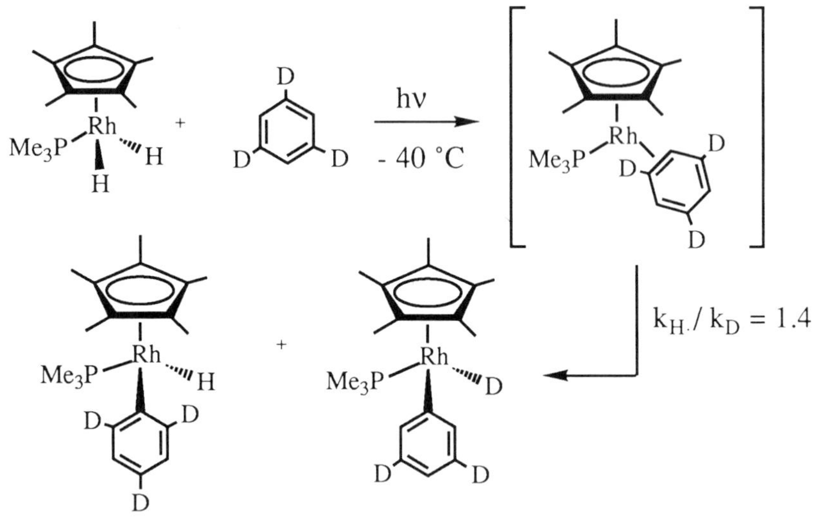

(8.24)

The intermediate implicated on the basis of these and several other experiments is an η^2-arene complex, as shown in eq. (8.24). Because the arene

C—H(D) bond is not broken in the formation of the arene complex, the only isotope effect expected is a secondary one, consistent with the small observed isotope effect of 1.05. The intramolecular isotope effect of 1.4 is observed for conversion of the η^2-arene complex into the phenyl hydride complex, a step that does involve breaking the C—H(D) bond.

An inverse kinetic isotope effect ($k_H/k_D = 0.51$) was observed [59] for the arene reductive elimination and dissociation reaction shown in eq. (8.25).

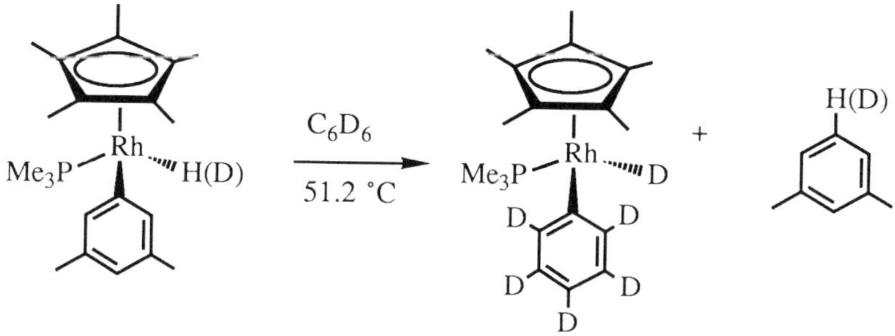

Me$_3$P–Rh(H(D)) ; C_6D_6, 51.2 °C → Me$_3$P–Rh–D (with C_6D_5) + H(D) arene (8.25)

By microscopic reversibility, an η^2-arene complex must be an intermediate in the elimination of arene from the aryl hydride complex, because it was established as an intermediate in the opposite process, formation of the rhodium aryl hydride complex from the free arene. This inverse kinetic isotope effect is similar to those discussed earlier in the formation of σ-complexes as intermediates in the reductive elimination of alkanes from alkyl hydride complexes.

Jones and Feher [59] also demonstrated that an equilibrium isotope effect was operative in the rhodium aryl hydride system, as shown in eq. (8.26). At 25°C, the occupation of H in the three isomers of $(C_5Me_5)(PMe_3)Rh(D)(C_6D_5H)$ was found in the statistical ratio of 2 : 2 : 1 (ortho : meta : para). However, occupation of the hydride site on Rh by H rather than D was 2.7 times larger than statistical, indicating a significant preference for D in the aromatic position.

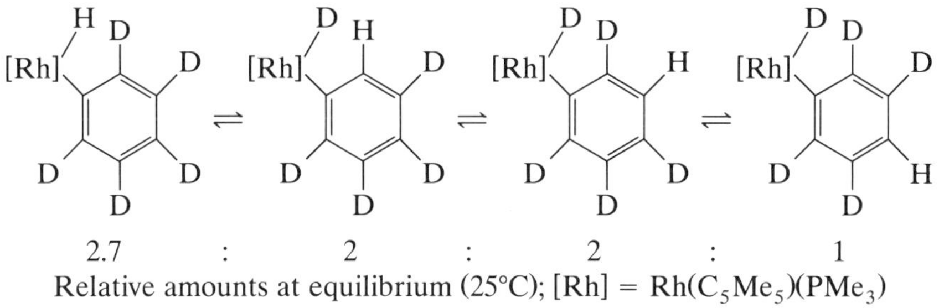

[Rh](H)(C_6D_5) ⇌ [Rh](D)(o-H) ⇌ [Rh](D)(m-H) ⇌ [Rh](D)(p-H)

2.7 : 2 : 2 : 1

Relative amounts at equilibrium (25°C); [Rh] = $Rh(C_5Me_5)(PMe_3)$

(8.26)

As found in several other examples, the preference for deuterium to occupy the site experiencing the larger force constant ($\nu_{CD} > \nu_{MD}$) determines the direction of the equilibrium isotope effect.

In addition to formation of metal hydride complexes by reaction of alkanes and arenes with low-valent metal complexes, recent studies have shown that vinyl hydride complexes resulting from oxidative addition of C—H bonds of unactivated alkenes to unsaturated metal centers can be observed in some cases. Stoutland and Bergman [61] found that reaction of the alkyl hydride complex $(C_5Me_5)(PMe_3)Ir(cyclohexyl)H$ with ethylene gives both the vinyl hydride complex and the π-ethylene complex [eq. (8.27)].

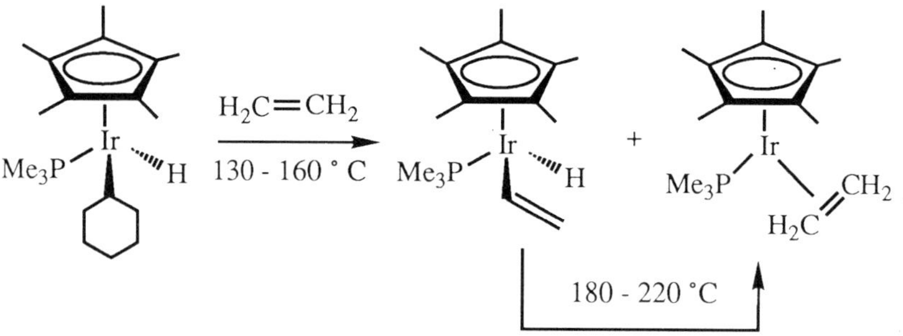

$H_2C{=}CH_2$, 130 - 160 °C; 180 - 220 °C (8.27)

An important conclusion established by this study is that the π complex is not an intermediate in the activation of the C—H bond to form the vinyl hydride complex; in fact, the vinyl hydride complex is converted to the alkene complex above 180°C.

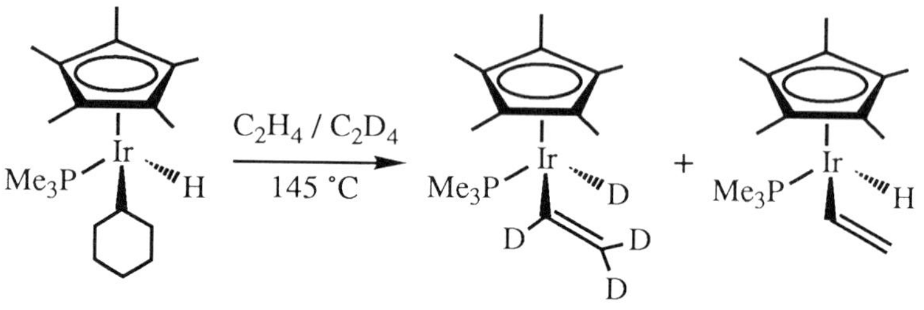

C_2H_4 / C_2D_4, 145 °C (8.28)

$$k_H/k_D = 1.49$$

The intermolecular isotope effect was determined by allowing $(C_5Me_5)(PMe_3)Ir(cyclohexyl)H$ to compete for reaction with C_2H_4 and C_2D_4. As shown in eq. (8.28), the kinetic isotope effect was found to be 1.49 ± 0.08 at 145°C. The intramolecular isotope effect was determined for the same system in a reaction with ethylene-d_2 [eq. (8.29)]. Because only one isotopomer of ethylene is present, the organometallic intermediate chooses (intramolecularly) whether to react with a C—H or a C—D bond. The intramolecular isotope effect determined from ethylene-d_2 was $k_H/k_D =$

1.18 ± 0.03.

$$(C_5Me_5)(PMe_3)Ir(H)(C_6H_{11}) \xrightarrow[145\ ^\circ C]{H_2C{=}CD_2} (C_5Me_5)(PMe_3)Ir(H)(CH{=}CD_2) + (C_5Me_5)(PMe_3)Ir(D)(CD{=}CH_2) \quad (8.29)$$

$$k_H/k_D = 1.18$$

An intriguiging observation is that (within experimental error) the *same* intramolecular isotope effect was obtained for all three isomers of ethylene-d_2 (1,1-d_2, *cis*-1,2-d_2, and *trans*-1,2-d_2). The data presently available for this system suggest the presence of rapidly equilibrating σ-complexes of ethylene as intermediates. The rapid equilibration is required by the fact that all three d_2-ethylenes gave similar isotope effects, and the presence of *some* intermediate is required by the fact that the intermolecular and intramolecular isotope effects are not the same. Comparisons of intermolecular versus intramolecular isotope effects also have been made for other reactions of Ir complexes with benzene [62]. Both the arene-activation experiments by Jones and Feher and the formation of the two products in the reaction of ethylene with the Ir complex show the value of careful studies of isotope effects, but there are notable differences in the chemistry. In the arene activation by $(C_5Me_5)(PMe_3)Rh$, an arene π complex was implicated (on the basis of isotope effects and other mechanistic information) as the intermediate in the formation of the aryl hydride. In contrast, the formation of the Ir vinyl hydride complex cannot proceed through formation of the π-complexed intermediate, because the π-bonded ethylene complex is the thermodynamically favored product. Instead, the unusual result was obtained that different transition states were required for formation of the vinyl hydride complex and the π-ethylene complex.

The σ-complexes established as intermediates in the reductive elimination of alkanes from alkyl hydride complexes must also be intermediates in the microscopic reverse, oxidative addition of a C—H bond to an unsaturated metal center. The $(C_5Me_5)(PMe_3)M$ (M = Rh and Ir) complexes utilized in arene and ethylene activation also react with saturated hydrocarbons. Reaction of $(C_5Me_5)(PMe_3)Rh(H)CH_2CMe_3$ with *n*-hexane (vs. *n*-hexane-d_{14}) gave $k_H/k_D = 1.2 \pm 0.1$ at -60°C [57], and photolysis of $(C_5Me_5)(PMe_3)IrH_2$ gave $k_H/k_D = 1.38$ in reaction with cyclohexane (vs. cyclohexane-d_{12}) to produce $(C_5Me_5)(PMe_3)Ir(cyclohexyl)H$ [63]. The small isotope effects observed in both of these cases are consistent with the presence of σ-alkane complexes as intermediates.

The three-center–two-electron bonding in these σ-alkane complexes is closely related to intramolecular agostic C—H····M interactions [2]. Hoff et al. [64] examined the influence of an agostic interaction by comparing the kinetics of the ligand displacement reaction shown in eq. (8.30) with the kinetics for the analogous tungsten complex containing $P(C_6D_{11})_3$ ligands.

$$[P(C_6H_{11})_3]_2W(CO)_3(py) + P(OMe_3) \longrightarrow [P(C_6H_{11})_3]_2W(CO)_3[P(OMe)_3] + py \quad (8.30)$$

The kinetic isotope effect observed for the overall reaction was normal ($k_H/k_D = 1.2$ at 25°C). It was suggested that the transition state for the reaction involved a W····H—C interaction that participated in displacement of the pyridine [eq. (8.31)], because no isotope effect would have been expected if the loss of pyridine were purely dissociative.

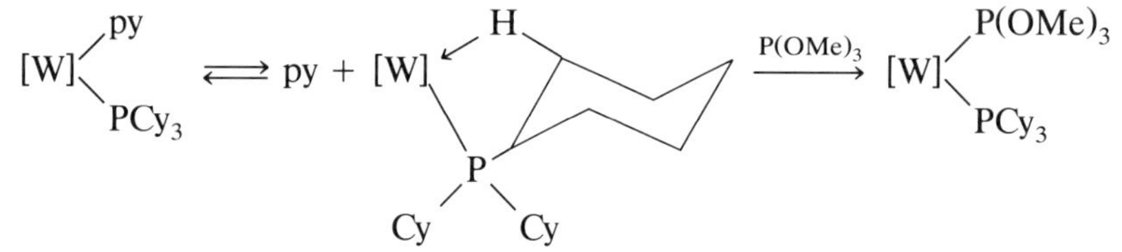

$$[W](py)(PCy_3) \rightleftharpoons py + [W]\cdots H\text{-}C_6H_{10}\text{-}PCy_2 \xrightarrow{P(OMe)_3} [W](P(OMe)_3)(PCy_3) \quad (8.31)$$

[W] = $(PCy_3)W(CO)_3$, Cy = cyclohexyl

Flood and co-workers [65] reported several kinetic isotope effects in their studies of C—H bond activation by $(PMe_3)_4OsH(CH_2CMe_3)$. In cyclohexane, thermolysis of this complex led to formation of a cyclometalated product formed by intramolecular oxidative addition of a C—H bond of a PMe_3 ligand (Scheme 8.8). Kinetic and mechanistic studies indicated this reaction occurred by two paths, one that was inhibited by PMe_3 and one that was

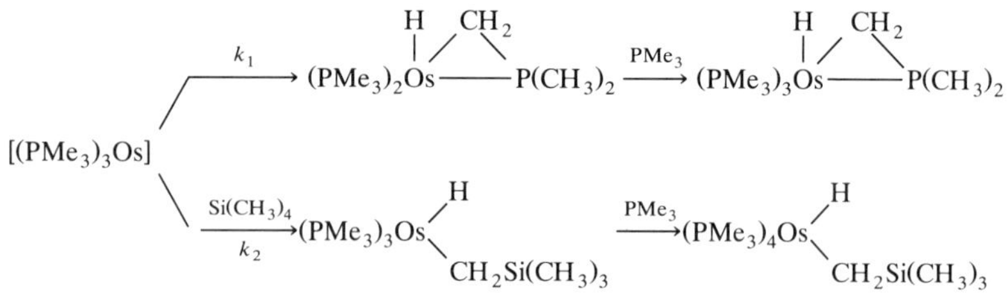

$$[(PMe_3)_3Os] \xrightarrow{k_1} (PMe_3)_2Os(H)(CH_2)P(CH_3)_2 \xrightarrow{PMe_3} (PMe_3)_3Os(H)(CH_2)P(CH_3)_2$$

$$[(PMe_3)_3Os] \xrightarrow[k_2]{Si(CH_3)_4} (PMe_3)_3Os(H)(CH_2Si(CH_3)_3) \xrightarrow{PMe_3} (PMe_3)_4Os(H)(CH_2Si(CH_3)_3)$$

Scheme 8.8

independent of $[PMe_3]$. An isotope effect was found for dissociation of PMe_3 from $(PMe_3)_4OsH(CH_2CMe_3)$; the rate constant for dissociation of $P(CH_3)_3$ from $(PMe_3)_4OsH(CH_2CMe_3)$ was higher ($k_H/k_D = 1.27 \pm 0.05$ at 65°C) than that for dissociation of $P(CD_3)_3$ from $[P(CD_3)_3]_4OsH(CH_2CMe_3)$. It was proposed that this was a steric isotope effect that was observed due to the slightly smaller size of $P(CD_3)_3$ compared to $P(CH_3)_3$, because the dissociation of the ligand was driven by steric crowding. Thermolysis of $(PMe_3)_4OsH(CH_2CMe_3)$ in $Si(CH_3)_4$ gave two products, the cyclometalated product resulting from intramolecular C—H activation, and $(PMe_3)_4OsH(CH_2SiMe_3)$ resulting from an intermolecular C—H activation. The rate of this reaction was the same in $Si(CH_3)_4$ as in $Si(CD_3)_4$. The product ratio was different in the deuterated versus nondeuterated solvent, however, suggesting that the products of intramolecular and intermolecular C—H activation resulted from a common intermediate. The intermediate implicated was $[Os(PMe_3)_3]$, which was suggested to result from dissociation of PMe_3 from $(PMe_3)_4OsH(CH_2CMe_3)$, followed by reductive elimination of neopentane. This intermediate could then undergo intramolecular oxidative addition of a C—H bond of PMe_3 (k_1 in Scheme 8.8) or intermolecular oxidative addition of a C—H bond of $Si(CH_3)_4$ (k_2 in Scheme 8.8). (The observed products would then result after reassociation of PMe_3.) The isotope effects for both of these steps were determined ($k_H/k_D = 2.2$ for k_1 and $k_H/k_D = 3.6$ for k_2, both at 80°C). A normal isotope effect also was observed for reaction of $(PMe_3)_4OsH(CH_2CMe_3)$ with benzene (vs. C_6D_6) to give $(PMe_3)_4OsH(C_6H_5)$. Normal isotope effects also were observed by Burk and Crabtree [66] and by Whitesides et al. [67] in other C—H activation reactions of benzene.

A study of the mechanism of formation of isobutane from $(C_5Me_5)_2Zr(H)CH_2CHMe_2$ by Bercaw et al. [68] also provided a convincing demonstration of the utility of isotope labeling experiments. Although the rates of formation of isobutane from $(C_5Me_5)_2Zr(H)CH_2CHMe_2$ and $(C_5Me_5)_2Zr(D)CH_2CDMe_2$ were essentially equal, a kinetic isotope effect of $k_H/k_D = 2.0$ was found in a comparison of the rates of decomposition of $(C_5Me_5)_2Zr(D)CH_2CDMe_2$ versus $[C_5(CD_3)_5]_2Zr(D)CH_2CDMe_2$. This unusual result, that a kinetic isotope effect is observed only upon deuteration of the methyl groups of the $C_5(CH_3)_5$ ring, is accounted for by the mechanism in Scheme 8.9. (CD_3 groups on the Cp* ring are not shown.) The first step, migration of a hydrogen from Zr to the Cp* ring, is a preequilibrium that should be subject to an equilibrium isotope effect. The rate-determining step was proposed to be oxidative addition of a C—H (C—D) bond of the Cp* ring to the unsaturated Zr intermediate. Reductive elimination from this complex then generates isobutane. One striking feature to emerge from this study is that although the overall process is a reductive elimination, the rate-determining step is oxidative addition (the microscopic reverse of reductive elimination). The overall observed kinetic isotope effect is influenced by both the equilibrium isotope effect (k_1/k_{-1}) and the kinetic isotope effect on the rate-determining step (k_2 in Scheme 8.9).

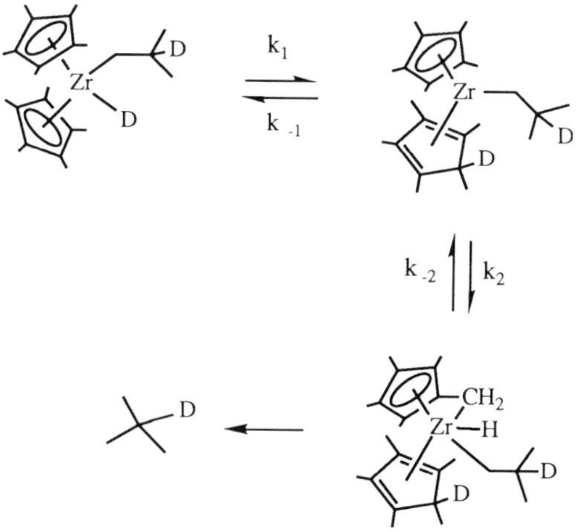

Scheme 8.9

In contrast to the previously discussed C—H bond activation reactions that typically proceed by even-electron pathways, Sherry and Wayland [69] found a stable rhodium radical complex that reacts with methane to form rhodium hydride and rhodium methyl complexes as shown in eq. (8.32),

$$2(\mathrm{TMP})\mathrm{Rh}\cdot + \mathrm{CH_4} \rightleftharpoons (\mathrm{TMP})\mathrm{RhH} + (\mathrm{TMP})\mathrm{RhCH_3} \qquad (8.32)$$

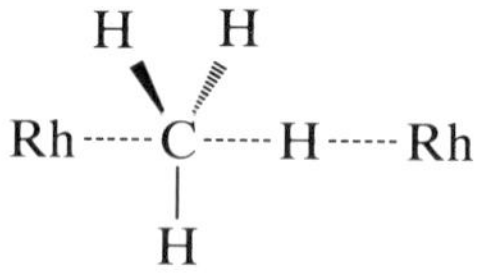

where TMP = tetramesitylporphyrin. The isotope effect found for reaction with CH_4 versus CD_4 was $k_H/k_D = 8.6 \pm 1.5$ at 23°C ($k_H/k_D = 5.0 \pm 1.0$ at 80°C). A termolecular, linear, four-center transition state [shown above] was suggested.

8.10. REACTIONS OF H_2 WITH METAL COMPLEXES

The addition of hydrogen to a metal center is a fundamentally important step in the hydrogenation and hydroformylation of olefins and other reactions. Chock and Halpern [70] reported an isotope effect of $k_H/k_D = 1.22$ at 30°C for the addition of H_2 (vs. D_2) to Vaska's complex, *trans*-$(PPh_3)_2IrCl(CO)$. More recently, San Filippo et al. [71] carried out a detailed investigation of the same reaction and found that the isotope effect ranged from $k_H/k_D =$

1.18 at 0°C to $k_H/k_D = 1.06$ at 30°C. The activation energy was higher for D_2 than for H_2 ($E_a^D - E_a^H = 0.38$ kcal mol^{-1}), and the ratio of Arrhenius preexponential factors was $A_H/A_D = 0.57$. Model calculations were carried out on this reaction. The mass and moment-of-inertia (MMI) term was unusually large (about 5.5), whereas the excitation term (EXC) was slightly lower than unity (about 0.8). (As noted in the introduction to this chapter, many calculations of expected isotope effects assume that the deviation of either of these terms from unity to be *negligible*.) The term due to zero-point energy (ZPE) was significantly less than unity (about 0.1). One of the conclusions reached in this experimental and theoretical study was that the combined effect of the MMI, EXC, and ZPE terms would predict an inverse isotope effect, contrary to the weak but normal isotope effect found experimentally. However, a substantial amount of hydrogen tunnelling was proposed to occur. The interpretation offered was that compensating factors accounted for the relatively small isotope effect for the overall reaction. Another conclusion of this study was that the transition state came early in the reaction (a reactant-like transition state) and involved side-on approach of H_2 (reminiscent of the now-familiar η^2-H_2 complexes) rather than end-on approach of H_2.

Wink and Ford [72] reported that flash photolysis of $(PPh_3)_2Rh(CO)Cl$ gives loss of CO and formation of $(PPh_3)_2RhCl$. This unsaturated complex has been proposed to be an intermediate in olefin hydrogenations by Wilkinson's catalyst, $(PPh_3)_3RhCl$. Flash photolysis studies carried out under H_2 and D_2 gave a value of $k_H/k_D = 1.5$ for addition of H_2 to $(PPh_3)_2RhCl$ at 23°C.

Lin and Marks [73] studied the hydrogenolysis of some metal alkyl complexes to give alkanes, and a kinetic isotope effect of $k_H/k_D = 2.5 \pm 0.4$ was found at 30°C for reaction of $(C_5Me_5)_2Th(CH_2CMe_3)(OCH_2CMe_3)$ with H_2 (vs. D_2). A four-center transition state involving rate-limiting H—H cleavage was suggested; furthermore, it was argued on the basis of other evidence that a preequilibrium involving a dihydrogen complex was not involved. The polar four-center transition state that was postulated is similar to that suggested for the hydrogenolysis of *n*-octyllithium by H_2 to produce octane. This temperature dependence of k_H/k_D for the hydrogenolysis of this alkyllithium compound by H_2 (vs. D_2) was found by Vitale and San Filippo [74] to range from $k_H/k_D = 1.64$ (at 50°C) to $k_H/k_D = 1.39$ (at 80°C). On the basis of $A_H/A_D = 0.24$ and $E_a^D - E_a^H = 1.25$ kcal mol^{-1} it was suggested that a small amount of tunnelling was operative in this hydrogenolysis reaction.

An inverse isotope effect of $k_H/k_D = 0.85$ was reported by Wochner and Brintzinger [75] for hydrogenolysis (H_2 vs. D_2) of the Zr—C bond of $Cp^*{}_2Zr(Cl)CH_2CMe_3$ to give neopentane and $Cp^*{}_2Zr(Cl)H$. The proposed mechanism involved coordination of H_2 to Zr, migration of H to a Cp* ring (compare to Scheme 8.9), and delivery of the H from the ring to the alkyl ligand to complete the hydrogenolysis.

8.11. REDUCTIVE ELIMINATION OF H_2 FROM MH_2

The kinetics of reductive elimination of H_2 from $(PMe_3)_2PtH_2$ have been reported by Packett and Trogler [76]. Comparison of the rate of decomposition of $(PMe_3)_2PtH_2$ (vs. PtD_2) at 21°C in THF gave an inverse isotope effect of $k_H/k_D = 0.45 \pm 0.10$. The inverse isotope effect was interpreted in terms of a late transition state involving Pt—H bond cleavage in the rate-determining step. The stretching frequency of the H—H bond exceeds that of the Pt—H bond, and a late transition state would therefore involve formation of an intermediate having higher force constants than the starting materials. It is possible that the transition state involves formation of an η^2-H_2 complex in a preequilibrium step rather than a single-step reductive elimination of H_2.

An inverse isotope effect of $k_H/k_D \approx 0.4$ was observed by Moore et al. [77] in the hydrogenation of norbornadiene to norbornene by $[(acetone)_2(PPh_3)_2IrH_2]^+$. The inverse isotope effect was rationalized by proposing an inverse equilibrium isotope effect on the rate-limiting step involving transfer of H from Ir to the alkene bound to Ir.

In contrast to these inverse isotope effects for loss of H_2 from MH_2 species, the reaction of CO with the cationic complex $[IrH_2(CO)_2(PPhMe_2)_2]^+$ to produce H_2 and $[Ir(CO)_3(PPhMe_2)_2]^+$, studied by Mays et al. [78], was found to be faster than the same reaction for the IrD_2 complex ($k_H/k_D = 2.1 \pm 0.3$ at 20°C).

Deutsch and Eisenberg [79] found a normal isotope effect ($k_H/k_D = 1.6 \pm 0.2$ at 35°C) for reductive elimination of H_2 from the Ir complex shown in eq. (8.33). This complex is an unusual example of an organometallic complex containing an alkyl group and two hydrides bonded to a single metal.

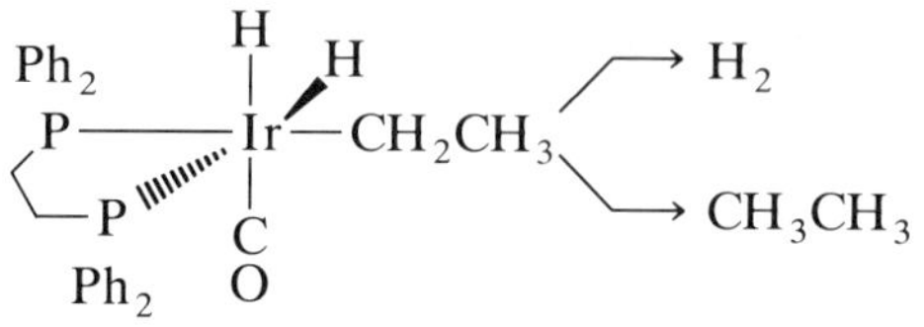

$$(Ph_2P\frown PPh_2)Ir(H)_2(CO)(CH_2CH_3) \longrightarrow H_2 \quad ; \quad \longrightarrow CH_3CH_3 \tag{8.33}$$

An especially interesting aspect of this study is that it was possible to observe competitive reversible elimination of H_2 and irreversible ethane elimination from the same complex. Activation parameters and isotope effects for both processes were reported. The isotope effect for elimination of ethane from (diphos)(CO)$IrEtH_2$ (vs. $IrEtD_2$) was $k_H/k_D = 2.4 \pm 0.1$ at 35°C. It was suggested that there was little C—H (or H—H) formation in the transition state for these reductive eliminations. The reactant-like transition state inferred for this reductive elimination of H_2 would appear to be in conflict with the proposal [71] of a reactant-like transition state for oxidative addition of H_2 to Vaska's complex, *trans*-$(PPh_3)_2IrCl(CO)$. Despite the overall similarity of these two Ir systems, there are some differences that could possibly affect the details of the reactions. The phosphines in (diphos)(CO)IrEtH_2 are constrained to a cis geometry by chelation, whereas the two PPh_3 ligands are trans in both $(PPh_3)_2IrCl(CO)$ and the dihydride formed from it.

8.12. FORMATION AND REACTIONS OF $M(\eta^2\text{-}H_2)$ COMPLEXES

Along with recent evidence that dihydrogen complexes $[M(\eta^2\text{-}H_2)]$ may be intermediates in the formation of stable metal dihydrides (MH_2) from reaction of metal complexes with H_2, a few examples of kinetic isotope effects on the formation and reactions of these $M(\eta^2\text{-}H_2)$ complexes have been reported. Grevels et al. [80] found a kinetic isotope effect of $k_H/k_D = 1.9$ on the rate of formation of $(CO)_5Cr(H_2)$. This dihydrogen complex was detected as a transient intermediate from the reaction of H_2 (D_2) with $(CO)_5Cr(C_6H_{12})$, which has a weakly bound cyclohexane ligand.

Hoff and co-workers [81] studied displacement of the η^2-H_2 ligand of Kubas' tungsten dihydrogen complex shown in eq. (8.34). The kinetic analysis was complicated by the fact that the dihydrogen complex $[M(\eta^2\text{-}H_2)]$ was in equilibrium with the dihydride (MH_2).

$$[P(C_6H_{11})_3]_2W(CO)_3(\eta^2\text{-}H_2) + py \rightarrow H_2 + [P(C_6H_{11})_3]_2W(CO)_3(py) \tag{8.34}$$

The dihydrogen complex reacted with pyridine about an order of magnitude faster than the dihydride, so that the initial rate of the reaction was fast due to the consumption of the dihydrogen complex, followed by a slower rate as the $MH_2 \rightleftharpoons [M(\eta^2\text{-}H_2)]$ equilibrium was reestablished. A normal isotope effect ($k_H/k_D = 1.7 \pm 0.1$ at 25°C) was found for loss of H_2 (D_2), but a negligible isotope effect ($k_H/k_D = 1.08 \pm 0.04$) was determined for the conversion of MH_2 to $M(\eta^2\text{-}H_2)$.

8.13. INSERTION OF CO_2 INTO AN M—H BOND

Sullivan and Meyer [82] found an inverse isotope effect for insertion of CO_2 into the Re—H bond of *fac*-(bipy)$(CO)_3$ReH to give *fac*-(bipy)$(CO)_3$-ReOC(=O)H. The kinetic isotope effect (Re—H vs. Re—D) was $k_H/k_D = 0.55 \pm 0.05$ in CH_3CN and was similar in THF and acetone. Although the isotope effects were similar in the different solvents, the rate of the insertion reaction was strongly solvent-dependent, with higher rates observed in solvents of higher dielectric constant. A polar four-center transition state involving hydride transfer from Re to CO_2 was suggested. The decarboxylation of a chiral rhenium formate complex was studied by Merrifield and Gladysz [83]. They reported $k_H/k_D = 1.55 \pm 0.19$ at 112°C for decomposition of Cp(NO)(PPh_3)ReOC(=O)H to give CO_2 and Cp(NO)(PPh_3)ReH.

8.14. INSERTION OF CO INTO AN M—H BOND

The insertion of CO into an M—H bond to give a formyl complex is often postulated in connection with the hydrogenation of CO by metal hydrides, but is seldom observable. Moloy and Marks [84] found that some $Cp^*_2ThH(OR)$ complexes containing bulky alkoxide ligands gave spectroscopically characterized formyl complexes, as shown in eq. (8.35). The equilibrium isotope effect measured (Th—H versus Th—D) at −78°C was $K_H/K_D = 0.31$.

$$Cp^*_2Th(H)(OR) + CO \underset{k_{-1}}{\overset{k_1}{\rightleftharpoons}} Cp^*_2Th(\eta^2\text{-}\ddot{O}\text{=}C\text{—}H)(OR) \longleftrightarrow Cp^*_2Th(\eta^2\text{-}O\text{—}\!:\!C\text{—}H)(OR) \qquad (8.35)$$

$$OR = OCH(CMe_3)_2,\ OCD(CMe_3)_2$$

This inverse equilibrium isotope effect is expected in view of the relatively low Th—H stretching frequency ($\nu_{ThH} = 1355\ cm^{-1}$) in the hydride starting material compared to the higher C—H stretching frequency of the formyl product (ν_{CH} not directly observed but suggested to be $\sim 2800\ cm^{-1}$). The kinetic isotope effects on the elementary steps [k_1 and k_{-1}, eq. (8.35)] were determined for this reaction at −56°C by ^1H NMR line-shape measurements. The results of these experiments indicated that $k_H/k_D = 2.8 \pm 0.4$ for the forward step (k_1) and $k_H/k_D = 4.1 \pm 0.5$ for the reverse step (k_{-1}). Thus the overall inverse *equilibrium* isotope effect is the result of normal *kinetic* isotope effects for each of the elementary steps, with a larger normal isotope effect on the reverse rate than on the forward rate.

8.15. α-ELIMINATION AND α-INSERTION REACTIONS

Although β-elimination has been studied extensively, fewer detailed mechanistic studies have been reported for α-elimination. Kinetic and mechanistic studies reported by Bercaw et al. [85] have shown that, in some cases, α-elimination is much faster kinetically than β-elimination, even though the latter may be favored thermodynamically. A kinetic isotope effect of k_H/k_D

= 0.46 at 140°C was found for the rearrangement of $Cp^*_2Ta(H)(\eta^2\text{-}CH_2O)$ to $Cp^*_2Ta(O)CH_3$ [eq. (8.36)].

$$Cp^*_2Ta(H)(\eta^2\text{-}CH_2O) \rightleftharpoons [Cp^*_2Ta{-}OCH_3] \longrightarrow Cp^*_2Ta(=O)CH_3 \qquad (8.36)$$

This isotope effect was interpreted as the result of an inverse thermodynamic isotope effect on the preequilibrium formation of the proposed intermediate, $[Cp^*_2Ta(OCH_3)]$.

Thermolysis of $Cp^*_2Ta(H)(=CH_2)$ gives methane and a mixture of bis-metalated Ta complexes [85] [eq. (8.37)]. As in the preceding case, the inverse kinetic isotope effect observed (k_H/k_D = 0.43 at 80°C) for thermolysis of $Cp^*_2Ta(H)(=CH_2)$ versus $Cp^*_2Ta(D)(=CD_2)$ was interpreted in terms of an inverse isotope effect on a preequilibrium in which the Ta—H bond was ruptured and a C—H bond (of significantly higher stretching frequency) was formed.

$$Cp^*_2Ta(H)(=CH_2) \rightleftharpoons [Cp^*_2Ta{-}CH_3] \longrightarrow CH_4 + [\text{bis-metalated Ta complex}] \rightarrow \rightarrow \qquad (8.37)$$

A particularly intriguing aspect of this study was that it was possible to determine the isotope effect on the forward rate constant (of a very closely related reaction) by monitoring the coalescence of the disastereotopic methyl groups of the C_5Me_4Ph ligand on the reaction shown in eq. (8.38).

$$(C_5Me_5)(C_5Me_4Ph)Ta(H)(=CH_2) \rightarrow [(C_5Me_5)(C_5Me_4Ph)Ta{-}CH_3] \qquad (8.38)$$

The normal kinetic isotope effect on this forward rate constant (k_H/k_D = 2.0 ± 0.6 at 60°C), coupled with the inverse isotope effect for the related overall reaction [eq. (8.37)] suggests that the kinetic isotope effect on the reverse rate (which has not been observed directly) would exhibit an even larger normal isotope effect.

8.16. PROTONATION REACTIONS OF ORGANOMETALLIC COMPLEXES

Heinekey, Michel, and Schulte [86] studied the protonation of $[Cp^*Ir(CO)]_2(\mu\text{-}CH_2)$, which was established by X-ray crystallography to have a trans arrangement of Cp* groups. The first observable product of protonation is *cis*-$[Cp^*Ir(CO)]_2(\mu\text{-}CH_2)(\mu\text{-}H)$, which slowly rearranges to the trans isomer. When the protonation of $[Cp^*Ir(CO)]_2(\mu\text{-}CH_2)$ was carried out at −88°C using the deuterated acid $D_2C(SO_2CF_3)_2$, the equilibrium shown in eq. (8.39) was established.

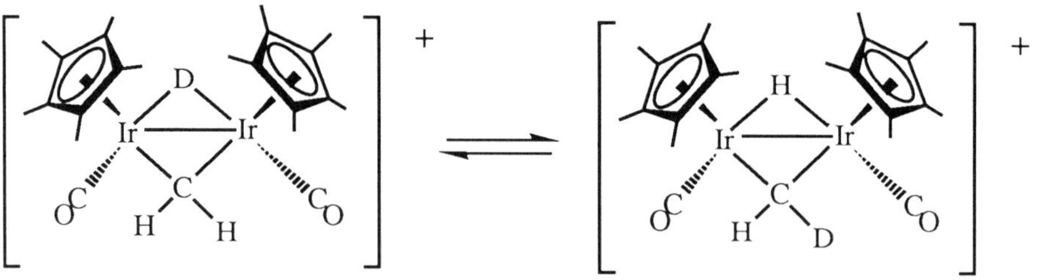

(8.39)

As found in the preceding cases, the equilibrium favored D in the site that had a higher force cosntant, in this case favoring D in the bridging methylene group instead of the bridging hydride position. The equilibrium constant observed at −88°C was $K_{eq} = 6 \pm 0.5$, which is considerably higher than the $K_{eq} = 2$ that would be expected on the basis of statistics in the absence of an isotope effect. Because rapid H–D scrambling between hydride and methylene sites occurred at this temperature, experiments at lower temperatures were performed in order to establish the kinetic site of protonation. It was suggested that the reaction proceeded by initial protonation of the methylene carbon, to give an intermediate with an agostic methyl group, which then formed $[Cp^*Ir(CO)]_2(\mu\text{-}CH_2)(\mu\text{-}H)$ by oxidative addition of a C—H bond to Ir. A kinetic isotope effect of $k_H/k_D = 4.1 \pm 0.5$ was found at −123°C for oxidative addition of a C—H (vs. C—D) bond of this agostic methyl intermediate to the Ir.

Casey, Gable, and Roddick [87] reported that isomerization of *trans*-$[Cp(CO)Fe]_2(\mu\text{-}CO)(\mu\text{-}CH_2)$ to an equilibrium mixture of the cis and trans isomers is catalyzed by acid. Protonation at the iron (either directly or by intramolecular proton transfer following initial protonation at a bridging carbonyl) was implicated by their kinetic studies as an important step in the isomerization. The extremely large isotope effect observed ($k_H/k_D = 105 \pm 20$ at −38°C) for isomerization catalyzed by CF_3CO_2H (vs. CF_3CO_2D) is strongly indicative of proton tunnelling.

Whitesides and Neilan [88] found a large isotope effect ($k_H/k_D = 27$) in the acid-catalyzed cis–trans isomerization of the iron diene complex shown in eq. (8.40). The tritium isotope effect ($k_H/k_T = 104 \pm 6$) was also determined for this reaction. Protonation at iron was suggested to be the rate-determining step in this isomerization.

$$\text{(Ph)CH=CH–CH=CH(Ph)}\,Fe(CO)_3 \xrightarrow{CF_3CO_2H} \text{Ph–CH=CH–CH=CH–Ph}\,Fe(CO)_3 \quad (8.40)$$

Electrophilic cleavage reactions of metal–carbon bonds by acid can exhibit large isotope effects that are indicative of proton tunnelling. The isotope effect for cleavage of the Fe–C bond of the substituted benzyl complex p-$FC_6H_4CH_2Fe(CO)_2(C_5H_5)$ was $k_H/k_D = 15–20$ for cleavage by CF_3CO_2H (vs. CF_3CO_2D) [89]. Cleavage of the Mn—C bond of p-$BrC_6H_4Mn(CO)_5$ by HBF_4 (vs. DBF_4) also had a large ($k_H/k_D \approx 20$) isotope effect [90]. In both of these cases protonation at the metal is indicated.

8.17. ISOTOPIC EFFECTS ON SPECTRAL PROPERTIES OF METAL HYDRIDES

Most of the effects caused by isotopic substitution of H for D discussed earlier were observed to affect the reaction kinetics and/or equilibrium constants. Isotopic substitution can, in some cases, also have dramatic effects on spectroscopic properties. Prediction of expected isotope effects for metal hydrides requires knowledge of M—H and M—D stretching frequencies. Although it often is possible to obtain this information from IR spectroscopy, in some cases interactions of normal modes for M—H stretches with other vibrations can alter their observed frequency. When two IR bands of similar energy (and appropriate symmetry) are present in the same molecule, they can undergo a resonance interaction in which the frequency of the higher-energy band is raised and the frequency of the lower-energy band is lowered. The magnitude of the interaction is inversely proportional to the difference in energy of the (unperturbed) vibrations. Additionally, the weak band "borrows" intensity from the stronger band. Because the frequencies of M—H stretches in metal hydrides often are similar to the frequencies of CO stretches in metal carbonyls, this effect frequently has been observed in metal carbonyl hydrides. An early observation of this effect was made by Vaska [91], who found that the ν_{CO} band of $(PPh_3)_2IrH_2(CO)Cl$ was observed at 1982 cm^{-1} in the dihydride but increased in energy (to 2023 cm^{-1}) in the IrD_2 complex. In this dihydride complex, the H trans to CO has a stretching frequency of 2100 cm^{-1} that undergoes resonance with the CO, resulting in a lowering of the observed frequency of the ν_{CO} band.

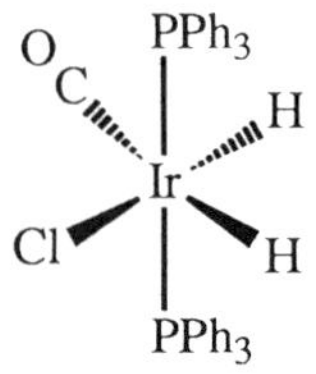

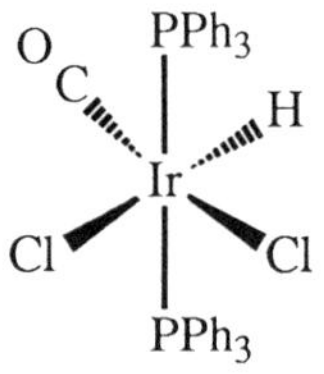

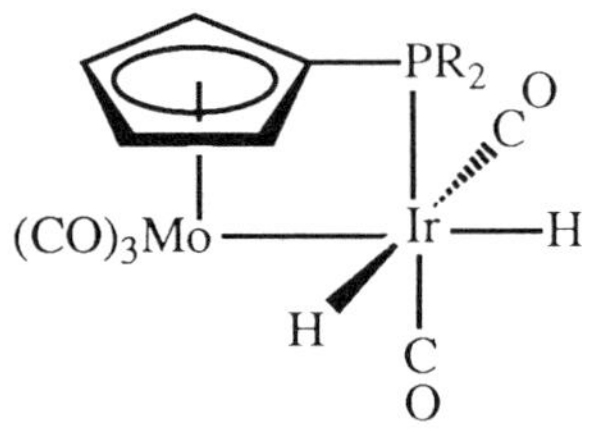

In the dideuteride, however, the ν_{MD} frequency (1497 cm^{-1}) is too far removed in frequency to interact significantly with the ν_{CO} band, so the IR band of the CO is no longer affected by the resonance interaction. Two important points are that the "unperturbed" frequency of the ν_{CO} band is observed in the dideuteride (not the dihydride), and the interaction is far more pronounced when the H and CO are trans to each other rather than cis. The ν_{CO} band of *trans*-$(PPh_3)_2Ir(CO)HCl_2$ appears at the same position in the dihydride as the dideuteride, because in this complex the H is trans to Cl, not CO. Kaesz et al. [92] found that the ν_{CO} for the CO trans to H appears at 2005 cm^{-1} in $HRe(CO)_5$ but shifts to 2000 cm^{-1} in $DRe(CO)_5$. The calculated interaction force constant between the trans ν_{CO} and $\nu_{Re\text{-}H}$ ($\sim$ 0.22 mdyn $Å^{-1}$) was almost 400 times greater than that between the cis ν_{CO} and $\nu_{Re\text{-}H}$. Casey, Bullock, and Nief [93] were able to assign the $\nu_{Ir\text{-}H}$ and ν_{CO} for both Ir and Mo carbonyl stretches for the bimetallic complex shown above, by comparing spectra of the dihydride and dideuteride. The ν_{CO} for the Ir—CO trans to H shifted to 21 cm^{-1} higher energy in the dideuteride, whereas the ν_{CO} for the Ir—CO cis to H shifted only 12 cm^{-1} to higher energy, and the highest energy ν_{CO} for Mo—CO shifted only 5 cm^{-1}. Other examples of metal carbonyl hydrides that exhibit observable shifts of the ν_{CO} bands upon deuteration of the hydride positions include $H_2Os(CO)_4$ [94], (diphos)(CO)$IrEtH_2$ [79], (diphos)(CO)$IrBrH_2$ [95] and (diphos)(CO)Ir(CN)H_2 [95].

Poliakoff and co-workers [96] pointed out a caveat in the potential use of similar perturbations in the study of metal carbonyl complexes containing dihydrogen ligands. Changes in the ν_{CO} bands may not be diagnostic of whether the hydrogen is bonded as a terminal hydride or as a dihydrogen ligand, because shifts in the IR spectra are expected in either case. The ν_{CO} bands could experience vibrational coupling with ν_{MH} bands in terminal hydride complexes, and with ν_{DD} bands in $M(\eta^2\text{-}D_2)$ complexes.

Crabtree and co-workers [97] found that the isotopic shift in the ^1H NMR spectra upon deuteration of rhenium polyhydrides was about 0.008 ppm upfield per deuterium. Thus a solution of $(PCy_3)_2ReH_{7-x}D_x$ ($x = 0$–6) showed seven evenly spaced singlets in the hydride region of the ^1H NMR spectrum (^{31}P decoupled) [97]. Similarly, an upfield isotope shift of -0.025

ppm per deuterium was found for [tris(pyrazolyl)borate]$ReH_{6-x}D_x$ ($x = 0$–5) at 298 K. The fact that the isotopic shifts in these and several related L_2ReH_7 complexes exhibited little or no temperature dependence was interpreted as evidence that these complexes were "classical" polyhydrides and not dihydrogen complexes. In contrast, experiments with partially deuterated samples of $[(PMe_2Ph)_3ReH_2(\eta^2\text{-}H_2)]^+$ showed a substantial preference for deuterium occupation of the η^2-H_2 site [97c]. The explanation for this effect is analogous to that discussed previously for the preference of D to occupy terminal positions (with H in bridging site) in partially deuterated agostic complexes [eqs. (8.6) and (8.7)].

Another spectroscopic change that occurs upon substitution of D for H is the observation of J_{HD} as an aid in the identification of dihydrogen complexes $M(\eta^2\text{-}H_2)$. Although the value of J_{HH} is generally small in classical dihydrides, such values can be quite large (but not directly observable) in dihydrogen complexes. Thus the observation of J_{HD} values in the range of 20–35 Hz (which would correspond to large values of J_{HH}) has been used as one of several criteria for the assignment of dihydrogen complexes. Heinekey and co-workers [98] observed large and highly temperature-dependent J_{HH} values in a series of IrH_3 complexes. For example, J_{HH} for $[(C_5H_5)(AsPh_3)IrH_3]^+$ changes from 376 Hz at −97°C to 570 Hz at −77°C. Although J_{HD} values were not resolved in partially deuterated samples, J_{HT} values of only 24 and 29 Hz were observed by ^{3}T NMR of a partially tritiated sample. These J_{HT} values are far smaller than would be predicted based on the known J_{HH} values. It was proposed that the surprisingly large J_{HH} couplings were actually exchange couplings of quantum mechanical origin [99].

8.18. CONCLUSIONS

Studies of isotope effects in transition metal hydride chemistry have been quite useful in mechanistic studies. Recent studies in C—H activation and dihydrogen complexes have relied on isotope effects to help answer important questions, and these are areas where use of isotope effects are expected to play a prominent role in the future. The use of isotopically labelled alkyl hydride complexes was instrumental in several studies that implicated σ-alkane complexes as important intermediates in reductive elimination of alkanes; these σ-complexes also explain the H–D exchange of alkyl and hydride sites in alkyl hydride complexes. Inverse isotope effects often are observed in organometallic reactions, especially when an M—H bond is ruptured and a C—H bond is formed. In many cases, these inverse isotope effects are not interpreted as an inverse isotope effect on an elementary step but instead as being reflective of an inverse thermodynamic isotope effect on a preequilibrium. Large isotope effects indicative of tunnelling have been observed in reactions involving protonation at the metal.

ACKNOWLEDGMENT

Research at Brookhaven National Laboratory is carried out under contract DE-AC02-76CH00016 with the US Department of Energy and is supported by its Division of Chemical Sciences, Office of Basic Energy Sciences. M. B. thanks Drs. Carol Creutz, George Gould, Jack Norton, and Mark Andrews for helpful discussions and comments.

REFERENCES

1. Rosenberg, E. *Polyhedron* **1989**, *8*, 383–405.
2. For a review of agostic (CH····M) interactions, see Brookhart, M.; Green, M. L. H.; Wong, L.-L. *Prog. Inorg. Chem.* **1988**, *36*, 1–124.
3. (a) Bigeleisen, J. *J. Chem. Phys.* **1949**, *17*, 675–678. (b) Bigeleisen, J.; Wolfsberg, M. *Adv. Chem. Phys.* **1958**, *1*, 15–76. (c) Bigeleisen, J. *Pure Appl. Chem.* **1964**, *8*, 217–223.
4. (a) Wolfsberg, M.; Stern, M. J. *Pure Appl. Chem.* **1964**, *8*, 225–242. (b) Stern, M. J.; Wolfsberg, M. *J. Chem. Phys.* **1966**, *45*, 2618–2629, 4105–4124. (c) Wolfsberg, M. *Acc. Chem. Res.* **1972**, *5*, 225–233. (d) Weston, R. E. *Science* **1967**, *158*, 332–342.
5. Westheimer, F. H. *Chem. Rev.* **1961**, *61*, 265–273.
6. Melander, L.; Saunders, W. H. *Reaction Rates of Isotopic Molecules*; Wiley: New York, 1980.
7. (a) German, E. D.; Kuznetsov, A. M.; Dogonadze, R. R. *J. Chem. Soc., Faraday Trans. 2* **1980**, *76*, 1128–1146. (b) German, E. D.; Kuznetsov, A. M. *J. Chem. Soc., Faraday Trans. 2* **1980**, *76*, 2203–2212.
8. Edidin, R. T.; Sullivan, J. M.; Norton, J. R. *J. Am. Chem. Soc.* **1987**, *109*, 3945–3953.
9. More O'Ferrall, R. A. *J. Chem. Soc. (B)* **1970**, 785–790.
10. For consideration of a weak-interaction model for proton transfers between transition metals, see Creutz, C.; Sutin, N. *J. Am. Chem. Soc.* **1988**, *110*, 2418–2427.
11. For discussions of tunnelling see the following: (a) Bell, R. P. *The Tunnel Effect in Chemistry*; Chapman and Hall: London, 1980. (b) Bell, R. P. *Chem. Soc. Rev.* **1974**, *3*, 513–544. (c) Lewis, E. S. In *Proton-Transfer Reactions*; Caldin, E., Gold, V., Eds. Chapman and Hall: London, 1975.
12. For recent theoretical work using the application of time-dependent perturbation theory to tunnelling and the application of this theory to organic reactions involving tunnelling, see Siebrand, W.; Wildman, T. A.; Zgierski, M. Z. *J. Am. Chem. Soc.* **1984**, *106*, 4083–4089 and 4089–4096.
13. For calculations related to tunnelling and Arrhenius plot curvature, see Stern, M. J.; Weston, R. E. *J. Chem. Phys.* **1974**, *60*, 2803–2807.
14. Bell, R. P. *The Proton in Chemistry*; Cornell University Press: Ithaca, NY, 1959; Chapter 11.
15. Schneider, M. E.; Stern, M. J. *J. Am. Chem. Soc.* **1972**, *94*, 1517–1522.
16. Stern, M. J.; Weston, R. E. *J. Chem. Phys.* **1974**, *60*, 2808–2814.
17. (a) Brunton, G.; Griller, D.; Barclay, L. R. C.; Ingold, K. U. *J. Am. Chem. Soc.* **1976**, *98*, 6803–6811. (b) Brunton, G.; Gray, J. A.; Griller, D.; Barclay, L. R. C.; Ingold, K. U. *J. Am. Chem. Soc.* **1978**, *100*, 4197–4200.
18. Kwart and co-workers have reported that some H transfers in 1,5-sigmatropic rearrangements exhibit temperature-*independent* isotope effects (and large values

of $A_H/A_D \sim 5$), and they have suggested that these parameters characterize nonlinear transition states for H transfer: (a) Kwart, H.; Brechbiel, M. W.; Acheson, R. M.; Ward, D. C. *J. Am. Chem. Soc.* **1982**, *104*, 4671–4672. (b) Kwart, H. *Acc. Chem. Res.* **1982**, *15*, 401–408. However, this proposal has been criticized, and model calculations lend no support to Kwart's proposal of the relationship of transition-state geometry to observed isotope effects. See McLennan, D. J.; Gill, P. M. W. *J. Am. Chem. Soc.* **1985**, *107*, 2971–2972; *Isr. J. Chem.* **1985**, *26*, 378–386. For additional information on the relationship of isotope effects to transition-state geometry and to A_H/A_D values, see references 9 and 16.

19. Roecker, L.; Meyer, T. J. *J. Am. Chem. Soc.* **1987**, *109*, 746–754.
20. (a) Calvert, R. B.; Shapley, J. R.; Schultz, A. J.; Williams, J. M.; Suib, S. L.; Stucky, G. D. *J. Am. Chem. Soc.* **1978**, *100*, 6240–6241. (b) Schultz, A. J.; Williams, J. M.; Calvert, R. B.; Shapley, J. R.; Stucky, G. D. *Inorg. Chem.* **1979**, *18*, 319–323. (c) Calvert, R. B.; Shapley, J. R. *J. Am. Chem. Soc.* **1978**, *100*, 7726–7727.
21. For a review of isotopic effects on NMR chemical shifts, see Hansen, P. E. *Ann. Rep. NMR Spectrosc.* **1983**, *15*, 105–234.
22. (a) Brookhart, M.; Lamanna, W.; Humphrey, M. B. *J. Am. Chem. Soc.* **1982**, *104*, 2117–2126. (b) Brookhart, M.; Lamanna, W.; Pinhas, A. R. *Organometallics* **1983**, *2*, 638–649.
23. (a) Ittel, S. D.; Van-Catledge, F. A.; Jesson, J. P. *J. Am. Chem. Soc.* **1979**, *101*, 6905–6911. (b) Holmes, S. J.; Clark, D. N.; Turner, H. W.; Schrock, R. R. *J. Am. Chem. Soc.* **1982**, *104*, 6322–6329. (c) Casey, C. P.; Fagan, P. J.; Miles, W. H. *J. Am. Chem. Soc.* **1982**, *104*, 1134–1136. (d) Dawkins, G. M.; Green, M.; Orpen, A. G.; Stone, F. G. A. *J. Chem. Soc., Chem. Commun.* **1982**, 41–43. (e) Cree-Uchiyama, M.; Shapley, J. R.; St. George, G. M. *J. Am. Chem. Soc.* **1986**, *108*, 1316–1317. (f) Dutta, T. K.; Vites, J. C.; Jacobsen, G. B.; Fehlner, T. P. *Organometallics* **1987**, *6*, 842–847. (g) Crabtree, R. H.; Holt, E. M.; Lavin, M.; Morehouse, S. M. *Inorg. Chem.* **1985**, *24*, 1986–1992. (h) Bleeke, J. R.; Kotyk, J. J.; Moore, D. A.; Rauscher, D. J. *J. Am. Chem. Soc.* **1987**, *109*, 417–423.
24. Sweany, R. L.; Halpern, J. *J. Am. Chem. Soc.* **1977**, *99*, 8335–8337.
25. Sweany, R. L.; Comberrel, D. S.; Dombourian, M. F.; Peters, N. A. *J. Organomet. Chem.* **1981**, *216*, 57–63.
26. Roth, J. A.; Wiseman, P.; Ruszala, L. *J. Organomet. Chem.* **1983**, *240*, 271–275.
27. Ungváry, F.; Markó, L. *Organometallics* **1982**, *1*, 1120–1125.
28. Bullock, R. M.; Rappoli, B. J.; Samsel, E. G.; Rheingold, A. L. *J. Chem. Soc., Chem. Commun.* **1989**, 261–263.
29. (a) Bullock, R. M.; Samsel, E. G.; *J. Am. Chem. Soc.* **1987**, *109*, 6542–6544. (b) Bullock, R. M.; Samsel, E. G. *J. Am. Chem. Soc.* **1990**, *112*, 6886–6898.
30. Connolly, J. W. *Organometallics* **1984**, *3*, 1333–1337.
31. Wassink, B.; Thomas, M. J.; Wright, S. C.; Gillis, D. J.; Baird, M. C. *J. Am. Chem. Soc.* **1987**, *109*, 1995–2002.
32. Shackleton, T. A.; Baird, M. C. *Organometallics* **1989**, *8*, 2225–2232.
33. Jacobsen, E. N.; Bergman, R. G. *J. Am. Chem. Soc.* **1985**, *107*, 2023–2032.
34. Leusink, A. J.; Budding, H. A.; Drenth, W. *J. Organomet. Chem.* **1967**, *9*, 295–306.
35. Melander, L. *Acta Chem. Scand.* **1971**, *25*, 3821–3826.
36. (a) Nalesnik, T. E.; Freudenberger, J. H.; Orchin, M. *J. Mol. Catal.* **1982**, *16*, 43–49. (b) Orchin, M. *Ann. N.Y. Acad. Sci.* **1983**, *415*, 129–134.
37. (a) Moody, A. E. Ph.D. Thesis, Colorado State University, Fort Collins, CO, 1989. (b) Eisenberg, D. C.; Kristjánsdóttir, S. S.; Lawrie, C. J. C.; Moody, A. E.; Norton, J. R. 198th National American Chemical Society Meeting, Miami Beach, Florida, September 10–15, 1989, INOR 128. (c) Eisenberg, D. C.; Norton, J. R. *Isr. J. Chem.*, in press.

38. (a) Vlcek, A., Jr.; Gray, H. B. *J. Am. Chem. Soc.* **1987**, *109*, 286–287. (b) Vlcek, A., Jr.; Gray, H. B. *Inorg. Chem.* **1987**, *26*, 1997–2001. (c) Harvey, E. L.; Stiegman, A. E.; Vlcek, A., Jr.; Gray, H. B. *J. Am. Chem. Soc.* **1987**, *109*, 5233–5235. (d) Roundhill, D. M.; Gray, H. B.; Che, C.-M. *Acc. Chem. Res.* **1989**, *22*, 55–61.
39. Doherty, N. M.; Bercaw, J. E. *J. Am. Chem. Soc.* **1985**, *107*, 2670–2682.
40. Halpern, J.; Okamoto, T.; Zakhariev, A. *J. Mol. Catal.* **1976**, *2*, 65–68.
41. Evans, J.; Schwartz, J.; Urquhart, P. W. *J. Organomet. Chem.* **1974**, *81*, C37–C39.
42. Ikariya, T.; Yamamoto, A. *J. Organomet. Chem.* **1976**, *120*, 257–284.
43. Burger, B. J.; Thompson, M. E.; Cotter, W. D.; Bercaw, J. E. *J. Am. Chem. Soc.* **1990**, *112*, 1566–1577.
44. McCarthy, T. J.; Nuzzo, R. G.; Whitesides, G. M. *J. Am. Chem. Soc.* **1981**, *103*, 3396–3403.
45. Ozawa, F.; Ito, T.; Yamamoto, A. *J. Am. Chem. Soc.* **1980**, *102*, 6457–6463.
46. Komiya, S.; Morimoto, Y.; Yamamoto, A.; Yamamoto, T. *Organometallics* **1982**, *1*, 1528–1536.
47. Brainard, R. L.; Whitesides, G. M. *Organometallics* **1985**, *4*, 1550–1557.
48. Kazlauskas, R. J.; Wrighton, M. S. *J. Am. Chem. Soc.* **1982**, *104*, 6005–6015.
49. Faller, J. W.; Anderson, A. S. *J. Am. Chem. Soc.* **1970**, *92*, 5852–5860.
50. (a) Foley, P.; DiCosimo, R.; Whitesides, G. M. *J. Am. Chem. Soc.* **1980**, *102*, 6713–6725. (b) Foley, P.; Whitesides, G. M. *J. Am. Chem. Soc.* **1979**, *101*, 2732–2733. (c) DiCosimo, R.; Moore, S. S.; Sowinski, A. F.; Whitesides, G. M. *J. Am. Chem. Soc.* **1982**, *104*, 124–133. (d) Moore, S. S.; DiCosimo, R.; Sowinski, A. F.; Whitesides, G. M. *J. Am. Chem. Soc.* **1981**, *103*, 948–949. (e) Brainard, R. L., Miller, T. M., Whitesides, G. M. *Organometallics* **1986**, *5*, 1481–1490.
51. McDade, C.; Bercaw, J. E. *J. Organomet. Chem.* **1985**, *279*, 281–315.
52. Abis, L.; Sen, A.; Halpern, J. *J. Am. Chem. Soc.* **1978**, *100*, 2915–2916.
53. Michelin, R. A.; Faglia, S.; Uguagliati, P. *Inorg. Chem.* **1983**, *22*, 1831–1834.
54. Hackett, M.; Ibers, J. A.; Whitesides, G. M. *J. Am. Chem. Soc.* **1988**, *110*, 1436–1448.
55. (a) Bullock, R. M.; Headford, C. E. L.; Kegley, S. E.; Norton, J. R. *J. Am. Chem. Soc.* **1985**, *107*, 727–729. (b) Bullock, R. M.; Headford, C. E. L.; Hennessy, K. M.; Kegley, S. E.; Norton, J. R. *J. Am. Chem. Soc.* **1989**, *111*, 3897–3908.
56. Buchanan, J. M.; Stryker, J. M.; Bergman, R. G. *J. Am. Chem. Soc.* **1986**, *108*, 1537–1550.
57. Periana, R. A.; Bergman, R. G. *J. Am. Chem. Soc.* **1986**, *108*, 7332–7346.
58. Parkin, G.; Bercaw, J. E. *Organometallics* **1989**, *8*, 1172–1179.
59. Jones, W. D.; Feher, F. J. *J. Am. Chem. Soc.* **1986**, *108*, 4814–4818; *Acc. Chem. Res.* **1989**, *22*, 91–100.
60. Gould, G. L.; Heinekey, D. M. *J. Am. Chem. Soc.* **1989**, *111*, 5502–5504.
61. Stoutland, P. O.; Bergman, R. G. *J. Am. Chem. Soc.* **1985**, *107*, 4581–4582; *J. Am. Chem. Soc.* **1988**, *110*, 5732–5744.
62. (a) McGhee, W. D.; Bergman, R. G. *J. Am. Chem. Soc.* **1988**, *110*, 4246–4262. (b) McGhee, W. D.; Hollander, F. J.; Bergman, R. G. *J. Am. Chem. Soc.* **1988**, *110*, 8248–8443.
63. Janowicz, A. H.; Bergman, R. G. *J. Am. Chem. Soc.* **1983**, *105*, 3929–3939.
64. Gonzalez, A. A.; Zhang, K.; Hoff, C. D. *Inorg. Chem.* **1989**, *28*, 4285–4290.
65. (a) Harper, T. G. P.; Shinomoto, R. S.; Deming, M. A.; Flood, T. C. *J. Am. Chem. Soc.* **1988**, *110*, 7915–7916. (b) Desrosiers, P. J.; Shinomoto, R. S.; Flood, T. C. *J. Am. Chem. Soc.* **1986**, *108*, 7964–7970. (c) Shinomoto, R. S.; Desrosiers, P. J.; Harper, T. G. P.; Flood, T. C. *J. Am. Chem. Soc.* **1990**, *112*, 704–713.
66. Burk, M. J.; Crabtree, R. H. *J. Am. Chem. Soc.* **1987**, *109*, 8025–8032.
67. Brainard, R. L.; Nutt, W. R.; Lee, T. R.; Whitesides, G. M. *Organometallics* **1988**, *7*, 2379–2386.

68. McAlister, D. R.; Erwin, D. K.; Bercaw, J. E. *J. Am. Chem. Soc.* **1978**, *100*, 5966–5968.
69. Sherry, A. E.; Wayland, B. B. *J. Am. Chem. Soc.* **1990**, *112*, 1259–1261.
70. Chock, P. B.; Halpern, J. *J. Am. Chem. Soc.* **1966**, *88*, 3511–3514.
71. Zhou, P.; Vitale, A. A.; San Filippo, J.; Saunders, W. H. *J. Am. Chem. Soc.* **1985**, *107*, 8049–8054.
72. Wink, D. A.; Ford, P. C. *J. Am. Chem. Soc.* **1987**, *109*, 436–442.
73. Lin, Z.; Marks, T. J. *J. Am. Chem. Soc.* **1987**, *109*, 7979–7985.
74. Vitale, A. A.; San Filippo, J. *J. Am. Chem. Soc.* **1982**, *104*, 7341–7343.
75. Wochner, F.; Brintzinger, H. H. *J. Organomet. Chem.* **1986**, *309*, 65–75.
76. Packett, D. L.; Trogler, W. C. *Inorg. Chem.* **1988**, *27*, 1768–1775; *J. Am. Chem. Soc.* **1986**, *108*, 5036–5038.
77. Howarth, O. W.; McAteer, C. H.; Moore, P.; Morris, G. E. *J. Chem. Soc., Dalton Trans.* **1984**, 1171–1180.
78. Mays, M. J.; Simpson, R. N.; Stefanini, F. P. *J. Chem. Soc.* (*A*) **1970**, 3000–3002.
79. Deutsch, P. P.; Eisenberg, R. *J. Am. Chem. Soc.* **1990**, *112*, 714–721.
80. Church, S. P.; Grevels, F.-W.; Hermann, H.; Schaffner, K. *J. Chem. Soc., Chem. Commun.* **1985**, 30–32.
81. Zhang, K.; Gonzalez, A. A.; Hoff, C. D. *J. Am. Chem. Soc.* **1989**, *111*, 3627–3632. Gonzalez, A. A.; Zhang, K.; Nolan, S. P.; de la Vega, R. L.; Mukerjee, S. L.; Hoff, C. D. *Organometallics* **1988**, *7*, 2429–2435.
82. Sullivan, B. P.; Meyer, T. J. *Organometallics* **1986**, *5*, 1500–1502.
83. Merrifield, J. H.; Gladysz, J. A. *Organometallics* **1983**, *2*, 782–784.
84. Moloy, K. G.; Marks, T. J. *J. Am. Chem. Soc.* **1984**, *106*, 7051–7064.
85. Parkin, G.; Bunel, E.; Burger, B. J.; Trimmer, M. S.; Van Asselt, A.; Bercaw, J. E. *J. Mol. Catal.* **1987**, *41*, 21–39.
86. Heinekey, D. M.; Michel, S. T.; Schulte, G. K. *Organometallics* **1989**, *8*, 1241–1246.
87. Casey, C. P.; Gable, K. P.; Roddick, D. M. *Organometallics* **1990**, *9*, 221–226.
88. Whitesides, T. H.; Neilan, J. P. *J. Am. Chem. Soc.* **1975**, *97*, 907–908.
89. Anderson, S. N.; Cooksey, C. J.; Holton, S. G.; Johnson, M. D. *J. Am. Chem. Soc.* **1980**, *102*, 2312–2318.
90. Motz, P. L.; Sheeran, D. J.; Orchin, M. *J. Organomet. Chem.* **1990**, *383*, 201–212.
91. Vaska, L. *J. Am. Chem. Soc.* **1966**, *88*, 4100–4101.
92. Braterman, P. S.; Harrill, R. W.; Kaesz, H. D. *J. Am. Chem. Soc.* **1967**, *89*, 2851–2855.
93. Casey, C. P.; Bullock, R. M.; Nief, F. *J. Am. Chem. Soc.* **1983**, *105*, 7574–7580.
94. L'Eplattenier, F.; Calderazzo, F. *Inorg. Chem.* **1967**, *6*, 2092–2097.
95. Johnson, C. E.; Eisenberg, R. *J. Am. Chem. Soc.* **1985**, *107*, 3148–3160. Fisher, B. J.; Eisenberg, R. *Organometallics* **1983**, *2*, 764–767. Johnson, C. E.; Fisher, B. J.; Eisenberg, R. *J. Am. Chem. Soc.* **1983**, *105*, 7772–7774.
96. Jackson, S. A.; Hodges, P. M.; Poliakoff, M.; Turner, J. J.; Grevels, F.-W. *J. Am. Chem. Soc.* **1990**, *112*, 1221–1233.
97. (a) Hamilton, D. G.; Luo, X.-L.; Crabtree, R. H. *Inorg. Chem.* **1989**, *28*, 3198–3203. (b) Luo, X.-L.; Crabtree, R. H. *J. Am. Chem. Soc.* **1990**, *112*, 4813–4821. (c) Luo, X.-L.; Crabtree, R. H. *J. Am. Chem. Soc.* **1990**, *112*, 6912–6918.
98. (a) Heinekey, D. M.; Payne, N. G.; Schulte, G. K. *J. Am. Chem. Soc.* **1988**, *110*, 2303–2305. (b) Heinekey, D. M.; Millar, J. M.; Koetzle, T. F.; Payne, N. G.; Zilm, K. W. *J. Am. Chem. Soc.* **1990**, *112*, 909–919.
99. (a) Zilm, K. W.; Heinekey, D. M.; Millar, J. M.; Payne, N. G.; Demou, P. *J. Am. Chem. Soc.* **1989**, *111*, 3088–3089. (b) Jones, D. H.; Labinger, J. A.; Weitekamp, D. P. *J. Am. Chem. Soc.* **1989**, *111*, 3087–3088. (c) Zilm, K. W.; Heinekey, D. M.; Millar, J. M.; Payne, N. G.; Neshyba, S. P.; Duchamp, J. C.; Szczyrba, J. *J. Am. Chem. Soc.* **1990**, *112*, 920–929.

CHAPTER 9

Acidity of Hydrido Transition Metal Complexes in Solution

S. SÓLEY KRISTJÁNSDÓTTIR and JACK R. NORTON

Department of Chemistry, Colorado State University, Ft. Collins, Colorado 80523

9.1. INTRODUCTION

Historically, the acidic nature of transition metal hydrides was appreciated but received little quantitative study. In 1932 Feigl and Krumholz [1] mentioned their "... Annahme, daß der Eisencarbonylwasserstoff sauren Charakter besitzt [hypothesis, that iron carbonyl hydride has acidic character]", and in 1937 Coleman and Blanchard [2] estimated the aqueous pK_a of $HCo(CO)_4$ and $H_2Fe(CO)_4$. Between 1953 and 1965 Hieber and co-workers [3] did measure the aqueous pK_a of some carbonyl hydride complexes. However, writing in 1971 after summarizing the few data then available, Schunn [4] noted that there was still "a definite need for the measurement of the acidity functions of a systematic series of metal hydrides under controlled, identical conditions."

The *thermodynamic acidity* (equilibrium extent of deprotonation) of transition metal hydrides plays an important role in many of their reactions. For example, the regiospecificity of the hydrocobaltation of olefins with electron-withdrawing substituents [reactions (9.1) and 9.2)] depends upon the pH of the solution. Under conditions where the cobalt hydride remains protonated, the α-substituted cobaloxime is formed [eq. (9.1)]; in media alkaline enough to deprotonate the hydride, the β-substituted cobaloxime is formed instead [eq. (9.2)] [5].

$$\text{CH}_2{=}\text{CH–CO}_2\text{Et} \xrightarrow[\text{(neutral)}]{\text{Co(dmgH)}_2\text{–H}_2\text{–py}} \text{H–CH}_2\text{–CH([Co])–CO}_2\text{Et} \tag{9.1}$$

$$\text{CH}_2{=}\text{CH–CO}_2\text{Et} \xrightarrow[\text{(alkaline)}]{\text{Co(dmgH)}_2\text{–H}_2\text{–py}} \text{[Co]–CH}_2\text{–CH(H)–CO}_2\text{Et} \tag{9.2}$$

$$[\text{Co}] = \text{Co(dimethylglyoximatoH)}_2\text{py}$$

The ability of transition metal hydrides to serve as catalysts is also affected by such deprotonation equilibria. The anionic ruthenium hydride **1**, which Grey, Pez, and co-workers [6] found to be a catalyst for the hydrogenation of ketones, aldehydes, and activated esters [eq. (9.3)], offers a good example. Linn and Halpern [7] showed that, in the presence of H_2, **1** is rapidly converted to $[RuH_3(PPh_3)_3]^-$ (**2**) and that **2** is a strong base, capable of deprotonating cyclohexanol in THF [eq. (9.4)].

$$\text{cyclohexanone} \xrightarrow[\mathbf{1},\ \mathbf{2},\ \text{or}\ \mathbf{3}]{H_2} \text{cyclohexanol} \tag{9.3}$$

H, H, PPh_3, PPh_3, PPh_2, Ru

1

$$\underset{\mathbf{2}}{[RuH_3(PPh_3)_3]^-} + \text{cyclohexanol (OH)} \underset{K_{eq}=0.13}{\rightleftharpoons} \underset{\mathbf{3}}{RuH_4(PPh_3)_3} + \text{cyclohexoxide } (O^-) \tag{9.4}$$

An induction period, during which the rate of hydrogenation builds up to a limiting value, is observed when **1** or **2** is used as a catalyst for the hydrogenation of cyclohexanone. However, when the protonated form **3** is employed, there is no induction period and the limiting rate of hydrogenation is observed immediately. The active form is thus **3**. The anionic polyhydrides **1** and **2** are converted to **3** by proton transfer from the hydrogenated product, and the hydrogenation reaction is thus autocatalytic when the catalyst is added as **1** or **2**.

The *kinetic acidity* of a transition metal hydride (the rate at which a proton is removed from its M—H bond) also can affect its chemistry. A good example is the reaction of the hydride complex $CpRe(NO)(PPh_3)H$ (**4**) with *n*-BuLi–tetramethylethylenediamine in tetrahydrofuran (THF) at $-78°C$ [eq. (9.5)] [8]. The Cp ring hydrogens are kinetically more acidic than the hydrogen ligand on the Re, so lithiation initially occurs on the ring, to give the kinetic product **5**; however, the hydrogen on the Re has greater thermodynamic acidity, so **5** eventually rearranges to the more stable anion **6**. The result of methylation varies with the conditions: methylation prior to rearrangement occurs on the Cp ring of **5** and gives **7**, whereas methylation after rearrangement to **6** occurs at rhenium and gives **8**.

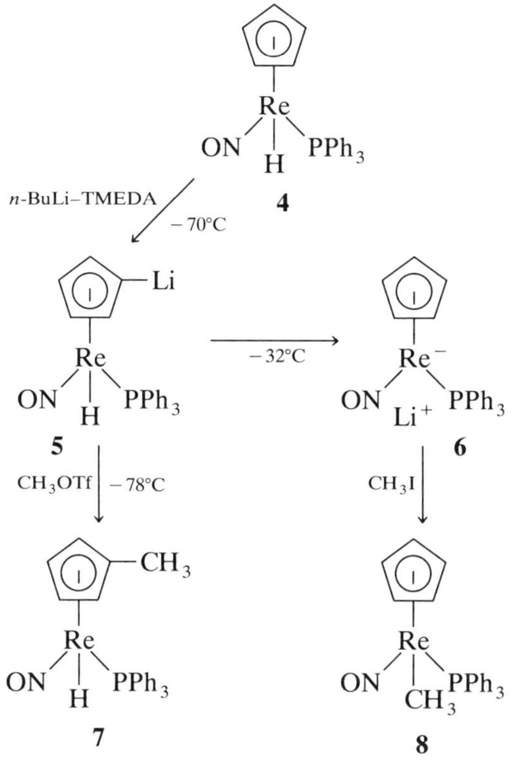

Re
ON H PPh_3
4
n-BuLi–TMEDA
−70°C
Li
Re
ON H PPh_3
5
−32°C
Re^-
ON Li^+ PPh_3
6
(9.5)
CH_3OTf −78°C
CH_3I
CH_3
Re
ON H PPh_3
7
Re
ON CH_3 PPh_3
8

A general explanation for the low kinetic acidity of hydrogen bound to a transition metal was first advanced by Walker, Kresge, Ford, and Pearson [9] in 1979: "Presumably slow rates are the result of the substantial structural and electronic rearrangements which accompany removal of the proton." An idea of how large these rearrangements are (and a comparison with those that accompany removal of a proton from nitrogen) can be gleaned from considering the changes that occur during reaction (9.6).

$$Me_3N + HCo(CO)_4 \rightarrow [Me_3NH][Co(CO)_4] \qquad (9.6)$$

The structures of Me_3N (electron diffraction [10]), $HCo(CO)_4$ (electron diffraction [11]), and $[Me_3NH][Co(CO)_4]$ (X-ray [12–16]) are shown in Figure 9.1. Because the lone pair in Me_3N is stereochemically active, the nitrogen in Me_3N is tetrahedral before and after H^+ transfer. In contrast, the pair of d electrons that receive the proton as we go from d^{10} $[Co(CO)_4]^-$ to d^8 $HCo(CO)_4$ are *not* stereochemically active, and there is a considerable difference in geometry about the cobalt between the distorted trigonal

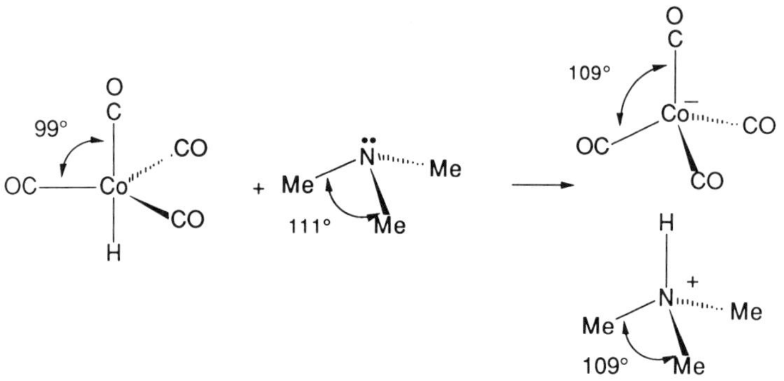

FIGURE 9.1

bipyramid in $HCo(CO)_4$ and the tetrahedral arrangement found [12–16] in $[Co(CO)_4]^-$.

The size of the electronic rearrangement that occurs as a transition metal is deprotonated can be inferred from the change in ν_{CO} for associated carbonyl ligands. For example, the principal ν_{CO} decreases by over 100 cm^{-1} between $HCo(CO)_4$ and $[Co(CO)_4]^-$, showing that most of the charge that arises from deprotonation is delocalized onto the carbonyl ligands.

The fact that these rearrangements are larger for transition metals than for nitrogen suggests that (at least for a *constant thermodynamic driving force*) proton transfers involving transition metals will be much slower than those involving nitrogen. Recent work in our own research group [13, 17] as well as in the Ford–Pearson research group [9, 18] has provided considerable evidence that this prediction is correct.

Oxygen is similar to nitrogen in that both elements generally have stereochemically active lone pairs; proton transfer reactions involving either element thus have relatively low barriers and are fast. In contrast, the rearrangements that a metal hydride must undergo during deprotonation are like those that must occur during the deprotonation of carbon acids [19]. A classic example of such a rearrangement occurs during the deprotonation of nitromethane in eq. (9.7). The resulting carbanion is better described by structure **A** than by structure **B**, so a sizable barrier results from the need to rehybridize the carbon from which the proton has been removed and to delocalize the resulting negative charge onto the nitro group. Reaction (9.7) thus occurs slowly even with very strong bases.

$$CH_3NO_2 + \text{base} \rightarrow \underset{\textbf{A}}{H_2C{=}\overset{+}{N}(O^-)_2} \qquad (\text{not } {}^-CH_2NO_2) \; \textbf{B} \tag{9.7}$$

It is worth emphasizing that these kinetic barriers are important only if we are comparing the rates of proton transfer reactions with similar equilibrium constants. We will therefore discuss the measurement and interpretation of the thermodynamic acidities of transition metal hydrides before we resume our discussion of the rates of their proton transfer reactions.

9.2. THERMODYNAMIC ACIDITY

As already mentioned, some early carbonyl hydride pK_a measurements were made in water [3], but this solvent is not suitable for general use. Most transition metal hydrides are either insoluble in water or react with it, and the range of pK_a values that can be *measured* in water is limited by its relatively high ion product (10^{-14}) and the levelling effect it thus exerts on strong acids and bases. It is nevertheless desirable to *refer* pK_a values measured in other solvents to an aqueous standard, and this will be done later.

Complications similar to those in water arise in methanol. Although the Pearson–Ford group has determined the pK_a of several hydrides in that solvent [9, 18], the deprotonation equilibria have been complicated (at least in mononuclear cases) by ion pairing involving the carbonylmetallate anions, methoxide ion, and alkali metal cations.

Tetrahydrofuran, a common solvent in organometallic synthesis, dissolves most transition metal hydrides without reacting with them and thus might seem to be a suitable solvent for measuring their pK_a values. Indeed, Miholova and Vlcek [20] published a preliminary report in 1971 on the qualitative acidities of several common carbonyl hydrides in THF (as determined by polarographic and IR techniques). However, *quantitative* equilibrium studies in THF are complicated by ion pairing. Even delocalized organic anions exhibit extensive ion pairing in THF [21] (although the pK_a values of the corresponding hydrocarbon acids can be determined indirectly in that solvent [22]). The formation of contact ion pairs involving carbonylmetallate anions in THF is common and has been studied extensively; several distinct types of these ion pairs have been identified [23].

Finally, there have been scattered hydride pK_a determinations in less-common solvents or solvent mixtures (e.g., the pK_a of $[Cp_2ReH_2]^+$ has been measured [24] in 60% dioxane–water). These solvent systems have typically been chosen for their convenience for the system at hand and the results are difficult to compare with literature acidity data in more conventional media.

We believe acetonitrile is the best solvent for measuring the pK_a values of transition metal hydrides. As acetonitrile is both a weak acid and a weak base [25], with an ion product that is approximately 10^{-44} [25c], pK_a values up to 36 [25d] can be measured directly in it. It solvates cations effectively enough to keep them from forming contact ion pairs with carbonylmetallate anions [26] and therefore leaves such anions symmetrically solvated. Finally, pK_a values in acetonitrile can be found in the literature for many organic acids [27–30], including the protonated forms of simple amines [28], of 1,8-diaminonaphthalene derivatives [29], and of several recently synthesized classes of strong neutral nitrogen bases [30].

Few transition metal hydrides are acidic enough in acetonitrile to make direct measurement of the pK_a equilibrium [eq. (9.8)] practical. Instead we generally have measured the position of a proton-transfer equilibrium involving the hydride and a base of known strength capable of effecting partial deprotonation [eq. (9.9)]. As Beer's law holds for carbonyl hydrides and carbonylmetallate anions at concentrations up to 0.1 *M* [31], the concentrations of carbonyl-containing species are conveniently determined by IR spectroscopy from the absorbance of carbonyl modes and the independently measured absorptivities of the pure compounds. (A typical spectrum, used in determining the equilibrium constant for the deprotonation of $HMn(CO)_4(PEtPh_2)$ by pyrrolidine [32], is shown in Figure 9.2.) When the bands due to the acid and its conjugate base overlap, the concentrations of both species can be calculated from the absorbances at two or more points by

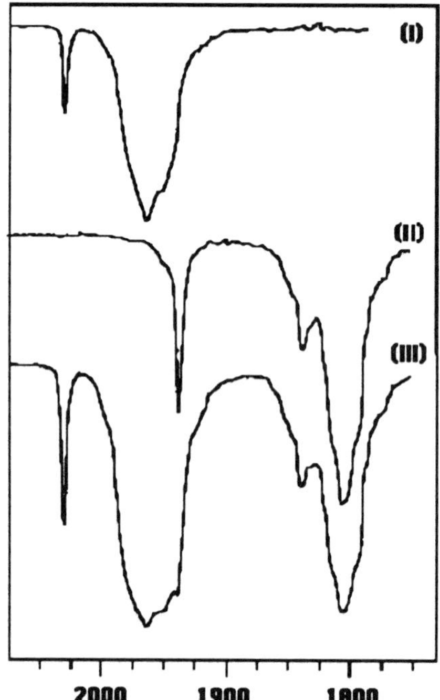

FIGURE 9.2
The carbonyl stretching region in the IR of the species involved in the equilibrium between $HMn(CO)_4PEtPh_2$ and pyrrolidine in CH_3CN: (I) $HMn(PEtPh_2)(CO)_4$; (II) $[Mn(PEtPh_2)(CO)_4]^-$; (III) $HMn(PEtPh_2)(CO)_4 + [Mn(PEtPh_2)(CO)]^-$.

overdetermined-least-squares methods [17c, 32].

$$\mathrm{M{-}H} \underset{\mathrm{CH_3CN}}{\overset{K_a}{\rightleftharpoons}} \mathrm{M^- + CH_3CNH^+} \tag{9.8}$$

$$\mathrm{M{-}H + B} \overset{K_{eq}}{\rightleftharpoons} \mathrm{M^- + BH^+} \tag{9.9}$$

If there is no association between the base B and its protonated form BH^+, the concentration of BH^+ equals that of M^-, and the concentration of B remaining can be calculated from the amount of base initially added (B_{tot}). If there is association between B and BH^+ [eq. (9.10)], $[BH^+]$ must be calculated from $[M^-]$ and $[B_{tot}]$ according to eq. (9.11) [13a, 17, 32].

$$\mathrm{B + BH^+ \rightleftharpoons BHB^+} \qquad K_f = \frac{[\mathrm{BHB^+}]}{[\mathrm{B}][\mathrm{BH^+}]} \tag{9.10}$$

$$[\mathrm{BH^+}]^2 + \left\{[\mathrm{B}]_{\mathrm{total}} - 2[\mathrm{M^-}] + \frac{1}{K_f}\right\}[\mathrm{BH^+}] - \frac{[\mathrm{M^-}]}{K_f} = 0 \tag{9.11}$$

The stoichiometry of eqs. (9.9) and (9.10) then gives $[BHB^+]$ and [B]. The pK_a of the transition metal hydride is calculated from K_{eq} and the known pK_a of BH^+ by eq. (9.12) [13a, 17, 32].

$$pK_a^{\mathrm{MH}} = pK_{eq} + pK_a^{\mathrm{BH^+}} \tag{9.12}$$

The fact that acetonitrile is a good solvent for electrochemistry also makes it useful for the *indirect* determination of pK_a values. Ryan, Tilset, and Parker [33] have measured the oxidation potential [eq. (iii) in Scheme 9.1] of several transition metal hydrides in acetontrile and have thus been able to derive the acidity of the corresponding 17-electron hydride cation

$$\mathrm{M{-}H \rightleftharpoons M^- + H^+} \qquad \Delta G^\circ = 2.301RT\, pK_a(\mathrm{M{-}H}) \tag{i}$$

$$\mathrm{M^- \rightleftharpoons M\cdot + e^-} \qquad \Delta G^\circ = FE^\circ_{\mathrm{ox}}(\mathrm{M^-}) \tag{ii}$$

$$\mathrm{M{-}H^{\cdot+} + e^- \rightleftharpoons M{-}H} \qquad \Delta G^\circ = -FE^\circ_{\mathrm{ox}}(\mathrm{M{-}H}) \tag{iii}$$

$$\mathrm{M{-}H^{\cdot+} \rightleftharpoons M\cdot + H^+} \qquad \Delta G^\circ = 2.301RT\, pK_a(\mathrm{M{-}H^{\cdot+}}) \tag{iv}$$

$$pK_a(\mathrm{M{-}H^{\cdot+}}) = pK_a(\mathrm{M{-}H}) + 23.06/1.37\left[E^\circ_{\mathrm{ox}}(\mathrm{M^-}) - E^\circ_{\mathrm{ox}}(\mathrm{M{-}H})\right] \tag{v}$$

Scheme 9.1
Determination of the pK_a of hydride cation radicals.

Table 9.1
Thermodynamic Acidities (pK_a values) of Transition Metal Hydrides in Water, Methanol, or Acetonitrile

Compound	Solvent		
	H_2O	MeOH	CH_3CN
$HV(CO)_6$	Strong [35, 36]		
$HV(CO)_5PPh_3$	6.8 [35]		
$HV(CO)_4(dppe)$[a]			< 17.5[a] [37]
$HV(\eta^6\text{-}C_6Me_3H_3)(CO)_3$	< 13 [38]		
$HCrCp(CO)_3$	5.4[b] [18b]	6.4[c] [18c]	13.3 [17a]
$HCrCp(CO)_3^{\cdot +}$			−9.5[d] [33]
$HMoCp(CO)_3$	> 5 [39] 6.2[b] [18b]	7.2[c] [18c]	13.9 [17a]
$HMoCp(CO)_3^{\cdot +}$			−6.0[d] [33]
$HMoCp^*(CO)_3$			17.1 [13a]
$HMoCp^*(CO)_3^{\cdot +}$			−2.5[d] [33]
$HMo(\eta^3\text{-}L)(CO)_3^{+}$[e]	2.4 [40]		
$HMo(\eta^3\text{-}L)(NO)_2^{+}$[e]	4.9 [40]		
$HWCp(CO)_3$	8.0[b] [18b]	9.0[c] [18c]	16.1 [17a]
$HWCp(CO)_3^{\cdot +}$			−3.0[d] [33]
$HWCp(CO)_2PMe_3$			26.6 [13a]
$H_2WCp(CO)_2PMe_3^{+}$			> 9 [33]
$HWCp(CO)_2PMe_3^{\cdot +}$			5.1[d] [33]
$HW(\eta^3\text{-}L)(CO)_3^{+}$[e]	2.9 [40]		
$HMn(CO)_5$	7.1 [3b]		14.2 [41]
$HMn(CO)_4(PPh_3)$			20.4 [32]
$HMn(CO)_4(PEtPh_2)$			21.6 [32]
$(\eta^3\text{-}C_6H_8\cdots H\cdots)Mn(CO)_3$[f]			22.2 [32]
$HMn(CO)_2\eta^6\text{-}C_6H_6)$			26.8 [32]
$HMnCp(CO)_2(SiCl_3)$			~ 18.5 [42]
$HRe(CO)_5$	Very weak [3d]		21.1 [13a]
$CpRe(CO)_2H_2$			23.0 [32]
$Cp_2ReH_2^{+}$	5.5[g] [24]		
$ReH_4(PMe_2Ph)_4^{+}$			25.3 [43]
$ReH_4(CO)(PMe_2Ph)_3^{+}$			< 19 [44]
$H_3Re_3(CO)_{12}$	3 [45]		
$[H_2Re_3(CO)_{12}]^-$	10 [45]		
$[HRe_3(CO)_{12}]^{2-}$	25 [45]		
$H_2M(PF_3)_4$ (M = Fe, Ru, Os)	Weak [46]		
$H_2Fe(CO)_4$	4.0 [47]	6.9 [18b]	11.4 [13a]
$[HFe(CO)_4]^-$	12.7 [47]		
$Fe(H_2)H(dmpe)_2^{+}$[h]		~ 12 [48]	
$HFeCp(CO)_2$			19.4 [13a]
$HFeCp^*(CO)_2$			26.3 [13a]
$HFeCp(CO)(SiCl_3)_2$			2.6 [42]
$(\mu\text{-}H)_2Fe_3(CO)_9(\mu_3\text{-}P\text{-}4\text{-}C_6H_4OCH_3)$			9.0 [32]
$(\mu\text{-}H)_2Fe_3(CO)_9(\mu_3\text{-}P\text{-}t\text{-}Bu)$			11.4 [32]
$(\mu\text{-}H)Fe_3(CO)_9(\mu_3\text{-}S\text{-}C_6H_{11})$			16.9 [32]
$H_2Ru(CO)_4$			18.7 [13a]
$HRuCp(CO)_2$			20.2 [13a]
$CpRu(dmpe)(H_2)^{+}$[h]			17.6 [49]
$H_4Ru_4(CO)_{11}P(OCH_3)_3$			12.4 [32]
$H_4Ru_4(CO)_{10}[P(OCH_3)_3]_2$			15.4 [32]
$H_2Os(CO)_4$		15.2 [18b]	20.8 [17a]

Table 9.1
Thermodynamic Acidities (pK_a values) of Transition Metal Hydrides in Water, Methanol, or Acetonitrile (Continued)

Compound	Solvent				
	H_2O		MeOH	CH_3CN	
$H(CH_3)Os(CO)_4$				23.0	[17a]
$H_2Os_2(CO)_8$				20.4	[17a]
$HM(PF_3)_4$ (M = Co, Rh, Ir)	Strong[i]	[50]			
$HCo(CO)_4$	Strong	[3c]	~ 1.[j] [3a]	8.4	[13a]
$HCo(CO)_3PPh_3$	7.0	[3c]		15.4	[13a]
$HCo(CO)_3P(OPh)_3$	4.9	[3c]		11.4	[13a]
$HCo(CO)_2(PPh_3)_2$	Very weak	[3e]			
$[HCo(CN)_5]^{3-}$	20	[51]			
$HCo(dmgH)_2P(n\text{-}Bu)_3$	10.5[k]	[52]			
$HRh(CO)Cl(PPh_3)_2{}^+$			1.8 [53]		
$HRh(dppe)^{2+}$ [l]			~ 1 [54]		
$HRh(dmgH)_2PPh_3$	9.5[k]	[55]			
$HRh(en)_2OH^+$	~ 12	[56]			
$[Rh_{13}(CO)_{24}H_3]^{2-}$				11.0	[17c]
$[Rh_{13}(CO)_{24}H_2]^{3-}$				16.5	[17c]
$[Cp^*Rh(\mu\text{-}PMe_2)_2(\mu\text{-}H)RhCp^*]^+$			< 10[m] [57]		
$HIr(CO)Cl(PPh_3)_2{}^+$			2.1 [53]		
$HIr(CO)Br(PPh_3)_2{}^+$			2.6 [53]		
$HIr(CO)I(PPh_3)_2{}^+$			2.8 [53]		
$HM(PEt_3)_3{}^+$ (M = Ni, Pd, Pt)			Weaker than EtOH[n] [58]		
$HNi(Ph_2PCH_2CH_2PPh_2)_2{}^+$			2.6 [59]		
$HNi[PPh(OEt)_2]_4{}^+$			2.0 [35]		
$HNi[P(OEt)_3]_4{}^+$				13	[59]
$HNi[P(OMe)_3]_4{}^+$			1.7 [35]	13.4	[60]
$HPd[P(OMe)_3]_4{}^+$			1.0 [60]	10.6	[60]
$HPt[P(OMe)_3]_4{}^+$			10.2 [60]	18.5	[60]
$Cp(PMe_3)_2Ru{=}C{=}C(\underline{H})CMe_3{}^+$				20.8	[61]
$Cp(PMe_3)(CO)Fe{=}C{=}C(\underline{H})CMe_3{}^+$				13.6	[61]
$(CO)_5Cr{=}C(OMe)C\underline{H}_3$	12.3	[62]			
trans-$[Mo({=}N\underline{H})F(dppe)_2]^+$ [l]			12.7 [63]		
trans-$[W({=}N\underline{H})F(dppe)_2]^+$ [l]			15.5 [64]		

[a] dppe = 1,2-bis(diphenylphosphino)ethane; the compound is deprotonated by NEt_3 (pK_a = 18.4) in CH_3CN.

[b] Measured in a 70:30 MeOH–H_2O mixture; the apparent pK_a of this mixture is 14.2 [65].

[c] Measured in a 70:30 MeOH–H_2O mixture and corrected to pure MeOH.

[d] From Scheme 9.1.

[e] L = 1,4,7-triazacyclononane.

[f] The complex contains a three-center–two-electron interaction involving the 4-endo-CH bond and the Mn [66].

[g] Measured in a 60% dioxane–water mixture.

[h] dmpe = 1,2-bis(dimethylphosphino)ethane.

[i] Measured in a water–acetone mixture.

[j] Measured in a MeOH–H_2O mixture.

[k] Measured in a 50:50 MeOH–H_2O mixture; dmgH = dimethylglyoxime monoanion.

[l] dppe = 1,2-bis(dipheynlphosphino)ethane.

[m] Deprotonated by KCN [pK_a(H_2O) = 9.21].

[n] Measured in an EtOH–H_2O mixture.

Table 9.2
Estimated pK_a Values[a] for Proton Transfer Reactions in Various Solvents

Compound	Observation	pK_a(H_2O) and Solvent	Refs.
$H_2Ir(PMe_3)_4{}^+$	$HIr(PMe_3)_4 + CH_2(CN)_2 \rightarrow [H_2Ir(PMe_3)_4][CH(CN)_2]^-$	> 12; THF	[67]
$[HFe(CNBu^t)_5]^+$	$Fe(CNBu^t)_5 + H_2Os(CO)_4 \rightarrow [HFe(CNBu^t)_5][HOs(CO)_4]$[b]	> 13; pentane	[68]
$[H_3Os(PMe_3)_4]^+$	*cis*-$H_2Os(PMe_3)_4 + NH_4{}^+ \rightarrow H_3Os(PMe_3)_4{}^+$	> 9.24; MeOH	[69]
$[H_3Fe(PMe_3)_4]^+$	*cis*-$H_2Fe(PMe_3)_4 + NH_4{}^+ \not\rightarrow$	< 9.24	
$[Os(bpy)(PPh_3)_2(CO)H_2]^{2+}$	$[Os(bpy)(PPh_3)_2(CO)H]^+ + CF_3CO_2H \rightleftharpoons [Os(bpy)(PPh_3)_2(CO)H_2]^{2+} + CF_3COO^-$; deprotonated by Et_2O	≈ −3; CF_3CO_2H	[70, 71]
$HReCp(NO)PPh_3$	H—Re bracketed between 26 and 30 Cp H pK_a estimated at 35.9 in THF	≈ 35; THF ≈ 42; THF	[8]
$HWCp(NO)_2$	$HWCp(NO)_2 + NEt_3 \rightarrow$	> 10; hexane	[72]

$[HIrCp^*(CO)_2]^+$	$HIrCp^*(CO)_2{}^+$ is deprotonated by NEt_3	< 11; CH_3CN	[73]
$[HIr(\eta^3\text{-}HBPz^*_3)(CO)_2]^+$	$HIr(\eta^3\text{-}HBPz^*_3)(CO)_2{}^+$ deprotonated by DBU but not by NEt_3[c, d]	≈ 14; CH_2Cl_2	[73]
$[Rh\{\eta^2\text{-}HBPz^*_2(Pz^*\underline{H})\}(CO)_2]^+$	$Rh\{\eta^2\text{-}HBPz^*_2(Pz^*\underline{H})\}(CO)_2{}^+ + NEt_3 \rightarrow Rh(\eta^3\text{-}HBPz^*_3)(CO)_2$[c]	< 11; CH_2Cl_2	[73]
$RuH_4(PPh_3)_3$	$RuH_3(PPh_3)_3{}^- + c\text{-}C_6H_{11}OH \rightleftharpoons RuH_4(PPh_3)_3 + c\text{-}C_6H_{11}O^-$ ($K_{eq} = 0.13$ in THF)	16; THF	[7]
$(CO)_3Ni{=}C(Bu)O\underline{H}$	$(CO)_3Ni{=}C(Bu)O\underline{H}$ + $PhCH_2C({=}O)CHLiPh \rightleftharpoons (CO)_3Ni{=}C(Bu)OLi + (PhCH_2)_2C{=}O$	16.3; THF	[74]
$Cp(CO)_2Mo(\mu\text{-}PPh_2)(\mu\text{-}H)Mn(CO)_4$	Fully deprotonated by KOH–EtOH	< 15; EtOH	[75]

[a] Aqueous pK_a values are estimated without any correction for dielectric or other solvent effects from the pK_a of the other partner in the proton transfer reaction.

[b] Estimated from the approximate aqueous pK_a of $H_2Os(CO)_4$ in Table 9.3.

[c] Pz^* = 3,5-dimethylpyrazol-1-yl.

[d] DBU = 1,8-diazabicyclo[5.4.0]undec-7-ene.

radicals. The pK_a of the parent 18-electron hydride (i) and the oxidation potential of the 18-electron conjugate base M^- (ii) combine with (iii) to complete a thermodynamic cycle for reaction (iv) and, thus, to allow the calculation of the pK_a of the cation radical by eq. (v).

The purity of the acetonitrile is critical for obtaining accurate pK_a values, particularly with the more acidic hydrides. We have found the most effective purification procedures to be (a) distillation from alkaline $KMnO_4$, followed by distillation from $KHSO_4$, fractional distillation from CaH_2, and vacuum transfer from P_4O_{10} or (b) distillation from $CuSO_4$, passage through a column of highly activated alumina, and distillation from that alumina [13a, 34].

We have tabulated all acetonitrile pK_a measurements on transition metal hydrides in Table 9.1. We have also attempted to compile in Table 9.1 all methanol and water pK_a values *measured directly in those solvents*. (A few hydride pK_a values obtained in related solvents have been included in the methanol and water columns.) A few pK_a values for carbon acid ligands are also given.

In some cases limits can be placed (Table 9.2) on the aqueous pK_a values of transition metal hydrides from their behavior in other solvents. Relative acidities can be deduced from the fact that certain proton transfer reactions do or do not occur, and absolute acidities can be estimated from these relative acidities if the aqueous pK_a of one partner in the proton transfer reaction is known.

Scant quantitative data are available on the acidities of vanadium, niobium, and tantalum carbonyl hydrides. Attempts at the protonation of $[V(CO)_6]^-$ in the presence of a proton acceptor solvent A do not yield $HV(CO)_6$ but result in the formation of $[HA][V(CO)_6]$. Treatment of $[V(CO)_6]^-$ with HCl in an anhydrous solvent causes oxidation of the vanadium center, yielding the paramagnetic $V(CO)_6$ and H_2 [36]. Addition of HCl to $[Nb(CO)_6]^-$ or $[Ta(CO)_6]^-$ under anhydrous conditions or in aqueous solution yields the chloro bridged dimers $[H(THF)_2][M_2(CO)_8Cl_3]$ rather than the corresponding tantalum or niobium carbonyl hydrides [76].

Phosphine substitution increases the stability of these complexes. Hydride complexes $HV(CO)_{6-n}L_n$ can be isolated: $HV(CO)_5PPh_3$ is known [35], and several diphosphine-substituted vanadium hydride complexes can be made from water and the corresponding monoanion [77]. The hydrides $HM(CO)_{6-n}L_n$ (where M = Nb [78] or Ta [79], $n > 1$) can be synthesized by running a solution of the corresponding carbonylmetallate anions [80] through a silica gel column, implying that the $[M(CO)_4L_2]^-$ anions are basic enough to be protonated by silica gel.

9.2.1. Estimated Aqueous pK_a Values from Acetonitrile pK_a Measurements

Although acetonitrile appears to be the most useful solvent for the *measurement* of transition metal hydride pK_a values, most of the acidity data in the literature are referenced to the aqueous scale. We have therefore decided to calculate approximate aqueous pK_a values from our acetonitrile pK_a values. Unfortunately no general equation relating acidities in acetonitrile to those in water [81, 82] can be written. Early work in the Kolthoff [27] and Coetzee [28a, 28b] groups showed the CH_3CN–H_2O difference to be fairly constant within a class of acids of similar structure (alkylammonium cations, pyridinium cations, phenols, or carboxylic acids) but to be different among different classes of acids. These workers concluded that there was no simple quantitative correlation between pK_a values in acetonitrile and those in water.

After a study of acidities in water, acetonitrile, and DMSO, Kolthoff, Chantooni, and Bhowmik [83] estimated the proton acceptor power of water to exceed that of acetonitrile by about 7.7 pK units. Subsequent measurements [84] have confirmed that the free energy of transfer of H^+ from acetonitrile to water ($\Delta G_{tr}^{H^+}$) is about -11 kcal mol^{-1} (-8 pK units). If we consider the definition of pK_a [eq. (9.13)], we can see that the difference between the pK_a of an acid in water (pK_a^W) and the pK_a of the same acid in acetonitrile (pK_a^{AN}) will be given by eq. (9.14) [85]:

$$\mathrm{HA} \overset{K_a}{\rightleftharpoons} \text{solvent } \mathrm{H^+} + \mathrm{A^-} \tag{9.13}$$

$$\mathrm{p}K_a^{W} = \mathrm{p}K_a^{AN} + \frac{\Delta G_{tr}^{H^+}(\mathrm{AN} \rightarrow \mathrm{W})}{2.301RT} + \frac{\Delta G_{tr}^{A^-}(\mathrm{AN} \rightarrow \mathrm{W}) - \Delta G_{tr}^{HA}(\mathrm{AN} \rightarrow \mathrm{W})}{2.301RT} \tag{9.14}$$

The $\Delta G_{tr}^{H^+}$ term in eq. (9.14) is constant regardless of the nature of the acid. However, the $\Delta G_{tr}^{A^-} - \Delta G_{tr}^{HA}$ term, which describes the effect of the change in solvent on the solvation energy *difference* between A^- and HA, can vary from one class of acid to another—which explains why the CH_3CN–H_2O pK_a difference varies from one class of acid to another.

Fortunately, the $\Delta G_{tr}^{A^-} - \Delta G_{tr}^{HA}$ term in eq. (9.14) *is* small for large delocalized anions. There is ample evidence that the solvation energies of organic anions do not depend appreciably on the solvent [86–89]. Kanabus-Kaminska, Gilbert, and Griller [90] concluded that "...for large organic anions, solvation energies [vary] little in going from one polar solvent to another" and that the major effect of a solvent change on an R—H deprotonation equilibrium is "attributable to the very small charged species, the proton".

We feel justified in assuming that the difference $\Delta G_{tr}^{A^-} - \Delta G_{tr}^{HA}$ can be neglected for the relatively large (and extensively delocalized) anions resulting from the deprotonation of transition metal acids. If so, the CH_3CN–H_2O ΔpK_a for transition metal hydrides will be equal to the $\Delta G_{tr}^{H^+}$ term of eq. (9.14) (about -8 pK units) only. The differences between the CH_3CN and

H_2O pK_a values for the four transition metal hydrides in Table 9.1 that have been examined in both solvents are consistent with this prediction. For example, the pK_a of $H_2Fe(CO)_4$ is 11.4 in CH_3CN but 4.0 in H_2O ($\Delta = -7.4$ units); that of $HMn(CO)_5$ is 14.2 in CH_3CN but 7.1 in H_2O ($\Delta = -7.1$); that of $HCo(CO)_3(P(OPh)_3)$ is 11.4 in CH_3CN but 4.95 in H_2O ($\Delta = -6.45$); and that of $HCo(CO)_3PPh_3$ is 15.4 in CH_3CN but 6.96 in H_2O ($\Delta = -8.44$). The average ΔpK_a for all four cases is -7.35 units.

If we combine this average ΔpK_a with the Kolthoff estimate (7.7 pK units) [83] of the proton acceptor power of water relative to that of

Table 9.3
Thermodynamic (Water Scale) Acidities of Common Transition Metal Hydrides (Continued)

Compound	pK_a(H_2O)
$HV(CO)_6$	Strong
$HV(CO)_5PPh_3$	6.8[a]
$HV(CO)_4$(dppe)[e]	< 10[b]
$HCrCp(CO)_3$	5.4[a]; 5.8[b]
$HMoCp(CO)_3$	6.2[a]; 6.4[b]
$HMoCp^*(CO)_3$	9.6[b]
$HWCp(CO)_3$	8.0[a]; 8.6[b]
$HWCp(CO)_2PMe_3$	19.1[b]
$HMn(CO)_5$	7.1[a]; 6.7[b]
$HMn(CO)_4(PPh_3)$	12.9[b]
$HMn(CO)_4(PEtPh_2)$	14.1[b]
$(\eta^3\text{-}C_6H_8\cdots H\cdots)Mn(CO)_3$[f]	14.7[b]
$HMn(CO)_2(\eta^6\text{-}C_6H_6)$	19.3[b]
$HRe(CO)_5$	Very weak[a]; 13.6[b]
$CpRe(CO)_2H_2$	15.5[b]
$Cp_2ReH_2^+$	5.5[a]
$H_3Re_3(CO)_{12}$	3[a]
$[H_2Re_3(CO)_{12}]^-$	10[a]
$[HRe_3(CO)_{12}]^{2-}$	25[a]
$MH(H_2)(dppe)_2^+$ (M = Fe, Ru, Os)[e]	> 10[c] [92]
$H_2Fe(CO)_4$	4.4[a]; 3.9[b]
$HFe(CO)_3NO$	< 3[d] [93]
$HFeCp(CO)_2$	11.9[b]
$HFeCp^*(CO)_2$	18.8[b]
$(\mu\text{-}H)_2Fe_3(CO)_9(\mu_3\text{-}P\text{-}4\text{-}C_6H_4OCH_3)$	1.5[b]
$(\mu\text{-}H)_2Fe_3(CO)_9(\mu_3\text{-}P\text{-}t\text{-}Bu)$	3.9[b]
$(\mu\text{-}H)Fe_3(CO)_9(\mu_3\text{-}S\text{-}C_6H_{11})$	9.4[b]
$H_2Ru(CO)_4$	11.2[b]
$HRuCp(CO)_2$	12.7[b]
$CpRu(H_2)(dmpe)^+$[e]	10.1[b]
$CpRu(H)_2(dppp)^+$[e]	8.4[c] [92]
$CpRu(H)_2(PPh_3)_2^+$	8.3[c] [92]
$CpRuH_2(dppe)^+$[e, g]	7.3[c, g] [92]

Table 9.3
Thermodynamic (Water Scale) Acidities of Common Transition Metal Hydrides (Continued)

Compound	$pK_a(H_2O)$
$CpRu(H)_2(dppm)^{+}$ [e]	7.1[c] [92]
$H_4Ru_4(CO)_{11}[P(OCH_3)_3]$	4.9[b]
$H_4Ru_4(CO)_{10}[P(OCH_3)_3]_2$	7.9[b]
$H_2Os(CO)_4$	13.3[b]
$H(CH_3)Os(CO)_4$	15.5[b]
$H_2Os_2(CO)_8$	12.9[b]
$HCo(CO)_4$	Strong[a]; 0.9[b]
$HCo(CO)_3(PPh_3)$	7.0[a]; 7.9[b]
$HCo(CO)_3P(OPh)_3$	4.9[a]; 3.9[b]
$HCo(CO)_2(PPh_3)_2$	Very weak[a]
$[HCo(CN)_5]^{3-}$	20[a]
$HRh(en)_2OH^+$	~ 12[a]
$[Rh_{13}(CO)_{24}H_3]^{2-}$	3.5[b]
$[Rh_{13}(CO)_{24}H_2]^{3-}$	9.0[b]
$HNi[P(OMe)_3]_4{}^+$	5.9[b]
$HPd[P(OMe)_3]_4{}^+$	3.1[b]
$HPt[P(OMe)_3]_4{}^+$	11[b]
$Cp(PMe_3)_2Ru{=}C{=}C(\underline{H})CMe_3{}^+$	13.3[b]
$Cp(PMe_3)(CO)Fe{=}C{=}C(\underline{H})CMe_3{}^+$	6.1[b]
$(CO)_5Cr{=}C(OMe)C\underline{H}_3$	12.3[a]

[a] Data from the H_2O column of Table 9.1.

[b] Estimated using eq. (9.15) from data in the CH_3CN column of Table 9.1.

[c] Estimated using eq. (9.18) from data measured in CH_2Cl_2.

[d] Estimated from the stability of $TlFe(CO)_3NO$ and $TlFe(CO)_2(NO)P(OC_6H_4Cl)_3$ compared to that of $TlFe(CO)_2(NO)P(C_6H_5)_3$.

[e] dmpe = $PMe_2CH_2CH_2PMe_2$, dppe = $PPh_2CH_2CH_2PPh_2$, dppp = $PPh_2CH_2CH_2CH_2PPh_2$, dppm = $PPh_2CH_2PPh_2$.

[f] The complex contains a three-center 2-electron interaction involving the 4-endo-CH bond and the Mn [66].

[g] A rapidly interconverting 1:2 mixture of the η^2-H_2 and dihydride forms is present in solution, so the measured pK_a (said [92] to be "the pK_a of the $CpRu(H_2)(dppe)^+$ form") is higher than the pK_a of either form for the reasons given in section 9.2.3.

acetonitrile, we obtain eq. (9.15) as a reasonable way of estimating the pK_a in water of a transition metal hydride from its pK_a measured in acetonitrile. Although inherently imprecise, eq. (9.15) should be accurate enough to allow the comparison of M—H pK_a values with other acidity data in the literature. Consideration of the four measured hydride ΔpK_a values mentioned previously suggests that the error in estimates from eq. (9.15) is ± 1 pK unit.

$$pK_a(H_2O) = pK_a(CH_3CN) - 7.5 \qquad (9.15)$$

For all the acetonitrile-measured transition metal hydride pK_a values listed in Table 9.1, an estimate from eq. (9.15) of the corresponding aqueous pK_a value is given in Table 9.3. Table 9.3 also contains all the hydride pK_a values that have been *measured* in water.

9.2.2. Estimated Aqueous pK_a Values from the Use of Tertiary Phosphines as an Internal Secondary Standard in CH_2Cl_2

There are cases when neither water nor acetonitrile is a feasible solvent for pK_a measurements. For example, the pK_a of most dihydrogen complexes [91] cannot be measured in acetonitrile because of the donor ability of that solvent; the CH_3CN can displace the coordinated H_2.

A useful way of estimating the aqueous pK_a values of these complexes is due to Jia and Morris [92]. The pK_a values of HPR_3^+ (where R = Cy, *n*-Bu) are known in H_2O by extrapolation from measurements in CH_3NO_2, and these cations can be used as secondary internal standards. The equilibrium constants for a succession of proton transfer equilibria like eqs. (9.16) and (9.17) can be measured easily in other solvents such as CH_2Cl_2, and the acidity of a cation $M(H_2)^+$ can thus be related to that of HPR_3^+.

$$M(H_2)^+ + M'\text{—}H \overset{K_1}{\rightleftharpoons} M\text{—}H + M'(H_2)^+ \tag{9.16}$$

$$M'(H_2)^+ + PR_3 \overset{K_2}{\rightleftharpoons} M'\text{—}H + HPR_3^+ \tag{9.17}$$

Just as the CH_3CN–H_2O ΔpK_a is fairly constant within a class of acids of similar structure (e.g., it is about 7.6 pK units for all alkylammonium cations [28b]), the CH_2Cl_2–H_2O ΔpK_a should be fairly constant for all large cations. The *relative* acidity of $M(H_2)^+$ and HPR_3^+ thus should be about the same in CH_2Cl_2 and in H_2O, and we can estimate the aqueous pK_a of the hydride cationic dihydrogen complex from the aqueous pK_a of HPR_3^+ by using

$$pK_a([M(H_2)]^+) = pK_a(HPR_3^+) - \log(K_1K_2) \tag{9.18}$$

We have attempted to compile all transition metal pK_a values measured or estimated in water in Table 9.3.

9.2.3. Uncertainty in pK_a Measurements Involving Tautomeric Equilibria

When a tautomeric equilibrium relates two or more acids that have the same conjugate base, pK_a measurement becomes complex. An important example is offered by the many dihydrogen complexes that exist in equilibrium with the corresponding dihydrides [eq. (9.19)].

$$\overset{+}{M} \leftarrow \overset{H}{\underset{H}{|}} \overset{K_3}{\rightleftharpoons} \overset{+}{M}\begin{matrix} \diagup H \\ \diagdown H \end{matrix} \tag{9.19}$$

If $K_a(H_2)$ is the equilibrium constant for the deprotonation of the dihydrogen complex in eq. (9.20) and $K_a(2H)$ is the equilibrium constant for the deprotonation of the dihydride *to the same conjugate base* in eq. (9.21), the less abundant (and thus more energetic) species in eq. (9.19) *must* be the stronger acid.

$$M(H_2)^+ + B \rightleftharpoons MH + BH^+ \quad K_a(H_2) \tag{9.20}$$

$$M(H)_2{}^+ + B \rightleftharpoons MH + BH^+ \quad K_a(2H) \tag{9.21}$$

If equilibrium (9.19) is maintained so rapidly that we can only measure $T = [M(H_2)^+] + [M(H)_2{}^+]$, a measurement of the acidity of the mixture will give an apparent K_a value (K_a^{app}) less than $K_a(H_2)$ [eq. (9.22)] and less than $K_a(2H)$ [eq. (9.23)]. The resulting apparent pK_a thus will be greater than the pK_a of the weaker acid (the more-abundant tautomeric form).

$$K_a^{app} = \frac{[MH][BH^+]}{[T][B]} = \frac{[MH][BH^+]}{\{[M(H_2)^+] + [M(H)_2{}^+]\}[B]} = \frac{1}{1+K_3}K_a(H_2) \tag{9.22}$$

$$K_a^{app} = \frac{1}{(1/K_3)+1}K_a(2H) = \frac{K_3}{1+K_3}K_a(2H) \tag{9.23}$$

The discrepancy between the apparent pK_a and the actual pK_a of the more-abundant tautomer approaches zero as either tautomer begins to dominate eq. (9.19). The correction is maximal (0.3 pK units) when $K_3 = 1$. The pK_a of whichever tautomer is less-abundant can be calculated from eq. (9.24):

$$pK_a(H_2) = pK_a(2H) - \log K_3 \tag{9.24}$$

9.2.4. Trends in Thermodynamic Acidities

Not surprisingly, the acidity of a hydride complex decreases if a π-acceptor ligand is replaced by a σ-donor ligand. Tables 9.1 and 9.3 imply that pK_a increases in the order $HV(CO)_6 < HV(CO)_5PPh_3 < HV(CO)_4(dppe)$ and show the following series of increasing pK_a:

$$HCo(CO)_4 < HCo(CO)_3(P(OPh)_3) < HCo(CO)_3(PPh_3);$$
$$HW(CO)_3Cp < HW(CO)_2(PMe_3)Cp;$$
$$HMn(CO)_5 < HMn(CO)_4(PPh_3) < HMn(CO)_4(PEtPh_2).$$

The same trend is evident in qualitative observations on protonation equilibria. For example, in CF_3COOH–CH_2Cl_2 the extent of protonation of the metal center increases in the order $CpMn(CO)_3 < CpMn(CO)_2PPh_3$ and in the order $(\eta^5\text{-}C_5Et_5)Mn(CO)_3 < (\eta^5\text{-}C_5Et_5)Mn(CO)_2PPh_3$ [94]. For cobalt centers, base strength decreases in the order $CpCo(PMe_3)_2 > CpCo(P(OR)_3)_2 > CpCo(CO)_2$ [95].

Replacement of Cp by Cp* also decreases acidity. Tables 9.1 and 9.3 show that pK_a increases in the order $HFe(CO)_2Cp < HFe(CO)_2Cp^*$ and $HMo(CO)_3Cp < HMo(CO)_3Cp^*$.

The acidity of analogous hydrides usually decreases down a column in the Periodic Table. Tables 9.1 and 9.3 show the following series of increasing pK_a:

$$HCr(CO)_3Cp < HMo(CO)_3Cp < HW(CO)_3Cp;$$
$$HMn(CO)_5 < HRe(CO)_5;$$
$$H_2Fe(CO)_4 < H_2Ru(CO)_4 < H_2Os(CO)_4;$$
$$HFe(CO)_2Cp < HRu(CO)_2Cp.$$

The same trend is again evident in qualitative observations on protonation equilibria. For example, in CF_3COOH–CH_2Cl_2 the extent of protonation of the metal center increases in the order $CpV(CO)_3PPh_3 < CpNb(CO)_3PPh_3$ and in the order $CpMn(CO)_2PPh_3 < CpRe(CO)_2PPh_3$ [96]. Ruthenocene is a stronger base then ferrocene [97], and $[CpRu(CO)_2]_2$ is a stronger base in acetic acid than $[CpFe(CO)_2]_2$ [98]. Although the trinuclear carbonyl clusters $Os_3(CO)_{12}$ and $Ru_3(CO)_{12}$ both can be protonated by trifluoroacetic acid in CH_2Cl_2 to give the edge-bridging $[HM_3(CO)_{12}]^+$ clusters [99], the osmium cluster is five times more basic than its ruthenium analog [100].

The pK_a values of $HNi[P(OMe)_3]_4^+$, $HPd[P(OMe)_3]_4^+$, and $HPt[P(OMe)_3]_4^+$ in Tables 9.1 and 9.3 depart from this pattern, with the palladium hydride being the most acidic. There also have been suggestions that $HRh(CO)_4$ is a stronger acid than $HCo(CO)_4$ and $HIr(CO)_4$. In 1975 Imjanitow [101] argued that, on the basis of their ability to catalyze the pinacol rearrangement, the acidity of rhodium carbonyl hydrides exceeded that of cobalt carbonyl hydrides. In 1981 Vidal and Walker [102] observed these hydrides by high-pressure IR spectroscopy and suggested the acidity order $HRh(CO)_4 > HCo(CO)_4 > HIr(CO)_4$ on the basis of three qualitative observations:

1. The preparation of these hydrides by the protonation of $[M(CO)_4]^-$ in tetraglyme–toluene required different acids for different metals.
2. $[Co(CO)_4]^-$ could be seen along with $HCo(CO)_4$ when the latter was generated in the presence of *N*-methylmorpholine, whereas $[Ir(CO)_4]^-$ was not observed when $HIr(CO)_4$ was generated under the same conditions.
3. The intensity of the IR spectra of the various hydrides $HM(CO)_4$ in dodecane was affected to different extents by different amines.

A protonation–deprotonation equilibrium clearly never was achieved for $HRh(CO)_4$ and $HIr(CO)_4$ in the Vidal–Walker experiments, and thus the HRh > HCo > HIr acidity order is questionable; in any case, no quantitative data were obtained. As local density calculations [103] support the acidity order HCo > HRh > HIr, more definitive experiments are desirable.

There is some evidence that bridging hydride ligands are more acidic than terminal ones—but the comparison is imprecise because significant structural differences necessarily are involved. For example, the pK_a values of $(\mu_2\text{-H})_4Ru_4(CO)_{11}[P(OMe)_3]$ and $(\mu_2\text{-H})_4Ru_4(CO)_{10}[P(OMe)_3]_2$ [18, 32] are lower than that of $H_2Ru(CO)_4$. Just as in carbanion chemistry, where cyclopentadiene (which gives rise to a stable delocalized anion) is more acidic than cycloheptatriene (which gives rise to a relatively unstable localized anion), what is important is not the nature of the hydride ligand but the extent to which the resulting anion is delocalized. Polynuclear complexes with bridging hydride ligands are *usually* acidic because they *usually* give extensively delocalized anions. Polynuclear complexes with terminal hydride ligands that give localized anions upon deprotonation [e.g., $H_2Os_2(CO)_8$ in eq. (9.25)] [17a] are not notably more acidic than related mononuclear complexes [e.g., $H_2Os(CO)_4$]. Such logic predicts that $H_2Fe_2(CO)_8$, which has terminal hydride ligands [104] but gives rise to a symmetric delocalized hydride upon deprotonation [eq. (9.26)] [105], will prove to be much more acidic than $H_2Fe(CO)_4$.

$$(CO)_4Os(H){-}Os(H)(CO)_4 \rightarrow [(CO)_4Os(H){-}Os(CO)_4]^{\ominus} + H^+ \tag{9.25}$$

$$(CO)_4Fe(H){-}Fe(H)(CO)_4 \rightarrow \left[(CO)_4Fe(\mu\text{-H})(\mu\text{-CO})_2Fe(CO)_4\right]^- + H^+ \tag{9.26}$$

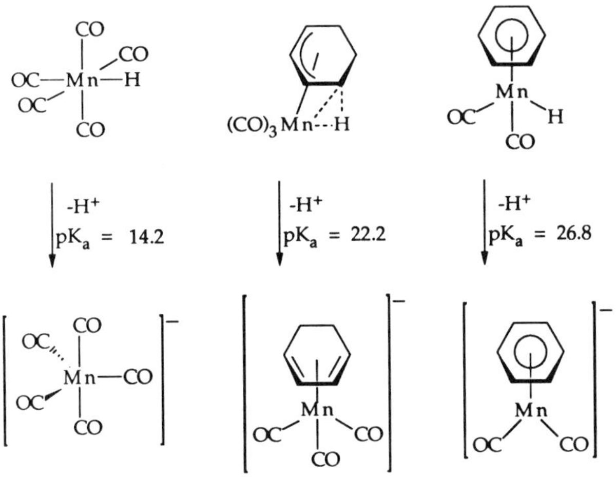

FIGURE 9.3

Similarly, in the series of compounds in Figure 9.3, the pK_a is determined by the stability of the anion rather than by the nature of the hydride ligand removed as a proton. Comparison of the pK_a of $HMn(CO)_5$ with that of $HMn(CO)_2(C_6H_6)$ shows that the pK_a increases by about 4 units for each CO ligand replaced by an olefin, implying a pK_a of about 22.5 for the hypothetical diene hydride complex $HMn(CO)_3(\eta^4\text{-}C_6H_8)$. This is about equal to the pK_a of the actual complex $(\eta^3\text{-}C_6H_8 \cdots H \cdots)Mn(CO)_3$, despite the fact that the proton is removed from the agostic C····H····Mn bond of the latter [66] rather than from a normal Mn—H bond. The acidity of transition metal hydrides appears to be determined more by the *stability of the conjugate base* rather than by the *nature of the hydrogen removed.*

Extrapolation of the CH_3CN pK_a values for the removal of successive protons from the interstitial sites in $[Rh_{13}(CO)_{24}H_{5-n}]^{n-}$ ($n = 2, 3$) suggests that the hypothetical $Rh_{13}(CO)_{24}H_5$ should be a fully dissociated strong acid in acetonitrile [17c]. The difference between successive pK_a values is comparable to that observed as successive protons are removed from H_3PO_4 in water and is larger than that observed as successive protons are removed from another large polynuclear anion, $[H_3PMo_{12}O_{40}]^{4-}$, in water [106].

9.2.5. Relation between Thermodynamic Acidities and M—H Bond Strengths

We have seen that the ΔpK_a between CH_3CN and H_2O varies little from one transition metal hydride to another and arises largely from the free energy transfer of H^+ between these two solvents, and that the ΔpK_a

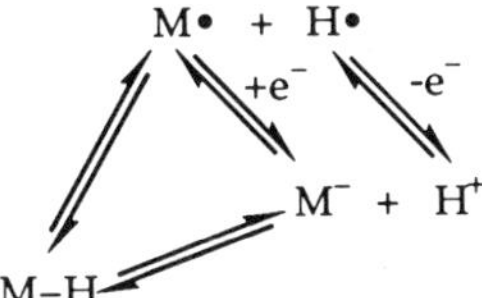

FIGURE 9.4
The acidity of MH in the gas phase depends on the bond dissociation energy of M—H and the electron affinity of M·.

between CH_2Cl_2 and H_2O is fairly constant for large catonic hydrides. The implication is that solvation has little effect on hydride *relative* acidities.

In fact, these relative acidities do not appear to vary even in the gas phase. Approximate gas-phase ΔH values for $H—M \rightarrow H^+ + M^-$ can be determined by studying the ion–molecule reactions of H—M with bases for which ΔH is known. Stevens Miller and co-workers [107, 108] found that ΔH increases between $HCo(PF_3)_4$ and $HIr(PF_3)_4$ and in the order

$$HCo(CO)_4 < HMn(CO)_5 \sim H_2Fe(CO)_4 < HRe(CO)_5.$$

Both series parallel the order of increasing pK_a values in Tables 9.1 and 9.3.

A glance at Figure 9.4 shows that the acidity of a hydride M—H in the gas phase is a function of the M—H bond dissociation energy (BDE) and of the electron affinity of the metalloradical M·.

Some gas-phase M—H BDE values are available [109], but little is known about M· electron affinity values [110]. There are few if any transition metal hydrides for which a known value of the gas-phase acidity (ΔH for $H—M \rightarrow H^+ + M^-$ in Figure 9.4) can be compared with independently determined values of both the M—H BDE and the M· electron affinity.

Analysis of the factors that influence acidity in solution is more practical. There the electron affinity of M· can be replaced by the easily measured potential of the $M\cdot/M^-$ couple, the gas-phase acidity of M—H can be replaced by the easily measured pK_a values in Tables 9.1 and 9.3, and the gas-phase M—H BDE can be replaced by its solution equivalent. The known gas-phase M—H bond strengths are probably reasonable estimates of the corresponding solution values [109], but, more importantly, there are a number of ways in which solution M—H bond strengths can be directly determined.

Calorimetry has been used by several groups, notably that of Hoff, to obtain solution M—H BDE values [111]—although the accuracy of the results inevitably depends upon the accuracy of the other bond strengths (e.g., M—M BDE values) that enter into the calculation. A few bond strengths also have been determined by kinetic methods [112]—but these methods are not applicable to many hydrides and depend on assumptions about the energetics of the same reactions in the reverse direction.

$$\text{M—H(sol)} \rightleftharpoons \text{M}^-\text{(sol)} + \text{H}^+\text{(sol)} \qquad \text{(i)}$$

$$\text{M}^-\text{(sol)} \rightleftharpoons \text{M}\cdot\text{(sol)} + e^- \qquad \text{(ii)}$$

$$\text{H}^+\text{(sol)} + e^- \rightleftharpoons 0.5\ \text{H}_2\text{(g)} \qquad \text{(iii)}$$

$$0.5\text{H}_2\text{(g)} \rightleftharpoons \text{H}\cdot\text{(g)} \qquad \text{(iv)}$$

$$\text{H}\cdot\text{(g)} \rightleftharpoons \text{H}\cdot\text{(sol)} \qquad \text{(v)}$$

$$M\text{—}H(sol) \rightleftharpoons \text{M}\cdot\text{(sol)} + \text{H}\cdot\text{(sol)}$$

$$\Delta\text{G}^\circ = \textit{bond dissociation free energy}$$

Scheme 9.2
Thermodynamic cycle for bond strength determination in solution.

The most useful M—H bond strengths for our purposes are those obtained by Tilset and Parker [113] from the solution equivalent (Scheme 9.2) of the thermodynamic cycle in Figure 9.4. Similar cycles for carbon acids were first written by Breslow [114] and, after refinement by Nicholas and Arnold [115], were used by Bordwell and co-workers [116] to determine C—H bond strengths in DMSO. ΔG° for step (i) in acetonitrile is available from the pK_a value of the hydride in the appropriate column of Table 9.1. ΔG° for steps (ii) and (iii), the reversible potential for the oxidation of the metal anions versus the hypothetical hydrogen electrode in acetonitrile, can in principle be obtained by cyclic voltammetry [117–120]. The free energies of dissociation of H_2 [step (iv)] and of solvation of H· [step (v)] complete the cycle, yielding the free energy for bond dissociation.

The bond dissociation free energy is easily converted to a bond dissociation enthalpy by assuming that S°(M—H) ≈ S°(M·) and that ΔS for M—H → M· + H· is therefore equal to S°(H·) in acetonitrile. However, S°(H·) in acetonitrile equals the entropy changes in steps (iv) and (v) of Scheme 9.2, so only the enthalpy changes for steps (iv) and (v) affect the bond dissociation enthalpies. The eventual expression for the M—H bond dissociation enthalpy is

$$\text{BDE (kcal mol}^{-1}) = 1.37\ \text{p}K_a + 23.06 E^\circ_{\text{ox}}(\text{M}^-) + 58.3 \qquad (9.27)$$

The M—H bond dissociation enthalpies that have been determined from eq. (9.27) are listed in Table 9.4 along with recent values determined by other methods. A recently determined [121] value of the Sn—H bond strength is included for comparison.

Table 9.4
M—H Bond Dissociation Enthalpies

Compound	Values from Eq. (9.27) (kcal mol^{-1})	Values from other methods (kcal mol^{-1})
$HCr(CO)_3Cp$	62	61[a]; 61.5 ± 1[a] [111e]
$HCr(CO)_3Cp^*$ ($Cp^* = \eta^5$-C_5Me_5)		62.3 ± 1[a] [111e]
$HCr(CO)_2PPh_3Cp$		59.8 ± 1[a] [111e]
$HCr(CO)_2PEt_3Cp$		59.9 ± 1[a] [111e]
$HCr(CO)_2P(OMe)_3Cp$		62.7 ± 1[a] [111e]
$HMo(CO)_3Cp$	70	65 ± 6[a] [111a]; 66 ± 8[a] [111b]; 66 ± 1.5[a] [111c]; ≤ 65[b]
$HMo(CO)_3Cp^*$	69	
$HW(CO)_3Cp$	73	≤ 67[b]
$HW(CO)_2PMe_3Cp$	70	
$HMn(CO)_5$	68	60[c] [107]; 63[d] [112]; ≤ 65[b]
$HMn(CO)_4PPh_3$	69	
$HMn(CO)_4PEtPh_2$	71 [119]	
$HRe(CO)_5$	75	
$HFe(CO)_2Cp$	58	
$HRu(CO)_2Cp$	65	
$H_2Fe(CO)_4$	68	
$H_2Os(CO)_4$		78[a] [122]
$HCo(CO)_4$	67	56[e]; ≤ 63[b]
$HCo(CO)_3P(OPh)_3$	66	
$HCo(CO)_3PPh_3$	65	
Bu_3Sn—H		73.7 ± 2.0 [121]

[a] Calorimetric.

[b] Upper limits estimated [113] from published activation parameters [123–125] for reactions of these hydrides with various conjugated olefins.

[c] Difference in the appearance potential for $^+Mn(CO)_5$ between $HMn(CO)_5$ and $\cdot Mn(CO)_5$ [101].

[d] Kinetic.

[e] Thermochemical, from various measurements of ΔH for the hydrogenolysis of $Co_2(CO)_8$, using a Co—Co BDE of 15 kcal mol^{-1}; for a discussion see [109].

As the result of a suggestion by Halpern, Tilset and Parker [113] estimated upper limits to several M—H bond strengths from the $\Delta H^\ddagger$ values for H· transfer from these hydrides to various conjugated olefins [123–125]. In all cases, the M—H BDE values calculated from eq. (9.27) exceed these kinetic upper limits (listed with ≤ signs in the right-hand column of Table 9.4) by several kilocalories per mole. Furthermore, BDE for H—$Mo(CO)_3Cp$ from eq. (9.27), 70 kcal mol^{-1}, is greater than the calorimetrically determined strength of this bond: Hoff and co-workers [111a–111c] obtained the same result (66 kcal mol^{-1}) for the H—$Mo(CO)_3Cp$ BDE in three different ways [126].

Other results in the literature are more consistent with the exact M—H BDE values from eq. (9.27) that are given in Table 9.4. Hoff has recently obtained values of the H—$Cr(CO)_3Cp$ BDE in two different ways: 61 kcal mol^{-1} from the heat of hydrogenation of the chromium dimer $[CpCr(CO)_3]_2$ [111c, 111e]; 61.5 kcal mol^{-1} from the heat of hydrogenation of $[Cp^*Cr(CO)_3]_2$ (where $Cp^* = \eta^5$-C_5Me_5) combined with the measured ΔH for H· transfer to ·$Cr(CO)_3Cp$ from H—$Cr(CO)_3Cp^*$ [111e]. Both determinations use Cr—Cr bond strengths derived from *directly measured* monomer–dimer equilibrium constants [111e, 127] and thus should be particularly accurate, and both values are relatively close to the BDE for H—$Cr(CO)_3Cp$ from eq. (9.27) (62 kcal mol^{-1}). Furthermore, Brown et al. [128] found that H· is abstracted from Bu_3SnH by ·$Re(CO)_5$ when the latter is generated from $Re_2(CO)_{10}$ by flash photolysis. The exact BDE value for H—$Re(CO)_5$ from eq. (9.27) given in Table 9.4 (75 kcal mol^{-1}, compared to 73.7 kcal mol^{-1} for Bu_3Sn—H) is consistent with this observation, but a lower H—$Re(CO)_5$ BDE would be inconsistent with it.

It remains uncertain whether the BDE values given by eq. (9.27) are systematically a little higher than the true values. Plausible sources of error include underestimation of the enthalpy of solvation of H· and neglect of ionic strength effects [129]. However, the eq. (9.27) bond strengths in Table 9.4 are accurate enough to rationalize many known reactions, and it is instructive to compare them with measured thermodynamic acidities.

9.2.6. Relative Influence of M—H Bond Strengths and Oxidation Potentials on Thermodynamic Acidities

Examination of Table 9.4 shows that there is little change in M—H BDE values when a π-acceptor ligand is replaced by a σ-donor ligand, or when Cp is replaced by Cp*. For example, there is less than 3 kcal mol^{-1} variation in the H—Cr bond strengths of $HCr(CO)_3Cp$, $HCr(CO)_3Cp^*$, and $HCr(CO)_2LCp$; the H—Mo bond strengths of $HMo(CO)_3Cp$ and $HMo(CO)_3Cp^*$ are essentially equal, and so are the H—Mn bond strengths of $HMn(CO)_5$ and $HMn(CO)_4L$. Local density calculations [103] show that substitution of PH_3 has a negligible (< 3 kcal mol^{-1}) effect on the H—Co BDE of $HCo(CO)_4$.

The increase in pK_a when a π-acceptor ligand is replaced by a σ-donor ligand thus occurs largely because the $M\cdot/M^-$ potential becomes more negative. This increase can be estimated in some cases when it has not been measured. Several groups have developed ways of predicting the potentials of metal complexes from their structure and substituents [130–134]. A particularly simple model, due to Pickett and Pletcher [131], calculates the potentials for substituted carbonyl complexes $M(CO)_{x-n}(L)_n^{y\pm}$ from eq. (9.28),

Table 9.5
Ligand Inductive Parameters for a Few Common Ligands [133, 134]

Ligand	CO	NO	$P(OC_6H_5)_3$	$P(C_6H_5)_3$	MeCN	NH_3
$\left[\frac{\delta E^\circ}{dn}\right]_L$	0	−0.08	−0.19	−0.36	−0.54	−0.80

where A is a constant dependent on the reference and solvent used, $[\delta E^\circ/dn]_L$ is a parameter established empirically for each ligand, and y is the charge on the complex

$$E^\circ_{ox} = A + n\left[\frac{\delta E^\circ}{dn}\right]_L + 1.48y \tag{9.28}$$

Values of $[\delta E^\circ/dn]_L$ for a few common ligands are given in Table 9.5.

Equation (9.28) can be used to estimate of the change in acidity caused by ligand substitution in a hydride complex with a six-coordinate conjugate base. When combined with eq. (9.27) and applied to compounds with the same M—H bond and therefore the same BDE, it gives eq. (9.29) for the effect of substitution on pK_a. Equation (9.29) correctly predicts the pK_a difference between $IICo(CO)_4$ and $HCo(CO)_3P(OPh)_3$ (calculated 3.2 observed 3.0) even though the coordination number of cobalt in the corresponding anions is not six.

$$\Delta pK_a = \frac{23.06}{1.37}\Delta\left[\frac{\delta E^\circ}{dn}\right]_L \tag{9.29}$$

Equations (9.28) and (9.29) work because oxidation potentials reflect the energy of the highest occupied molecular orbital (HOMO), and the HOMO energy is a simple additive function of ligand effects [134]. Sarapu and Fenske [135] have shown that there is an excellent correlation between the oxidation potentials of substituted manganese carbonyl cations $Mn(CO)_{6-x}(CNMe)_x^{\;+}$ and their calculated (nonempirical Fenske–Hall) HOMO energies.

All of this suggests that the acidities of a series of homometallic hydrides should increase as the HOMO energies of their conjugate bases decrease. However, Bursten and Gatter [136] argued that "relative acid strength in a series of related complexes ... is directly related to the magnitudes of both the HOMO/second-highest orbital gap and the HOMO/LUMO gap". Although the physical basis for this proposal is not apparent, these Fenske–Hall molecular orbital calculations of the HOMO–LUMO gap do reproduce general trends in the acidity of hydride complexes [136].

When we compare the hydride complexes of *different* transition metals, variation in M—H BDE values becomes an important cause of variation in acidity. All the bond strengths in Table 9.4 increase ($Cr < Mo < W$; $Mn < Re$; $Fe < Ru$) down a column in the Periodic Table, and this increase is a major cause of the increase in pK_a down a column. In some cases, the corresponding oxidation potentials are known to vary little: the potentials of $[CpMo(CO)_3]^-$ and $[CpW(CO)_3]^-$ are equal [113], and those of $[Mn(CO)_5]^-$ and $[Re(CO)_5]^-$ differ by only 123 mV [113, 137]. In other cases, however, varying the metal has a substantial effect on both the bond strength and the oxidation potential, and both influence the pK_a. A similar combination of factors determines gas-phase acidities: Ziegler, in view of his local density calculations on carbonyl hydride complexes, has concluded that M—H bond strengths *as well as* M· electron affinities are responsible for the increase in proton affinity down a triad [103b].

In summary, we can offer two generalizations that, if not universally valid, are certainly useful: (1) altering the ligands of a hydride complex H—M changes its acidity only by affecting the HOMO of M^-, whereas (2) altering the metal in H—M changes its acidity by affecting the H—M bond strength as well.

9.3. KINETIC ACIDITY

We have already seen that there are substantial electronic and structural changes when transition metal complexes are protonated and deprotonated and that the resulting high barriers cause the proton transfer reactions of transition metals to be slower than those of oxygen or nitrogen. However, the wide variety of ways in which hydrides can be coordinated to transition metals means that the barriers to their removal as protons vary considerably, providing a challenge to our understanding of the factors that influence proton transfer rates.

9.3.1. Equations Relating Proton Transfer Rates to Equilibrium Constants

If "the members of [a] series of weak bases or acids are not too dissimilar in structure and large steric effects are not present" [138], the relation between the rate constants (k) and the equilibrium constants (K) for their proton transfer reactions with a single partner can be described by the Brønsted equation [eq. (9.30)] [139]. The Brønsted coefficient α usually lies between 0 (for very exothermic reactions) and 1 (for very endothermic reactions). Although attempts have been made to infer transition-state structure from the value of α [19, 140], recent work on proton transfers involving carbon acids has shown that no simple relationship exists [141].

$$\log k = \alpha \log K + C \tag{9.30}$$

Proton transfer reactions have also been treated [19, 142] by the general equation first derived by Marcus for electron transfer reactions [143]. In eq. (9.31), $\Delta G_0^\ddagger$ is the intrinsic barrier—the barrier the transfer would face if it were thermoneutral and $\Delta G°$ were zero. Taking the derivative of $\Delta G^\ddagger$ by

ΔG° [eq. (9.32)] gives the Brønsted parameter α as a function of ΔG° and $\Delta G_0^\ddagger$. Equation (9.32) predicts that a Brønsted plot will be linear only if the intrinsic barrier $\Delta G_0^\ddagger$ is large relative to ΔG°.

$$\Delta G^\ddagger = \left(1 + \frac{\Delta G^\circ}{4\,\Delta G_0^\ddagger}\right)^2 \Delta G_0^\ddagger \tag{9.31}$$

$$\alpha = \frac{\partial\,\Delta G^\ddagger}{\partial\,\Delta G^\circ} = \frac{1 + \Delta G^\circ/4\,\Delta G_0^\ddagger}{2} \tag{9.32}$$

9.3.2. Agreement of Eq. (9.32) with Experiment; Intrinsic Barriers for Proton Transfer Reactions Involving Mononuclear Hydrides

We have measured the rate constants in CD_3CN at 25°C for proton transfer from $HW(CO)_3Cp$, $HMn(CO)_5$, $H_2Fe(CO)_4$, and $HCo(CO)_4$ to a series of parasubstituted anilines [13b]. For all four hydrides, the resulting Brønsted plots (Figure 9.5) are linear, as expected from eq. (9.32) for large intrinsic barriers $\Delta G_0^\ddagger$. Extrapolation of the measured values of $\log k_{H^+}$ to thermoneutrality ($pK_{eq} = 0$) gives intrinsic barriers of 9.3 kcal mol^{-1} for $CpW(CO)_3H$, 10.9 kcal mol^{-1} for $HMn(CO)_5$, 10.7 kcal mol^{-1} for $H_2Fe(CO)_4$, and 10.7 kcal mol^{-1} for $HCo(CO)_4$. The slopes α of these Brønsted plots are about 0.5 but, as predicted by eq. (9.32), increase [$HCo(CO)_4 < HMn(CO)_5 \sim H_2Fe(CO)_4 < HW(CO)_3Cp$] as pK_{eq} and ΔG° increase.

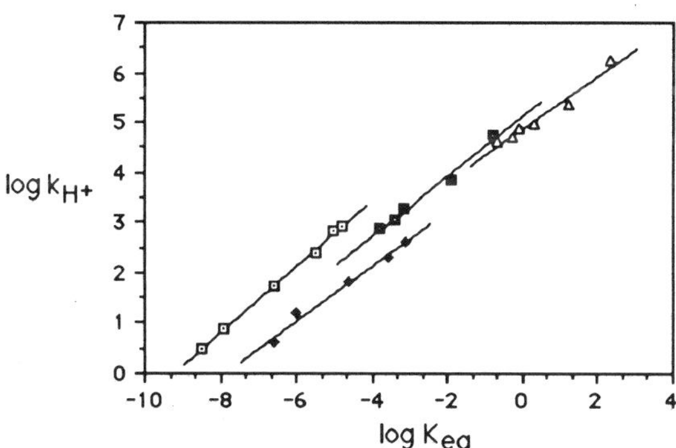

FIGURE 9.5
Brønsted plots showing $\log k_{H^+}$ (where k_{H^+} is the rate constant) as a function of $\log K_{H^+}$ (where K_{H^+} is the equilibrium constant) in CD_3CN at 25°C for H^+ transfer onto various para-substituted anilines from the following: $HW(CO)_3Cp$ (□); $HMn(CO)_5$ (♦); $H_2Fe(CO)_4$ (■); and $HCo(CO)_4$ (△). (Reference 13b.)

9.3.3. Variation of Intrinsic Barrier and Proton Transfer Rate with Type of Hydride Ligand

Steric congestion can increase the intrinsic barrier to proton transfer by forcing the transfer to occur at a relatively large separation of the reactants [155]. For example, amine bases deprotonate $[HMo(CO)_2(dppe)_2]BF_4$ very slowly [144], and the exact rate depends on the steric bulk around the nitrogen rather than on the thermodynamic basicity of the amine. Small anions X^- catalyze this deprotonation by ferrying the proton from the metal to the amine. The rate depends on the identity ($F^- > Cl^- > OAc^- > Br^- > I^-$) and concentration of the X^- catalyst but not on the concentration or identity of the amine (which now serves only as a thermodynamic proton sink). $\Delta G^\ddagger$ is 22 kcal mol^{-1} when X = Cl, and the intrinsic barrier $\Delta G_0^\ddagger$ [which can be estimated from eq. (9.31) as 20 kcal mol^{-1}] is much greater than that found for the uncrowded hydrides in Figure 9.5.

Intrinsic barriers also appear to be larger for proton transfer reactions involving nonterminal hydride ligands. The edge-bridging hydrides of $H_4Ru_4(CO)_{11}(P(OMe)_3)$ have a lower pK_a (Tables 9.1 and 9.3) than the terminal ones of $H_2Ru(CO)_4$, but are transferred to aniline more than 10^4 times more slowly [32]. Steric factors must be part of the reason that deprotonation of bridging hydrides is slow, but the very delocalization of charge that stabilizes their conjugate bases also obliges them to undergo extensive structural and electronic rearrangement as proton transfer occurs. There is evidence [62, 145, 146] that carbon acids that yield highly conjugated anions upon deprotonation have relatively large intrinsic barriers, although the picture is clouded by the difficulty [19a, 147] in determining accurate intrinsic barriers for carbon acids.

The deprotonation of interstitial hydrides is also much slower than would be expected from their thermodynamic acidity. Although the thermodynamic acidities of $[H_3Rh_{13}(CO)_{24}]^{2-}$ and $H_2Fe(CO)_4$ are almost equal (Tables 9.1 and 9.3), the rate constant for deprotonation of $[H_3Rh_{13}(CO)_{24}]^{2-}$ by aniline in acetonitrile (1.2×10^{-3} M^{-1} s^{-1}) is 10^7 times slower than that for deprotonation of $H_2Fe(CO)_4$ (5.4×10^4 M^{-1} s^{-1}) under the same conditions [17c]. The intrinsic barrier $\Delta G_0^\ddagger$ for the removal by aniline of a proton from an interstitial site of $[H_3Rh_{13}(CO)_{24}]^{2-}$ is 21.3 kcal mol^{-1}, about twice the value typical of mononuclear hydrides.

9.3.4. Rates of Metal – Metal Proton Transfer

When a proton is transferred from a metal hydride to an oxygen or nitrogen base, the intrinsic barrier is largely due to the need for structural and electronic reorganization of the metal hydride partner; we saw in the introduction to this chapter that little reorganization is needed as a proton is transferred onto an oxygen or nitrogen lone pair. When a proton is transferred between two metals, the intrinsic barrier reflects the need for structural and electronic reorganization of both partners. If we treat the transfer of a proton from metal A to metal B in eq. (9.33) like an electron-transfer cross reaction, the intrinsic barrier $\Delta G_0^\ddagger$ for eq. (9.33) should be the average [148] of $\Delta G_{AA}^\ddagger$ [the barrier to the AH/A$^-$ proton self-exchange in eq. (9.34)] and $\Delta G_{BB}^\ddagger$ [the barrier to the BH/B$^-$ proton self-exchange in eq. (9.34)].

$$\mathrm{A{-}H} + \mathrm{B}^- \underset{k_{BA}}{\overset{k_{AB}}{\rightleftharpoons}} \mathrm{A}^- + \mathrm{B{-}H} \qquad K_{AB} = \frac{k_{AB}}{k_{BA}} \tag{9.33}$$

$$\mathrm{A{-}H} + \mathrm{A}^- \underset{k_{AA}}{\overset{k_{AA}}{\rightleftharpoons}} \mathrm{A}^- + \mathrm{A{-}H} \qquad \mathrm{B{-}H} + \mathrm{B}^- \underset{k_{BB}}{\overset{k_{BB}}{\rightleftharpoons}} \mathrm{B}^- + \mathrm{B{-}H} \tag{9.34}$$

Substitution of this expression for $\Delta G_0^\ddagger$ into eq. (9.31) and acceptance of the restriction that $\Delta G^\circ < 4\,\Delta G_0^\ddagger$ (so that the term quadratic in $\Delta G^\circ/4\,\Delta G_0^\ddagger$ can be neglected [149, 150]) lead to eq. (9.35). Restatement of eq. (9.35) in terms of the corresponding rate and equilibrium constants gives eq. (9.36), a cross relation like that familiar for electron transfer.

$$\Delta G_{AB}^\ddagger = \tfrac{1}{2}\left[\Delta G_{AA}^\ddagger + \Delta G_{BB}^\ddagger + \Delta G^\circ\right] \tag{9.35}$$

$$k_{AB} = \sqrt{k_{AA} k_{BB} K_{AB}} \tag{9.36}$$

The cross-reaction rate constants we have measured [151] for proton transfer among the group VI metals (Table 9.6) agree very well with those predicted by substituting the known self-exchange rate [13b] and equilibrium [17a] constants into eq. (9.36). The particularly good agreement between calculated and observed rate constants for the Mo–W cross reaction may reflect the minimal difference in geometry [152–154] between $CpMo(CO)_3H$ and $CpW(CO)_3H$.

The observed self-exchange rates in Table 9.6 agree reasonably well with those calculated from the structures and vibrational frequencies of $HM(CO)_3Cp$ and $M(CO)_3Cp^-$ by Creutz and Sutin [155], who have used a weak-interaction model including barriers resulting from changes in geometry. Similar calculations predict $H_2M(CO)_4/[HM(CO)_4]^-$ self-exchanges to be slower than the $HM(CO)_3Cp/[M(CO)_3Cp]^-$ ones—as is observed. This

Table 9.6
Observed and Calculated[a] Rate Constants for Group VI $HMCp(CO)_3 - MCp(CO)_3^-$ Proton Exchange Reactions in CH_3CN at 25°C

	Metal Anion k_{exch} (M^{-1} s^{-1})		
Metal Hydride	$KCrCp(CO)_3$	$KMoCp(CO)_3$	$KWCp(CO)_3$
$HCrCp(CO)_3$	1.8×10^4	1.66×10^4 (2.6×10^4)	3.12×10^4 (1.5×10^5)
$HMoCp(CO)_3$		2.5×10^3	8.5×10^3 (1.4×10^4)
$HWCp(CO)_3$			6.5×10^2

[a] Calculated values are given in parentheses.

weak-interaction model predicts that proton transfer rate constants will decrease "extremely rapidly with reactant separation", and thus also rationalizes the experimental observation that self-exchange between sterically congested hydrides (e.g., $HMo(CO)_2(dppe)_2^+$ [147b] $HCr(tripod)_2^+$, where tripod = $MeC(CH_2PMe_2)_3$ [156]) and their conjugate bases is extremely slow.

9.3.5. Why Do the Proton Transfer Reactions of Transition Metal Hydrides Fit Theory So Well?

There is every reason to believe that all our measured rate constants belong to the *proton transfer step* itself. The isotope effects k_H/k_D observed in the group VI self-exchange reactions [the k_H/k_D's for the $H(D)MCp(CO)_3/MCp(CO)_3^-$ exchange are 3.6, 3.7, and 3.7 for Cr, Mo, and W, respectively] [13b] are those expected for true proton transfer reactions with linear symmetric transition states. Although similar isotope effects might be observed if H· transfer were rate-determining, these results rule out the possibility that the sequence in eq. (9.37) (rate-determining electron transfer followed by fast H· transfer) is masquerading as a proton transfer reaction.

$$M^- + H{-}M \xrightarrow[e^-\ \text{transfer}]{\text{rate-determining}} M^{-}\cdot + H{-}M\cdot \xrightarrow[\text{fast}]{H\cdot\ \text{transfer}} M{-}H + M^- \qquad (9.37)$$

Further, and very importantly, proton transfer from transition metal hydrides cannot be preceded by the formation of the sort of hydrogen-bonded intermediates [eq. (9.38)] that have bedeviled the study of other types of proton transfer reactions [19, 142, 145]. The H—M bond in transition metal hydrides is polarized $H^{\delta-}{-}M^{\delta+}$ [157] and should be unable to form a B····H—M hydrogen bond [158]. We have evidence that this prediction is valid: The H—M stretching bands of $HRe(CO)_5$ and even the highly acidic $HCo(CO)_4$ are easily observed by Raman spectroscopy, and their frequency and shape do not change in the presence of an excess of bases strong enough to effect partial deprotonation [159].

$$B + H{-}M \nrightarrow B\cdots H{-}M \qquad (9.38)$$

9.3.6. A Kinetic Acidity Series for Transition Metal Hydrides

It is instructive to plot kinetic acidity versus thermodynamic acidity for transition metal hydrides as a class. Figure 9.6 displays the relationship between the rate and equilibrium constants (listed in Table 9.7) for proton transfers to aniline from most of the common mononuclear hydrides [13b].

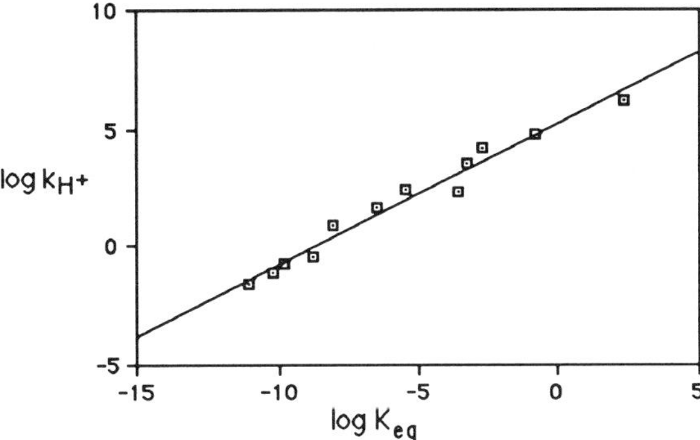

FIGURE 9.6
Plot of log k_{H^+} (where k_{H^+} is the rate constant for proton transfer to aniline in CD_3CN at 25°C) versus log K_{eq} for mononuclear hydrides. The line is the least-sqaures fit to the data for all hydrides. (Data from Tables 9.1 and 9.7.)

Table 9.7
Rate Constants for Proton Transfer to Aniline in CD_3CN at 25°C

MH	k_{H^+} (M^{-1} s^{-1})	Ref.
$HCo(CO)_4$	17,000,000	[13b]
$H_2Fe(CO)_4$	54,000	[13b]
$HCrCp(CO)_3$	17,000	[13b]
$HMoCp(CO)_3$	3,900	[13b]
$MWCp(CO)_3$	250	[13b]
$HMn(CO)_5$	210	[13b]
$HMoC_5Me_5(CO)_3$	49	[13b]
$H_2Ru(CO)_4$	8	[13b]
$HFeCp(CO)_2$	0.4	[13b]
$HMn(CO)_4PPh_3$	0.18	[160]
$H_2Os(CO)_4$	0.08	[13b]
$HMn(CO)_4P(EtPh_2)$	0.025	[160]
$HWCp(CO)_2PMe_3$	$< 10^{-3}$	[13b]

Because the transition state geometry inevitably changes along with the thermodynamic driving force, Figure 9.6 is not a Brønsted plot; nevertheless the points in it fall reasonably close to a straight line. The relationship between rate and thermodynamic driving force is thus reasonably constant for proton transfer from these hydrides to aniline, suggesting that the relative rates in Table 9.7 can be used to diagnose proton transfer mechanisms in other reactions.

9.3.7. Use of Kinetic Acidity Series to Identify Proton Transfer Reactions

The relative rates at which the M = Cr, Mo, and W hydrides react in eq. (9.39) [13b] agree with the relative rates for the same hydrides in the diagnostic kinetic acidity series in Table 9.7. This finding and the observed solvent effects and activation parameters are all consistent with the proton transfer mechanism shown.

$$Cp_2ZrMe_2 + H\text{—}M(CO)_3Cp \longrightarrow \left[\begin{array}{c} H \\ Cp_2\overset{+}{Zr}\cdots CH_3 \quad \overset{-}{M}(CO)_3Cp \\ | \\ CH_3 \end{array} \right] \xrightarrow{-CH_4}$$

$$\begin{array}{c} Cp_2\overset{+}{Zr} \leftarrow O{=}C{=}\overset{-}{M}(CO)_2Cp \\ | \\ CH_3 \end{array} \qquad (9.39)$$

Similar reasoning has helped Bullock [161] establish a proton transfer mechanism for reaction (9.40). The cationic methylvinylidene complex $Cp(PMe_3)_2Ru{=}C{=}C(H)CMe_3{}^+$ precipitates when toluene solutions of $Cp(PMe_3)_2Ru\text{—}C{\equiv}C\text{—}CMe_3$ and HM [where M = $Cr(CO)_3Cp$, $Mo(CO)_3Cp$, $W(CO)_3Cp$] are mixed, and the rate law in eq. (9.41) is obeyed when reaction (9.40) is carried out in acetone or acetonitrile. The more-acidic $HMo(CO)_3Cp$ reacts about 10 times faster than $HW(CO)_3Cp$—about the ratio expected from Table 9.7 for a rate-determining proton transfer.

$$CpRu(PMe_3)_2\text{—}C{\equiv}C\text{—}CH_3 + MH \xrightarrow{k_1} [CpRu(PMe_3)_2{=}C{=}C(H)\text{—}CH_3]^+ + M^- \qquad (9.40)$$

$$\frac{d[Cp(Me_3P)_2Ru{=}C{=}C(H)CMe_3{}^+]}{dt} = k_1[Cp(Me_3P)_2Ru\text{—}C{\equiv}C\text{—}CMe_3][HM] \qquad (9.41)$$

The CH_3CN pK_a of $Cp(PMe_3)_2Ru{=}C{=}C(H)CMe_3{}^+$ (given in Table 9.1) makes $\Delta G = -6.4$ kcal mol^{-1} in eq. (9.40) if HM = $HW(CO)_3Cp$.

Despite this thermodynamic driving force, k_1 for $HW(CO)_3Cp$ is only 10 times faster than the rate constant (k_{H^+}) for the uphill ($\Delta G \sim +7.5$ kcal mol^{-1}) proton transfer to aniline from $HW(CO)_3Cp$. The relatively sluggish H^+ transfer to $Cp(PMe_3)_2Ru{-}C{\equiv}C{-}CMe_3$ must be the result of an intrinsic barrier to protonation at its alkynyl carbon far larger than that to protonation at the nitrogen of aniline.

Kinetic acidity arguments also have been used to rule out a proton transfer mechanism for the addition of hydrides to the vinylidene-bridged dimer in eq. (9.42) [162]. In contrast to the pattern seen for phosphine-substituted hydrides in Table 9.7, the rate constants for eq. (9.42) remain almost unchanged as L varies from CO to PPh_3 to PMe_3, a pattern inconsistent with that of the kinetic acidities in Table 9.7 but consistent with the pattern of the M—H bond dissociation enthalpies in Table 9.4 (which change little when CO is replaced by a phosphine). The change in the mechanism for $HMo(CO)_3Cp$ from reaction (9.40) (formation of a vinylidene ligand by H^+ transfer) to reaction (9.42) (reaction of a vinylidene ligand by H· transfer) illustrates how slight a variation in substrate can persuade a transition metal hydride to switch mechanisms.

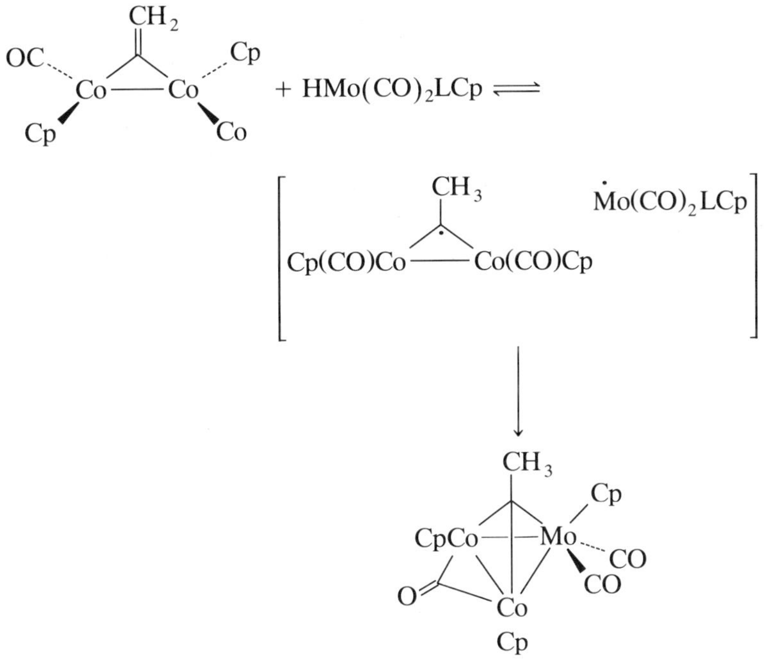

CH_2
OC Cp
Co—Co + $HMo(CO)_2LCp$ ⇌
Cp Co

[CH_3 $\dot{M}o(CO)_2LCp$
Cp(CO)Co—Co(CO)Cp] (9.42)

↓

CH_3
Cp
CpCo—Mo CO
O CO
Co
Cp

9.3.8. Kinetic Acidity of Ligands

The slow deprotonation rates typical of carbon acids are retained when they are coordinated to a metal. The removal of H^+ from the carbon α to the carbene ligand of the chromium complex $(CO)_5Cr{=}C(OCH_3)(CH_3)$ by

OH^- or piperidine is remarkably slow for an acid with an aqueous pK_a of 12.3 [62]. The intrinsic barrier to deprotonation by piperidine is higher for $(CO)_5Cr{=}C(OCH_3)(CH_3)$ than for any carbon acids except nitroalkanes.

Proton transfer is even slow at *oxygen* ligands of certain types. The oxo ligand in $Re(O)(OH)(MeC{\equiv}CMe)_2$ is connected to rhenium by a *triple* bond, and considerable rearrangement must occur as it is protonated to a hydroxyl ligand. As a result, the tautomerism in eq. (9.43) is quite slow unless it is catalyzed by a strong acid [163].

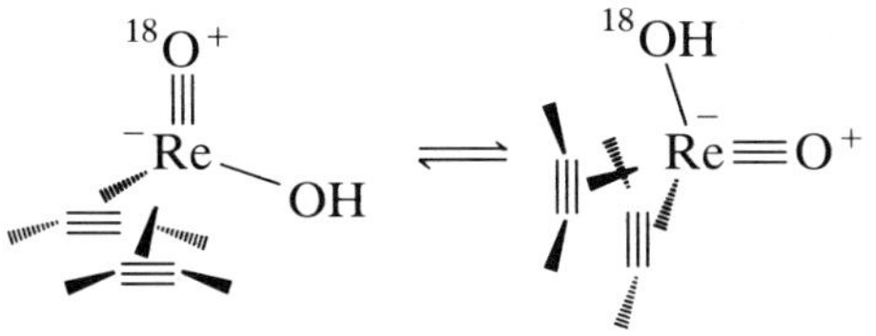

(9.43)

9.3.9. Competing Protonation Sites

9.3.9.1. BRIDGING VERSUS TERMINAL

The fact that intrinsic barriers are larger for proton transfer reactions involving nonterminal hydride ligands means that the kinetic barrier to putting a proton into a bridging site is greater than the barrier to putting one into a terminal site. When $Cp(CO)_2(\mu\text{-}PPh_2)WPt(CO)(PPh_3)$ is treated with HBF_4 [164], protonation initially occurs at a terminal position on tungsten, to give the kinetic product **9a** [eq. (9.44)]. Over a period of 2 weeks at 25°C, **9a** isomerizes to the more stable hydride-bridged dinuclear complex **10a**.

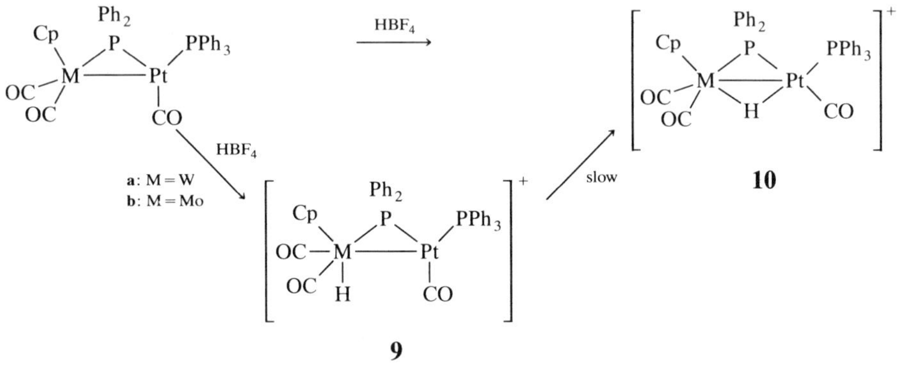

a: M = W
b: M = Mo

9

10

(9.44)

In the corresponding molybdenum system, the barriers to protonation and deprotonation are smaller, as one would expect from the kinetic acidity trends in Table 9.7. Thus, when $Cp(CO)_2(\mu\text{-}PPh_2)MoPt(CO)(PPh_3)$ is protonated, the intermediate terminal molybdenum hydride **9b** is not observed and the hydride-bridged cation **10b** is the first observable product.

In some cases the terminal hydride initially formed cannot rearrange to the more stable hydride-bridged product without a catalyst. The cation **11** in eq. (9.45) is stable in solution, but addition of halide ions catalyzes its isomerization ($Cl^- > Br^- > I^- \gg F^-$) to the hydride-bridged cation **12** [165]. The ineffectiveness of fluoride ion shows that this isomerization is not simply a base-assisted proton transfer. A kinetic study in the presence of Cl^- has found the rate law in eq. (9.46), where $k_{obs} = k_{Cl^-}K_{Cl^-}$. The high barrier to proton transfer from the terminal to the bridging site thus can be overcome by the alternate pathway to the thermodynamically favored product shown in eq. (9.45).

$$[\mathrm{Cp(H)(NO)Re}(\mu\text{-}\mathrm{PCy_2})\mathrm{Pt(PPh_3)_2}]^+\ (\mathbf{11}) + \mathrm{Cl^-} \underset{}{\overset{K_{Cl^-}}{\rightleftharpoons}} \mathrm{Cl^-}\cdots[\mathrm{Cp(H)(NO)Re}\rightarrow\mathrm{Pt}(\mu\text{-}\mathrm{PCy_2})(\mathrm{PPh_3})_2]^+ \xrightarrow{+\mathrm{Cl^-},\ k_{Cl^-}} [\mathrm{Cp(Cl)(H)(NO)Re}(\mu\text{-}\mathrm{PCy_2})\mathrm{PtCl(PPh_3)_2}]^- \xrightarrow[\text{fast}]{-2\mathrm{Cl^-}} [\mathrm{Cp(ON)Re}(\mu\text{-}\mathrm{PCy_2})(\mu\text{-}\mathrm{H})\mathrm{Pt(PPh_3)_2}]^+\ (\mathbf{12}) \quad (9.45)$$

$$-\frac{d[\mathbf{11}]}{dt} = k_{obs}[\mathbf{11}][\mathrm{Cl}^-]^2 \quad (9.46)$$

9.3.9.2. METAL VERSUS LIGAND

The small intrinsic barrier typical for protonation at oxygen or nitrogen often causes protonation of an oxygen or nitrogen ligand to be more facile than protonation of the metal even if the latter is thermodynamically favored. The kinetic site of protonation [eq. (9.47)] of the $[(\mu\text{-}H)M_3(CO)_{11}]^-$ clusters **13** (where M = Fe [166] for **13a**, Ru [167] for **13b**, and Os [168] for **13c**) is always a carbonyl oxygen. The fate of the resulting O-protonated species **14** depends on the identity of the metal. When **13c** (M = Os) is protonated, the final product, **15c**, is formed rapidly even at $-80°C$ [168]. With iron (**13a**), **14a** is the only product observed at low temperatures; upon warming up to $-30°C$, it decomposes to $Fe_3(CO)_{12}$ (**15a** may be thermodynamically unstable relative to **14a**) [166]. The behavior of **13b** is intermediate: protonation with CF_3SO_3H yields **14b**, which rearranges to **15b** in approximately 20 s at room temperature [167b]. The rearrangement of **14** to **15** shows a large kinetic isotope effect k_H/k_D (which may be related to the large k_H/k_D values that have been reported [169] for the protonation of several polynuclear metal carbonyls).

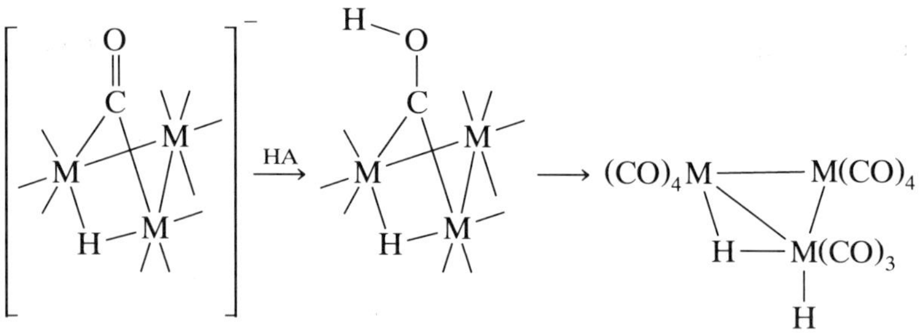

$$\mathbf{13} \xrightarrow{HA} \mathbf{14} \longrightarrow \mathbf{15} \qquad (9.47)$$

The kinetic site of protonation of the related $[Ru_3(NO)(CO)_{10}]^-$ (**16**) is a *nitrosyl* oxygen rather than a carbonyl one [eq. (9.48)]. When HA is a sufficiently strong acid (e.g., CF_3SO_3H) only the kinetic product **17** is observed. Although **17** is stable at room temperature when $CF_3SO_3^-$ is the strongest base present, Brønsted bases such as X^-, NO_3^-, or $CF_3CO_2^-$ tautomerize **17** to the thermodynamically favored **18** [170]. It follows that the conjugate acid of one of these bases eventually should give **18** when it is used to protonate **16**, and **18** is indeed the eventual product when **16** is treated with CF_3CO_2H. A similar mechanism (initial deprotonation back to **13**) may effect the rearrangement of **14** to **15** in eq. (9.47).

$$\underset{\mathbf{16}}{[Ru_3(CO)_{10}NO]^-} \underset{A^-}{\overset{\text{fast}\;HA}{\rightleftharpoons}} \underset{\mathbf{17}}{Ru_3(CO)_{10}NOH}$$

$$\mathbf{16} \xrightarrow[\text{slow}]{HA} \underset{\mathbf{18}}{HRu_3(CO)_{10}NO} \qquad (9.48)$$

The carbonyl oxygens in $CpM(CO)_{3-x}L_x$ and (arene)$M(CO)_3$ form hydrogen bonds with HCl and $(CF_3)_3COH$ even though protonation eventually occurs at the metal [158b]. Although observation of a COH or NOH species has been rare [167b, 170], it is arguable that the kinetic site of protonation is a carbonyl or nitrosyl oxygen for all carbonyl and nitrosyl complexes, even mononuclear ones.

Electrophiles attack the rhodium cyanide complexes LRh(σ-CN) [where L = $N(CH_2CH_2PPh_2)_3$ or $P(CH_2CH_2PPh_2)_3$] at either the metal center or the cyanide nitrogen. CF_3SO_3Me gives $LRh(CNMe)^+$, whereas CF_3SO_3H gives the cis hydrido cyanide complex $LRh(H)(CN)^+$ [171]. The kinetically favored site for both additions is probably the nitrogen of the cyanide ligand, with $LRh(CNH)^+$ subsequently rearranging to $LRh(H)(CN)^+$ and $LRh(CNMe)^+$ not being able to rearrange to $LRh(Me)(CN)^+$.

Protonation at sulfur also seems to be kinetically favored over protonation at a transition metal. Addition of either hard or soft alkylating agents to the iron thiolato complexes $[RSFe(CO)_3L]^-$ (**19**) results in the formation of neutral thioether complexes, and the product of the protonation in eq. (9.49) is the coordinated thiol complex $(HSR)Fe(CO)_3L$ [172].

$$(\text{at } -78°\text{C}) \quad RSFe(CO)_4^- \underset{NEt_3}{\overset{HBF_4}{\rightleftharpoons}} (HSR)Fe(CO)_4 \qquad (R = Me, Et) \tag{9.49}$$

$$\underset{\mathbf{19a}}{PhSFe(CO)_3P(OEt)_3^-} \xrightarrow{HBF_4} \underset{\mathbf{20b}}{(H)(PhS)Fe(CO)_3P(OEt)_3} \tag{9.50}$$

In contrast, protonation of **19a**, with R = Ph and L = $P(OEt)_3$, yields a thiolate hydride complex **20a** [eq. (9.50)]; initial formation of metal-bound thiol has not been observed. For **19b**, R = Me, L = PEt_3, the initial product of protonation at −78°C is an η^2-thiol (**21**) with an agostic S · · · H · · · Fe interaction; **21** then isomerizes to the thiolato hydride form **20b** [eq. (9.51)]. It is arguable that initial protonation always occurs at either sulfur or the S—Fe bond, giving either an η^1 or an η^2 thiol complex as the kinetic product.

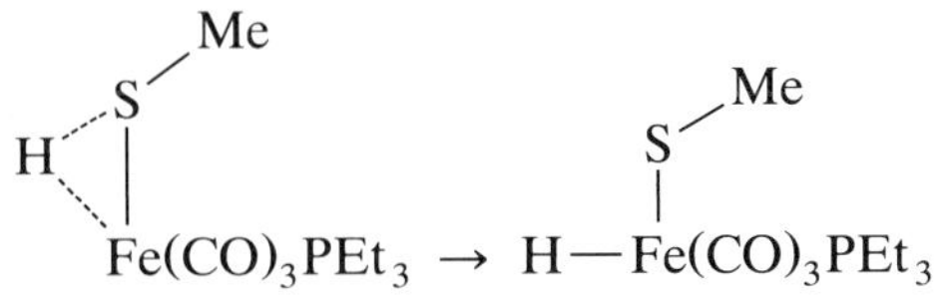

$$\underset{\mathbf{21}}{(\eta^2\text{-H}\cdots S(Me))Fe(CO)_3PEt_3} \rightarrow \underset{\mathbf{20b}}{H{-}Fe(SMe)(CO)_3PEt_3} \tag{9.51}$$

Because a large intrinsic barrier is typical for protonation at a carbon ligand as well as for protonation at a transition metal, carbon and metal sites can compete kinetically for protonation and deprotonation. We saw in the introduction to this chapter that the hydrogens of the Cp ligand in the rhenium complex $CpRe(NO)(PPh_3)H$ are more kinetically acidic than the hydride ligand.

On the other hand, the metal appears to be the kinetically favored site for protonation of the vinylidene complex *trans*-$Re{=}C{=}CHPh(Cl)(dppe)_2$ (**22**). With Et_3NH^+ the equilibrium constant K_4 for protonation of **22** to **23** [eq. (9.52)] is greater than 4. The mechanism of the conversion of the kinetic product **23** to the thermodynamic product, the alkylidyne complex **24**, has been examined in detail [173].

$$\underset{\mathbf{22}}{Re{=}C{=}C\langle} \underset{B}{\overset{K_4,\ BH^+}{\rightleftharpoons}} \underset{\mathbf{23}}{H{-}\overset{+}{Re}{=}C{=}C\langle} \qquad \mathbf{22} \xrightarrow{k_2,\ BH^+} \mathbf{24}; \quad \mathbf{23} \xrightarrow{k_3} \mathbf{24}: \ \overset{+}{Re}{\equiv}C{-}C{\langle}{-}H \qquad \begin{array}{l} Re = ReCl(dppe)_2 \\ B = NEt_3 \end{array} \tag{9.52}$$

Both the k_2 path and the k_3 path in eq. (9.52) convert significant amounts of **23** to **24**. The k_2 (base-catalyzed) path in eq. (9.52) involves slow reprotonation of **22**, related to **23** by the fast equilibrium K_4, at the thermodynamic (carbon) site; the k_3 path in eq. (9.52) can be thought of as an intramolecular proton transfer [173b]. These two paths epitomize the mechanisms (base-catalyzed or intramolecular) that can be imagined for the tautomerization of *any* kinetic protonation product to the corresponding thermodynamic product. There is seldom enough evidence to tell which one is operating. *Probably* the tautomerizations in eqs. (9.44), (9.45), and (9.51) are intramolecular; the one in eq. (9.48) has been shown to be base-catalyzed; the one in eq. (9.47), and the tautomerization of $LRh(CNH)^+$, could plausibly occur by either mechanism.

The tautomerization of a carbon acid to a transition metal hydride has been proposed as a necessary step in the elimination of benzene from the ruthenium nitrosoethyl complex **25** [eq. (9.53)] [174]. In the absence of a base, elimination is slow. When a neutral base such as DBU is used, the rate of disappearance of **25** increases linearly with base concentration. Apparently base abstraction of the acidic methylene proton from the nitrosoethyl ligand is the rate-determining step; protonation at the metal center gives the ruthenium hydride **26**, and Ph–H elimination gives the product oximate complex **27**. Apparently the tautomerization of **25** to **26** is not possible by an intramolecular mechanism and can occur only by a base-catalyzed mechanism [175].

$$\underset{\mathbf{25}}{Cp^*Ru(\text{—}N(=O)CH_2CH_3)(Ph)(PPhMe_2)} \xrightarrow{\text{cat. base}} \underset{\mathbf{26}}{Cp^*Ru(\text{—}N(\rightarrow O)=CHCH_3)(Ph)(H)(PPhMe_2)} \xrightarrow[+L]{\text{fast, } -PhH} \underset{\mathbf{27}}{Cp^*Ru(\text{—}N(\rightarrow O)=CHCH_3)(PPhMe_2)(L)} \quad (9.53)$$

9.3.10. Protonation at Metal versus Formation of an η^2-H_2 Ligand from a Hydride Ligand

The acidity of molecular hydrogen is enhanced enormously when it is coordinated [91]. Evidence is offered by the recent observation that in the presence of base $Ru(OEP)(THF)_2$ (where OEP = octaethylporphyrinato dianion) catalyzes exchange between H_2 and D_2O [reaction (9.54)] [176]. The observation that $Os(OEP)(\eta^2\text{-}H_2)$ is stable and can be formed by the reversible protonation of $[Os(OEP)H]^-$ suggests that eq. (9.54) occurs by the mechanism in eq. (9.55).

$$(THF)_2(Ru) + KOD + D_2O + H_2 \xrightarrow[50°\,C]{THF} H_2/HD/D_2 \quad (9.54)$$

$$(THF)_2(Ru) \xrightarrow{+H_2} (H\text{—}H)(Ru)(THF) \xrightarrow{OD^-} [(H)(Ru)(THF)]^- \xrightarrow{D_2O} (H\text{—}D)(Ru)(THF) \xrightarrow{-HD} (THF)_2(Ru) \quad (9.55)$$

The kinetic acidity of η^2-H_2 ligands [91] appears to exceed that of normal hydride ligands. A 1 : 6 mixture of the cationic dihydride **28a** and the cationic dihydrogen complex **28b** is formed when CpRu(dmpe)H [where dmpe = 1,2-bis(dimethylphosphino)ethane] (**28**) is protonated [49, 177]. Interconversion of these two tautomers is slow, but saturation transfer experiments at 298 K have measured k_4 in eq. (9.55) as 9.0×10^{-3} s^{-1}. The pK_a of the dihydrogen complex **28b** has been measured as 17.6 in acetonitrile (Table 9.1); the pK_a of **28a** must be lower for the reasons given in Section 9.2.3 and can be calculated as 16.8 from eq. (9.24). When the conjugate base **28** is present in a solution of **28a** and **28b**, saturation of the Cp signal of **28** results in measurable magnetization transfer only to **28b**; in the presence of Et_3N, the signal of **28b** broadens whereas that of **28a** does not. The *kinetic acidity* of the dihydrogen complex **28b** thus exceeds that of the dihydride **28a** ($k_{H_2} \gg k_{2H}$).

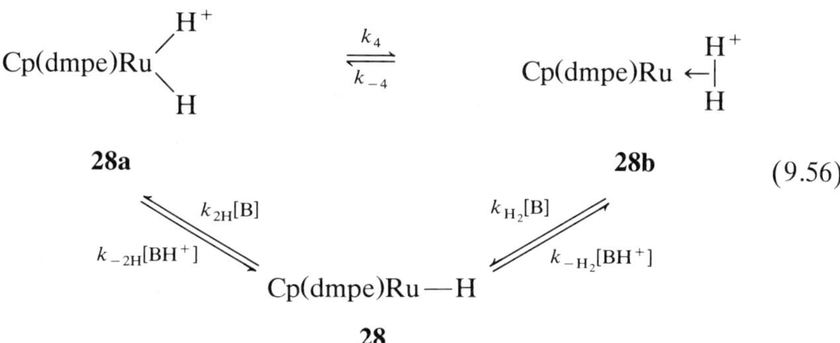

(9.56)

Because **28a** and **28b** are both related to **28** by deprotonation equilibria with similar equilibrium constants (the preceding pK_a values show that **28b** is a slightly weaker acid than **28a**), the fact that $k_{H_2} \gg k_{2H}$ means that k_{-H_2} must be greater than k_{-2H}. The protonation of the M—H bond in **28** to an η^2-H_2 complex is thus faster than the protonation of the same complex at the metal.

The regiochemistry of the protonation of $Cp^*_2WH_2$ suggests that this result may be general. Treatment of $Cp^*_2WH_2$ with D^+ yields an unexpected mixture of isotopomers [178]. The isomer (**29a**) where deuterium occupies the central position predominates (~ 90%), although there is a small amount (~ 10%) of the laterally deuteriated isotopomer (**29b**). Because there is no W-localized central orbital [179], Parkin and Bercaw [178] explained the kinetic preference for central protonation by charge-controlled formation of an intermediate η^3-H_3 or an ($\eta^2 H_2$)(H) complex [eq. (9.57)]. [An intramolecular path, perhaps involving an $(\eta^2\text{-}H_2)(H)$ complex, eventually equilibrates **29a** and **29b**.]

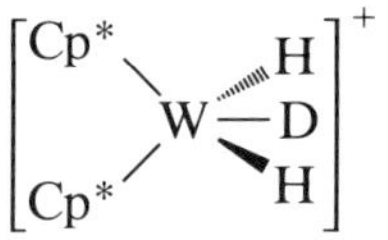

29a 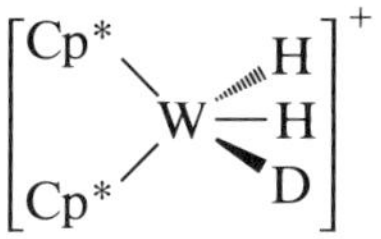 **29b**

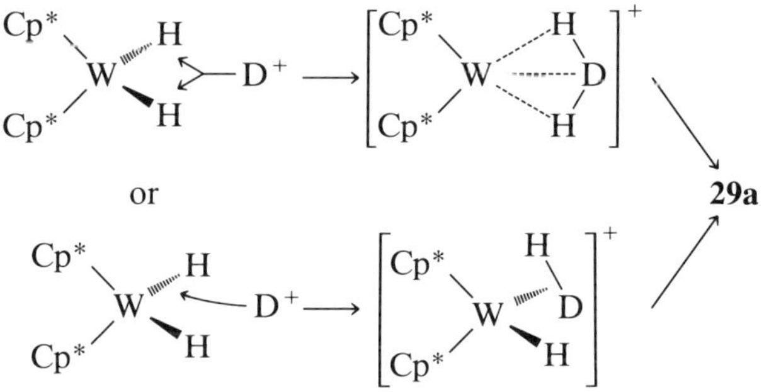

(9.57)

Some curious results in the literature can be explained by postulating that the protonation of an M—H bond to an $\eta^2\text{-}H_2$ complex is faster than the protonation of the corresponding metal M. The binuclear complex **30** evolves H_2 and forms the dication **31** when dissolved in strong acid [reaction (9.58)], whereas the closely related binuclear complex **32** gives the stable diprotonated derivative **33** under the same conditions [reaction (9.59)] [180].

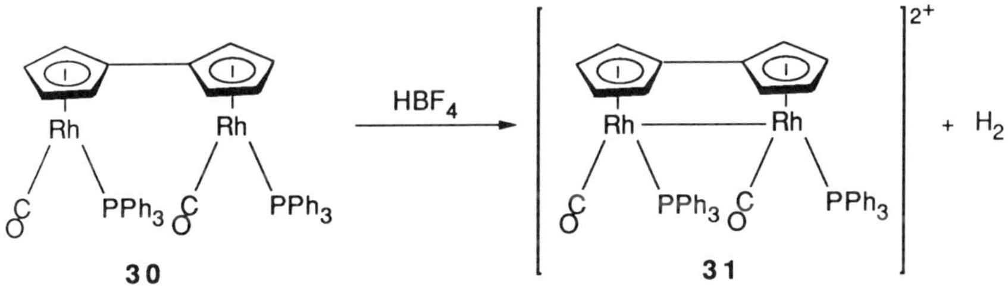

(9.58)

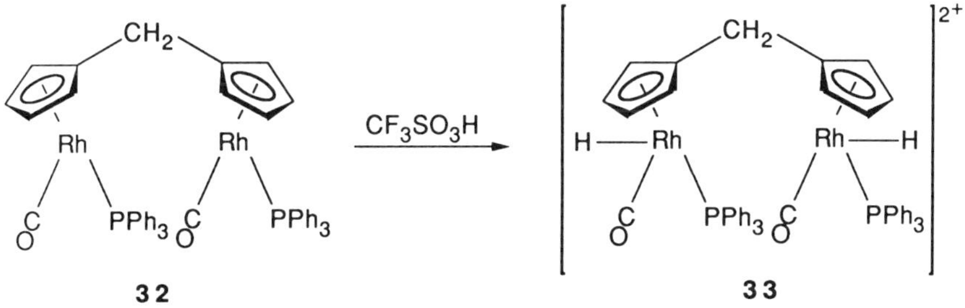

(9.59)

Several H_2 evolution reactions like eq. (9.58) are known [181]. Generally, it has been assumed that they proceed through diprotonated complexes like **33** [181, 182], but this assumption is unlikely to be correct: (a) No diprotonated binuclear complex like **33** has ever been observed to evolve H_2. (b) The yield of H_2 is often less than 100% and is acid-dependent [182b]. A more plausible explanation involves the mechanism in eq. (9.60), with **35** being the kinetic product of the protonation of **34**. The H_2 ligand in **35** can either be replaced by the remote metal center to give **36** or rearrange via **34** to the thermodynamically stable diprotonated complex **37**. The product ratio will be determined by the competition between the acid-independent (**35** → **36**) path and the acid-dependent (**35** → **37**) path.

$$[L_nM\frown ML_n(H)]^+ \ (\mathbf{34}) \xrightleftharpoons[\text{fast}]{H^+} [L_nM\frown ML_n(\leftarrow H{-}H)]^{2+} \ (\mathbf{35}) \xrightarrow[\text{slow}]{-H_2} [L_nM\frown\!\!-\!\!-ML_n]^{2+} \ (\mathbf{36})$$

$$\mathbf{34} \xrightarrow[\text{slow}]{H^+} [(H)L_nM\frown ML_n(H)]^{2+} \ (\mathbf{37}) \qquad (9.60)$$

The ease with which an M—H bond is protonated to an η^2-H_2 complex suggests that the protonation of an M—C bond to an η^2-alkane complex may also be facile. Protonation of d^0 alkyls presumably occurs at an M—C bond, raising the possibility of a σ-complex [183, 184] intermediate [eq. (9.61)] in protonolysis reactions such as reaction (9.39). Such σ complexes could even be intermediates in the formation of alkyl hydrides by the protonation of d^n alkyls ($n \geq 2$).

$$Cp_2Zr(CH_3)_2 + H^+ \longrightarrow Cp_2\overset{+}{Zr}(CH_3)(\leftarrow H{-}CH_3) \qquad (9.61)$$

ACKNOWLEDGMENTS

We are grateful to Prof. Branka Ladanyi, Dr. Norman Sutin, Dr. Carol Creutz, and Prof. A. K. Rappé for assistance on theoretical points, and to Prof. John Brauman for helpful discussions of acidity in nonaqueous solvents. We would like to acknowledge the contributions of everyone in the Norton group who has worked on the acidity of transition metal hydrides: Richard F. Jordan (who started it all), Robin T. Edidin, Jeffrey M. Sullivan, Eric J. Moore, Anne E. Moody, and Rolf T. Weberg. This work has been supported by the National Science Foundation, most recently under grant CHE-88-19760.

REFERENCES

1. Feigl, F.; Krumholz, P. *Monatsh. Chem.* **1932**, *59*, 314.
2. Coleman, G. W. Unpublished work mentioned in Blanchard, A. A. *Chem. Rev.* **1937**, *21*, 3.
3. (a) Hieber, W.; Hübel, W. *Z Elektrochem.* **1953**, *57*, 235. (b) Hieber, W.; Wagner, G. *Z. Naturforschung* **1958**, *B13*, 339. (c) Hieber, W.; Lindner, E. *Chem. Ber.* **1961**, *94*, 1417. (d) Beck, W.; Hieber, W.; Braun, G. *Z. Anorg. Allg. Chem.* **1961**, *308*, 23. (e) Hieber, W.; Duchatsch, H. *Chem. Ber.* **1965**, *98*, 2933.
4. Schunn, R. A. In *Transition Metal Hydrides*; Muetterties, E. L., Ed.; Marcel Dekker: New York, 1971; Chapter 5.
5. (a) Schrauzer, G. N.; Windgassen, R. J. *J. Am. Chem. Soc.* **1967**, *89*, 1999. (b) Bhandal, H.; Pattenden, G. *J. Chem. Soc., Chem. Commun.* **1988**, 1110. (c) Pattenden, G. *Chem. Soc. Rev.* **1988**, *17*, 361.
6. (a) Grey, R. A.; Pez, G. P.; Wallo, A. *J. Am. Chem. Soc.* **1981**, *103*, 7536–7542. (b) Grey, R. A.; Pez, G. P.; Wallo, A.; Corsi, J. *Ann. N.Y. Acad. Sci.* **1983**, *415*, 235–243.
7. Linn, D. E.; Halpern, J. *J. Am. Chem. Soc.* **1987**, *109*, 2969–2974.
8. Crocco, G. L.; Gladysz, J. A. *J. Am. Chem. Soc.* **1988**, *110*, 6110–6118.
9. Walker, H. W.; Kresge, C. T.; Ford, P. C.; Pearson, R G. *J. Am. Chem. Soc.* **1979**, *101*, 7428.
10. Beagley, B.; Hewitt, T. G. *Trans. Faraday Soc.* **1968**, *64*, 2561.
11. McNeil, E. A.; Scholer, F. R. *J. Am. Chem. Soc.* **1977**, *99*, 6243.
12. Calderazzo, F.; Fachinetti, G.; Marchetti, F.; Zanazzi, P. *J. Chem. Soc., Chem. Commun.* **1981**, 181. For simplicity, we show a tetrahedral geometry for $[Co(CO)_4]^-$ rather than the slightly distorted structure it has in the $[Me_3NH][Co(CO)_4]$ ion pair in the solid state. In polar solvents $[Co(CO)_4]^-$ is generally a free tetrahedral ion. In acetonitrile (the solvent used in our own work) the IR spectrum shows no departure from tetrahedral geometry for $[Co(CO)_4]^-$ with a protonated aniline [13] bis(triphenylphosphine)iminium (PPN^+) [14, 15] various electron-acceptor cations [15], or even Na^+ [15, 16] as the counterion. Even with Tl^+, only weak association (i.e., contact ion pair formation) occurs [14].
13. (a) Moore, E. J.; Sullivan, J. M.; Norton, J. R. *J. Am. Chem. Soc.* **1986**, *108*, 2257. (b) Edidin, R. T.; Sullivan, J. M.; Norton, J. R. *J. Am. Chem. Soc.* **1987**, *109*, 3945.

14. Schramm, C.; Zink, J. I. *J. Am. Chem. Soc.* **1979**, *101*, 4554.
15. Bockman, T. M.; Kochi, J. K. *J. Am. Chem. Soc.* **1989**, *111*, 4669.
16. Edgell, W. F.; Barbetta, A. *J. Am. Chem. Soc.* **1974**, *96*, 415.
17. (a) Jordan, R. F.; Norton, J. R. *J. Am. Chem. Soc.* **1982**, *104*, 1255. (b) Jordan, R. F.; Norton, J. R. *ACS Symp. Ser.* **1982**, *198*, 403. (c) Weberg, R. T.; Norton, J. R. *J. Am. Chem. Soc.* **1990**, *112*, 1105.
18. (a) Pearson, R. G.; Ford, P. C. *Comments Inorg. Chem.* **1982**, *1*, 279. (b) Walker, H. W.; Pearson, R. G.; Ford, P. C. *J. Am. Chem. Soc.* **1983**, *105*, 1179. (c) Pearson, R. G. *Chem. Rev.* **1985**, *85*, 41.
19. (a) Kresge, A. J. *Chem. Soc. Rev.* **1973**, *2*, 475. (b) Bell, R. P. *The Proton in Chemistry*; Cornell University Press: Ithaca, NY, 1973. (c) Reutov, O. A.; Beletskaya, I. P.; Butin, K. P. *CH-Acids*; Pergamon: New York, 1978. (d) *Comprehensive Chemical Kinetics*; Bamford, C. H., Tipper, C. F. H., Eds.; Elsevier: Amsterdam, 1977; Vol. 8, Chapter 2 (Hibbert, F.), Chapter 3 (Crooks, J. E.). (e) Simmons, E. L. *Prog. React. Kinet.* **1977**, *8*, 161. (f) Albery, W. J. *Ann. Rev. Phys. Chem.* **1980**, *31*, 227. (g) Stewart, R. *The Proton: Applications to Organic Chemistry*; Academic: New York, 1985.
20. Miholova, D.; Vlcek, A. *Proc. Conf. Coord. Chem., 3rd* **1971**, 221.
21. Hogen-Esch, T. E.; Smid, J. *J. Am. Chem Soc.* **1966**, *88*, 318.
22. Kaufman, M. J.; Gronert, S.; Streitwieser, A., Jr. *J. Am. Chem. Soc.* **1988**, *110*, 2829, and previous papers in the series.
23. (a) Darensbourg, M. Y.; Darensbourg, D. J.; Burns, D.; Drew, D. A. *J. Am. Chem. Soc.* **1976**, *98*, 3127. (b) Darensbourg, M. Y.; Darensbourg, D. J.; Barros, H. L. C. *Inorg. Chem.* **1978**, *17*, 297. (c) Darensbourg, M. Y.; Jimenez, P.; Sacket, J. R.; Hanckel, J. M.; Kump, R. L. *J. Am. Chem. Soc.* **1982**, *104*, 1521. (d) Darensbourg, M. Y.; Hanckel, J. M. *Organometallics* **1982**, *1*, 82. (e) Darensbourg, M. Y.; Hanckel, J. M. *J. Organomet. Chem.* **1981**, *217*, C9. (f) Darensbourg, M. Y.; Barros, H.; Borman, C. J. *J. Am. Chem. Soc.* **1977**, *99*, 1647.
(g) Darensbourg, M. Y.; Barros, H. L. C. *Inorg. Chem.* **1979**, *18*, 3786.
24. Green, M. L. H.; Pratt, L.; Wilkinson, G. *J. Am. Chem. Soc.* **1958**, 3916.
25. (a) Kolthoff, I. M.; Chantooni, M. K., Jr. *J. Phys. Chem.* **1968**, *72*, 2270. (b) Bykova, L. N.; Petrov, S. I. *J. Anal. Chem. USSR Engl. Transl.* **1972**, *27*, 965. (c) Whereas references 25a and 25b report upper limits to the ion product of acetonitrile ($K_{auto} < 10^{-33}$), Schwesinger [30d] suggested that the actual value of K_{auto} for acetonitrile is close to 10^{-44} based on the extrapolation of the acidity of CH_3CN in DMSO to acetonitrile solvent (Schwesinger, R., private communication). (d) With stronger bases [$pK_a(BH^+) > 35–36$] rapid condensation of the solvent takes place [30d].
26. No contact ion pair formation between Na^+, K^+, PPN^+, or any other polynuclear cation and any organometallic anion has ever been observed in CH_3CN [13, 15–17, 23c]. There is weak interaction between Tl^+ and $[Co(CO)_4]^-$ in CH_3CN [14].
27. (a) Kolthoff, I. M.; Chantooni, M. K., Jr. *J. Am. Chem. Soc.* **1970**, *92*, 7025. (b) Kolthoff, I. M.; Chantooni, M. K., Jr.; Bhowmik, S. *J. Am. Chem. Soc.* **1966**, *88*, 5430.
28. (a) Coetzee, J. F. *Prog. Phys. Org. Chem.* **1967**, *4*, 45. (b) Coetzee, J. R.; Padmanabhan, G. R. *J. Am. Chem. Soc.* **1965**, *87*, 5005. (c) Cauquis, G.; Deronzier, A.; Serve, D.; Vieil, E. *J. Electroanal. Chem. Interfacial Electrochem.* **1975**, *60*, 205.
29. Kurasov, L. A.; Pozharskii, A. F.; Kuz'menko, V. V. *J. Org. Chem. USSR* **1983**, *19*, 759.
30. (a) Schwesinger, R. *Chimia* **1985**, *39*, 269–272. (b) Schwesinger, R. *Angew. Chem., Int. Ed. Engl.* **1987**, *26*, 1164. (c) Schwesinger, R.; Mißfeldt, M.; Peters,

K.; von Schnering, H. G. *Angew. Chem., Int. Ed. Engl.* **1987**, *26*, 1165–1167. (d) Schwesinger, R.; Schlemper, H. *Angew. Chem., Int. Ed. Engl.* **1987**, *26*, 1167–1169.
31. Sullivan, J. M. Transition metal hydride complexes: The relationship between thermodynamic and kinetic acidity; Ph.D. Thesis, Colorado State University, 1986.
32. Kristjánsdóttir, S. S.; Moody, A. E.; Weberg, R. T.; Norton, J. R. *Organometallics* **1988**, *7*, 1983.
33. Ryan, O. B.; Tilset, M.; Parker, V. D. *J. Am. Chem. Soc.* **1990**, *112*, 2618–2626, and references therein.
34. (a) *Recommended Methods for Purification of Solvents*; Coetzee, J. K., Ed.; Pergamon: New York, 1982. (b) Kiesele, H. *Anal. Chem.* **1980**, *52*, 2230.
35. Hieber, W.; Winter, E.; Scubert, E. *Chem. Ber.* **1962**, *95*, 3070–3076.
36. Calderazzo, F.; Pampaloni, G.; Vitali, D. *Gazz. Chim. Ital.* **1981**, *111*, 455.
37. Davison, A.; Ellis, J. E. *J. Organomet. Chem.* **1972**, *36*, 131.
38. Davison, A.; Reger, D. L. *J. Organomet. Chem.* **1970**, *23*, 491–496.
39. Fischer, E. O.; Hafner, W.; Stahl, H. O. *Z. Anorg. Allg. Chem.* **1955**, *282*, 47.
40. Chadhuri, P.; Wieghardt, K.; Tsai, Y.-H.; Krüger, C. *Inorg. Chem.* **1984**, *23*, 427–432.
41. Kristjánsdóttir, S. S.; Norton, J. R. Unpublished work.
42. Jetz, W.; Graham, W. A. G. *Inorg. Chem.* **1971**, *10*, 1647.
43. Kristjaśnsdóttir, S. S.; Loendorf, A.; Norton, J. R. Unpublished work.
44. Luo, X.-L.; Crabtree, R. H. *J. Chem. Soc., Chem. Commun.* **1990**, 189–190.
45. Kaesz, H. D. *Chem. Br.* **1973**, *9*, 344.
46. Kruck, Th.; Prasch, A. *Z. Anorg. Allg. Chem.* **1969**, *371*, 1.
47. Gamembeck, F.; Krumholz, P. *J. Am. Chem Soc.* **1971**, *93*, 1909.
48. Estimated from the reported equilibrium data in EtOH in Baker, M. V.; Field, L. D.; Young, D. J. *J. Chem. Soc., Chem. Commun.* **1988**, 546–548.
49. Chinn, M. S.; Heinekey, D. M. *J. Am. Chem. Soc.* **1987**, *109*, 5865–5867.
50. Kruck, T.; Lang, W.; Derner, N.; Stadler, M. *Chem. Ber.* **1968**, *101*, 3816.
51. Lim, H. S.; Anson, F. C. *Inorg. Chem.* **1971**, *10*, 103.
52. Schrauzer, G. N.; Holland, R. J. *J. Am. Chem. Soc.* **1971**, *93*, 1505.
53. Pearson, R. G.; Kresge, C. T. *Inorg. Chem.* **1981**, *20*, 1878.
54. Halpern, J.; Riley, D. P.; Chan, A. S. C.; Pluth, J. J. *J. Am. Chem. Soc.* **1977**, *99*, 8055.
55. Ramasami, T.; Espenson, J. H. *Inorg. Chem.* **1980**, *19*, 1846.
56. Gillard, R. D.; Heaton, B. T.; Vaughan, D. H. *J. Chem. Soc., A* **1970**, 3126.
57. Klingert, B.; Werner, H. *J. Organomet. Chem* **1987**, *333*, 119–128.
58. (a) Gerlach, D. H.; Kane, A. R.; Parshall, G. W.; Jesson, J. P.; Meutterties, E. L. *J. Am. Chem. Soc.* **1971**, *93*, 3543. (b) Schunn, R. A. *Inorg. Chem.* **1976**, *15*, 208.
59. (a) Tolman, C. A. *Inorg. Chem.* **1972**, *11*, 3128. (b) Tolman, C. A. *J. Am. Chem. Soc.* **1970**, *92*, 4217–4222.
60. Pearson, R. Private communication.
61. Bullock, M. R. *J. Am. Chem. Soc.* **1987**, *109*, 8087–8089.
62. Gandler, J. R.; Bernasconi, C. F. *Organometallics* **1989**, *8*, 2282–2284.
63. Henderson, R. A.; Davies, F.; Dilworth, J. R.; Thorneley, R. N. F. *J. Chem. Soc., Dalton Trans.* **1981**, 40.
64. Henderson, R. A. *J. Chem. Soc., Dalton Trans.* **1983**, 51.
65. Koskikallio, J. *Suom. Kemistil. B* **1957**, *30*, 111, 115.
66. Brookhart, M.; Lamanna, W.; Pinhas, A. R. *Organometallics* **1983**, *2*, 638.
67. Behr, A.; Herdtweck, E.; Herrmann, W. A.; Keim, W.; Kipshagen, W. *Organometallics* **1987**, 2307–2313.
68. Bassett, J.-M.; Farrugia, L. J.; Stone, F. G. A. *J. Chem. Soc., Dalton Trans.* **1980**, 1789–1790.

69. Werner, H.; Gotzig, J. *Organometallics* **1983**, *2*, 547–549.
70. Sullivan, B. P.; Caspar, J. V.; Johnson, S. R.; Meyer, T. J. *Organometallics* **1984**, *3*, 1241.
71. Sullivan, B. P.; Lumpkin, R. S.; Meyer, T. J. *Inorg. Chem.* **1987**, *26*, 1247.
72. Legzdins, P.; Martin, D. T. *Inorg. Chem.* **1979**, *18*, 1250–1254.
73. Ball, R. G.; Ghosh, C. K.; Hoyano, J. K.; McMaster, A. D.; Graham, A. G. *J. Chem. Soc., Chem. Commun.* **1989**, 341–342.
74. Simunic, J. L.; Pinhas, A. R. *Inorg. Chem.* **1989**, *28*, 2400.
75. Casey, C. P.; Bullock, R. M. *Organometallics* **1984**, *3*, 1100.
76. Calderazzo, F.; Pampaloni, G.; Zanazzi, P. F. *J. Chem. Soc., Chem Commun.* **1982**, 1304–1305.
77. These reactions may be driven by the insolubility of the resulting hydride in water, as $[V(CO)_4(diars)]^-$, where diars = *o*-phenylenebisdimethylarsine, is stable in water at ambient temperature, and $HV(CO)_4(dmpe)$ is deprotonated quantitatively by $[R_4N]OH$ in methanol. Ellis, J. E.; Faltynek, R. A. *J. Organomet. Chem.* **1975**, *93*, 205.
78. Bachmann, K.; Rehder, D. *J. Organomet. Chem.* **1984**, 177.
79. Rehder, D.; Fornalczy, M.; Otmann, P. *J. Organomet. Chem.* **1987**, *331*, 207.
80. Davison, A.; Ellis, J. E. *J. Organomet. Chem.* **1971**, *31*, 239–247.
81. Hillery and Cohen [82] published an equation relating the acetonitrile and water pK_a values of carboxylic acids and a few mineral acids. The $pK_a(CH_3CN)$ values they used for H_2SO_4 and $HClO_4$ are incorrect (see reference 43 in reference 13a). The $pK(H_2O)$ values they used for the mineral acids were "apparent" acidity values derived from leaving group abilities.
82. Hillery, P. S.; Cohen, L. A. *J. Am. Chem. Soc.* **1983**, *105*, 2760–2770.
83. Kolthoff, I. M.; Chantooni, M. K., Jr.; Bhowmik, S. *J. Am. Chem. Soc.* **1968**, *90*, 23–28.
84. (a) Kolthoff, I. M.; Chantooni, M. K., Jr. *J. Phys. Chem.* **1972**, *76*, 2024. (b) Marcus, Y.; Kamlet, M. J.; Taft, R. W. *J. Phys. Chem.* **1988**, *92*, 3613.
85. A discussion of solvent effects on acid–base equilibria can be found in Reichardt, C. *Solvent and Solvent Effects in Organic Chemistry*, 2d ed.; VCH: Weinheim, 1988; Chapter 4.
86. (a) Wayner, D. D. M.; McPhee, D. J.; Griller, D. *J. Am. Chem. Soc.* **1988**, *110*, 132–137. (b) Stevenson, G. R.; Hashim, R. T. *J. Phys. Chem.* **1986**, *90*, 3217.
87. Kolthoff et al. [83] concluded that there is no appreciable energy change associated with the transfer of picrate ion from acetonitrile to water.
88. ΔG_{tr} from acetonitrile to DMSO is less than 2 kcal mol^{-1} for substituted benzylic carbanions: Sim, B. A.; Griller, D.; Wayner, D. D. M. *J. Am. Chem. Soc.* **1989**, *111*, 754–755.
89. Although the tabulated differences in the CH_3CN/H_2O ΔpK_a among different classes of acids suggest that there are real differences in solvation among the conjugate bases of these different classes of acids, it has been argued that ΔG_{tr} from acetonitrile to water is negligible even for the relatively small anion H^-: Griller, D.; Martinho Simões, J. A.; Mulder, P.; Sim, B. A.; Wayner, D. D. M. *J. Am. Chem. Soc.* **1989**, *111*, 7872–7876.
90. Kanabus-Kaminska, J. M.; Gilbert, B. C.; Griller, D. *J. Am. Chem. Soc.* **1989**, *111*, 3311.
91. Kubas, G. J. *Acc. Chem. Res.* **1988**, *21*, 120. Kubas, G. J. *Comments Inorg. Chem.* **1988**, *7*, 17–40. Crabtree, R. H.; Hamilton, D. G. *Adv. Organomet. Chem.* **1988**, *28*, 299.
92. Jia, G.; Morris, R. H. *Inorg. Chem.* **1990**, *29*, 582.
93. Pedersen, S. E.; Robinson, W. R. *Inorg. Chem.* **1975**, *14*, 2365–2371.
94. Lokshin, B. V.; Ginzburg, A. G.; Setkina, V. N.; Kursanov, D. N.; Nemirovskaya, I. B. *J. Organomet. Chem.* **1972**, *37*, 347–353.

95. Werner, H. *Pure Appl. Chem.* **1982**, *54*, 177.
96. Lokshin, B. V.; Pasinsky, A. A.; Kolobova, N. E.; Anisimov, K. N.; Makarov, Yu. V. *J. Organomet. Chem.* **1973**, *55*, 315–319.
97. Cerichelli, G.; Illuminati, G.; Ortaggi, G.; Guiliani, A. M. *J. Organomet. Chem.* **1977**, *127*, 357.
98. Harris, D. C.; Gray, H. B. *Inorg. Chem.* **1975**, *14*, 1215.
99. (a) Knight, J.; Mays, M. J. *J. Chem. Soc. A* **1970**, 711–714. (b) Koridze, A. A.; Kizas, O. A.; Astakhova, N. M.; Petrovskii, P. V.; Grishin, Y. K. *J. Chem. Soc., Chem. Commun.* **1981**, 853–855. (c) White, J. W.; Wright, W. *J. Chem. Soc. A* **1971**, 2843–2847.
100. Delley, B.; Manning, M. C.; Ellis, D. E.; Berkowitz, J.; Trogler, W. C. *Inorg. Chem.* **1982**, *21*, 2247–2253.
101. Imjanitow, N. S. *Hung. J. Ind. Chem.* **1975**, *3*, 331.
102. Vidal, J. L.; Walker, W. E. *Inorg. Chem.* **1981**, *20*, 249.
103. (a) Ziegler, T. *Organometallics* **1985**, *4*, 675. (b) Ziegler, T.; Tschinke, V.; Becke, A. *J. Am. Chem. Soc.* **1987**, *109*, 1351.
104. Inferred from the ^{1}H chemical shift reported by Krusic, P. J.; Jones, D. J.; Roe, D. C. *Organometallics*, **1986**, *5*, 456.
105. (a) Chin, H. B.; Bau, R. *Inorg. Chem.* **1978**, *17*, 2314. (b) Collman, J. P.; Finke, R. G.; Matlock, P. L.; Wahren, R.; Komoto, R. G.; Brauman, J. I. *J. Am. Chem. Soc.* **1978**, *100*, 1119.
106. Fruchart, J.-M.; Souchay, P. *Compt. Rend. C* **1968**, *266*, 1571.
107. Stevens Miller, A. E.; Kawamura, A. R.; Miller, T. M. *J. Am. Chem. Soc.* **1990**, *112*, 457–458.
108. Unpublished work of Stevens Miller and Beauchamp quoted in reference 107 and in Martinho Simões, J. A.; Beauchamp, J. L. *Chem. Rev.* **1990**, *90*, 629–688.
109. A recent review discusses the determination of M—H bond strengths in the gas phase and elsewhere: Martinho Simões, J. A.; Beauchamp, J. L. *Chem. Rev.* **1990**, *90*, 629–688. Relatively few M—H BDE values have been directly determined *in the gas phase*, and many depend upon assumptions regarding other bond strengths.
110. Meckstroth, W. K.; Ridge, D. P. *J. Am. Chem. Soc.* **1985**, *107*, 2281, and references therein.
111. (a) Landrum, J. T.; Hoff, C. D. *J. Organomet. Chem.* **1985**, *282*, 215. (b) Hoff, C. D. *J. Organomet. Chem.* **1985**, *282*, 201. (c) Nolan, S. P.; de la Vega, R. L.; Hoff, C. D. *J. Organomet. Chem.* **1986**, *315*, 187. (d) Nolan, S. P.; Hoff, C. D.; Stoutland, P. O.; Newman, L. J.; Buchanan, J. M.; Bergman, R. G.; Yang, G. K.; Peters, K. S. *J. Am. Chem. Soc.* **1987**, *109*, 3143 (see also Stoutland, P. O.; Bergman, R. G.; Nolan, S. P.; Hoff, C. D. *Polyhedron* **1988**, *7*, 1429). (e) Kiss, G.; Zhang, K.; Mukerjee, S. L.; Hoff, C. D.; Roper, G. C. *J. Am. Chem. Soc.* **1990**, *112*, 5657. (f) Bruno, J. W.; Marks, T. J.; Morss, L. R. *J. Am. Chem. Soc.* **1983**, *105*, 6824. (g) Schock, L. E.; Marks, T. J. *J. Am. Chem. Soc.* **1988**, *110*, 7701. (h) Calhorda, M. J.; Dias, A. R.; Minas da Piedade, M. E.; Salema, M. S.; Martinho Simões, J. A. *Organometallics* **1987**, *6*, 734. (i) Dias, A. R.; Martinho Simões, J. A. *Polyhedron* **1988**, *7*, 1531. (j) Wayland, B. B. *Polyhedron* **1988**, *7*, 1545.
112. For an example see Billmers, R.; Griffith, L. L.; Stein, S. E. *J. Phys. Chem.* **1986**, *90*, 517.
113. Tilset, M.; Parker, V. D. *J. Am. Chem. Soc.* **1989**, *111*, 6711. In the original paper, $E°$ values relative to the standard hydrogen electrode in acetonitrile were used instead of $E°$ values relative to the standard hydrogen electrode in water. Correction by $\Delta G°$ for the transfer of H^+ from water to acetonitrile was

therefore unnecessary. Removing this error (Tilset, M.; Parker, V. D. *J. Am. Chem. Soc.* **1990**, *112*, 2843) adds 8 kcal mol^{-1} to the values originally published and gives the values in Table 9.4.

114. (a) Breslow, R.; Balasubramanian, K. *J. Am. Chem. Soc.* **1969**, *91*, 5182. (b) Breslow, R.; Chu, W. *J. Am. Chem. Soc.* **1973**, *95*, 411. (c) Jaun, B.; Schwarz, J.; Breslow, R. *J. Am. Chem. Soc.* **1980**, *102*, 5741.
115. Nicholas, A. M. de P.; Arnold, D. R. *Can. J. Chem.* **1982**, *60*, 2165.
116. Bordwell, F. G.; Cheng, J. P.; Harrelson, J. A., Jr. *J. Am. Chem. Soc.* **1988**, *110*, 1229, and references therein. Although these results agree well with known gas-phase C—H BDE values, Bordwell and co-workers suggested in a more recent publication (Bordwell, F. G.; Harrelson, J. A., Jr.; Satish, A. V. *J. Org. Chem.* **1989**, *54*, 3101) that this good agreement may be the result of fortuitous cancellation of errors.
117. Tilset and Parker [113] used irreversible potentials from derivative cyclic voltammetry and used known or estimated rate constants for M· dimerization to estimate the corresponding reversible potentials. The results agree well with those directly measured by rapid scan cyclic voltammetry for $[Mn(CO)_5]^-$ and $[CpFe(CO)_2]^-$ by Pugh and Meyer [118] and for $[Mn(CO)_4(PPh_2Et)]^-$ by Kristjánsdóttir, Elliot, and Norton [119], although they do not agree as well with the rapid-scan $[Mn(CO)_5]^-$ and $[Mn(CO)_4PPh_3]^-$ results of Kuchynka and Kochi [120].
118. Pugh, J. R.; Meyer, T. J. *J. Am. Chem. Soc.* **1988**, *110*, 8245.
119. Kristjánsdóttir, S. S.; Elliott, C. M.; Norton, J. R. Unpublished work.
120. Kuchynka, D. J.; Kochi, J. K. *Inorg. Chem.* **1989**, *28*, 855. There is a discrepancy of over 200 mV in the reversible potential of $[Mn(CO)_5]^-$ and a discrepancy of about 100 mV in the reversible potential of $[Mn(CO)_4PPh_3]^-$.
121. Burkey, T. J.; Majewski, M.; Griller, D. *J. Am. Chem. Soc.* **1986**, *108*, 2218.
122. Calderazzo, F. *Ann. N.Y. Acad. Sci.* **1983**, *415*, 37.
123. Sweany, R. L.; Halpern, J. *J. Am. Chem. Soc.* **1977**, *99*, 8335.
124. Sweany, R. L.; Comberrel, D. S.; Dombourian, M. F.; Peters, N. A. *J. Organomet. Chem.* **1981**, *216*, 57.
125. Ungváry, F.; Markó, L. *Organometallics* **1982**, *1*, 1120.
126. However, all three numbers depend upon the same kinetically determined Mo—Mo bond strength (from Amer, S.; Kramer, G.; Poë, A. *J. Organomet. Chem.* **1981**, *209*, C28).
127. McLain, S. J. *J. Am. Chem. Soc.* **1988**, *110*, 643.
128. Hanckel, J. M.; Lee, K. W.; Rushman, P.; Brown, T. L. *Inorg. Chem.* **1986**, *25*, 1852.
129. For a detailed analysis of possible errors in the determination of M—H bond strengths by eq. (9.27), see Eisenberg, D. C.; Norton, J. R. *Isr. J. Chem.* **1991**, *31*, in press.
130. Treichel, P. M.; Mueh, H. J.; Bursten, B. E. *Isr. J. Chem.* **1977**, *15*, 253.
131. Pickett, C. J.; Pletcher, D. *J. Organomet. Chem.* **1975**, *102*, 327.
132. Chatt, J.; Kan, C. T.; Leigh, J.; Pickett, C. J.; Stanley, D. R. *J. Chem. Soc., Dalton Trans.* **1980**, 2032.
133. Lever, A. B. P. *Inorg. Chem.* **1990**, *29*, 1271–1285.
134. Substituent effects on the electrochemistry of organometallic compounds have been reviewed: Bursten, B. E.; Green, M. R. *Prog. Inorg. Chem.* **1988**, *36*, 419.
135. Sarapu, A. C.; Fenske, R. F. *Inorg. Chem.* **1971**, *10*, 38.
136. (a) Bursten, B. E.; Gatter, M. G. *J. Am. Chem. Soc.* **1984**, *106*, 2554–2558. (b) Bursten, B. E.; Gatter, M. G. *Organometallics* **1984**, *3*, 895–899. (c) Bursten, B. E.; Gatter, M. G. *Organometallics* **1984**, *3*, 941–943.
137. The electron-impact fragmentation pattern of $(CO)_5MnRe(CO)_5$ suggests that the gas-phase electron affinities of $\cdot Mn(CO)_5$ and $\cdot Re(CO)_5$ are approximately

equal [110], although Ziegler's local density calculations [103] predict a significant difference.

138. Stewart, R. [19g], p. 269.
139. Brønsted, J. N.; Pedersen, K. J. *Z. Phys. Chem.* (*Leipzig*) **1924**, *108*, 185.
140. (a) Bell, R. P. *Proc. Roy. Soc. London. Series A* **1936**, *154*, 414. (b) Evans, M. G.; Polanyi, M. *Trans. Faraday Soc.* **1936**, *32*, 1340. (c) Leffler, J. E. *Science* (*Washington D.C.*) **1953**, *117*, 340. (d) Hammond, G. S. *J. Am. Chem. Soc.* **1955**, *77*, 334.
141. (a) Kreevoy, M. M.; Lee, I.-S. H. *J. Am. Chem. Soc.* **1984**, *106*, 2550. (b) Agmon, N. *J. Am. Chem. Soc.* **1980**, *102*, 2164. (c) Pross, A. *J. Org. Chem.* **1984**, *49*, 1811.
142. (a) Marcus, R. A. *J. Phys. Chem.* **1968**, *72*, 891. (b) Cohen, A. O.; Marcus, R. A. *J. Phys. Chem.* **1968**, *72*, 4249. (c) Marcus, R. A. *J. Am. Chem. Soc.* **1969**, *91*, 7225. (d) Marcus, R. A. *Faraday Symp. Chem. Soc.* **1975**, *10*, 60.
143. (a) Marcus, R. A. *J. Chem. Phys.* **1968**, *43*, 679. (b) Newton, T. W. *J. Chem. Educ.* **1968**, *45*, 571.
144. (a) Darensbourg, M. Y.; Ludvig, M. M. *Inorg. Chem.* **1986**, *25*, 2894. (b) Hanckel, J. M.; Darensbourg, M. Y. *J. Am. Chem. Soc.* **1983**, *105*, 6979.
145. (a) Bernasconi, C. F. *Tetrahedron* **1985**, *41*, 3219. (b) Bernasconi, C. F. *Acc. Chem. Res.* **1987**, *20*, 301.
146. See reference 142b and comments in Raycheba, J. M. T.; Geier, G. *Inorg. Chem.* **1979**, *18*, 2486.
147. Saunders, W. H., Jr. *J. Phys. Chem.* **1982**, *86*, 3321.
148. Such an assumption is plausible during *electron* transfer because "the force field from one reactant does not influence the other" [143a], but it is less clear *a priori* that it is true during *proton* transfer (a reaction that obliges the reactants to come relatively close to each other). Marcus remarked while discussing proton transfers in 1968 that "additivity might be expected to hold best if neither [intrinsic barrier] is near zero" [142b] and later used a simple BEBO model to show that it is not seriously in error even when one intrinsic barrier is substantially larger than the other [142d].
149. The intrinsic barriers to the deprotonation of transition metal hydrides are much larger than the values of $\Delta G°$ associated with the small equilibrium constants we are considering, and thus, as with methyl transfers [150], the quadratic term can be neglected. Marcus noted in 1968 [142a] that the application of such cross relations to proton transfers was probably limited to $|\Delta G°/4\,\Delta G_0^\ddagger| < 1$ even when the quadratic term was included (in the form of the f_{12} factor in his eq. 4).
150. Lewis, E. S. *J. Phys. Chem.* **1986**, *89*, 3756. (Lewis does state that proton transfers are a case where a contribution from the quadratic term is possible.)
151. Kristjánsdóttir, S. S.; Norton, J. R. *J. Am. Chem. Soc.* **1991**, in press.
152. These metals have essentially equal covalent radii, as calculated from the M–Cl distances in $CpM(CO)_3Cl$ (1.50 and 1.51 Å for W and Mo, respectively) [153], so their M—H distances should be about equal. The covalent radius of Cr is considerably shorter (1.30 Å [154], as calculated from $[CpCr(CO)_3]_2SnCl_2$) so the Cr—H bond should be shorter.
153. Bueno, C.; Churchill, M. R. *Inorg. Chem.* **1981**, *20*, 2197.
154. Stephens, F. S. *J. Chem. Soc., Dalton Trans.* **1975**, 230.
155. Creutz, C.; Sutin, N. *J. Am. Chem. Soc.* **1988**, *110*, 2418.
156. Thaler, E.; Folting, J. C.; Caulton, K. G. *Inorg. Chem.* **1987**, *26*, 374.
157. Chen, H.-W.; Jolly, W. L.; Kopf, J.; Lee, T. H. *J. Am. Chem. Soc.* **1979**, *101*, 2607. See also calculations by Antovic, D.; Davidson, E. R. *J. Am. Chem. Soc.* **1987**, *109*, 977, and references therein.
158. (a) Neutron diffraction has shown that the O—H ligand in *cis*-$[IrH(OH)(PMe_3)_4]PF_6$ bends toward the hydride ligand, indicating that the latter

is serving as a *base* in the formation of a hydrogen bond: Stevens, R. C.; Bau, R.; Milstein, D.; Blum, O.; Koetzle, T. F. *J. Chem. Soc., Dalton Trans.* **1990**, 1429–1432. (b) IR studies in liquid xenon have shown that, in the presence of HCl or $(CF_3)_3COH$, the carbonyl oxygen in $CpM(CO)_{3-x}L_x$ or $AreneM(CO)_3$ complexes serves as the hydrogen bond acceptor even though protonation eventually occurs at the metal: Lokshin, B. V.; Kazarian, S. G.; Ginzburg, A. G. *J. Mol. Struct.* **1988**, *174*, 29.

159. Kristjánsdóttir, S. S.; Norton, J. R.; Moroz, A.; Sweany, R. L.; Whittenburg, S. L. *Organometallics* **1991**, in press.
160. Kristjánsdóttir, S. S.; Norton, J. R. Unpublished results.
161. Bullock, R. M. *J. Am. Chem. Soc.* **1987**, *109*, 8087–8089.
162. Jacobsen, E. N.; Bergman, R. G. *J. Am. Chem. Soc.* **1985**, *107*, 2023.
163. Erikson, T. K. G.; Mayer, J. M. *Angew. Chem. Int. Ed. Engl.* **1988**, *27*, 1527.
164. Powell, J.; Sawyer, J. F.; Smith, S. J. *J. Chem. Soc., Chem. Commun.* **1985**, 1312–1313.
165. Powell, J.; Sawyer, J. F.; Stainer, M. V. R. *J. Chem. Soc., Chem. Commun.* **1985**, 1314–1316.
166. Hodali, H. A.; Shriver, D. F.; Ammlung, C. A. *J. Am. Chem. Soc.* **1978**, *100*, 5239.
167. (a) Keister, J B. *J. Organomet. Chem.* **1980**, *190*, C36. (b) Nevinger, L. R.; Keister, J. B.; Maher, J. *Organometallics* **1990**, *9*, 1900.
168. (a) Pribich, D. C.; Rosenberg, E. *Organometallics* **1988**, *7*, 1741. (b) Rosenberg, E. *Polyhedron* **1989**, *8*, 383–405.
169. (a) Mays, M. J.; Simpson, R. N. F. *J. Chem. Soc. A* **1968**, 1444. (b) Knight, J.; Mays, M. J. *J. Chem. Soc. A* **1970**, 711.
170. (a) Stevens, R. E.; Guettler, R. D.; Gladfelter, W. L. *Inorg. Chem* **1990**, *29*, 451. (b) Gladfelter, W. L.; Stevens, R. E. *J. Am. Chem. Soc.* **1982**, *104*, 6454.
171. Bianchini, C.; Laschi, F.; Orraviani, M. F.; Peruzzini, M.; Zanello, P.; Zanobini, F. *Organometallics* **1989**, *8*, 893.
172. Darensbourg, M. Y.; Liaw, W.-F.; Riordan, C. G. *J. Am. Chem. Soc.* **1989**, *111*, 8051.
173. (a) Carvalho, M. F. N. N.; Henderson, R. A.; Pombeiro, A. J. L.; Richards, R. L. *J. Chem. Soc., Chem. Commun.* **1989**, 1796. (b) Such "intramolecular" paths can also involve base catalysis by the solvent. The previous reference says that the k_3 pathway for reaction (9.52) "appears to be intramolecular", although the solvent is THF.
174. Chang, J.; Seidler, M. D.; Bergman, R. G. *J. Am. Chem. Soc.* **1989**, *111*, 3258–3271.
175. Rosenberg [168b] has established both base-catalyzed and intramolecular mechanisms for a tautomerization (Adams, R. D.; Golembeski, N. M. *J. Am. Chem. Soc.* **1979**, *101*, 2579) in which a proton moves from an osmium to a *more basic* ligand nitrogen.
176. Collman, J. P.; Wagenknecht, P. S.; Hembre, R. T.; Lewis, N. S. *J. Am. Chem. Soc.* **1990**, *112*, 1294.
177. Chinn, M. S.; Heinekey, D. M. *J. Am. Chem. Soc.* **1990**, *112*, 5166–5175.
178. Parkin, G.; Bercaw, J. E. *J. Chem. Soc., Chem. Commun.* **1989**, 255–257.
179. (a) Lauher, J. W.; Hoffmann, R. *J. Am. Chem. Soc.* **1976**, *98*, 1729. (b) Petersen, J. L.; Lichtenberger, D. L.; Fenske, R. F.; Dahl, L. F. *J. Am. Chem. Soc.* **1975**, *97*, 6433. (c) Brintzinger, H. H.; Lohr, L. L., Jr.; Tang Wong, K. L. *J. Am. Chem. Soc.* **1975**, *97*, 5146.
180. Bitterwolf, T. E.; Spink, W. C.; Rausch, M. D. *J. Organomet. Chem.* **1989**, *363*, 189–195.
181. (a) Bitterwolf, T. E.; Ling, A. C. *J. Organomet. Chem.* **1973**, *57*, C17–C18. (b) Bitterwolf, T. E. *J. Organomet. Chem.* **1983**, *252*, 305.

182. (a) Hillman, M.; Michaile, S.; Feldberg, S. W.; Eisch, J. J. *Organometallics* **1985**, *4*, 1258–1263. (b) Michaile, S.; Hillman, M.; Eisch, J. J. *Organometallics* **1988**, *7*, 1059–1065.

183. Some evidence for a d^0 η^2-H_2 complex has been reported: Henderson, R. A. *J. Chem. Soc., Chem. Commun.* **1987**, 1670.

184. (a) Buchanan, J. M.; Stryker, J. M.; Bergman, R. G. *J. Am. Chem. Soc.* **1986**, *108*, 1537, and references therein. (b) Periana, R. A.; Bergman, R. J. *J. Am. Chem. Soc.* **1986**, *108*, 7332–7346, 7346–7355, and references therein. (c) Brookhart, M.; Green, M. L. H.; Wong, L.-L. *Prog. Inorg. Chem.* **1988**, *36*, 1–124. (d) Bullock, R. M.; Headford, C. E. L.; Hennessy, K. M.; Kegley, S. E.; Norton, J. R. *J. Am. Chem. Soc.* **1989**, *111*, 3897. (e) Parkin, G.; Bercaw, J. E. *Organometallics* **1989**, *8*, 1172. (f) Gould, G. L.; Heinekcy, D. M. *J. Am. Chem. Soc.* **1989**, *111*, 5502.

CHAPTER 10

Nucleophilic Reactivity of Transition Metal Hydrides

JAY A. LABINGER

Division of Chemistry and Engineering 139-74, California Institute of Technology Pasadena, CA 91125

10.1. DEFINITION OF HYDRIDIC CHARACTER

When is a hydride a hydride?

The term "metal hydride" implies, if not an ionic compound, at least considerable polarization of the covalent bond in the direction M^+H^- and consequent nucleophilic character of the hydrogen. Although this is in fact characteristic of many (by no means all!) main-group metal hydrides, it has long been known that it is *not* a general description of the behavior of most transition metal hydrides. The majority of transition metal hydrides do not exhibit reactivity typical of boron and aluminum hydride reagents, such as reduction of carbonyl compounds, nucleophilic displacement, and so forth.

There are several reasons for seeking patterns of hydridic or nucleophilic reactivity among the transition metal hydrides. Besides the fundamental desire to understand the factors that govern the nature of this important class of compound, there are potential practical applications of immense importance. The extensive body of reduction techniques using main-group hydrides is probably fully adequate for the vast majority of laboratory-scale problems that arise, but the reagents are usually too expensive for large-scale commercial application. Because these reagents cannot be formed directly from H_2 under reasonably mild conditions, there is no possibility of catalytic reduction. Transition metal hydrides, in contrast, *are* frequently formed and/or regenerated under H_2, thus opening up the possibility of catalytic reduction *if* we can determine how to reproduce the reactivity of their main-group cousins.

In this chapter we explore some of the reactions and trends that have been found; as we shall see, this topic is relatively undeveloped. An issue that must be addressed first is how to recognize and, especially, to quantify hydridic character. One might expect that a "hydridicity" scale should be definable, completely analogous to acidity scales that have been developed recently for transition metal hydrides. A bit of reflection, though, makes it clear that the situation is much less straightforward.

Acidity is basically defined by eq. (0.10.1): the ability of the compound in question to transfer a proton to a standard base B:. Granted, what is straightforward in principle may be much less so in practice; complications of differential solvation, kinetic versus thermodynamic limitations, and so forth, all complicate the issue (see Chapter 9). However, an analogous equation (10.2) to define hydridic character suffers from (in addition to the preceding complications) a more fundamental defect: If MH satisfies the 18-electron rule, then so does M^-, but *not* M^+! The product of eq. (10.2) will almost never be a stable species as written (for that matter, the number of stable, coordinatively unsaturated acceptors A is extremely small as well), making it virtually impossible to find equilibria from which hydridic character could be reduced.

$$MH + B\colon \rightleftarrows M^- + BH^+ \tag{10.1}$$

$$MH + A \rightleftarrows M^+ + AH^- \tag{10.2}$$

One might hope to establish at least a *relative* scale using reactions such as eq. (10.3), where L is a weakly bound ligand, so that the influence of M—L bonding on the position of the equilibrium would not be a major perturbation. Unfortunately, the fast majority of such reactions tend to give not H^- transfer but formation of stable, hydrogen-bridged, bimetallic compounds, so even this type of quantitation probably is not available. Hence, we will have to fall back on more phenomenological tests: to look for the sort of nucleophilic reactivity we are trying to achieve and to see if any correlations with the nature of the transition metal hydride complex may be deduced.

$$MH + M'L^+ \rightleftarrows ML^+ + M'H \tag{10.3}$$

10.2. TESTS OF NUCLEOPHILICITY

A number of reactions might be used to assay nucleophilic or hydridic character. Possibly the most clearcut would be a nucleophilic substitution by the metal hydride:

$$M{-}H + RX \rightarrow RH + M^+ + X^- \tag{10.4}$$

Here we still have important ambiguities, now of a mechanistic nature. A reaction with the stoichiometry of eq. (10.4) *may* proceed by a simple S_N2 pathway, but alternatives such as radical routes (10.5) or nucleophilic attack by metal (10.6) are also possible and not necessarily readily distinguishable. For example, stereochemistry at R could be used to discern between (10.4) and (10.5), but not between (10.4) and (10.6).

$$\begin{aligned} R\cdot + MH &\rightarrow RH + M \\ M\cdot + RX &\rightarrow R\cdot + MX \end{aligned} \qquad (10.5)$$

$$MH + RX \rightarrow X^- + M(H)(R)^+ \rightarrow M^+ + RH \qquad (10.6)$$

Another class of reaction that characterizes nucleophilicity is addition of H^- to *polar* unsaturated bonds (aldehydes and ketones; acids and esters; metal carbonyls and CO_2; imines; etc.). Again there is the possibility that reduction of carbonyl compounds might involve single-electron pathways; this has been found even for main-group reagents [1]. Furthermore, there is the issue of prior coordination of the unsaturated substrate; if that is a factor, then the favorability of a reaction such as eq. (10.7) might depend more on the strength of the M—O bond of the product than on any character of the M—H bond of the reagent [2]. Thus, although CO_2 might seem to be a good candidate for revealing nucleophilicity of hydride, the range of transition metal hydrides that insert CO_2 is immense [3]; it is unlikely that any useful patterns can be deduced therefrom. Similarly, catalytic hydrogenation of a polar unsaturated substrate does not by itself imply nucleophilicity, especially if the same transition metal hydride is a catalyst for hydrogenation of nonpolar bonds.

$$MH + R_2C{=}O \rightarrow M(H)\,(\eta^2\text{-}O{=}CR_2) \rightarrow M{-}OCHR_2 \qquad (10.7)$$

A couple of other potential test reactions include abstraction of H^- by a hydride acceptor, such as trityl cation, and the tendency to undergo protonolysis to give H_2. There are few examples and no detailed studies of the first and, again, complicating single-electron pathways may intervene. As for the second, most transition metal hydrides will react with a sufficiently strong acid; for example, the moderately acidic $HMn(CO)_5$ reacts with neat triflic acid to give H_2 [4]. Issues such as formation of η^2-H_2 complexes and their stability can come into play as well.

One can also imagine using ground-state criteria to assess hydridic behavior: The degree of polarization as M^+H^- ought to be related in some way to reactivity. One could use theoretical and/or spectroscopic methods to attempt to determine the charge at H. Here we have the dual problem of not only assessing the nature of the bonding but also determining how it does in fact relate to reactivity. For example, Sweany has used vibrational spectroscopy on matrix-isolated $M(CO)_n$ and $HM(CO)_n$ to probe charge distribution; the latter exhibit higher C—O stretching force constants, suggesting that negative charge is withdrawn onto the hydrogen atom. However, this tendency is virtually *constant* over the series M = Mn, Re, Fe, Co, which strongly implies that the well-known high acidity of $HCo(CO)_4$ does not say anything about ground-state $(OC)_4Co^-H^+$ polarization [5]. X-ray photoelectron spectroscopy (XPS) studies lead to the same conclusion [6]. Another alternative is the ability of the transition metal hydride to function as a donor *through hydrogen* to a Lewis acid, as in eq. (10.8); qualitative trends for simple hydridometal carbonyls appear generally opposite to acidity orders [7].

$$M—H + A \rightleftarrows M—H \cdots A \tag{10.8}$$

The bottom line, if there is one, is that we should not expect any single, simple, unambiguous method for identifying or measuring nucleophilicity in transition metal hydrides. The approach we will take is to try to identify potential patterns of nucleophilic character and to examine them from as many of the previously mentioned points of view as possible.

If one were to take a rather simplistic approach to prediction of hydridic behavior, one might come up with the following:

1. If nucleophilic reactivity is related to H^- character, then transition metal hydrides with net negative charge, or with more strongly donating ligands, should be more nucleophilic. (This predicts a trend *opposite* to that of acidity, which seems not unreasonable.)
2. If we are looking for reactivity resembling that of main-group hydrides, then perhaps transition metal hydrides with d^0 configurations would be the best place to look. A possible extension would be to concentrate on the left-hand end of the transition series in the Periodic Table. This also may be a trend complementary to that of acidity: Moore, Sullivan, and Norton [8] concluded that acidity generally increases on moving from left to right (and, less consistently, from bottom to top) within the transition series.

In fact, both of these somewhat naive predictions work at least to some extent, although there is not yet enough information to determine the generality of, and exceptions to, these "rules." In the remainder of this chapter we will examine some of the chemistry in each of the preceding areas, as well as several examples that do not really fall into either.

10.3. HYDRIDIC CHARACTER OF ANIONIC TRANSITION METAL HYDRIDES

In 1981 Pez and co-workers described a rational approach to the design of a homogeneous catalyst for hydrogenation of polar unsaturates, such as ketones, esters, and nitriles, based on just this notion: that anionic transition metal hydrides should have nucleophilic character and hence should be particularly reactive toward these substrates. Syntheses of two such complexes are shown in eqs. (10.9) and (10.10). Several indicators of hydridic character were indeed present: Compound **1** reacts readily with HCl and

MeI; also, the low M—H stretching frequencies and the slight solubility in completely nonpolar solvents suggested strong $Ru\text{–}H^-\text{—}K^+$ ion pairing [9]. A related osmium complex, $K^+[OsH_3L_3]^-$, is also soluble in aromatics, and the X-ray structure reveals strong ion pairing via $H^-\text{—}K^+$ interactions [10].

$$(PPh_3)_3RuHCl + K^+C_{10}H_8{}^- \xrightarrow{Et_2O}$$

$$K^+\left[(PPh_3)_2H_2Ru\text{–}C_6H_4PPh_2\right]^- \cdot C_{10}H_8 \cdot Et_2O \qquad (10.9)$$

1

$$((PPh_3)_2RuHCl)_2 + K^+[\text{naphthalene}]^- \rightarrow K^+{}_2[(PPh_3)_3(PPh_2)Ru_2H_4]^{2-} \qquad (10.10)$$

2

In fact, both **1** and **2** were found to be moderately good (pre)catalysts for hydrogenation of polar substrates. For example, acetone is hydrogenated in the presence of **1** at 85°C, 6 atm H_2, over 16 h; **2** is about four times more active. This contrasts with the behavior of non-anionic ruthenium hydrides such as $RuHCl(PPh_3)_3$, which is only slightly active for acetone hydrogenation although it is an excellent olefin hydrogenation catalyst. Nitriles can be hydrogenated at comparable rates and esters somewhat more slowly. Another indication that nucleophilic character is the key to this reactivity is the fact that hexafluoroacetone (as well as fluorinated esters) is significantly more reactive than the unfluorinated analogs. The mechanism proposed for catalysis is based upon eq. (10.7) [11].

Unfortunately, although the successful design of a catalyst would appear to validate the initial premise, it does not stand up to closer mechanistic examination, as demonstrated subsequently by Halpern and co-workers. Kinetic studies show that the active catalyst is *not* anionic $RuH_3(PPh_3)_3{}^-$ (**3**), which would be formed by hydrogenation of precatalyst **1** (or can be synthesized independently [12]); rather, it is *neutral* $RuH_4(PPh_3)_3$ (**4**), which is (reversibly) formed from the anion by eq. (10.11). This can be seen most clearly from the fact that catalysis by the anionic hydride **3** requires an induction period, during which small amounts of alcohol are formed and serve to protonate **3** to **4**; if **4** is used, there is no induction period. The catalytic activity of **4** was ascribed to the facile displacement of H_2 by even a weak ligand such as a ketone, which in turn is attributed to the formulation of **4** as $RuH_2(\eta^2\text{-}H_2)(PPh_3)_3$, and apparently has nothing to do with nucleophilic character [13]. Similarly, iridium polyhydride complexes such as $IrH_5(PPr^i{}_3)_2$ hydrogenate activated esters such as $CF_3CO_2CH_2CF_3$, with no indication of any anionic species being involved [14].

$$\underset{\mathbf{3}}{RuH_3L_3{}^-} + ROH \rightleftarrows \underset{\mathbf{4}}{RuH_4L_3} + RO^- \qquad (10.11)$$

Although this result appears disappointing, evidence for nucleophilic behavior *has*, in fact, been found for a group of anionic hydridometal carbonyl complexes. Most of the work here comes from extensive studies by Darensbourg and her group. We will highlight key results here; for more details, see a recent review [15] on the general topic of anionic transition metal hydrides and their chemistry.

The group VI metal hydrides $[HM(CO)_4L]^-$ (where M = Cr, Mo, W and L = CO, $P(OMe)_3$, PR_3) were found to reduce aldehydes and ketones to the corresponding alcohols, with an additional acidic species needed to complete reaction (10.12). The reaction is believed to proceed via hydride transfer to give intermediate **5**, which can be detected spectroscopically under certain conditions [16].

$$R_2C{=}O + HM(CO)_4L^- + HX \rightarrow R_2CHOH + [M(CO)_4L] + X^- \qquad (10.12)$$

$$R_2C{=}O + HM(CO)_4L^- \rightarrow \underset{\mathbf{5}}{R_2CHO{-}M(CO)_4L} \xrightarrow{H^+} R_2CHOH \qquad (10.13)$$

The following support a nucleophilic hydride transfer mechanism:

1. Reactivity toward organic substrates follows the general order RCOCl > RCHO > R_2CO, typical of hydride reagents.
2. The reactivity order

$$[HW(CO)_4(P(OMe)_3)]^- > [HW(CO)_5]^- > [HCr(CO)_5]^-$$

corresponds to what might have been anticipated for hydridic transfer: The poorer π-acceptor ligand favors negative charge localization at H, whereas W > Cr is the opposite of the acidity order mentioned earlier.
3. No exchange with or inhibition by free CO can be detected, suggesting that generation of a vacant site and coordination of the organic substrate prior to reaction are not taking place.

Under certain conditions, these reductions may be run catalytically under H_2, although activities are not very high. $[M(CO)_5(OAc)]^-$ functions as a catalyst precursor for hydrogenation of cyclohexanone and benzaldehyde to the corresponding alcohols (THF, 125°C, 700 psig H_2, ca. 10 turnovers per day); the proposed mechanism (Scheme 10.1) involves hydrogenolysis of the

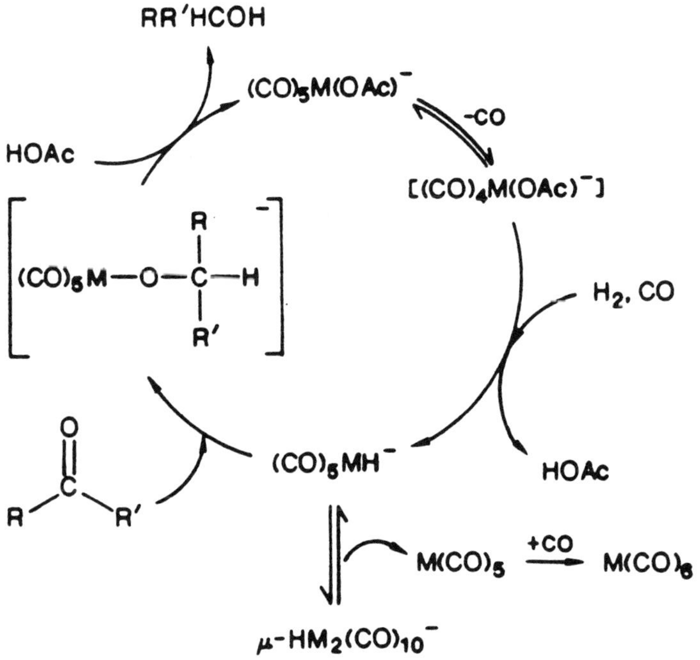

Scheme 10.1

M—O bond to produce both the nucleophilic anionic hydride and the acidic co-reagent. Catalysis also is achieved starting with the dimeric hydride anions, $[(\mu\text{-H})M_2(CO)_{10}]^-$ [17].

Other reactions that may be diagnostic for nucleophilic behavior also have been reported, including the following: protonolysis [which makes the choice of the acid co-reagent in eq. (10.12) crucial—with the more active hydrides, stronger HX gives decomposition instead of reduction of substrate] [16]; formation of adducts with, and even transfer of hydride to [eq. (10.14)] [18] Lewis acids; ion pairing through M—H, as manifested by counterion effects upon IR and NMR spectroscopic results [19]; and reactivity with alkyl halides.

$$2HCr(CO)_5^- + AlMe_3 \rightarrow HAlMe_3^- + HCr_2(CO)_{10}^- \qquad (10.14)$$

The last item demands attention because, as noted earlier, it is often difficult to distinguish between nucleophilic and radical pathways. Several lines of evidence support the former for at least some of the systems; for example, the relative reactivities shown in Table 10.1 suggest a nucleophilic path for the most hydridic (according to the previously mentioned studies) $HW(CO)_4(P(OMe)_3)^-$, whereas some participation of a different pathway is

Table 10.1
Relative Reactivities of Anionic Hydrides with Alkyl Halides [20]

Hydride	Bu^nBr	Bu^sBr	Bu^tBr
$HCr(CO)_5^-$	≡ 1	0.9	1.8
$HW(CO)_5^-$	1.8	1.0	1.6
$HCr(CO)_4(P(OMe)_3)^-$	17	1.1	0.5
$HW(CO)_4(P(OMe)_3)^-$	28	2.4	0.1
$CpV(CO)_3H^-$	1.2		
$HRu(CO)_4^-$	0.6		
$HFe(CO)_4^-$	0	0	0
$HFe(CO)_3(P(OMe)_3)^-$	0.01		

indicated for $HM(CO)_5^-$. Likewise, reaction of the hexenyl bromide in eq. (10.15) gives no cyclized product **6**—a useful marker for radical intermediates—with $HW(CO)_4(P(OMe)_3)^-$, but does afford around 10–30% **6** with $HM(CO)_5^-$ [20].

$$\text{CH}_2{=}\text{CH(CH}_2)_4\text{Br} + \text{HM}^- \rightarrow \text{CH}_2{=}\text{CH(CH}_2)_3\text{CH}_3 + \underset{\mathbf{6}}{\text{C}_5\text{H}_9\text{CH}_3} + \text{M} + \text{Br}^- \quad (10.15)$$

A few studies have been reported with complexes from other groups of the Periodic Table. $HFe(CO)_3(PR_3)^-$ reacts (slowly) with RX, but these are thought to proceed via initial attack at the metal center [eq. (10.6)] rather than at the hydride; calculations indicate a lower negative charge density at H for this anion than for the group VI examples [21]. This *could* reflect dependence upon Periodic Table position, a point that will be discussed at length in the following section. On the other hand, $CpV(CO)_3H^-$ reduces alkyl halides, and only a small amount of cyclized product was observed in eq. (10.15); but this was ascribed to the rapid capture of radical rather than to a substantial nucleophilic pathway; *all* the reaction is thought to proceed via chain mechanism (10.5) [22]. It is rather striking that, in spite of the move to the left in the Periodic Table plus the replacement of some of the CO ligands with the strong donor Cp ligand, $CpV(CO)_3H^-$ appears to be *less* nucleophilic than the group VI anions. $CpRe(CO)_2H^-$ likewise exhibits no hydridic character [23].

The dianions $HM(CO)_5^{2-}$ (where M = V, Nb, Ta) are sufficiently strong hydride donors to transfer H^- to $Fe(CO)_5$, giving the known formyl compound $Fe(CO)_4(CHO)^-$ [eq. (10.16)] [24]. $HRu(CO)_4^-$ does not react with $Fe(CO)_5$ but will transfer H^- to the more electrophilic $CpRe(CO)_2(NO)^+$ [eq. (10.17)]; it has been suggested that nucleophilic character is responsible for some of the homogeneous CO reduction chemistry exhibited by ruthenium carbonyl systems [25]. (It should be noted, though, that a nucleophilic pathway for these reactions has by no means been unequivocally demonstrated.) A cluster ruthenium hydride anion exhibits apparently strong hydride donor properties [eqs. (10.18) and (10.19)], but only under a CO atmosphere [26]; these may reflect the high stability of $Ru_3(CO)_{12}$ more than anything else.

$$HM(CO)_5^{2-} + Fe(CO)_5 \xrightarrow[\text{fast}]{\text{THF, }-50^\circ\text{C}} M(CO)_5(THF)^- + Fe(CO)_4(CHO)^- \quad (10.16)$$

$$HRu(CO)_4^- + CpRe(CO)_2(NO)^+ \rightarrow CpRe(CO)(NO)(CHO) \quad (10.17)$$

$$K^+HRu_3(CO)_{11}^- + Ph_3C^+BF_4^- + CO \rightarrow Ph_3CH + Ru_3(CO)_{12} + KBF_4 \quad (10.18)$$

$$K^+HRu_3(CO)_{11}^- + CO \rightarrow KH + Ru_3(CO)_{12} \quad (10.19)$$

Besides the limited applications of nucleophilic anionic transition metal hydrides to catalytic reduction [$HFe(CO)_4^-$ also catalyzes reductions of aldehydes, using water–CO or H_2 as reductant; but, as seen before, it is not at all clear that any nucleophilic reactivity is involved], the hydridic group VI transition metal hydrides have synthetically useful properties. For example, the strong preference for reacting with acyl halides allows the transformation of eq. (10.20) to proceed with good selectivity [27]. These complexes also exchange readily with weak deuterioacids, such as MeOD, D_2O, or DOAc, which makes them potentially valuable deuterating agents; preparation of deuterated main-group hydride reagents is more difficult and expensive.

$$BrCH_2CH_2CH_2COCl + HW(CO)_4(P(OMe)_3)^- \rightarrow BrCH_2CH_2CH_2CHO \quad (10.20)$$

10.4. HYDRIDIC CHARACTER OF EARLY TRANSITION METAL HYDRIDES

The first well-characterized transition metal hydrides with d^0 electronic configurations were the zirconium(IV) complexes $(Cp_2ZrH_2)_n$ and $(Cp_2ZrHCl)_n$ [28, 29]. It was quickly recognized that these species, and other group IV hydrides [30] that followed, resemble main-group hydrides in their reactivity: rapid decomposition by water and other proton sources, reduction of ketones, and so forth. Extensive nucleophilic chemistry of $(Cp_2ZrHCl)_n$

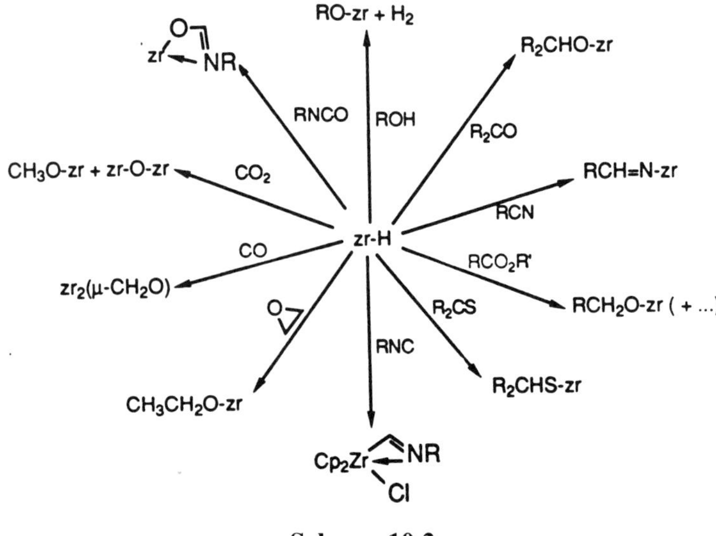

Scheme 10.2

has been documented [31]; some examples are summarized in Scheme 10.2. Similar reactivity has been established for group III [32] (including lanthanide [33] and actinide [34]) hydride complexes. Compound **7**, which was designed to act as a "hydride sponge" and does in fact remove H^- from HBR_3^-, was reported to take H^- from $(Cp_2ZrHCl)_n$ as well, although no details were given [35].

Me_2B BMe_2

7

One particularly interesting manifestation of nucleophilic behavior is the ability to transfer hydride to coordinated CO. As implied earlier, promising applications of homogeneous catalysis to CO hydrogenation are exceedingly rare, and it has been speculated that a large kinetic barrier to insertion of CO into an M—H bond may be responsible. Nucleophilic attack on coordinated CO might be an alternate route into CO—H_2 chemistry [36, 37], and several intriguing reports of CO reduction effected by group IV systems have appeared [38–40]. No conclusive demonstration of intermolecular, nucleophilic attack versus intramolecular insertion has been achieved, but several lines of evidence tend to support the former (although perhaps ironically, the only case where intramolcular insertion *is* explicitly supported, by means of crossover experiments, involves an actinide hydride, eq. (10.21) [41]). In

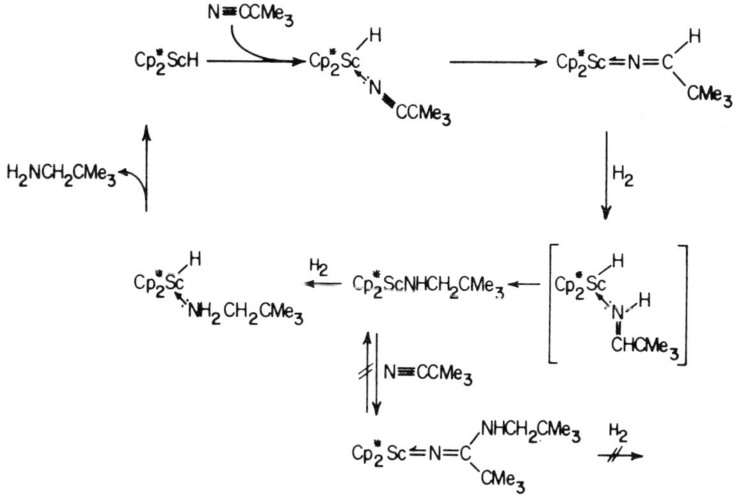

Scheme 10.3

particular, $Cp^*_2ZrH_2$ transfers hydride to a variety of metal carbonyls, as illustrated in eq. (10.22) [42].

$$Cp^*_2ThH(OR) + CO \rightarrow Cp^*_2Th(\eta^2\text{-}CHO)(OR) \qquad (10.21)$$

$$Cp^*_2ZrH_2 + Cp_2NbR(CO) \rightarrow Cp^*_2ZrH\big(OCH{=}NbRCp_2\big) \qquad (10.22)$$

Unfortunately, there are no examples of catalysis, either of CO hydrogenation or of reduction of organic carbonyl complexes. The difficulty is that these species are *too much* like main-group hydrides: They form extremely stable M—O bonds as a consequence of the high electropositivity of the metals. Hence, closing a catalytic cycle by hydrogenolysis of an M—O bond probably will not be possible. The restriction is not so severe when it comes to M—N bonds, and catalysis based on nucleophilic reactivity may well be achievable. For example, Cp^*_2ScH catalyzes nitrile hydrogenation according to Scheme 10.3; here hydrogenolysis of the Sc—N bond *is* sufficiently favorable for catalysis to proceed [43].

What happens as we move away from the extreme left end of the transition series? Is nucleophilic character unique to d^0 configurations, or is it a more general consequence of periodicity that will still be present, albeit attenuated, in groups V, VI, and so on? This question was probed systematically, using reactivity toward ketones as a test for nucleophilicity; as shown in Table 10.2, there does appear to be a distinct trend, at least for the metallocene derivatives shown here: Hydridic character falls off smoothly on progressing from group IV to group VII [44]. That direct nucleophilic attack

Table 10.2
Qualitative Reactivity of Hydride toward Ketones [44]

Complex	Reactivity with $(CH_3)_2C{=}O$	Reactivity with $(CH_3)(CF_3)C{=}O$
$(Cp_2ZrH_2)_n$	Fast/25°	
Cp_2NbH_3	Slow/25°	Fast/25°
Cp_2MoH_2	No reaction/78°	Slow/25°
Cp_2ReH	No reaction/78°	No reaction/78°

is involved is implied by (1) much higher reactivities of the more electrophilic fluorinated ketone and (2) the fact that coordinatively saturated complexes react readily under conditions where ligand exchange or replacement (and hence generation of a potential vacant site for substrate coordination) are known to be very slow.

Attempts to exploit the moderated nucleophilicity of group V transition metal hydrides for CO hydrogenation have been only partially successful: The reaction of Cp_2NbH_3 with $Fe(CO)_5$ gives (by NMR) a carbene-like intermediate [eq. (10.23)] similar to those seen with Zr hydrides [eq. (10.22)], but no net CO reduction is achieved [45]. The same hydride reacts with $Cr(CO)_6$, but only at elevated temperatures; ethane derived from CO reduction *is* formed, but only stoichiometrically; a niobium–oxygen bonded product again forms irreversibly [46]. Chemistry related to CO hydrogenation has been achieved with less-nucleophilic hydrides, as in eqs. (10.24) and (10.25) [47]; the strong Zr—O bonds formed undoubtedly have much to do with the favorability of these reactions. (A related reaction has been used as the basis for a systematic approach to quantitate nucleophilicity; see Section 10.5.) It does not appear likely that the goal of *catalytic* CO reduction using nucleophilic early transition metal hydrides will be achievable.

$$Cp_2NbH_3 + Fe(CO)_5 \rightarrow Cp_2H_2Nb{-}OCH{=}Fe(CO)_4 \quad (10.23)$$

$$Cp_2MoH_2 + Cp_2ZrMe(COMe) \rightarrow Cp_2ZrMe(OEt) \quad (10.24)$$

$$Cp_2ReH + Cp_2ZrMe(COMe) \rightarrow Cp_2MeZr(OCHMeReCp_2) \quad (10.25)$$

Not surprisingly, the nucleophilicity of early transition metal hydrides depends upon the nature of the ligands as well. For example, the monocyclopentadienyl complex $Cp^*Ta(PMe_3)_2H_3Cl$ reacts considerably more rapidly with both MeOH and $Me_2C{=}O$ than does Cp_2NbH_3 [48]; likewise, the phosphine hydride $Ta(dmpe)_2H_5$ is rapidly decomposed by EtOH [49]. Apparently, phosphines are better than Cp at promoting hydridic character. In the other direction, $CpMo(CO)_3H$ shows no reactivity toward trifluoracetone, whereas Cp_2MoH_2 does [44]; it also exhibits *acidic* character toward an "amphoteric" reagent [eq. (10.26)] rather than hydridic character (such as formation of an H—Al bonded adduct) [50].

$$CpMo(CO)_3H + Ph_2PNBu^tAlMe_2 \rightarrow [CpMo(CO)_3]^-[AlMe_2NBu^tPPh_2H]^+ \quad (10.26)$$

We should note that, even though the general trends favoring acidity and nucleophilicity proceed oppositely, it is by no means impossible for both modes of reactivity to appear in the same system. For example, Cp_2MoH_2, although moderately hydridic, can be deprotonated by a sufficiently strong base, such as $Me_3P{=}CH_2$ [51]. Conversely, the weak acid $CpMo(CO)_3H$ gives up hydride to Ph_3C^+ [52], as does $CpOs(CO)_2H$ [53]. A more striking illustration is the behavior of $Cp_2MH_3^+$ (where M = Mo, W), which would be expected to be considerably more acidic than the parent neutral Cp_2MH_2 and indeed is deprotonated readily by weak bases; but it may also be more *hydridic* than Cp_2MH_2: The cations reduce acetone to isopropanol at room temperature, whereas the neutrals are unreactive! It was suggested, though, that this may instead reflect the possibility of coordinating the substrate via eq. (10.27) [54].

$$Cp_2MH_3^+ \rightleftarrows Cp_2MH(\eta^2\text{-}H_2)^+ \xrightarrow[-H_2]{Me_2C{=}O} Cp_2MH(Me_2C{=}O)^- \rightarrow \cdots \quad (10.27)$$

Other tests for hydridic character in these early transition metal hydrides are less clearcut. For example, the reactions of $Cp_2NbH(CO)$ with Bu^nBr and Bu^tBr proceed at about the same rate, suggesting radical rather than (or in addition to) nucleophilic behavior [55]. A number of adducts with Lewis acids have been synthesized, and many appear to involve coordination via the hydride [56]; however, AlH_nX_{3-n} forms adducts more readily with Cp_2ReH than Cp_2MH_2 (where M = Mo, W), although the latter appear more nucleophilic according to other tests. This may involve direct Re—Al rather than ReH—Al bonding and may have more to do with steric factors than with the nature of the metal hydride [57].

Of course, the most important application of $(Cp_2ZrHCl)_n$ is in organic synthesis via hydrozirconation of *nonpolar* substrates, olefins and acetylenes. Does nucleophilic character play any role here? The accepted explanation for the favorability of hydrozirconation lies in the relative instability of the η^2-olefin intermediate, a consequence of the d^0 configuration [58]. Doherty and Bercaw [59] concluded that the electronic effects on the rate of insertion in Cp^*_2NbH(olefin) complexes imply that the hydrogen migrates more or less as H^-; however, similar effects are found for a late transition metal hydride (RhH_2ClL_3) as well [60]. Hence, there may not be much correlation between facility of insertion and hydridic character. Crabtree [61] suggested that the latter Rh hydride *is* relatively hydridic, on the grounds of the regiochemistry of insertion [eq. (10.28) and (10.29)]: Formation of the α-cyanoethyl isomer is more consistent with migration as H^-. A similar experiment was used to infer hydridic character for $CpFe(CO)(PPh_3)H$ [62].

$$RhHCl_2L_3 + CH_2{=}CHCN \rightarrow L_3Cl_2Rh{-}CH(CN)CH_3 \quad (10.28)$$

$$IrH(CO)L_3 + CH_2{=}CHCN \rightarrow L_2(CO)(CH_2{=}CHCN)Ir{-}CH_2CH_2CN \quad (10.29)$$

Olefin hydrogenation that apparently does depend on nucleophilic character has been reported recently [eq. (10.30)] [63]. The stoichiometry is related to the reductions of aldehydes by anionic group VI hydrides plus acid [eq. (10.12)], but here the first step appears to be protonation of the (electron-rich) olefin, followed by trapping of the carbonium ion by the metal hydride. The generality of this method is not yet known; a transition metal hydride with relatively *low* nucleophilic character may be required, as the carbonium ion is a strong hydride acceptor, whereas a more-hydridic metal hydride will react preferentially with the strong acid; an electron-rich olefin is probably also essential.

$$Me_2C{=}CMe_2 + CpMo(CO)_3H + HOTf \rightarrow$$

$$\cdot\ Me_2CHCHMe_2 + CpMo(CO)_3OTf \quad (10.30)$$

10.5. OTHER EXAMPLES OF NUCLEOPHILIC HYDRIDES

The complex $CpW(NO)_2H$ is formally isoelectronic to $CpFe(CO)_2H$, which has been seen to have little if any hydridic character; also, the NO ligand is generally considered at least as good a π-acceptor as CO. It is rather surprising, then, that this complex is readily protonolyzed, with formation of H_2 [eq. (10.31)] [64]. Bursten and Gatter [65, 66] presented a theoretical explanation. First, the energy separation between the highest occupied molecular orbital (HOMO) and the lowest unoccupied molecular orbit (LUMO) of a carbonylmetallate anion, $M(CO)_n{}^-$, was shown to correlate well with the acidity of the conjugate acid $HM(CO)_n$. This was justified on the grounds that the separation ΔE^{CB} will be large when the anion is electronically stable. The corresponding test for nucleophilicity would, at first sight, seem to be ΔE^{CB} for the cation $M(CO)_n{}^+$; because those species are not stable, though, $M(CO)_n(HCN)^+$ was used as a model for the solvated cation, and solvation energies were found to dominate to the extent that no

Table 10.3
ΔE^{HS} Values for Hydride Complexes [65, 66]

Complex	Model Anion for Calculation	ΔE^{HS} (eV)	Hydridic?[a]
$(Cp_2ZrH_2)_n$	Cp_2TiH^-	6.99	Strongly
Cp_2NbH_3	$Cp_2VH_2^-$	6.59	Moderately
$Cp_2NbH(CO)$	$Cp_2V(CO)^-$	1.54	Moderately
Cp_2MoH_2	Cp_2CrH^-	0.05	Weakly
$CpCr(NO)_2H$	$CpCr(NO)_2^-$	2.41	Moderately
$CpCr(CO)_3H$	$CpCr(CO)_3^-$	0	No
$CpFe(CO)_2H$	$CpFe(CO)_2^-$	0.71	No

[a]Reactivity toward ketones and/or protonolysis.

trends appear at all. As an alternative, the separation between the HOMO and the next-lower orbital in the anion was calculated; a large value for this parameter (termed ΔE^{HS}) would reflect a propensity to lose an electron pair and to produce a relatively stable cation. As can be seen in Table 10.3, a comparison of these quantities accords with the fact that $CpW(NO)_2H$ is somewhat hydridic, whereas $CpW(CO)_3H$ and $CpFe(CO)_2H$ are not. Correlation with trends shown in Table 10.2 is also found.

$$CpW(NO)_2H + HOTs \xrightarrow[\text{THF}]{25^\circ\text{C, 30 min}} CpW(NO)_2OTs + H_2 \quad (10.31)$$

Is it possible to draw any general conclusions from this one compound, or is such a result highly specific to the particular structure involved? Kundel and Berke [67] noted that another hydrido(nitrosyl)tungsten complex, $HW(CO)_2(NO)(P(OPr^i)_3)_2$, which has a completely different detailed structure, *also* exhibits nucleophilic character: protonolysis by weak acids such as phenol; reduction (with an acid co-reagent) of aldehydes. Berke proposes that the oxidation state of the metal is crucial: These are formally W(0) (taking the nitrosyl as NO^+), as are the anionic $HW(CO)_4L^-$ reagents whose chemistry is quite similar [67]. More examples of this class will be needed before any useful conclusions can be reached.

Another apparent manifestation of hydridic character that might not have been expected from the considerations discussed previously is the copper hydride $(CuH(PPh_3))_6$, which undergoes conjugate addition to conjugated enones [68] and also serves as catalyst (or precatalyst) for hydrogenation of enones, either to the corresponding ketone or all the way to the alcohol, depending upon conditions [69]. We have discussed previously whether such catalysis can be taken as a meaningful indicator of nucleophilic character; one indication that it *is* in the present case is the fact that simple olefins are *not* hydrogenated. In contrast, most previously known homogeneous catalysts for ketone hydrogenation, such as $RhL_2H_2^+$ [70], are also

Table 10.4
Relative Acidities and Nucleophilicities [72]

Complex	Thermodynamic Acidity[a]	Kinetic Acidity[b]	Kinetic Nucleophilicity[c]
$HRe(CO)_5$	2×10^{-8}	4×10^{-8}	139
$HMn(CO)_5$	2×10^{-2}	1×10^{-2}	100
$H_2Os(CO)_4$	3×10^{-8}	4×10^{-7}	58
$CpW(CO)_3H$	2×10^{-3}	1×10^{-2}	9
$CpW(CO)_2(PMe_3)$	5×10^{-14}	$< 10^{-7}$	5
$CpCr(CO)_3H$	$\equiv 1$	$\equiv 1$	$\equiv 1$

[a] From pK_a values.
[b] From rate of reaction with $PhNH_2$.
[c] From rate of reaction (10.32).

good olefin hydrogenation catalysts. In fact, previous work has suggested that simple copper hydrides, CuH and CuH^+, are hydridic in spite of their position in the Periodic Table [71]; coupled with the apparent ability of phosphine ligands to promote hydridic character, perhaps nucleophilic behavior here is not unreasonable.

We close with a study that attempts to quantitate nucleophilicity for a group of hydrides that should not be very nucleophilic at all according to the preceding criteria! Martin, Warner, and Norton [72] found that the rhenium acyl **8** is reduced by several metal carbonyl hydrides according to eq. (10.32); clean second-order kinetics are found, suggesting that the rate constants can be taken as a measure of the nucleophilicity of the hydride. These are collected in Table 10.4 [72]. As can be seen, none of the entries exhibits particularly strong nucleophilicity in any of the tests discussed previously; trends on changing ligands and Periodic Table position do not agree with previous findings either.

$$\underset{\mathbf{8}}{CH_3CH_2C({=}O)Re(CO)_4(CH_3CN)} + HM(CO)_nL \longrightarrow$$

$$CH_3CH_2CHO + (CH_3CN)(CO)_4Re{-}M(CO)_nL \quad (10.32)$$

It is not clear what to make of these results. Norton suggests that the initial interaction is not transfer of H^- to the carbonyl center, but rather formation of a three-center MHRe bond. if so, the relative reaction rates may reflect tendency to form such interactions, which may not be closely related to nucleophilicity in the sense we have been using. Furthermore, one might anticipate very important steric factors here, which may account for the fact that the cyclopentadienyl metal carbonyl hydrides are all substantially less reactive than the simple metal carbonyl hydrides; differences within the two groups are much less pronounced. Still, these findings must remind us that there is much we do not yet understand in this field.

10.6. CONCLUSION

Although we have not been able to come up with any unequivocal method for assessing nucleophilic character, based on the (often vague and sometimes contradictory) criteria available, several trends do stand out:

1. There is a clear correlation with position in the Periodic Table: Transition metal hydrides of groups III and IV are strongly nucleophilic, whereas those of the neighboring groups V and, to some extent, VI are less so.
2. Net negative charge and/or low formal oxidation state generally increase nucleophilic character, although the number of definitive examples is still rather small.
3. Good donor ligands promote nucleophilicity better than CO; there is some indication that phosphines may be better than Cp in this regard. On the other hand, the strong π-acceptor NO also seems to produce hydridic character in a couple of examples.

With regard to practical applications, catalytic reduction of carbonyl compounds cannot be carried out with the one class of transition metal hydrides that is reliably nucleophilic—the early metal hydrides—because the strong M—O bonds preclude catalysis. There are several promising indications that nucleophilic transition metal hydrides from groups further to the right may be exploited for such catalysis, but our understanding of the factors that control such behavior is still at a very preliminary stage.

REFERENCES

1. Ashby, E. C.; Goel, A. B.; DePriest, R. N. *J. Am. Chem. Soc.* **1980**, *102*, 7780.
2. For an example and some leading references, see Hoffman, D. M.; Lappas, D.; Wierda, D. A. *J. Am. Chem. Soc.* **1989**, *111*, 1531.
3. Darensbourg, D. J.; Kudaroski, R. A. *Adv. Organomet. Chem.* **1983**, *22*, 129, and references cited therein.
4. Trogler, W. C. *J. Am. Chem. Soc.* **1979**, *101*, 6459.
5. Sweany, R. L.; Owens, J. W. *J. Organomet. Chem.* **1983**, *255*, 327.
6. Chen, H. W.; Jolly, W. L.; Kopf, J.; Lee, T. H. *J. Am. Chem. Soc.* **1979**, *101*, 2607.
7. Richmond, T. G.; Basolo, F.; Shriver, D. F. *Organometallics* **1982**, *1*, 1624.
8. Moore, E. J.; Sullivan, J. M.; Norton, J. R. *J. Am. Chem. Soc.* **1986**, *108*, 2257.

9. Pez, G. P.; Grey, R. A.; Corsi, J. *J. Am. Chem. Soc.* **1981**, *103*, 7528.
10. Huffman, J. C.; Green, M. A.; Kaiser, S. L.; Caueton, K. G. *J. Am. Chem. Soc.* **1985**, *107*, 5115.
11. Grey, R. A.; Pez, G. P.; Wallo, A. *J. Am. Chem. Soc.* **1981**, *103*, 7536.
12. Chan, A. S. C.; Shieh, H. S. *J. Chem. Soc., Chem. Commun.* **1985**, 1379.
13. Linn, D. E., Jr.; Halpern, J. *J. Am. Chem. Soc.* **1987**, *109*, 2969.
14. Goldman, A. S.; Halpern, J. *J. Am. Chem. Soc.* **1987**, *109*, 7537.
15. Darensbourg, M. Y.; Ash, C. E. *Adv. Organomet. Chem.* **1987**, *27*, 1.
16. Gaus, P. L.; Kao, S. C.; Youngdahl, K.; Darensbourg, M. Y. *J. Am. Chem. Soc.* **1985**, *107*, 2428.
17. Tooley, P. A.; Ovalles, C.; Kao, S. C.; Darensbourg, D. J.; Darensbourg, M. Y. *J. Am. Chem. Soc.* **1986**, *108*, 5465.
18. Darensbourg, M. Y.; Slater, S. *J. Am. Chem. Soc.* **1981**, *103*, 5914.
19. Kao, S. C.; Darensbourg, M. Y.; Schenk, W. *Organometallics* **1984**, *3*, 871.
20. Kao, S. C.; Spillett, C. T.; Ash, C.; Lusk, R.; Park, Y. K.; Darensbourg, M. Y. *Organometallics* **1985**, *4*, 83.
21. Darensbourg, M. Y.; Ash, C. E.; Kao, S. C.; Silva, R.; Springs, J. *Pure Appl. Chem.* **1988**, *60*, 131.
22. Kinney, R. J.; Jones, W. D.; Bergman, R. G. *J. Am. Chem. Soc.* **1978**, *100*, 635, 7902.
23. Yang, G. K.; Bergman, R. G. *J. Am. Chem. Soc.* **1983**, *105*, 6500.
24. Warnock, G. F. P.; Ellis, J. E. *J. Am. Chem. Soc.* **1984**, *106*, 5016.
25. Dombek, B. D.; Harrison, A. M. *J. Am. Chem. Soc.* **1983**, *105*, 2485.
26. Bricker, J. C.; Nagel, C. C.; Shore, S. G. *J. Am. Chem. Soc.* **1982**, *104*, 1444.
27. Kao, S. C.; Gauss, P. L.; Youngdahl, K.; Darensbourg, M. Y. *Organometallics* **1984**, *3*, 1601.
28. James, B. D.; Nanda, R. K.; Wallbridge, M. G. H. *Inorg. Chem.* **1967**, *6*, 1979.
29. Kautzner, B.; Wailes, P. C.; Weigold, H. *J. Chem. Soc. Chem. Commun.* **1969**, 1105. Wailes, P. C.; Weigold, H. *J. Organomet. Chem.* **1970**, *24*, 405.
30. For a survey of group IV and V hydrides, see Toogood, G. E.; Wallbridge, M. G. H. *Adv. Inorg. Chem. Radiochem.* **1982**, *25*, 267.
31. Etievant, P.; Gautheron, B.; Tainturier, G. *Bull. Soc. Chim. France*, **1978**, 292.
32. St. Clair, M. A.; Santarsiero, B. D.; Bercaw, J. E. *Organometallics* **1989**, *8*, 17.
33. Evans, W. L.; Meadows, J. H.; Wayda, A. L.; Hunter, W. E.; Atwood, J. L. *J. Am. Chem. Soc.* **1982**, *104*, 2008.
34. Fagan, P. F.; Manriquez, J. M.; Maata, E. A.; Seyam, A. M.; Marks, T. J. *J. Am. Chem. Soc.* **1981**, *103*, 6550.
35. Katz, H. E. *J. Am. Chem. Soc.* **1985**, *107*, 1421.
36. Labinger, J. A. *ACS Adv. Chem. Series* **1978**, *167*, 149.
37. Caulton, K. G. *J. Mol. Catal.* **1981**, *13*, 71.
38. Huffman, J. C.; Stone, J. G.; Krusell, W. C.; Caulton, K. G. *J. Am. Chem. Soc.* **1977**, *99*, 5829.
39. Shoer, L. I.; Schwartz, J. *J. Am. Chem. Soc.* **1977**, *99*, 5831.
40. Wolczanski, P. T.; Bercaw, J. E. *Acc. Chem. Res.* **1980**, *13*, 121.
41. Fagan, P. J.; Moloy, K. G.; Marks, T. J. *J. Am. Chem. Soc.* **1981**, *103*, 6959.
42. Threlkel, R. S.; Bercaw, J. E. *J. Am. Chem. Soc.* **1981**, *103*, 2650.
43. Bercaw, J. E.; Davies, D. L.; Wolczanski, P. T. *Organometallics* **1986**, *5*, 443.
44. Labinger, J. A.; Komadina, K. H. *J. Organomet. Chem.* **1978**, *155*, C25.
45. Labinger, J. A.; Wong, K. S.; Scheidt, W. R. *J. Am. Chem. Soc.* **1978**, *100*, 3254.
46. Wong, K. S.; Labinger, J. A. *J. Am. Chem. Soc.* **1980**, *102*, 3652.
47. Marsella, J. A.; Caulton, K. G. *J. Am. Chem. Soc.* **1980**, *102*, 1747.
48. Mayer, J. M.; Bercaw, J. E. *J. Am. Chem. Soc.* **1982**, *104*, 2157.
49. Tebbe, F. N. *J. Am. Chem. Soc.* **1973**, *95*, 5823.

50. Grimmett, D. L.; Labinger, J. A.; Bonfiglio, J. N.; Masuo, S. T.; Shearin, E.; Miller, J. S. *Organometallics* **1983**, *2*, 1325.
51. Fiederling, K.; Grob, I.; Malisch, W. *J. Organomet. Chem.* **1983**, *255*, 299.
52. Beck, W.; Schloter, K. *Z. Naturforsch., B: Anorg. Chem., Org. Chem.* **1978**, *33*, 1214.
53. Hoyano, J. K.; May, C. J.; Graham, W. A. G. *Inorg. Chem.* **1982**, *21*, 3095.
54. Igarash, T.; Ito, T. *Chem. Lett.* **1985**, 1699.
55. Silva, R. *Abstracts of Papers*, 194th National Meeting of the American Chemical Society: Washington, DC, 1987; INOR 397.
56. Tebbe, F. N. *J. Am. Chem. Soc.* **1973**, *95*, 5412. Otto, E. E. H.; Brintzinger, H. H. *J. Organomet. Chem.* **1979**, *170*, 209.
57. Ishchenko, V. M.; Soloveichik, G. L.; Bulychev, B. M.; Sokolova, T. A. *Zh. Neorg. Khim.* **1984**, *29*, 119 (*Russ. J. Inorg. Chem.* **1984**, *29*, 66).
58. Schwartz, J.; Labinger, J. A.; *Angew. Chem., Int. Ed. Engl.* **1976**, *15*, 333.
59. Doherty, N. M.; Bercaw, J. E. *J. Am. Chem. Soc.* **1985**, *107*, 2671.
60. Halpern, J.; Okamoto, T. *Inorg. Chim. Acta* **1984**, *89*, L53.
61. Crabtree, R. H. In *Comprehensive Coordination Chemistry*; Wilkinson, G., Gillard, R. D., McCleverty, J. A., Eds.; Pergamon: New York, 1987; Vol. 2, p. 689.
62. Cariou, M.; Etienne, M.; Guerchais, J. E.; Kergoat, R.; Kubicki, M. M. *J. Organomet. Chem.* **1987**, *327*, 393.
63. Bullock, R. M.; Rappoli, B. J. *J. Chem. Soc., Chem. Commun.* **1989**, 1447.
64. Legzdins, P.; Martin, D. T. *Inorg. Chem.* **1979**, *18*, 1250.
65. Bursten, B. E.; Gatter, M. G. *Organometallics* **1984**, *3*, 895.
66. Bursten, B. E.; Gatter, M. G. *J. Am. Chem. Soc.* **1984**, *106*, 2554.
67. Kundel, P.; Berke, H. *J. Organomet. Chem.* **1987**, *335*, 353.
68. Mahoney, W. S.; Brestensky, D. M.; Stryker, J. M. *J. Am. Chem. Soc.* **1988**, *110*, 291.
69. Mahoney, W. S.; Stryker, J. M. *J. Am. Chem. Soc.* **1989**, *111*, 8818.
70. Schrock, R. R.; Osborn, J. A. *J. Chem. Soc., Chem. Commun.* **1970**, 567.
71. Pearson, R. G. *Chem. Rev.* **1985**, *85*, 41.
72. Martin, B. D.; Warner, K. E.; Norton, J. R. *J. Am. Chem. Soc.* **1986**, *108*, 33.

CHAPTER 11

Summary and Conclusions

A. DEDIEU

Laboratoire de Chimie Quantique, UPR 139 du CNRS, Université Louis Pasteur, 67000 Strasbourg, France

It is clear from reading the previous chapters that our understanding of the molecular properties and reactivity pattern of transition metal hydrides and transition metal hydride complexes has increased significantly during the last decade. Both theory and experiment have contributed to this advance, with an increasing degree of interplay.

For small systems, computational techniques have now reached a degree of precision such that they can assist efficiently solvent-free experiments, either in the gas phase or in matrices. This is the case, for instance, for the assignment of spectroscopic features or the determination of M—H bond strengths in transition metal hydrides (either the neutral or the positive cation). Similarly for homoleptic transition metal hydrides MH_n, for which new approaches and techniques are needed, confrontation of theory and experiment is particularly useful. The FeH_2 dihydride system on which experiments have been carried out (both in the gas phase and in rare-gas matrices) is a particularly representative example, in that reliable calculations have helped to assess the FeH bond strength and the H—Fe—H geometry (see Chapters 1–3). In a related fashion one may expect that the experimental characterization of dihydrogen complexes $M(H_2)$ will also take advantage of the various calculations carried out on such systems.

Both theory and experiment have also been of help in determining the nature of the bonding in these simple hydrides. It is now well-established that the bonding is largely dictated by the relative positioning of the atomic states of the metal and the art of the theoretician is therefore to choose a method that accounts properly for the separation of these states (Chapter 3). In fact, the route is now open for a systematic investigation of these systems, from both the theoretical and experimental sides.

The situation is less clear-cut, however, for transition metal hydride *complexes*. In the gas phase and to a lesser extent in matrices, information is still scarce and the results are often much more complex than those of simple hydrides. This is generally due to the involvement of the ligands in the bonding pattern. Solution studies have made available more systematic information, provided that the effects of the solvent have been unravelled. Current theoretical studies have not included solvent effects. The size of the systems under consideration also make them less amenable to very accurate determination of thermodynamic, kinetic, or spectroscopic data. Nevertheless, theory and theoretical calculations may be used in several ways to investigate these transition metal hydride complexes.

(1) They first provide a conceptual framework for the experimental observations. Thus the various $(\eta^2\text{-}H_2)ML_n$ structures have been rationalized (see Chapter 5) and the factors that control the η^2 geometry of H_2 with respect to the η^1 geometry have been delineated [1, 2]. A detailed picture of the oxidative addition mechanism is now emerging (Chapter 4). Hydride migration to a coordinated ligand, such as C_2H_4, CO_2, H_2CO, CO, and CH_2, has been successfully explained. For C_2H_4, CO_2, and H_2CO the corresponding transition state is a four-center one involving both π donation to σ^*_{M-H} and π back-donation from σ_{M-H} to π^*. On the other hand, for CO and CH_2 the transition state is a three-center transition state that experiences orbital interactions between the $1s_H$, the π^* orbital of the ligand and one appropriate d metal orbital (see Chapter 6). The photochemistry of transition metal hydride complexes can be rationalized easily (especially the competition between the homolytic M—H cleavage and the heterolytic loss of an ancillary ligand) by considering the evolution of the potential energy surfaces arising from the various electronic states of the metal (see Chapter 7).

(2) They permit the assessment of the M–H bonding. It is worth emphasizing here that the most recent and sophisticated calculations—in particular, those accounting correctly for the relative ordering of the electronic states of the metal—point to a predominantly *covalent* M—H bond. This does not preclude the possibility that HML_n systems may behave as proton donors or hydride donors in the presence of a base or an acid, respectively. Nevertheless, the M—H bond may be regarded as intrinsically covalent at least for middle and late transition metals [3, 4]. (It may be noted that the Sc—H bond of the Cl_2ScH model system also has been found to be covalent in the generalized valence bond (GVB) calculations of Rappé [5].) Experimental studies that are reported in Chapters 9 and 10 also underline the fact that the nature of the M–H bonding is not always related to the reactivity. As a consequence, the idea of competition between the covalent M—H bond and the other bonds present in a transition metal complex (either in its ground-state structure or in a structure corresponding to a transition state) is used more and more in a valence bond (VB) type of picture to rationalize experimental findings. This kind of analysis is, in fact, linked to the recognition of the importance of the various electronic states of the transition metal in the complex [6].

(3) Theory can propose and/or analyze transient species in catalytic cycles. Transition state structures are generally not amenable to precise structural and electronic characterization (although isotope-effect studies already have provided very useful information, see Chapter 8). The involvement, along a reaction path, of structures displaying specific interactions (e.g., agostic interaction or, more generally, any two-electron–three-center interactions; interactions in η^2-H_2 preequilibrium structures) can be assessed by carrying out the corresponding calculations.

At this stage it is important to summarize our knowledge concerning the techniques used in these calculations. One should mention at the outset that the methodology for accurate calculations on transition metal complexes in general and on transition metal hydride complexes in particular is not completely assessed yet, and that studies and progress will continue to be made in this direction [7]. Going beyond the Hartree–Fock approximation seems to be a requisite in most cases and even calculations carried out with the MP2 method or single-reference-based methods (methods that are commonly used in the organic realm) very often do not work. This essentially stems from the presence of a manifold of low-lying electronic states originating from the atomic states of the metal. This so-called near-degeneracy effect has to be taken into account, generally by using a multireference wave function. The problem is, of course, more accute for low valent complexes. Large saturated organometallic hydrides appear to be more tractable, as a result of reduced mixing of the atomic states. Thus, for such systems single-reference-based methods *may* give qualitative results.

The degree of geometry optimization to be used in the calculations is still a matter of debate. The most rigorous method is to carry out a full geometry optimization (by using the energy gradient technique) and at a level where the energy correlation is taken into account properly. Such a procedure is quite expensive in terms of computing time and generally is precluded by the size of transition metal hydride complexes (with several ligands in addition to the hydride). One may think of carrying out a partial optimization procedure based on discrete parabolic fittings. However, the neglect of some portions of the full potential-energy surface may lead to a misevaluation. The $HMn(CO)_5 \rightarrow Mn(CO)_4(CHO)$ insertion reaction, which has been studied quite extensively [8, 9], is particularly illustrative of this statement: The earlier study of Nakamura and Dedieu [8] did not consider the possibility of an η^2 coordination mode for the formyl intermediate that was later found to be lower in energy than the η^1 coordination mode [9]. In fact, the presence of an energy barrier for the $\eta^1 \rightarrow \eta^2$ interconversion and the choice of too smaller values for the increments of the parabolic interpolation procedure kept the search artificially in the region of the η^1 structure. To circumvent this kind of problem, a more popular technique therefore is to perform the geometry optimization with the gradient technique at the self-consistent field (SCF) level and then to carry out a calculation including electron correlation (at the

MP2, MC-SCF + CI, or MR-CI level). This procedure also may be somewhat erroneous or may lead at least to some inaccuracy. Consider, for instance, the CO_2 + $CuH(PH_3)_3$ insertion reaction where the effect of MP2 calculations is to shorten the C—H bond of the corresponding transition state by 0.3 Å relative to the SCF level [9].

Another problem illustrated by the same reaction is the use of an effective core potential (ECP) to reduce the size of the calculation: On going from the all-electron to the ECP calculation (at the RHF level), the C—H bond length in the transition state is reduced by 0.4 Å [10]. More generally, the choice of a proper ECP is far from being trivial, as stressed by Koga and Morokuma (see Chapter 6) in their analysis of hydride migration to the carbonyl ligand in $[HRh(CO)]^{n+}$. Bauschlicher and Hay (Chapters 3 and 4) have noticed the importance of treating explicitly the outer core electrons. This is also apparent from a recent article by Blomberg, Schüle, and Siegbahn [6], who, like Bauschlicher, have pointed out the importance of relativistic contributions, even for second-row transition metal hydride complexes.

Besides these ab-initio techniques, the extended Hückel (EH) method is still a valuable analysis tool. One should keep in mind, however, that this method probably overestimates the polarity of the M—H bond. The fact that this bond is consistently found to be strongly localized on the H atom (which is at odds with the results of ab-initio calculations based on a correlated wave function) practically excludes any discussion of the nucleophilicity or electrophilicity of the transition metal hydrides. Another drawback lies in the fact that any argument drawn from EH calculation rests heavily on molecular orbital theory and, as we have seen, the alternative VB-type analysis can be very useful. Despite these shortcomings and because of its inherent simplicity, this method remains (when checked against experiment or more-sophisticated calculations) invaluable for setting analogies, rationalizing structural data, or shedding light on reactivity patterns. All these are features of fundamental importance for a better synergy between theory and experiment.

It is appropriate at this point to turn to a more chemical viewpoint of transition metal hydride chemistry and to examine the remaining problems or at least the points that need some clarification. At the same time it may be useful to identify the fields of research that have emerged recently or that will be at the forefront in the coming years.

We already have mentioned the need for additional studies investigating the basic characteristics of the M—H bond, such as its bond strength (especially for the first-row transition metal atom), its vibrational features, and its propensity to cleave homolytically or heterolytically. Progress needs to be made also in the techniques used to assess the occurrence of the η^2 coordination of H_2. Sweany pointed out (Chapter 2) the complexity of the vibrational spectra in dihydrogen transition metal complexes. This complexity is far from being always understood. A recent report by Cotton and Luck [11] also underlines the difficulty of using the variable-temperature ^{1}H NMR longitudinal relaxation times (T_1) to assess whether a polyhydride molecule contains molecular dihydrogen ligands.

Paradoxically, much more is known about the acidic behavior of transition metal hydrides than about their hydridic behavior. In particular, the factors controlling the hydricity are far from being well-understood. There is currently no simple and unambiguous method either for identifying or for measuring the nucleophilicity, except for some correlation between the nucleophilic character and the position in the Periodic Table (see Chapter 10). This is so despite the desirability of determining this factor to design more-efficient catalytic reduction processes.

The assessment of the mechanisms of either thermal or photochemical reactions is still an active area of research. Isotope effect studies designed to probe the nature of the transition state in C—H bond activation reactions, oxidative addition, or insertion reactions have generally met with success. There may be some discrepancies, however, between theory and experiment for the structure of the transition state: For example, an η^2 geometry generally is put forward by experimental studies whereas the DVM-Xα calculations of Ziegler et al. [12] seem to favor an η^1 geometry with the hydrogen atom more tightly bound to the metal atom than the carbon atom. Isotope effects also may be used in conjunction with kinetic acidity series to identify proton transfer reactions. Despite this the actual mechanism of this transfer is far from being fully elucidated: Norton's contention is that it "cannot be preceded by the formation of the sort of hydrogen bonded intermediates that have bedeviled the study of other types of proton transfer reactions", due to the M($\delta+$)–H($\delta-$) polarization of the M—H bond. The fact that in Raman spectroscopy the frequency and shape of the M—H stretching bands of $HRe(CO)_5$ and $HCo(CO)_4$ do not change in the presence of an excess of bases (ones strong enough to effect partial deprotonation) supports this contention (see Chapter 9). This is in apparent contradiction with the theoretical results discussed previously, which pointed to an essentially covalent M—H bond. Furthermore, there are two reports [13, 14] of crystal structures involving $M\cdots H\cdots NR_3$ systems. In both cases, however, the N—H bond is quite short, so that the bonding may be described in terms of $M^-\cdots HNR_3^+$ ion pairs. Calculations should help to clarify this point.

Several aspects of the mechanism of competitive photochemical loss of H and of an ancillary ligand in transition metal hydride complexes have been unravelled through theoretical and matrix isolation studies (see Chapters 2 and 7). The latter technique has some distinct limitations: Volatility requirements of the parent complex have restricted the type of systems explored to those containing CO and/or cyclopentadienyl ligands. One also should be aware that the coordinatively unsaturated system formed by photolysis may well experience some sort of interaction with the matrix and thus may display a structure different from that of the gas phase. Clearly, gas-phase experiments are needed to allow a closer comparison with theoretical studies.

Studies in less conventional media may be also of help [15]. The photochemical and photophysical behavior of hydrido complexes with metal-to-ligand charge transfer (MLCT) excited states will also undoubtedly receive more attention [16]. One of the next steps in the theoretical approach will be to estimate the lifetime of such excited states and to evaluate the probability of nonradiative processes.

Studies related to dihydrogen and polyhydrogen systems form a new field of research that is expanding quite rapidly. There is a need for a variety of characterization techniques other than neutron diffraction (*vide supra*). In particular, one should better understand the spectral properties of $(H_2)ML_n$ complexes (see Chapter 2). The reactivity of such systems also deserves a lot of attention. It is important, for instance, to know the various requisites toward achieving a dihydrogen–dihydride equilibrium. Among the questions that have not yet received definite answers, one finds the mechanism of the H–D exchange in dihydrogen systems. The reaction of hydrogen with transition metal clusters and the possibility occurrence of binuclear or polynuclear polyhydrogen complexes is also a topic of growing interest. All these problems should benefit from the progress made in gas-phase experiments and in theoretical studies. Burdett, Eisenstein, and Jackson (Chapter 5) have underlined how simple arguments based on molecular orbital theory could already orient our thinking. Yet extending this to the quantitative level is far from being obvious. The determination of the vibrational frequencies and of their relative intensities remains a challenge in most cases, especially if there is mixing with the ligand modes (e.g., carbonyl modes). The accurate determination of the carbonyl frequencies is in itself a difficult problem [17]. The discrepancies found in ab-initio values for the relative energies of the tetrahydrogen $Cr(CO)_4H_4$ isomers (Table 5.1) illustrate that it is difficult to get reliable results for this class of systems without including geometry optimization and electron correlation. Clearly, more methodological work must be done before studies of this type can be extended to several metal atoms.

The "ab-initio" design of catalytic cycles is one of the ultimate goals—if not a dream—of theoretical organometallic chemistry [18]. As pointed out in Chapters 4 and 6, systematic theoretical studies of simple processes need to be undertaken in order to assess the performance of the various methods used in characterizing energy profiles and the geometries of stationary and transition states. It may be worthwhile to recall here that MP2 calculations, although effective in the organic field, fail to account properly for near-degeneracy correlation effects, which stand as probably one of the most difficult problems to be overcome.

On the other hand, however, one may expect that organic analogues will provide assistance—at least in the first instance—in performing theoretical studies on the solvation effects and dynamics of processes involving transition metal hydrides. Such studies are certainly among the greatest challenges of the years to come.

REFERENCES

1. Saillard, J.-Y.; Hoffman, R. *J. Am. Chem. Soc.* **1984**, *106*, 2006.
2. Bickelhaupt, M. F.; Baerends, E. J.; Ravenek, W. *Inorg. Chem.* **1990**, *29*, 350.
3. Low, J. J.; Goddard III, W. A. *J. Am. Chem. Soc.* **1984**, *106*, 6928.
4. Hay, P. J. *J. Am. Chem. Soc.* **1987**, *109*, 705.
5. Rappé, A. K. *J. Am. Chem. Soc.* **1987**, *109* 5605.
6. Blomberg, M. R. A.; Schüle, J.; Siegbahn, P. E. M. *J. Am. Chem. Soc.* **1989**, *111*, 6156.
7. Salahub, D. R.; Zerner, M. C. Eds. *The Challenge of d and f Electrons*; ACS Symposium Series 394; American Chemical Society: Washington, DC, 1989.
8. Nakamura, S.; Dedieu, A. *Chem. Phys. Lett.* **1984**, *111*, 243.
9. (a) Ziegler, T.; Verluis, L.; Tschinke, V. *J. Am. Chem. Soc.* **1986**, *108*, 612. (b) Axe, F. U.; Marynick, D. S. *Organometallics* **1987**, *6*, 572.
10. Sakaki, S.; Ohkubo, K. *Inorg. Chem.* **1989**, *28*, 2583.
11. Cotton, F. A.; Luck, R. L. *J. Am. Chem. Soc.* **1989**, *111*, 5757.
12. Ziegler, T.; Tschinke, V.; Fan, L.; Becke, A. D. *J. Am. Chem. Soc.* **1989**, *111*, 9177.
13. Calderazzo, F.; Fachinetti, G.; Marchetti, F.; Zanazzi, P. F. *J. Chem. Soc., Chem. Commun.* **1981**, 181.
14. Wehmann-Ooyevaar, I. C. M.; Grove, D. M.; Van der Sluis, P.; Spek, A. L.; Van Koten, G. *J. Chem. Soc., Chem. Commun.* **1990**, 1367.
15. Jobling, M.; Howdle, S. M.; Healy, M. A.; Poliakoff, M. *J. Chem. Soc., Chem. Commun.* **1990**, 1287 and references therein.
16. Megehee, E. G.; Meyer, T. J. *Inorg. Chem.* **1989**, *28*, 4084 and references therein.
17. Blomberg, M. R. A.; Brandemark, U.; Johansson, J.; Siegbahn, P.; Wennerberg, J. *J. Chem. Phys.* **1988**, *88*, 4324.
18. (a) Daniel, C.; Koga, N.; Han, J.; Fu, X. Y.; Morokuma, K. *J. Am. Chem. Soc.* **1988**, *110*, 3773. (b) Koga, N., Morokuma, K. In *The Challenge of d and f Electrons*; ACS Symposium Series 394; American Chemical Society: Washington, DC, 1989. (c) Versluis, L.; Ziegler, T.; Baerends, E. J.; Ravenek, W. *J. Am. Chem. Soc.* **1989**, *111*, 2018.

Index

G

H

N

V

W

X

Y

Z